DYNAMICS AND OPTIMAL CONTROL
OF ROAD VEHICLES

T0177486

Dynamics and Optimal Control of Road Vehicles

David J. N. Limebeer

Professor of Control Engineering and Fellow of New College
Department of Engineering Science
University of Oxford
Formerly Head of the Department of Electrical and Electronic
Engineering, Imperial College London

Matteo Massaro

Associate Professor of Applied Mechanics
Department of Industrial Engineering
University of Padova

OXFORD
UNIVERSITY PRESS

OXFORD
UNIVERSITY PRESS

Great Clarendon Street, Oxford, OX2 6DP,
United Kingdom

Oxford University Press is a department of the University of Oxford.
It furthers the University's objective of excellence in research, scholarship,
and education by publishing worldwide. Oxford is a registered trade mark of
Oxford University Press in the UK and in certain other countries

© David J. N. Limebeer and Matteo Massaro 2018

The moral rights of the authors have been asserted

First Edition published in 2018

All rights reserved. No part of this publication may be reproduced, stored in
a retrieval system, or transmitted, in any form or by any means, without the
prior permission in writing of Oxford University Press, or as expressly permitted
by law, by licence or under terms agreed with the appropriate reprographics
rights organization. Enquiries concerning reproduction outside the scope of the
above should be sent to the Rights Department, Oxford University Press, at the
address above

You must not circulate this work in any other form
and you must impose this same condition on any acquirer

Published in the United States of America by Oxford University Press
198 Madison Avenue, New York, NY 10016, United States of America

British Library Cataloguing in Publication Data
Data available

Library of Congress Control Number: 2018941696

ISBN 978–0–19–882571–5 (hbk.)
ISBN 978–0–19–882572–2 (pbk.)
DOI: 10.1093/oso/9780198825715.001.0001

Links to third party websites are provided by Oxford in good faith and
for information only. Oxford disclaims any responsibility for the materials
contained in any third party website referenced in this work.

Preface

The dynamics and control of road vehicles is a practical subject, but to truly master it, one must engage with a number of areas of theory.

In 1623 Galileo Galilei wrote about the importance of mathematics in the physical sciences in *The Assayer* (Italian: *Il Saggiatore*):

> Philosophy [i.e. physics] is described in this grand book [i.e. the universe], which while permanently open to our gaze cannot be understood unless one first learns the language, and reads the characters, in which it is composed. It is written in the language of mathematics, and its characters are triangles, circles, and other geometric figures without which it is humanly impossible to understand a single word.

There are hundreds (and probably thousands) of car- and motorcycle-related books available to the enthusiast. Almost all of these books are devoid of theoretical underpinning—our aim is to provide this underpinning. In our view the ready availability of powerful computers and computer-based modelling, analysis, and optimisation tools make the importance of understanding the basics no less important now than it was before the computer age.

The first focus of this book is classical mechanics and its use in building vehicle and tyre dynamics models. The second is optimal control, which is used to solve a range of minimum-time and minimum-fuel, and track-curvature reconstruction problems. As is known classically, all this material is bound together by the calculus of variations and stationary principles. Classical mechanics and optimal control are both beautiful subjects. We cannot hope to treat either fully, but we do hope to provide a thorough introduction to these topics and point the reader in the direction of the literature that will do these topics proper justice. We have supplemented our treatment of this material with a number of examples that were designed to highlight obscurities and subtleties in the theory. When determining the scope of this text, we were guided by the following principles:

- In order to achieve a proper understanding of the subject, the basics must be covered thoroughly. While we are the first to acknowledge the power and versatility of many contemporary computer codes, trying to subcontract understanding to a computer program that was written by a third party is foolish. In sympathy with the motto in [1], 'the purpose of models is insight, not numbers'.
- The theoretical underpinning must be accessible to both practising engineers and researchers. There is little point in offering analysis tools that, because of their complexity, are unlikely to find their way into everyday usage.
- We hope to expose fully the links between mechanics and optimal control, and how these links can be used in the modern automotive industry.

This is a book for senior undergraduates, graduates, academics, and practising automotive engineers. The assumed background includes: (a) linear differential equations; (b) linear algebra and matrix theory; (c) a introductory course in classical control theory that covers transfer functions and frequency response theory; (d) linear systems theory, including state-space system descriptions; and (e) basis mechanics and optimal control.

A particular strength of the book is its unified treatment of tyre, car, and motorcycle dynamics, and the application of optimal control to vehicle-related problems within a single text. These topics are usually treated independently and we believe that a present-day vehicle dynamicist should be familiar with all of these subject areas.

Chapter 1 is almost entirely discursive and covers the early history of road vehicles, outlining some of the important technological achievements that underpin the development of modern road-vehicular transport. The focus is on bicycles, motorcycles, and road cars; the history of steering mechanisms for four-wheeled vehicles is considered early on. We also discuss several early engine, suspension, and tyre developments.

Chapter 2 provides a comprehensive review of the classical mechanics required when building vehicle models. We review both vector-based methods and the variational approach to classical mechanics—we have made an effort to highlight the links between the two approaches. The connections between classical mechanics and mathematical physics are many and varied, but we would characterize our treatment as being philosophically aligned with the theoretical physicist, touching on modern geometric ideas. We study a wide range of illustrative examples with a particular focus on nonholonomic systems that include Čhaplygin's sleigh, rolling balls, and rolling discs. We also discuss briefly equilibria and stability, and the connection between time-reversal symmetry and dissipation.

Chapter 3 focuses on modern tyre modelling. While classical two-dimensional nonholonomic constraint models work reasonably well at very low speeds, these models are not acceptable in realistic applications. This chapter aims to explain the mechanisms and modelling issues related to the generation of tyre forces and moments. Physical models such as the brush and string models are used to clarify the basic concepts. Building upon these findings, the empirical models widespread in vehicle dynamics analyses are discussed. An overview of some of the advanced models currently used in the industry is also given.

In an attempt to 'sneak up' on modern vehicle modelling, Chapter 4 considers a number of precursory car and bicycle models, as well as an important oscillatory phenomenon commonly referred to as shimmy. We begin the chapter by considering a classical single-track car model that will be used to analyse such things as yaw stability, the effects of acceleration and braking, and some of the influences of cornering. After that we consider the Timoshenko–Young bicycle model that has a roll freedom. Bicycle stability is then considered with the help of a point-mass bicycle model. Shimmy, including gyroscopic shimmy, is considered in the concluding part of the chapter. The reader will also be introduced to the fundamental notions of *oversteer* and *understeer*. In a subsequent chapter the single-track models introduced in this chapter will be developed into fully fledged bicycle, motorcycle, and car models; shimmy-related phenomena will reappear in the context of both bicycle and motorcycle dynamics.

Chapter 5 provides a comprehensive analysis of the bicycle model that is based on Whipple's seminal 1899 paper [2]. This remarkable work contains, for the first time, a set of nonlinear differential equations that describe the general motion of a bicycle and rider. Since the appropriate computing facilities were not available at the time, Whipple's general nonlinear equations could not be solved and consequently were not pursued beyond simply reporting them. Instead, Whipple studied a set of linear differential equations that correspond to small motions about a straight-running trim condition at a given constant speed. In the context of the nineteenth century, stability properties could be evaluated using the Routh criterion. Some of the handling characteristics of bicycles are analysed using control-theoretic ideas. Extensions relating to accelerating and braking vehicles, tyred vehicles, and vehicles with flexible frames are also considered.

Chapter 6 deals with road-surface modelling, and vehicle suspension systems and their ride dynamics. A wide variety of car and motorcycle suspension configurations are now available. While most of these systems are 'very different' from each other, many of their important properties can nonetheless be analysed within a common ride dynamics framework. The chapter begins with an analysis of the simple two-degree-of-freedom single-wheel-station (quarter-car) model. We discuss the validity of this model, and study several of its properties including its mode shapes, its invariant equation and the associated interpolation conditions, its frequency-response characteristics, design compromises, and suspension component-value optimization. We then discuss the single-track suspension model that can be used for motorcycles or single-track car models. There are invariant equations, interpolation conditions, frequency-response characteristics and design compromises associated with this model too. In the last part of this chapter we study full-vehicle-suspension models, including their mode shapes, chassis warping, invariant equations, and (multivariable) interpolation constraints. Under certain known circumstances the half-car and full-car models decompose into isolated single-wheel-station systems. We conclude the chapter with some introductory remarks relating to suspension synthesis from a passive circuit-theoretic perspective.

Chapter 7 considers some of the modelling techniques used to replicate the important features of bicycle, car, motorcycle, and driver behaviours. Contemporary high-fidelity vehicle modelling goes beyond what can be accomplished using pencil-and-paper methods, and moves into the domain of computer-assisted modelling, a combination of symbolic, numerical, and manual methodologies. These 'advanced' vehicle models include a number of features that were neglected in models described in Chapters 4 and 5, including suspensions with anti-dive/-squat geometries, environmental influences such as aerodynamic effects, three-dimensional road geometries, and tyre models of the type described in Chapter 3.

Chapter 8 focuses on nonlinear optimal control and its applications. The chapter begins by introducing the fundamentals of optimal control and typical problem formulations. This is followed by the treatment of first-order necessary conditions including the Pontryagin minimum principle (PMP), and dynamic programming and the Hamilton–Jacobi–Bellman equation. Singular arcs and bang–bang controls are relevant in the solution of many minimum-time and minimum-fuel problems and so these issues are discussed with the help of examples that have been worked out in detail. The chapter then turns towards direct and indirect numerical methods suitable for large-scale optimal control problems. It concludes with an example relating to the calculation of an optimal track curvature estimate from global positioning system (GPS) data.

Chapter 9 considers the solution of minimum-time and minimum-fuel vehicular optimal control problems. These problems are posed as fuel usage optimization problems under a time-of-arrival constraint, or minimum-time problems under a fuel usage constraint. The first example considers three variants of a simple fuel usage minimization problem under a time-of-arrival constraint. The first variant is worked out theoretically, and serves to highlight several of the structural features of these problems; the other two more complicated variants are solved numerically. The second example is also a multi-stage fuel usage minimization problem under a time-of arrival constraint. More complicated track and vehicle models are employed; the problem is solved numerically. The third problem is a lap time minimization problem taken from Formula One and features a thermoelectric hybrid powertrain. The fourth and final problem is a minimum-time closed-circuit racing featuring a racing motorcycle and rider.

We believe that there is enough material in this text, when used as a lecture-content smorgasbord, to support several courses. Chapter 2 could be used to support (at least partially) a course in advanced classical mechanics—many other books could do the same. Chapter 3 could be used to support an advanced course in tyre dynamics. Parts of Chapters 4 to 7 could be used to support a course on advanced vehicle dynamics. Unlike many courses and books, this material is not restricted to four-wheeled vehicles, or single-track models of four-wheeled vehicles. Chapters 8 and 9 could be used to support a course on applied vehicular optimal control.

In writing this book, our aim was to produce a comprehensive and yet accessible text that emphasizes the theoretical aspects of vehicular modelling and control. In any exercise of this sort, the selection of material is bound to involve taste and compromise. We have made no attempt to review all the material that could be construed as being relevant to vehicular modelling and control. Rather, we have restricted our attention to work that we believe will be helpful to readers in developing their knowledge of the fundamentals of the subject, and to material that has played a role in educating us. In the case of well-established theory, we have made extensive use of classical texts, and have sought to illustrate subtle and obscure aspects of the material by example. Despite our best efforts, there is bound to be important work that has escaped our attention. To those authors, we offer our sincere apologies.

Acknowledgements

As with any major book-writing undertaking, we owe a debt of gratitude to many colleagues, collaborators, students, and postdocs, who shared many hours of their time discussing technical problems with us.

It is impossible to study vehicle dynamics in any serious way without some mastery of classical mechanics. We would like to thank Vittore Cossalter, Roberto Lot, Jaap Meijaard, and Robin S. Sharp who helped us to understand this material. These discussions were augmented by studying many of the classic books alluded to in Chapter 2.

Tyres play a fundamentally important role in the dynamics of all road vehicles. We learnt much of what we know about tyres from Vittore Cossalter, Hans Pacejka, and Robin Sharp as well as their review papers and books. Their research made a major contribution to Chapter 3.

Our interest in vehicle dynamics began with a fascination with bicycles, which then moved on to cars and motorcycles. We would like to acknowledge countless discussions with those already mentioned, as well as with Karl Åström, David J. Cole, Mont Hubbard, Ichiro Kageyama, Luca Leonelli, Jason K. Moore, Arend Schwab, Malcolm Smith, and Efstathios Velenis who all contributed, in one way or another, to the material in Chapters 4, 5, 6, and 7.

In the latter parts of the book we turn our attention to optimal control and its application to vehicle-related problems. We would like to thank Enrico Bertolazzi, Francesco Biral, Sina Ober-Blöbaum, and Anil V. Rao for sharing their knowledge of this topic with us and for contributing to the material in Chapters 8, and 9.

The people who really made our book-writing exercise worthwhile were our students and postdocs (past and present): Nicola Dal Bianco, Flavio Farroni, Peter Fussy, Edoardo Marconi, Mehdi Imani Masouleh, Gareth Pease, Giacomo Perantoni, Mark Pullin, James Sadauckas, Ingrid Salisbury, Amrit Sharma, Davide Tavernini, Anthony Tremlett, and Matteo Veneri, who gave this project more support and momentum than they will ever realize. Gareth Pease and Giacomo Perantoni made a substantial contribution to the examples in Chapter 9.

We would also like to thank Henry MacKeith for his careful copy-editing work.

David Limebeer
Matteo Massaro

Oxford
August 2018

Contents

1

A History of Road Vehicles

Road vehicles have continued to evolve over the last two centuries from their inception at the height of the industrial revolution in Europe. This chapter aims to outline some of the important technological achievements that underpin the development of modern road vehicular transport. It will hopefully also provide an insight into some of the key features of these complex machines. A complete history of two- and four-wheeled vehicle development will not be attempted, since there are many excellent books on this subject. These hundreds of books are mostly non-technical and come from a wide-ranging authorship. Further reading on the history of bicycles can be found in [3–5]. Since motorcycling history is not the focus of these texts, the reader may wish to consult [6–10] on motorcycle dynamics and design. The history of the motor car can be found in titles such as [11–15], which can be studied in combination with many more books and online resources. A repeat of most of this material is both unnecessary and a distraction from our primary focus, which is the dynamics, modelling, and control of road vehicles.

1.1 Prehistory

Rutted and potholed roads, exacerbated by steering deficiencies, made carriage journeys in eighteenth-century England notoriously dangerous. To alleviate the danger and discomfort of these journeys, Erasmus Darwin (Figure 1.1), grandfather of Charles Darwin, developed a steering mechanism that improved the comfort and stability of the horse-drawn carriage [16]. This mechanism, reinvented over fifty years later, was widely used in early motor cars.[1]

1.1.1 Darwinian steering

Darwin began trying to improve carriage steering in about 1759, within three years of starting his medical practice as a physician at Lichfield. Darwin focused on two defects in carriage steering design, which are easily appreciated by studying Figure 1.2.

The first defect relates to passenger comfort. As seen in Figure 1.2, steering was achieved by rotating the front axle about its centre. In this arrangement the front wheels had to be small, so that the axle could rotate the front wheels under the carriage

---

[1] In the early 1800s G. Lankensperger came up with the same idea, which was patented by Rudolph Ackermann in 1818. George Lankensperger's name seems to have been largely forgotten and this type of steering arrangement has become known in modern times as an implementation of 'Ackermann steering'.

Dynamics and Optimal Control of Road Vehicles. D. J. N. Limebeer and M. Massaro.
© D. J. N. Limebeer and M. Massaro 2018. Published in 2018 by Oxford University Press.
DOI: 10.1093/oso/9780198825715.001.0001

Figure 1.1: Erasmus Darwin (1731–1802); physician, philosopher, slave-trade aboli-
tionist, inventor, poet, and grandfather of Charles Darwin. Oil painting by Joseph
Wright of Derby, Derby Museum and Art Gallery.

chassis. Small front wheels meant that the carriage would rise and fall suddenly as it
passed over stones and potholes. This would produce jolts that could injure the horses,
damage the carriage, and cause discomfort to the traveller. The second defect relates
to the danger of overturning. In a conventional carriage, with a rotating front axle,
Darwin observed that the base on which the carriage rests changes from a rectangular
shape to a triangular one under sharp cornering (see Figure 1.2). Darwin's new design
overcame both defects. In the new steering arrangement the front axle is fixed with the
front wheels free to rotate on two short stub axles that are attached to the carriage's
chassis (see Figure 1.3). This design change allowed for larger front wheels, maintained
the 'four-legged stool' posture under turning, and meant that the wheels did not
need to be parallel when they turned. Darwin also observed that it would be better
if, in turning, the inside wheel turned more sharply than the outside wheel. In this

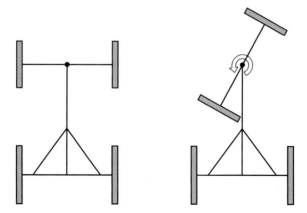

Figure 1.2: Carriage on a rectangular base in straight running (left); the base becomes
triangular under sharp cornering (right).

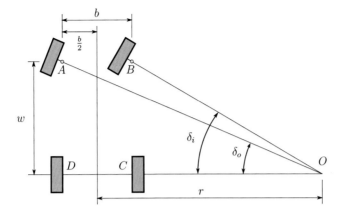

Figure 1.3: Erasmus Darwin's criterion for good steering.

arrangement the front wheels both run tangentially to the track so that there is no scuffing of the wheels on the road, the carriage is easier to pull, it is less likely to overturn, and there is less tyre wear. This begs the questions: Can Darwin's steering ideal be achieved in practice? And, if so, what mechanism should be used? Some clues to the answers to these questions can be found in surviving manuscripts from the 1766–9 time period, which include sketches of four-bar linkages [16].

In 1878 C. Jeantaud, a french car manufacturer, proposed a link mechanism that would achieve practically Darwinian (or Ackermann) steering. This mechanism consists of a four-bar linkage with the two track rods pointing towards the middle of the rear axle [15]. The arrangement was used on several early motor cars with rigid front axles. In order to see how it works, we will now analyse the linkage mechanism using modern kinematic techniques.

It is immediate from Figure 1.3 that

$$\cot \delta_i = \frac{r - \frac{b}{2}}{w} \qquad \text{and} \qquad \cot \delta_o = \frac{r + \frac{b}{2}}{w} = \frac{r - \frac{b}{2}}{w} + \frac{b}{w},$$

and so the condition for ideal Darwinian steering is

$$\delta_o = \cot^{-1}\left(\cot \delta_i + \frac{b}{w}\right). \tag{1.1}$$

In order to approximate (1.1) using a four-bar linkage, we begin by observing using Figure 1.4 (a) that

$$a_1 = a_3 + 2a_2 \sin \beta. \tag{1.2}$$

Next, we use Figure 1.4 (b) to establish that

$$\delta_0 = \beta + \sigma_1 - \frac{\pi}{2} \qquad\qquad \sigma_2 = \frac{\pi}{2} - \beta - \delta_i \tag{1.3}$$

under general steering conditions. We will consider δ_i the mechanism input, while δ_o is the mechanism output. Resolving the linkage displacements along the x- and y-axes

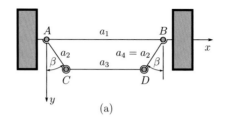

(a)

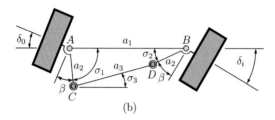

(b)

Figure 1.4: Four-bar steering mechanism.

respectively gives

$$a_3 \cos \sigma_3 = a_1 - a_2 \cos \sigma_1 - a_2 \cos \sigma_2 \tag{1.4}$$

$$a_3 \sin \sigma_3 = a_2 \sin \sigma_1 - a_2 \sin \sigma_2. \tag{1.5}$$

We can now square (1.4) and (1.5), and add them in order to eliminate the external angle σ_3. This gives

$$k_1 - k_2(\cos \sigma_1 + \cos \sigma_2) + \cos(\sigma_1 + \sigma_2) = 0, \tag{1.6}$$

where

$$k_1 = \frac{a_1^2 + 2a_2^2 - a_3^2}{2a_2^2} \qquad k_2 = \frac{a_1}{a_2}.$$

Equation (1.6) is sometimes called Freudenstein's equation in the mechanisms literature [17]. To see how the four-bar linkage steering mechanism works in practice, consider the case in which $w = 3.5\,\mathrm{m}$, $b = 1.5\,\mathrm{m}$, $\beta = \tan^{-1}(\frac{b}{2w})$, and $a_2 = 0.1\,\mathrm{m}$. It is immediate that $a_1 = b$, that a_3 can be computed from (1.2), and that the ideal steering angles on the inner and outer wheels are related by (1.1). The true steering angles, on the other hand, come from solving the nonlinear equations (1.3) and (1.6): for a given δ_i, σ_2 is computed using the right-hand part of (1.3), (1.6) is solved for σ_1, and δ_o is computed using the left-hand side of (1.3). Figure 1.5 shows that the four-bar steering arrangement provides a good approximation to ideal steering behaviour even at high near-wheel steering angles. Figure 1.6 shows the radius of turn and the steering angle error as a function of the near-wheel steering angle. It is observed that for the chosen steering geometry the steering angle error is less than 1.4° even in tight turns of radius $\leq 10\,\mathrm{m}$.

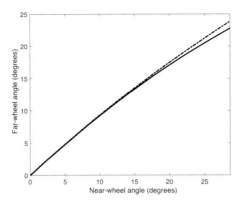

Figure 1.5: Darwinian steering angles. The dot-dash line shows the ideal Darwinian steering angles δ_i and δ_o (see Figure 1.3), while the solid line shows the angles achieved using a four-bar linkage.

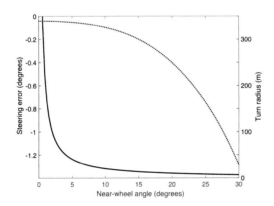

Figure 1.6: Steering angle and steering radius. The dotted line shows the steering angle error as a function of the near-wheel steering angle δ_i. The solid line shows the vehicle's radius of turn as a function of the near-wheel steering angle.

1.1.2 The differential

In 1827 Onésiphore Pecqueur, a watchmaker, took out a patent that was to become another key element in the development of the automobile. In order to appreciate his ground breaking invention, we consider a pair of wheels on an axle turning at ω rad/s.

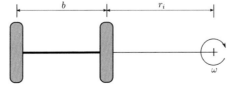

If both wheels have the same radius, and rotate without slipping,

$$\frac{\Omega_o}{\Omega_i} = 1 + \frac{b}{r_i}$$

where Ω_i and Ω_o are the angular velocities of the inner and outer wheels respectively. If these wheels are attached to a rotating axle, this speed difference, would result in slippage and tyre wear. Aware of this speed differential, Pecqueur used his watchmaking skills to invent the mechanical differential; one of the first automotive engineers was therefore a watchmaker. Unfortunately, Pecqueur never lived to see the vital role that the differential has played in the development of the motor car. His differential was not practically applied until 1869, after Beau de Rochas and Nikolaus Otto invented the internal combustion engine.

1.2   Birth of the motor car

It is said that success has a thousand fathers—the motor car is no exception. In the middle of the nineteenth century engineers from many countries were developing variants of the horseless carriage inspired by Nicolas-Joseph Cugnot's work on the first full-scale self-propelled mechanical vehicle (shown in Figure 1.7) [13, 14]. A variety of

Figure 1.7: The 1769 Cugnot fardier in the Musée des Arts et Métiers, Paris.

steam-powered road vehicles were used during the early part of the nineteenth century, including steam cars, steam buses, and steamrollers.

In Britain, sentiment against these contraptions led to the draconian 1865 'Red Flag Act', which required all road locomotives to travel at a maximum of 4 mph in the country and 2 mph in the city—as well as requiring a man carrying a red flag or lantern to walk in front of the vehicles to warn bystanders of its approach. It has been estimated that the Red Flag Act retarded Britain's work on the motor car by thirty years. The Red Flag Law was repealed in 1896, by which time the internal combustion engine was well into its infancy [11].

1.2.1   Early engines

In the early stages of the development of the motor car, the research focus was the engine. The Belgium engineer Jean Joseph Étienne Lenoir is credited with the invention of the first commercially successful internal combustion engine, which was a converted double-acting steam engine with slide valves to admit the air–fuel mixture and to discharge exhaust products [18]. Lenoir's engine was a two-stroke spark-ignition engine that used a mixture of coal gas and air. Though only about 4 per cent efficient, it was a durable machine, and by 1865 more than 400 were in use in France and 1,000 in Britain for low-power tasks such as pumping and printing. In 1862 Lenoir built the first automobile with an internal combustion engine. He had adapted his engine to run on liquid fuel and he required two to three hours to make a ten-kilometre trip.

Lenoir's engine interested Nikolaus August Otto, a German engineer, who observed that more power could be obtained by compressing the air–fuel mixture before feeding it into the cylinder [11]. Although the concept of the four-stroke engine, with the vital 'mixture compression prior to ignition' feature, had been invented and published in a pamphlet in 1862 by Alphonse Beau de Rochas [19], this paper remained unknown until 1884, when the patent filed by Otto in 1876 was invalidated. Otto was the first to make it practical. The four strokes are: (1) the downward intake—coal gas and air enter the piston chamber, (2) the upward compression stroke—the piston compresses the air–fuel mixture, (3) the downward power stroke—the fuel mixture is ignited by an electric spark, and (4) the upward exhaust stroke—the exhaust gas is discharged from the piston chamber. Otto perfected the four-stroke 'Otto cycle' and it was this power unit that opened the way to the motoring age.

1.2.2   Early road vehicles

In the 1860s Karl Benz bought a bicycle and convinced himself that vehicles for personal transport should have three, or preferably four wheels, and be power driven [11]. In the spring of 1885 Benz installed a two-stroke engine that he had developed into a three-wheeled vehicle, and in July 1886 the Benz motor car ran successfully on public roads. Benz's 1886 patent 'Fahrzeug mit Gasmotorenbetrieb (vehicle with gas-engine drive) [20] represents the first road vehicle designed and built specifically for mechanized transportation using an internal combustion engine. As shown in Figure 1.8, the Benz Motorwagen's innovative features included a four-stroke engine, three wire-spoke wheels, and two-passenger seating. Its 1,600-cc, three-quarter-horsepower engine produced a top speed of approximately 8 mph. In 1894 Enrico Bernardi developed the first Italian car (see Figure 1.9), which features one of the earliest jet carburettors and cam-operated valves.

At the time there were no car commercials and gasoline was a household cleanser sold at the local chemist. Even the best roads were rough, narrow tracks and the public was far from sold on the need for, or practicality of, motorized transportation. Benz's car would change all that, but not without a critical advertising campaign care of his wife Bertha. In 1888 Bertha appropriated a Motorwagen and drove 66 miles from Mannheim to Pforzheim to visit her mother. This was apparently the first long-distance trip ever attempted in a horseless carriage. Her 8-mph exploit alarmed the local citizens, while earning priceless publicity for the Motorwagen. She informed her husband of her safe arrival by telegram and returned home the next day. Bertha's journey also identified several refinements to make the car safer and more powerful, including brake improvements and an extra gear for climbing hills. Benz began selling

Figure 1.8: Replica of the Benz patented motor car. The original was the world's first gasoline-engined automobile. Unlike Daimler's motorized carriage, this motor car formed an autonomous entity of chassis and engine. Karl Benz designed the car as a three-wheeler, because he was not satisfied with the steering systems available for four-wheeled vehicles back in 1886. Mercedes-Benz Museum, Stuttgart,© Daimler AG.

cars that summer, and his model 3 Motorwagen was a star of the 1889 World Fair in Paris. Today, Bertha's famous road trip is celebrated every two years with an antique car rally along the same stretch of highway.

In 1893 Karl Benz achieved another crucial technical breakthrough when he registered a 'vehicle steering device with steering circles arranged at a tangent to the

Figure 1.9: 1894 Bernardi car, currently in the Museum of Machines 'Enrico Bernardi', University of Padova. It's 1.5–2.5 hp engine runs in the speed range 430–800 rpm. The car is still running.

Figure 1.10: The Benz Victoria motor car was the first vehicle with the double-pivot steering for which Carl Benz filed a patent in 1893. Mercedes-Benz Museum, Stuttgart, © Daimler AG.

wheels' with the Imperial German Patent Office (patent no. DRP 73 515, issued on 14 March 1894). His double-pivot steering system was used in the Victoria and Vis-à-Vis models—the first four-wheeled motor cars from Benz & Cie, which superseded the three-wheeled Patent Motor Car. Benz's steering arrangement is shown in Figure 1.10, where the front-wheel stub axles and four-bar linkage are clearly visible.

In 1872 Gottlieb Daimler was appointed the technical director to the firm of Otto and Langen, where he remained for ten years [11]. After a dispute with Otto in 1882, Daimler and Wilhelm Maybach set up their own company—Daimler was determined to improve on Otto's engine and put it on wheels. To do this he built an extension to the greenhouse at his villa in Cannstatt, which became his development laboratory. Daimler and Maybach concentrated on producing the first lightweight, high-speed, gasoline-fuelled engine. By 1885 they had developed an engine with a surface-mounted carburettor that vaporised the petrol and mixed it with air; this Otto-cycle engine produced a fraction of a kilowatt. This 'greenhouse emission', shown in Figure 1.11, was the 'Grandfather Clock' [12].

The Daimler-Maybach vehicle testbed was a two-wheeler, which emphasized the need for weight reduction and compact design; we will return to this topic in Section 1.4. In 1886 Daimler placed a Grandfather Clock in an American-made carriage and secured his place in automotive history as the inventor of the four-wheeled, four-stroke, gas-powered car. Still air-cooled, but now with an output of 1.1 hp, the engine was centrally located in front of the rear bench seat (see Figure 1.12). In 1887 the engine became water-cooled but still lacked an efficient radiator. The crucially new feature of this vehicle was its power transmission. Depending on the chosen ratio, the engine's belt pulley drove discs of different sizes on a layshaft. The power was transferred to gears on the rear wheels via pinions on both sides. Instead of a differential, a slip clutch was mounted on each side of the layshaft.

Figure 1.11: The Daimler-Maybach 'Grandfather Clock' single-cylinder engine was the world's first small, high-speed internal combustion engine to run on gasoline; it was light and yet powerful enough to drive a vehicle. Mercedes-Benz Museum, Stuttgart, © Daimler AG.

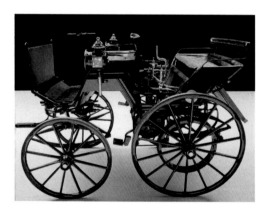

Figure 1.12: Daimler motor carriage, 1886—the first four-wheel vehicle driven by a high-speed combustion engine ('Grandfather Clock'). Mercedes-Benz Museum, Stuttgart, © Daimler AG.

1.2.3   Early car industry

The end of the First World War presented several great challenges for the German automotive industry. The collapse of the currency, the growing number of new automotive companies that had previously operated in the armaments sector, and the loss of important foreign markets all had a detrimental effect on its business. These problems, plus the penetration of the domestic market by foreign producers such as the Ford Motor Company, which were assisted by the Peace Treaty of Versailles, led to a severe structural crisis in the German automotive industry. As with other industries, companies strived to form larger conglomerates in order to achieve economies of scale. After initially entering into a joint venture with the aim of rationalizing production, the Daimler Motorengesellschaft and Karl Benz's Benz & Cie merged in 1926 to form Daimler-Benz AG.

As the stifling 1865 Red Flag Act mentioned earlier was hindering the development of the automobile in Britain, the first cars to run on British roads were of German or French origin. In 1896 the Red Flag Act was repealed and bicycle manufacturers such as Riley, Swift, and Sunbeam graduated to cars [14]. Herbert Austin, manager of the Wolseley Sheep Shearing Machine Company, was one of the early pioneers. The Austin Motor Company, based at Longbridge, was later to become one of the greatest car manufacturers in the world. In 1884 Frederick Henry Royce founded F. H. Royce and Company, making domestic electric fittings. In 1894 they started making dynamos and electric cranes. The company was re-registered in 1899 as Royce Ltd with a public share flotation. Following a decline in trade after the Second Boer War, and the arrival of increasing competition in cranes and dynamos from Germany and the United States, Royce began considering the motor car as a potential new product for the company. Royce imported a used Decauville in 1903, but was dissatisfied with its performance, and so redesigned and built a superior vehicle [14]. Three copies of the model were made, and, due to Royce's rigorous testing, were unusually reliable for the time. Later, Charles Rolls joined Royce and the first Silver Ghost was on road in 1906 [14]. Even in those early days, the engines could do over 30,000 km before requiring a service. A total of 7,874 Silver Ghost cars were produced from 1907 to 1926, and it is now one of the most valuable cars in the world (see Figure 1.13).

By the end of the nineteenth century the motor car industry had become an international business. In France, Armand Peugeot led the family business in its quest to produce its first motorized vehicle. In 1889, as a result of a collaboration with steam specialist Léon Serpollet, he produced the Serpollet-Peugeot—a steam-powered three-wheeler. By the following year, Armand had abandoned steam in favour of petrol, and had built Peugeot's first four-wheeler, a petrol-driven quadricycle with a Daimler engine [11]. By 1903, France was by far the most advanced motoring nation in Europe, producing half the world's output of cars—some 60,000 per year. The Italians were also active in the motor car industry from its inception. Fiat SpA, or Fabbrica Italiana Automobili Torino (Italian Automobile Factory of Turin), was founded in 1899, Lancia in 1906, Alfa Romeo in 1910, Maserati in 1914, Ferrari in 1947, Abarth in 1949, and Lamborghini in 1961. Italy's automotive industry is best known for its automobile designs, sports cars, and supercars. In 2014 a 1962 Ferrari 250 GTO was sold for $38,115,000 (£22,843,633) at Bonhams' Quail Lodge auction in Carmel, California.

Figure 1.13: Rolls-Royce Silver Ghost; centenary celebrations at the Midland Hotel Manchester in 2004. In 1907, the Silver Ghost was declared 'the Best Car in the World' after travelling from London to Glasgow twenty-seven times, covering 14,371 consecutive miles. Photograph by Malcolm Asquith.

The automotive industry in the United States began in the 1890s and evolved to become the largest in the world. It was, however, overtaken by Japan in the 1980s, and subsequently by China in 2008. The American motor vehicle industry began with hundreds of cottage-scale manufacturers, but by the end of the 1920s it was dominated by General Motors, Ford, and Chrysler [13]. Henry Ford began building cars in 1896 and founded his own company in 1903. The Ford Motor Company pioneered mass production, with the first conveyor-belt-based assembly line in 1913, producing the Model T. Henry Ford focused on delivering one product for the masses; one car, one colour, one price. Ford was also a pioneer in establishing foreign manufacturing facilities, with production facilities created in England in 1911, and Germany and Australia in 1925. The General Motors Corporation was founded in 1908 by William

Figure 1.14: 1962 Ferrari 250 GTO from the Ralph Lauren collection on display at the Boston Museum of Fine Arts.

Durant. The company initially acquired Buick, Oldsmobile, and Oakland (later to become Pontiac) in 1908. In 1909 GM acquired Cadillac, along with a number of other car companies and parts suppliers. GM also became an innovator in technology under the leadership of Charles F. Kettering. GM followed Ford by expanding overseas, and purchased England's Vauxhall Motors in 1925, Germany's Opel in 1929, and Australia's Holden in 1931. During the late 1920s, General Motors overtook Ford to become the largest car manufacturer in the world. After leaving GM in 1920, Walter Chrysler took control of the Maxwell Motor Company, and reorganized it into the Chrysler Corporation. He then acquired Dodge in 1927. This acquisition gave Chrysler the manufacturing facilities and dealer network that it needed to significantly expand its production and sales. In its early years Chrysler pioneered the first practical mass-produced four-wheel hydraulic brakes. Chrysler also pioneered rubber engine mounts to reduce vibration, oilite bearings, and crank shaft superfinishing.

1.3 Early bicycles

Early bicycles are notable for their variety of styles and the ingenuity in their design. Various interesting and occasionally amusing concepts were built to overcome inherent design problems that would eventually be solved by the advent of the 'safety' bicycle.

The first recorded instance of a two-wheeled velocipede was in 1817 when the German inventor Baron Karl von Drais (see Figure 1.15), inspired by the idea of skating without ice, invented the Draisine, or running machine [3, 4, 21], as shown in Figure 1.16. The initial impetus for Drais to develop the Draisine, and its subsequent short-lived success, was spurred by the need for horseless transport. In 1817 oats had become expensive, because of 'the year without a summer' in 1816 that followed the volcanic eruption of Mount Tambora in Indonesia the year before [21, 22]. Thus horses became expensive to keep, and it is said that to some they became more attractive as a food source than as a means of transport (although we have not found any definitive proof for this version of history, it remains plausible).

Figure 1.15: Karl Drais, the inventor of the first 'bicycle'.

Figure 1.16: The Draisine, patented in 1817 by Karl Drais as the Laufmaschine. Riders could steer the front wheel while resting their elbows on the front pad and pushing along with their feet. Image reproduced with permission of the Bicycle Museum of America, New Bremen, Ohio, USA.

The Draisine and its copies were nicknamed the dandy horse after the foppish men who rode them, or the hobby horse after the child's toy. Two in-line wheels were supported by a wooden frame, with the front wheel steerable about a vertical axis. Balance was ensured because the only method of propulsion was through the rider's feet contacting the ground. The original Draisine had a spoon brake on the rear wheel that pressed against the iron cladding of the wooden wheels (its iron tyres). All told, it had a combined weight of approximately 22 kg. The vertical steering axis is likely to have made it a fast-steering, jittery ride.

On 12 January 1818, Drais received his first patent for the Draisine from the state of Baden; a French patent was awarded a month later [4]. The Draisine is illustrated in Figure 1.16. In addition to the stuffed leather saddle, it featured a small stuffed rest,

Figure 1.17: The hobby horse, built in 1818, was designed by Denis Johnson imitating the original Draisine. The hobby horse was a 'walking machine' and was popular for approximately two years. From the cycles exhibits in the Science Museum, London.

on which the rider's arms were laid, to help maintain balance and comfort. Optional extras for the Draisine included an umbrella, a sail for windy days, a tandem version and a version with an extra seat in the front for ladies [4].

To popularize his machine, Drais travelled to France in October 1818, where a local newspaper praised his skilful handling of the Draisine as well as the grace and speed with which it descended a hill. The reporter also noted that the baron's legs had 'plenty to do' when he tried to mount his vehicle on muddy ground. Despite a mixed reception, the Draisine enjoyed a short period of European popularity.

The Draisine wasn't patented in England, and in late 1818 the Draisine was copied and improved by Denis Johnson, who used a steel frame to reduce its weight. Johnson removed the brakes—a retrogressive step. He began manufacturing his design, as shown in Figure 1.17, and it became popular in England. Despite the public's enduring desire for rider-propelled transportation, the Draisine was too flawed to survive as a viable contender, the lack of drive and Johnson's removal of the brakes being the most serious impediments.

1.3.1 Driven bicycles

Traditional credit for introducing the first pedal-driven two-wheeler, in approximately 1840, goes to the Scotsman Kirkpatrick MacMillan [3], although this is a matter of some controversy. MacMillan's design resembled a modified Draisine, but with the rear wheel driven by a crank rocker mechanism; a 'treadle' drive. It was an unusual, and arguably innovative, design because the concept of driving the rear wheel, which is the standard in modern bicycles, seems from this point to have been ignored and forgotten until the 'safety' was invented in 1884.

Another account has it that pedals were introduced in 1861 by the French coach builder Pierre Michaux when a customer brought a broken Draisine into his shop for repairs and Michaux instructed his son, Ernest, to affix cranks and pedals to it. In September 1894 a memorial was dedicated in honour of the Michaux machine. Figure 1.18 shows this vehicle, which weighed an unwieldy 27 kg and was known at the time as the velocipede, but is often referred to as the 'boneshaker'. Its nickname derived from its construction which, in combination with the cobblestone roads of the day, made for an uncomfortable ride.

Also laying claim to the invention of the driven bicycle was Pierre Lallement, who, in 1866, was granted a US patent [23] (see Figure 1.19) for a boneshaker-style bicycle, which it is reported he first made in 1863 [4]. The question as to whether Michaux, or Lallement, was first to have the 'driven-front-wheel' idea is unlikely to ever be answered.

The classic boneshaker had wooden, spoked, wheels with iron 'tyres' that would slip on loose surfaces. Its directly driven front wheel was tiring to pedal fast and rubbed against the rider's legs when cornering. Braking was achieved either through back-pedalling or through an ineffective spoon brake on the rear wheel applied by a cord connected to revolving handlebars or levers. The wheels ran on bronze axle bushes lubricated by reservoirs of whale oil. The long metal spring, onto which the leather saddle was attached, was the only suspension-like component, and was evidently not

Figure 1.18: The boneshaker, by Pierre Michaux et Cie of Paris, France, circa 1869. In the wake of the Draisine, the boneshaker represented the next major development in bicycle design. It was developed in France and achieved its greatest popularity in the late 1860s. Its most significant improvement over the Draisine was the addition of cranks and pedals to the front wheel. Different types of (not very effective) braking mechanisms were used, depending on the manufacturer. From the cycles exhibits in the Science Museum, London.

up to its task of absorbing road shocks. The steep steering axis was an improvement in stability terms over the Draisine, and is likely to have been marginally more stable.

Although 'velocipedomania' for the boneshaker design only lasted about three years (1868–70), the popularity of the machine is evidenced by the large number of surviving examples.

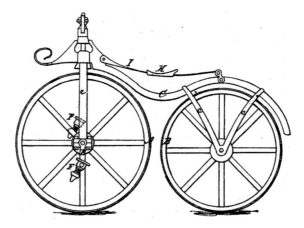

Figure 1.19: Pierre Lallement's patent drawing of his velocipede design from 1866 [23].

1.3.2   Transition to the ordinary

There were a number of notable improvements to the boneshaker design, including a kickstand and, of more interest to dynamicists, the addition of a 25 mm (1 inch) thick solid rubber tyre by Clément Ader in 1869 [4].

Also in 1869, wire-spoked wheels were introduced that became known as 'spider' wheels; they were considerably lighter than wooden wheels and were easier to make true. To alleviate the wheels rubbing against the rider's legs, a centrally hinged frame (rather than a front steering axis) was made, as were rear steering models, but neither proved successful [5]. Peyton and Peyton introduced a rear-driven bicycle, the 'improved bicycle' in 1869, which used a crank rocker mechanism to drive the rear wheel—much like the supposed MacMillan bicycle from 29 years earlier—as did McCall and other inventors in the same year. The rear treadle-driven design's complexity apparently prevented its widespread adoption [4].

Bearings were another important improvement made during this era. At first conical bushes were used, which could easily be adjusted for wear over time. By the late 1870s hard-wearing machine-made ball bearings were being manufactured, and were adopted in the bicycles of the day [4]. These spherical bearings reduced friction and wear, and increased axle rigidity, which undoubtedly improved riding stability.

Directly driven front wheels required the rider to pedal with a high cadence in order to reach high speeds. Internal front-hub gearing was used to overcome this problem in some designs, but it was costly, heavy, and complex. Unlike their wooden forebears, spider wheels offered the possibility of almost arbitrary wheel sizes, up to a radius equalling a person's leg length. This presented a simple solution to the problem of the necessarily high cadence; by increasing the wheel radius, high gearing was automatically achieved.

This led to boneshakers with larger-radius front wheels and smaller rear wheels, to keep the wheelbase from growing too large. These bicycles were know as 'transitionals', in that they heralded the era of the full-blown, 'high-wheelers', or 'ordinaries'.

1.3.3   The ordinary

By the mid 1870s bicycles had acquired large front wheels and small rear wheels as standard. The large radius of the front wheel provided a smoother ride in addition to its gearing advantage. The front wheel diameter eventually reached up to 1.5 m (60 inches) with a rear wheel diameter of around 0.4 m (16 inches), as shown in Figure 1.20. The seat was so far off the ground that a mounting step had to be positioned over the rear wheel to help the rider vault onto their saddle, which had to be done whilst moving in order to maintain balance. The vastly different wheel diameters earned it the nickname, 'penny-farthing' during its eventual demise.

Improvements in design continued; steel tubing replaced iron tubing, reducing the weight of bicycles to as little as 18 kg. Geared front axles combined with smaller wheels were repeatedly tried, but their weight consistently made them inferior to directly driven high-wheeled bicycles. H. Kellogg introduced suspension components into the high-wheeler [24] through an adjustable spring in the backbone.

The ordinary became popular worldwide as a viable means of commuting. The ordinary enjoyed its heyday in the 1880s. Thanks to its adjustable crank, and several

Figure 1.20: A classic ordinary ('penny-farthing') bicycle. From the cycles exhibits in the Science Museum, London.

other new mechanisms, it racked up record speeds of about 7 m/s (16 mph). But, as is often said, pride comes before a fall. The high centre of gravity and forward position of the rider made the penny-farthing difficult to mount and dismount, as well as dynamically challenging to ride. In the event that the front wheel hit any small obstacle, the entire machine rotated forward about its front axle, and the rider, with legs trapped under the handlebars, was unceremoniously propelled head-first into the road. Thus the term 'taking a header' came into being.

1.3.4   Rise of the safety bicycle

In 1876 Henry J. Lawson proposed a rear-driven bicycle that he called the 'safety bicycle'. It was driven by a treadle mechanism (crank-rocker) as had been tried before, but the other enhancements made to ordinarys could now be applied. This first iteration was not very successful as it was deemed too rickety [4], so in 1879 Lawson introduced what he called the 'bicyclette' [25], Figure 1.21, which was a safety with a chain-driven rear wheel. Chain drives were not new (they had been used on tricycles earlier), but their inclusion on bicycles was revolutionary. Other manufacturers of the time also introduced chain drives for the rear wheel of this new style of bicycle, but it could not immediately supplant the high-seated ordinary as the model of choice.

Modifications of the ordinary high-wheeler were made to improve its safety, which resulted in bicycles such as the Facile, which had pedal-driven rockers rather than cranks and allowed the rider to sit lower down and closer to the centre of the frame. The 'American Star' bicycle, patented in 1880 [27], Figure 1.22, swapped the ordinary's front and rear wheels for steering, leaving the rider seated above a large rear wheel driven by rockers and steering a small front wheel through a long steering column. The American Star did enjoy a period of popularity in the USA; aided by stunts such as that shown in Figure 1.23 by a bicycle club enthusiast, which demonstrated its

Figure 1.21: Lawson 'bicyclette', patented in 1879, with a follow-on US patent issued in 1886 [26]. This bicycle was the first commercial rear-wheel chain-driven bicycle, invented in 1879. Science Museum Group Collection.

advantages over the ordinary. However, the rider of the Star was still precariously high, and likely to 'pull a wheelie' and fall off the rear when encountering obstacles.

It wasn't until the Rover safety (see Figures 1.24 and 1.25) was introduced in 1884 that rear-wheel chain-driven bicycles achieved widespread public acceptance. To cement its place as the new state of the art, a professional rider was hired to set a new speed record over 100 miles (7 hours 5 minutes) [4, 5].

The original Rover safety (see Figure 1.24) had a remote steering arrangement, a vertical steering axis and a larger front wheel than rear. This was improved in a revision of the Rover (see Figure 1.25); it had a greater steering axis angle and the wheels were made equal in diameter (0.91 m), which improved its stability. The irony of the modern design's resemblance of the original boneshaker 'did not escape some astute observers' [4].

These early safety bicycles had the wheelbase reduced to as small a distance as

Figure 1.22: The American Star bicycle, patented in 1880, made use of front-wheel steering [27].

Figure 1.23: The American Star bicycle being ridden down the steps of the United States Capitol by Will Robertson of the Washington Bicycle Club in 1885 [4].

was feasible. The crank bracket was placed just in front of the rear wheel and the front wheel was placed so that it didn't interfere with the pedals when turning. The frame accommodated this design with either a curved seat post connecting the seat to the crank bracket, or with an open frame. As the designs of the safety improved, the wheelbase was extended by moving the crank bracket further forward, which allowed a straight tube to connect the seat to the crank bracket, reducing weight for a given stiffness. This design, known as the diamond frame, was first introduced in 1890 and remains the standard pattern for bicycles to this day; it is shown in Figure 1.26.

Figure 1.24: The original Rover safety bicycle, first introduced in 1884, was the first commercially successful rear-wheel chain-driven bicycle [3].

Figure 1.25: An improved version of the Rover safety bicycle. This 1890 model had direct steering and a raked front wheel. It set the basis for the standard bicycle pattern for years to come. Image reproduced with the permission of the Bicycle Museum of America, New Bremen, Ohio, USA.

1.3.5 Further developments

The diamond frame safety bicycle with pneumatic tyres has been the standard modern design for well over a century. That is not to say that further improvements have not been made.

Planetary hub gears, chain rings, and sprockets with derailleurs allowed an efficient power transfer from the rider's legs to the wheels. Folding bikes, first patented in 1893 [29], extended the utility of bicycles by allowing them to be carried and transported more easily. Bicycles were also made that were specifically designed for women—and later marketed for those with reduced mobility, too—which lacked the top crossbar, such as that patented by J. M. Starley in 1888 [30]. These models permitted mounting by stepping through rather than over, and were aptly named 'step-through' frames; see Figure 1.27 for an early example.

Calliper brakes and disc brakes vastly improved braking performance. Lightweight aluminium and carbon frames have reduced bicycle mass. Suspension components

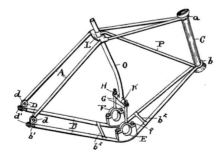

Figure 1.26: An early version of the diamond safety frame, the basic form of which is still used today [28].

Figure 1.27: Early example of a 'step-through' frame specifically designed for women.

including telescopic front forks and rear swinging arms, with springs and dampers, have improved rider comfort and handling, especially on off-road bicycles.

Recumbent bicycles have also been developed. These have a longer wheelbase than diamond framed bicycles and the rider is seated in a reclined position between the wheels, closer to the ground. This position can improve riding comfort and reduces air resistance due to the smaller profile. A recent record, set in 2009, for the longest distance covered by a bicycle in one hour was 90.7 km (56.4 miles) and the fastest human-powered vehicle record of 40.05 m/s (89.6 mph) was set in 2016; both were achieved on recumbent bicycles with enclosed fairings for reduced air resistance.

1.4 Powered two-wheeled vehicles

As early as 1818 a French caricaturist postulated the seemingly ludicrous idea of a Draisine with a steam engine [4]. However, within fifty years the concept had become reality.

1.4.1 First steam-powered machine

If one considers a wooden frame with two wheels and a steam engine a 'motorcycle', then the first machine of this type was probably American. In 1867 Sylvester Howard Roper demonstrated a motorcycle (Figure 1.28) at fairs and circuses in the eastern US. His machine was powered by a charcoal-fired, two-cylinder engine, whose connecting rods drove a crank on the rear wheel. The chassis of the Roper steam velocipede was based on that of the boneshaker bicycle.

Sylvester H. Roper, a pioneer of early powered bicycles, was possibly also the first person to die whilst riding a motorcycle. According to the Boston Globe,

Figure 1.28: Sylvester Roper steam motorcycle, 1867. This vehicle is powered by a two-cylinder steam engine that uses connecting rods fixed directly to the rear wheel. Division of Work and Industry, National Museum of American History, Smithsonian Institution, Washington, DC USA.

> The dramatic fatality occurred (June 1, 1896) yesterday morning at the new Charles River bicycle track, just across the Harvard Bridge on the Cambridge side. The deceased had for years enjoyed a reputation as an able mechanical engineer, who had perhaps been more identified with steam propulsion as applied to carriages and for general road use than any other man in New England. Ever since 1859 he has been at work on various contrivances for conveyances with steam as a motive power. ... The machine was cutting out a lively pace on the backstretch when the men seated near the training quarters noticed that the bicycle was unsteady. The forward wheel wobbled badly, and then suddenly the cycle was deflected from its course and plunged off the track into the sand, throwing the rider and overturning.

In 2002 Sylvester H. Roper was inducted into the American Motorcyclist Association's Hall of Fame.

1.4.2 First motorcycle

Gottlieb Daimler is considered by many to be the inventor of the first true motorcycle, or motor bicycle, since his machine was the first to employ an internal combustion engine. In 1885 Daimler and Maybach combined a Daimler engine with a bicycle, creating a machine with wooden-spoked front and rear wheels with iron tyres as well as a pair of smaller spring-loaded outrigger wheels, named the Reitwagen ('riding car', see Figure 1.29). This wooden motorcycle was designed to be a platform for engine demonstration rather than a practical vehicle. Nonetheless, it represents a true, internal combustion engine-powered motorcycle, and Maybach rode three kilometres from Cannstatt to Untertürkheim at up to 7.5 mph (12 km/h).

Figure 1.29: Daimler and Maybach motorcycle ('Reitwagen'); the 'Grandfather Clock' engine is visible. Gottlieb Daimler, who later teamed up with Karl Benz to form the Daimler-Benz Corporation, is credited with building the first motorcycle in 1885. Mercedes-Benz Museum, Stuttgart, © Daimler AG.

1.4.3 Production motorcycles

Jules-Albert de Dion and his engineering partner Georges Bouton began producing self-propelled steam vehicles in 1882. They filed a patent for a single-cylinder gasoline engine in 1890 and production started five years later. The de Dion Bouton engine, which was a small, lightweight, high-revving four-stroke 'single', used battery-and-coil ignition that replaced the troublesome hot-tube ignition system. The engine had a bore of 50 mm and a stroke of 70 mm, giving rise to a swept volume of 138 cc. De Dion Bouton also used this fractional kilowatt engine, which was widely copied by others including the Indian and Harley-Davidson companies in the USA, in road-

Figure 1.30: Hildebrand and Wolfmueller motorcycle. This machine, patented in 1894, was the first successful production motorcycle.

Figure 1.31: 2017 MotoGP motorcycles: Movistar Yamaha (Maverick Vinales and Valentino Rossi), Ducati Team (Andrea Dovizioso), Pull & Bear Aspar Team (Alvaro Bautista). Reproduced with permission of Dorna.

going tricycles. The De Dion Bouton engine is arguably the forerunner of all motorcycle engines.

The first successful production motorcycle was the Hildebrand and Wolfmueller, which was patented in Munich in 1894 (see Figure 1.30). The engine of this vehicle was a 1428 cc water-cooled, four-stroke parallel-twin, which was mounted low on the frame with cylinders in a fore-and-aft configuration. This machine produced less than 2 kW and had a top speed of approximately 10 m/s (22 mph). As with the Roper steamer, the engine's connecting rods were coupled directly to a crank on the rear axle. The Hildebrand and Wolfmueller, which was manufactured in France under the name Petrolette, remained in production until 1897.

In combination with advanced materials, modern tyres, sophisticated suspension systems, stiff and light frames, and the latest in brakes, fuels, and lubricants, these powerful engines have evolved to power present-day Grand Prix machines with straight-line speeds of approximately 100 m/s (220 mph); Figure 1.31 shows four 2017 MotoGP racing motorcycles.

At these elevated speeds, the need for efficient aerodynamics design becomes crucially important.

1.5 Tyres

The earliest tyres were bands of leather, and then hoops of iron, placed on to the wooden wheels of carts and wagons. In order to fit them, iron or steel tyres were heated in a forge fire, placed over the wheel and then quenched. This caused the tyre to contract and fit tightly on the wheel. The tyre tied the wheel segments together, providing a wear-resistant surface to the perimeter of the wheel.

Figure 1.32: R. W. Thomson's tyre [31]; see also Figure 4 in [32].

1.5.1 The first tyre

In 1847 R. W. Thomson patented, 'a new and useful improvement in carriage wheels' [32]. As shown in Figure 1.32, Thomson's tyre was made from a reinforced vulcanized rubber inner tube surrounded by a leather casing. Despite its demonstrable advantages, Thomson's invention was fifty years ahead of its time. In the 1840s not only were there no motor cars, but bicycles were only just starting to appear. This lack of demand together with the high production costs reduced this generation of pneumatic tyres to a mere curiosity.

In 1887 John Boyd Dunlop developed the first practical pneumatic or inflatable tyre. Unaware that Robert Thomson had already patented a design for a pneumatic tyre in 1847, in 1888 he patented a pneumatic tyre made of textile material and rubber that contained an inflatable rubber inner tube. As shown in Figure 1.33, Dunlop's invention was a combination of a tyre and a wheel. Competition events soon showed its superiority over solid tyres in terms of both comfort and rolling resistance.

In 1890 Thomson's patent was rediscovered and Dunlop's was deemed invalid. The surrounding controversy gave the new tyres enormous publicity, and bicycle race testing accelerated tyre development. Dunlop went on to market his design successfully, initially for bicycles, but the company that bore his name was soon making tyres for the motor cars that were emerging in the 1890s.

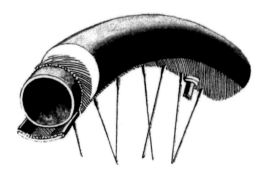

Figure 1.33: John Boyd Dunlop's tyre [31]; see also Figure 1 in [33].

1.5.2 Detachable tyres

The Dunlop tyre was both costly and inconvenient to repair and so Charles Kingston Welch patented his wire-bead design that we are familiar with today [34] (see Figure 1.34). The Welch invention enabled a tyre with wire beads to be mounted and demounted very simply from the wheel. In 1891 the Michelin brothers, Édouard and André, patented tyres that could be mounted or demounted by hand. In 1904, Firestone and Goodyear developed straight-sided wire-bead tyres and almost all the tyre manufacturers in the United States were following their manufacturing techniques by 1908.

In 1896 T. B. Jeffery patented the clincher rim, which is a thickened edge bead, or clinch, usually of hard rubber, moulded into the form of a hook that locked into a corresponding loop in the wheel rim. This arrangement ensures that the tyre is securely held when the tube is inflated [35]. Correspondingly, as the tyre pressure is reduced, the tyre becomes easier to remove (see Figure 1.35).

The first uses of pneumatic tyres on a four-wheeled vehicle was on L'Éclair, built by the Michelin brothers with 'Jeffery type' beaded tyres (1895). For the first time ever, a motor car was 'riding on air'.

Patent restrictions associated with Welch rim tyres, which were enforced by the Pneumatic Tyre Company (renamed Dunlop Rubber Company after 1900), meant that most four-wheel vehicle development (particularly in Europe) was based on beaded-edge tyres. The downside of 'clincher' or beaded-edge tyres is the possibility of rapid deflation following a sharp blow, particularly if the tyre is under-inflated. A 'blowout' or quick loss of air from a puncture or tear is a possibility on any inflated tyre, but this early method of fastening the tyre to the rim will result in the tyre being torn from the rim if tyre pressure is lost, because air pressure holds the two together.

While suitable for bicycles, where the tyre forces act predominantly through the diametral axis of the wheel, beaded-edge tyres are less suitable for cars, especially at higher speeds. When cornering at high speed, large lateral forces and significant load transfers tend to pull the outer (loaded) bead from its seating on the rim. The tyre 'dislodgement' problem was worsened by the rapid increase in vehicular weight resulting from the addition of multiple seating and more elaborate bodywork.

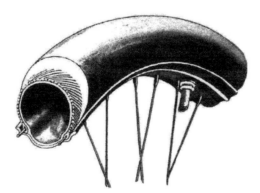

Figure 1.34: The Welch well-based rim [31]; see also Figure 1 in [34].

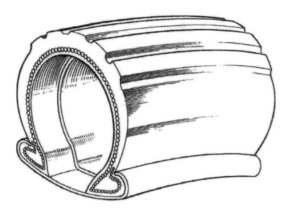

Figure 1.35: The Jeffery clincher [31]; see also Figure 1 in [35].

1.5.3 Tyre developments

The Pneumatic Tyre Company's aggressive protection of its 'Welch rim' patent meant that beaded-edge-equipped cars remained predominant, bringing with them prolonged dislodgement problems. Beaded-edge tyres continued to be fitted to new cars across Europe even after the Welch rim patent expired in 1904. Immediately prior to the First World War a movement away from beaded-edge to wired-on tyres began to take place, but this transition was slowed by the war effort. However, by the early 1920s the tide was turning and by 1925 beaded-edge tyres finally became obsolete.

In 1915 the Palmer Tyre Company pioneered the first rubberized cord fabric and made the first corded tyre. The tyre casings were built using sheets of cord material (typically nylon, rayon, or polyester), cut at an angle and laid across each other, with each ply completely separated from the next by its rubber coating. All the strands of cord in each sheet were laid parallel to each other and pressed into sheet rubber. With this design the sidewalls and tread are not differentiated, giving the tyre structure great rigidity and good steering characteristics. With a corded or cross-ply tyre the maximum width and the height of the tyre above the rim are approximately equal, giving the cross-ply tyre its characteristic high-profile appearance.

It is well known that if all cords in a tyre follow a path straight across (radial), instead of on the bias, the resulting tyre is too flexible and poor handling results. The radial cords are shown as '2' in Figure 1.36 and are located in the tyre's radial planes. Despite the initial carcass flexibility problem, radial ply tyres have a number of advantages including: (i) a soft ride due to their flexible side walls, (ii) reduced fuel consumption due to a lower rolling resistance, (iii) extended tyre life due to reduced heat generated within the tyre, and (iv) reduced road-induced vibration. To increase the tyre's stiffness it can be encased in additional belts oriented closer to the direction of travel, usually in the form of a toroidal spiral. These belts can be made of steel (hence the term steel-belted radial), polyester, or aramid fibers such as Twaron or Kevlar. The radial cords in the sidewall provide flexibility and ride comfort, while the rigid steel belts reinforce the tread region giving high mileage and performance. Radial ply tyres, introduced in 1950s, are now dominant.

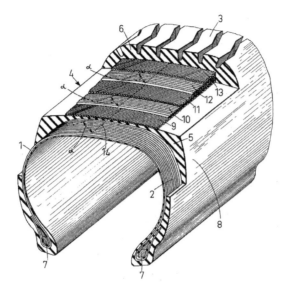

Figure 1.36: Radial ply tyre [39].

The first tubeless tyre was patented by E. B. Killen in 1928 [36], but this tyre appears to have been abandoned. In 1955, O. J. Patin, assignor to the Goodrich Company, applied for the patent 'rim for tubeless truck tires' [37]. By reinforcing the side walls, the company was able to combine the puncture-sealing features of inner tubes with an improved ride quality and superior air retention. While Goodrich awaited approval from the US Patent Office, the tubeless tyres underwent high-speed road testing, were put in service in a fleet of taxis, and were used by Ohio state police cars and a number of privately owned passenger cars. Within three years, the tubeless tyre became standard on most new automobiles [38].

2
Topics in Mechanics

2.1 Background

This book is not about mechanics, but those with a serious interest in vehicle dynamics must understand mechanics. Classical mechanics is a subject that has been studied for well over three hundred years, and over that time it has evolved into a broad and sophisticated subject that is now covered by hundreds of books and many thousands of research papers. We will not attempt to review this literature. Instead, we will point the reader in the direction of some of our preferred texts. In this way one should be able to discover any literature required.

A good undergraduate-level treatment of vector mechanics can be found in Synge and Griffith [40]. This book also provides an introductory treatment of generalized coordinates and constraints (chapter 10), and analytical mechanics (chapters 15 and 16). The chapter on the theory of vibrations (chapter 17) may also be of interest. We would recommend this book to those with an engineering background, especially if mechanics has not been a focal point of their prior education.

A more advanced set of books comprise those by Goldstein and his colleagues [41, 42], which are aimed primarily at graduate-level physicists. These books have more of an analytical mechanics flavour, and therefore focus on variational principles. Given the present context, the reader should be familiar with the material on classical mechanics (not relativity or quantum mechanics), with the treatment of rigid body rotations standing out as being particularly relevant.

The course notes on mechanics by Landau and Lifshitz [43] are a masterpiece of clarity and succinctness, but are probably not for those in the early stages of learning the subject. The focus is on analytical mechanics (rather than vector mechanics) and the treatment is from the perspective of the theoretical physicist; not dissimilar to Goldstein [41]. While there is certainly an overlap with other books, seeing the central ideas (the conservation of energy, momentum, and angular momentum) treated in a different way, and with different insights, is helpful. In the context of vehicle dynamics, chapters I, II, and VI are particularly relevant.

Lanczos' book [44] on variational principles in mechanics is a favourite of ours. This book provides an excellent overview of mechanics, and shows the relationships between vector mechanics ideas and d'Alembert's principle, virtual work, and the energy-based methods of Euler, Lagrange, and Hamilton. Lanczos' treatment of constraints (and the physical significance of the Lagrange multiplier) is excellent (chapter 5). To readers who feel they know mechanics already, we would respectfully suggest: 'read Lanczos anyway'.

Dynamics and Optimal Control of Road Vehicles. D. J. N. Limebeer and M. Massaro.
© D. J. N. Limebeer and M. Massaro 2018. Published in 2018 by Oxford University Press.
DOI: 10.1093/oso/9780198825715.001.0001

The connections between classical mechanics and mathematical physics are many and varied. Arnold's book [45] treats mechanics from a theoretical physicist's point of view with an emphasis on variational principles and analytical dynamics. All the basic problems in dynamics, including the theory of oscillations, the theory of rigid body motion, and the Hamiltonian formalism are examined. This book starts from the beginning, assuming only standard courses in differential and integral calculus, differential equations, geometry (vector spaces, vectors), and linear algebra (linear operators, quadratic forms). This is not an easy book for readers with an engineering background, but it offers a great deal for those willing to put in the effort. Many of the basic concepts are illustrated with beautiful sketches.

The main emphasis of Milne's book [46] is the power and utility of vector-based methods. The techniques of analytical mechanics were deliberately excluded in order to focus on elementary mathematical methods. However, this book also shows how difficult problems can be solved, with vectorial methods providing a clear insight into 'how systems move'. The book highlights the tremendous power of scalar and vector triple products. One theme of the book is the vector product $\mathbf{a} \times \mathbf{b}$ and its applications, with the identity $(\mathbf{a} \times \mathbf{b}) \times \mathbf{c} = \mathbf{b}(\mathbf{a} \cdot \mathbf{c}) - \mathbf{a}(\mathbf{b} \cdot \mathbf{c})$ playing a fundamental role. Several interesting examples are provided.

An introduction to the geometric approach to classical mechanics is given in [47]. This book provides the unifying viewpoint of Lagrangian and Hamiltonian mechanics in the language of differential geometry and fills a gap between traditional classical mechanics texts and advanced modern mathematical treatments of the subject. This book provides a comprehensive treatment of the ideas and methods of geometric mechanics, and is aimed at the graduate student, or the specialist who wants to refresh their knowledge of the formulations of Lagangrian and Hamiltonian mechanics.

All road vehicles have rolling wheels. As a result, nonholonomic constraints have a significant role to play in any study of vehicle dynamics, even though force-generating tyres are predominant in the modern literature. If one accepts this point of view, Neǐmark and Fufaev's book [48] on nonholonimc systems is certainly helpful. Apart from a masterful treatment of the general theory, it contains a detailed analysis of several examples of direct relevance. These include: Čhaplygin's sleigh, a disc rolling on a horizontal plane, the bicycle, and aircraft and road vehicle shimmy phenomena. Books with a more applied flavour include [49–52].

The chapter begins (Section 2.2) with the formulation of the equations of motion using the methods of Newton and Lagrange, which are techniques we employ throughout the book. Topics such as generalized coordinates, constraints, and kinetic and potential energy are discussed briefly. Some of the basics of conservation laws are discussed in Section 2.3. Hamilton's equations and Poisson brackets, and their relationship to conservation laws, are reviewed in Section 2.4.

A brief review of rotating reference frames using matrix-theoretic arguments and linear transformations is given in Section 2.5. This approach has found favour in the modern robotics literature [53], because it is well suited to computer-based computations. In our view this approach also clarifies several aspects of the classical theory. One of the key ideas is the use of skew-symmetric matrix representations of the vector cross product. Important also is the fact that rotations and cross products commute, by which we mean $\mathcal{R}(\mathbf{a} \times \mathbf{b}) = \mathcal{R}\mathbf{a} \times \mathcal{R}\mathbf{b}$ in which $\mathcal{R}$ represents a rotation matrix.

Equilibria, stability, and linearization are discussed briefly in Section 2.6; our simple treatment follows that given in [45].

Time-reversal symmetry, which is a property associated with many conservative systems, is discussed in Section 2.7.

The chapter contains numerous examples of varying levels of difficulty that serve two purposes. First, they illustrate several of the key ideas from mechanics that will be required in the later chapters. Second, they will provide a brief comparison between the ideas of vectorial and analytical mechanics. In some cases we will solve the problem using both vector mechanics ideas and Lagrange's equations, thereby providing this comparison. These examples have also been chosen to illustrate, in a simple context, some of the fundamental ideas required in the study of mechanics and vehicle dynamics.

The chapter concludes with three substantial examples. The first is a detailed analysis of the classical Chaplygin sleigh [48], which is given in Section 2.8. This system is conservative and nonholonomic, as well as being time-reversible. An intriguing feature of this system is that linearizations around stable equilibria are neither conservative nor time-reversible. An interesting question relates to the effect of linearization on properties such as energy conservation and time-reversibility.

In the second example we study a ball rolling on an inclined surface and then on an inclined turntable. This example is elegantly solved using vectorial methods and serves as a warning: 'Do not underestimate the subtleties that can arise in systems containing multiple rotating components.' The rolling ball is dealt with in Section 2.9.

The third example (given in Section 2.10) is the rolling disc, which is ubiquitous in the context of literature on nonholonomic dynamic systems. Nonholonomic rolling constraints are a feature of early bicycle models such as those introduced by Whipple [2] and Carvallo [54]; more will be said about this later. This example also illustrates the use of Euler angles and the use of linearization techniques in the study of stability. A facility with Euler angles is required, because bicycles yaw and roll through large angles, as well as pitch through smaller ones. In common with the bicycle, we will show that the rolling disc has speed-dependent stability properties.

While computers obviate the burden of performing tedious and often complicated calculations, they do not remove the need to understand mechanics, or to make decisions relating to analysis frameworks, reference frames, problem formulations, and context-dependent approximations.

We consider Sections 2.2, 2.3, 2.5, and 2.6 essential background that will be required to understand the remainder of the book. Section 2.4 on Hamiltonian mechanics can be skipped on a first reading. Sections 2.8, 2.9, and 2.10 will be of interest to those wishing to gain an understanding of rolling-contact mechanics and nonholonomic constraints. Section 2.7 looks at some of the deeper issues at the foundations of mechanics—this material can be skipped by those not interested in the more esoteric issues in classical mechanics.

2.2 Equations of motion

The equations of motion of a mechanical system can be derived using 'vectorial mechanics', or 'variational techniques (analytical mechanics)'. 'Vectorial mechanics' is based on forces and moments, and builds on Newton's three laws (published in 1687 in the *Principia Mathematica*). 'Analytical mechanics' is based on kinetic energy and work (sometimes replaceable by potential energy), and builds on the 'principle of stationary action', which was developed by Euler (1707–83), Lagrange (1736–13), Jacobi (1804–51), and Hamilton (1805–65). These analytical principles can all be derived from d'Alembert's (1717–83) principle, which is the dynamic analogue of the principle of virtual displacement for forces applied to a static system [44]. D'Alembert's principle itself makes use of Newton's second law.

We will now briefly review the basics of analytical and vector mechanics, since both are used to obtain the equations of motion of mechanical systems.

2.2.1 Inertial reference frame

In order to describe and analyse mechanical systems one requires a frame of reference to which positions can be referred. In general, the laws of motion take different forms in different reference frames. A coordinate system in which a free particle that is instantaneously at rest remains at rest, for all time, is called *inertial*. The laws of mechanics are the same at all moments in time in all inertial coordinate systems. Any coordinate system in uniform rectilinear motion relative to an inertial coordinate system is itself inertial. In the context of vehicle dynamics problems, any earth-fixed frame of reference will be treated as inertial, although this is not strictly true.

2.2.2 Newton's equations

Newton's laws of motion laid the foundations of classical mechanics. They describe the relationship between a body, the forces acting on the body, and the body's subsequent motion. They have been expressed in several different ways, but can be summarized as follows:

1. When viewed in an inertial reference frame, an object either remains at rest or continues to move at a constant velocity, unless acted upon by a force.
2. The vector sum $\sum F$ of all the external forces acting on an object is equal to the mass m of that object multiplied by the acceleration a of the object (bold characters are used to denote vector quantities).
3. When one body exerts a force on a second body, the second body simultaneously exerts a force equal in magnitude and opposite in direction on the first body.

A particle of mass m, moving with velocity v, is deemed to have *linear momentum* $P = mv$. It follows from the second law that

$$\frac{dP}{dt} = F, \tag{2.1}$$

in which F is the force acting on the particle. If r is the position of m relative to some (fixed) point O, then the particle's *angular momentum* is given by $H_O = r \times mv$. Premultiplication of (2.1) by r gives

$$\frac{dH_O}{dt} = r \times F = M_O, \tag{2.2}$$

where M_O is the *moment* of F around O.

In the case of a constellation of particles, we have

$$\boldsymbol{P} = \sum \boldsymbol{P}_i$$
$$= \sum m_i \dot{\boldsymbol{r}}_i$$
$$= \left(\sum m_i \right) \frac{d}{dt} \left(\frac{\sum m_i \boldsymbol{r}_i}{\sum m_i} \right)$$
$$= m \dot{\boldsymbol{r}}_G,$$

where m is the constellation's mass and $\boldsymbol{r}_G$ is the position of its mass centre. Thus

$$\boldsymbol{F} = \sum \boldsymbol{F}_i = \frac{d\boldsymbol{P}}{dt}$$
$$= m \ddot{\boldsymbol{r}}_G; \qquad (2.3)$$

for the constellation.

Now suppose that $\boldsymbol{r}_i = \bar{\boldsymbol{r}} + \bar{\boldsymbol{r}}_i$:

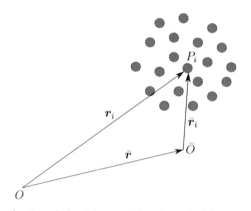

If $\boldsymbol{v}_i$ is the absolute velocity of the ith particle, there holds

$$\boldsymbol{H}_O = \sum \boldsymbol{r}_i \times m_i \boldsymbol{v}_i$$
$$= \sum (\bar{\boldsymbol{r}} + \bar{\boldsymbol{r}}_i) \times m_i \boldsymbol{v}_i$$
$$= \bar{\boldsymbol{r}} \times \sum m_i \boldsymbol{v}_i + \sum \bar{\boldsymbol{r}}_i \times m_i \boldsymbol{v}_i$$
$$= \bar{\boldsymbol{r}} \times \boldsymbol{P} + \boldsymbol{H}_{\bar{O}}, \qquad (2.4)$$

which is the constellation's angular momentum (about O). If $\bar{O}$ is the constellation's mass centre, (2.4) becomes

$$\boldsymbol{H}_O = \boldsymbol{r}_G \times \boldsymbol{P} + \boldsymbol{H}_G. \qquad (2.5)$$

We now suppose that the particles are the constituents of a *rigid body*. Suppose also that $\bar{O}$ is the origin of a moving reference frame which has translational velocity $\bar{\boldsymbol{v}}$ and angular velocity $\boldsymbol{\omega}$. In this case the absolute velocity of each particle is given by

$$v_i = \bar{v} + \omega \times \bar{r}_i.$$

This gives

$$
\begin{aligned}
H_{\bar{O}} &= \sum m_i \bar{r}_i \times v_i \\
&= \sum m_i \bar{r}_i \times (\bar{v} + \omega \times \bar{r}_i) \\
&= \sum m_i \bar{r}_i \times \bar{v} + \sum m_i(\bar{r}_i \times (\omega \times \bar{r}_i)) \\
&= m\bar{r}_G \times \bar{v} + \sum m_i(\omega(\bar{r}_i \cdot \bar{r}_i) - \bar{r}_i(\omega \cdot \bar{r}_i)) \\
&= m\bar{r}_G \times \bar{v} + \sum m_i\left((\bar{r}_i \cdot \bar{r}_i)I - \bar{r}_i\bar{r}_i^T\right)\omega \\
&= m\bar{r}_G \times \bar{v} + J_{\bar{O}}\omega,
\end{aligned}
\tag{2.6}
$$

where $J_{\bar{O}}$ is the *moment of inertia tensor* of the rigid body around $\bar{O}$. If the summation in (2.6) is replaced by a continuous integral the inertia tensor can be represented by a symmetric matrix of the form

$$
J_{\bar{O}} = \begin{bmatrix} J_{xx} & J_{xy} & J_{xz} \\ J_{xy} & J_{yy} & J_{yz} \\ J_{xz} & J_{yz} & J_{zz} \end{bmatrix} \quad J_{kk} = \int (i^2 + j^2)dm \quad J_{ij} = -\int ij\,dm \quad i \neq j.
\tag{2.7}
$$

If $\bar{O}$ is a stationary point there holds

$$H_{\bar{O}} = J_{\bar{O}}\omega.
\tag{2.8}$$

When $\bar{r}$ is coincident with the body's mass centre (i.e. $\bar{O} = G$), (2.6) becomes

$$
\begin{aligned}
H_{\bar{O}} &= m\bar{r}_G \times \bar{v} + J_G\omega \\
&= \bar{r}_G \times P + J_G\omega.
\end{aligned}
\tag{2.9}
$$

When $\bar{r}_G = 0$ the mass centre coincides with O, and

$$H_O = J_G\omega.
\tag{2.10}$$

We conclude this this brief summary of vector mechanics by finding the rate of change of angular momentum about a moving point that coincides instantaneously with a fixed point. Consider (2.4) in which point O is fixed, while point $\bar{O}$ is moving. Differentiating this equation with respect to time gives

$$\frac{dH_O}{dt} = \frac{dH_{\bar{O}}}{dt} + \dot{\bar{r}} \times P + \bar{r} \times \frac{dP}{dt}.$$

At the moment of coincidence, $\bar{r} = 0$, and $\dot{\bar{r}} = \bar{v}$ at the same instant. That is

$$M_O = \frac{dH_O}{dt} = \left.\frac{dH_{\bar{O}}}{dt}\right|_{O=\bar{O}} + \bar{v} \times P.
\tag{2.11}$$

The $\bar{v} \times P$ term is zero when point $\bar{O}$ is either fixed, or it coincides with G, or $\bar{v}$ is parallel to v_G, where v_G is the velocity of the mass centre. Thus the two rates of change of angular momentum in (2.11) are equal if the moving point is instantaneously at rest, or it is moving parallel to the mass centre, or when the mass centre is at rest.

In sum, the dynamics of a rigid body are governed by the translational equation (2.1) and the rotational equations given in (2.11). A comprehensive treatment of vector mechanics can be found in [46].

2.2.3   Properties of the inertia tensor

The inertia tensor (matrix) has some useful properties that we will now review briefly.

The *perpendicular axis theorem* applies to laminatae, that is, thin flat bodies lying in the plane. If J_x and J_y are the moments of inertia about the x- and y-axes, respectively, which lie in the laminate, then the moment of inertia about the z-axis is[1]

$$J_z = J_x + J_y. \tag{2.12}$$

The *parallel axis theorem* applies to general rigid bodies. Let J_G be the moment of inertia of a rigid body, with mass m, about an axis through its mass centre G. Then the moment of inertia around a parallel axis through a point O is given by

$$J_O = J_G + m \left(\boldsymbol{r}^T \boldsymbol{r} I - \boldsymbol{r}\boldsymbol{r}^T \right), \tag{2.13}$$

where I is the identity matrix and $\boldsymbol{r}$ is the position of G relative to O.

Since the inertia matrix is symmetric and non-negative, it has real non-negative eigenvalues and orthogonal eigenvectors. The eigenvalues are the principal moments of inertia and the (orthogonal) eigenvectors are the principal axes (of the inertia matrix). To reference the inertia tensor to alternative axes, one uses

$$J_F = \mathcal{R}^T J_I \mathcal{R}, \tag{2.14}$$

in which J_I and J_F are referenced to the initial and final coordinate systems, respectively, and $\mathcal{R}$ is the transformation between the two (see Section 2.5).

In most vehicle-related problems a vertical plane of symmetry can be assumed, with the pitch axis a principal axis that is perpendicular to the plane of symmetry. The other two axes lie in the plane of symmetry itself, with their locations fixed by experiment.

Suppose that in Figure 2.1 the roll moment of inertia is measured as J_ϕ, that the yaw moment of inertia is measured as J_ψ, and that the product of inertia is $J_{\phi\psi}$. The corresponding principal moments of inertia J_1 and J_2, and the inclination angle μ of the principal axis system, are given by

$$J_{1,2} = \frac{J_\phi + J_\psi \mp \sqrt{(J_\phi - J_\psi)^2 + 4J_{\phi\psi}^2}}{2} \qquad \mu = \frac{1}{2} \arctan\left(\frac{-2J_{\phi\psi}}{J_\phi - J_\psi} \right). \tag{2.15}$$

In the case that $J_{\phi\psi} > 0$ and $J_\phi < J_\psi$, a typical motorcycle scenario, μ is positive as shown in Figure 2.1.

---

[1] $J_z \leq J_x + J_y$ in the case of a thick laminate.

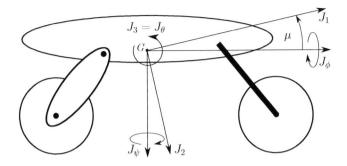

Figure 2.1: Inertia axis systems for a symmetric road vehicle.

Example 2.1 Suppose a motorcycle has a mass $m^M = 200$ kg, with its centre of mass a distance $b^M = 0.75$ m from the rear ground-contact point and at a height $h^M = 0.55$ m above the ground. When a rider of mass $m^R = 80$ kg sits on the saddle, his centre of mass is located at $b^R = 0.60$ m from the rear contact point and $h^R = 0.90$ m from the ground. Supposing the motorcycle has inertia matrix J^M and the rider has inertia matrix J^R with respect to their centres of mass (the x-axis points forwards, the z-axis points downwards, and the y-axis points to the rider's right)

$$
J^M = \begin{bmatrix} 10 & 0 & 2 \\ 0 & 45 & 0 \\ 2 & 0 & 35 \end{bmatrix} \qquad
J^R = \begin{bmatrix} 6.5 & 0 & 0 \\ 0 & 6.5 & 0 \\ 0 & 0 & 4.0 \end{bmatrix}.
$$

Compute the position of the centre of mass, and the inertia matrix of the motor-cycle–rider combination. Using the definition of the centre of mass and (2.13), one obtains

$$
m^{M+R} = 280 \qquad b^{M+R} = 0.71 \qquad h^{M+R} = 0.65 \qquad
J^{M+R} = \begin{bmatrix} 23.5 & 0 & -1 \\ 0 & 59.8 & 0 \\ -1 & 0 & 40.3 \end{bmatrix},
$$

where J^{M+R} is defined with respect to the whole-system mass centre. The principal moments of inertia in Figure 2.1 of the motorcycle–rider assembly are $J_1 = 23.4$ kg m^2 and $J_2 = 40.3$ kg m^2, while the inclination angle is $\mu = -3.4$ degrees.

2.2.4   Generalized coordinates

In order to describe the configuration of a system of N rigid bodies as many as $6N$ coordinates might be needed; this number will reduce if the motions of these bodies is restricted in some way. Any $n \leq 6N$ quantities $q_1, q_2, \cdots, q_n$ that completely de-termine the configuration of the system are called *generalized coordinates*, with the *generalized velocities* given by $\dot{q}_1, \dot{q}_2, \cdots, \dot{q}_n$ in which the 'dot' denotes the derivative with respect to time. Taken together, the generalized coordinates and the generalized velocities can be used to determine the *state* of the system; a system of N rigid bodies may require as many as $12N$ states to describe its motion. The relationships between the generalized coordinates, the generalized velocities, and the generalized accelerations are called the *equations of motion*. The solution of the mechanical problem can be thought of as a curve described by a point C with coordinates $q_1, q_2, \cdots, q_n$ moving on an n-dimensional manifold in the configuration space. Given the n initial generalized coordinates $q_1, q_2, \cdots, q_n$, and the n initial generalized velocities $\dot{q}_1, \dot{q}_2, \cdots, \dot{q}_n$, the solution is obtained by integration.

The $2n$-dimensional space with coordinates $q_1, q_2, \cdots, q_n$ and $\dot{q}_1, \dot{q}_2, \cdots, \dot{q}_n$ is called the *phase space*. When time is included, the resulting $2n + 1$ dimensional space is called the *state space*. If the system's motion starts at some point P in the state space, its subsequent motion is determined by the *equations of motion*. The equations of motion provide a set transformations

$$q_i(t) = f_i(q_1, q_2, \cdots, q_n; \dot{q}_1, \dot{q}_2, \cdots, \dot{q}_n; t)$$
$$\dot{q}_i(t) = g_i(q_1, q_2, \cdots, q_n; \dot{q}_1, \dot{q}_2, \cdots, \dot{q}_n; t),$$

called the *phase flow*, this name comes from the analogy with fluid flows [44].

2.2.5 Constraints

Constraints on the generalized coordinates are called *holonomic* constraints. Constraints that can only be given as relationships between differentials of the coordinates (and not as finite relationships between the coordinates themselves) are called *nonholonomic* constraints (after Hertz). In addition, constraints can be independent of time or time-dependent. These constraints are called *scleronomic* and *rheonomic* respectively (after Bolzmann). The four possibilities can be summarized as follows

$$
\begin{array}{ccc}
 & \text{holonomic} & \text{nonholonomic} \\
\text{scleronomic} & \left\{ \begin{array}{c} f_1(q_1, \cdots, q_n) = 0 \\ \vdots \\ f_m(q_1, \cdots, q_n) = 0 \end{array} \right. & \left\{ \begin{array}{c} F_{11}dq_1 + \cdots + F_{1n}dq_n = 0 \\ \vdots \\ F_{m1}dq_1 + \cdots + F_{mn}dq_n = 0 \end{array} \right. \\[3em]
\text{rheonomic} & \left\{ \begin{array}{c} f_1(q_1, \cdots, q_n, t) = 0 \\ \vdots \\ f_m(q_1, \cdots, q_n, t) = 0 \end{array} \right. & \left\{ \begin{array}{c} F_{11}dq_1 + \cdots + F_{1n}dq_n + F_{1t}dt = 0 \\ \vdots \\ F_{m1}dq_1 + \cdots + F_{mn}dq_n + F_{mt}dt = 0. \end{array} \right.
\end{array}
$$

If the motion is restricted by m constraints, these constraints must be appended to the equations of motion, thus increasing by m the number of equations (and variables) governing the system's behaviour. The additional variables are the constraint forces in the case of the vectorial mechanics framework and Lagrange multipliers in the case of the variational mechanics framework. In the case of holonomic constraints only $n - m$ of the qs and $n - m$ of the $\dot{q}$s can be defined arbitrarily, with the remainder determined by the constraint equations and their derivatives. In the case of nonholonomic constraints, n of qs and $n - m$ of $\dot{q}$s can be defined arbitrarily, with the remainder determined by the constraints equations.

Holonomic constraints of the form

$$f(q_1, \cdots, q_n) = 0 \tag{2.16}$$

can be 'disguised' as nonholonomic by writing

$$0 = \frac{\partial f}{\partial q_1} \dot{q}_1 + \cdots + \frac{\partial f}{\partial q_n} \dot{q}_n$$
$$= F_1 dq_1 + \cdots + F_n dq_n. \tag{2.17}$$

The converse, however, is not necessarily true, because the kinematic constraint may not be integrable. Constraint (2.17) can only be integrated if there exists an *integrating factor* $\beta(q)$ such that

$$\beta(q) F_j(q) = \frac{\partial h(q)}{\partial q_j}, \qquad j = 1, \cdots, n, \tag{2.18}$$

for some function $h(q)$. Conversely, if $\beta(q) F(q)$ is the gradient of $h(q)$, then (2.17) is integrable. Schwartz's theorem allows (2.18) to be replaced by

$$\frac{\partial(\beta F_k)}{\partial q_j} = \frac{\partial(\beta F_j)}{\partial q_k} \qquad j, k = 1, \cdots, n, \tag{2.19}$$

which does not involve $h(q)$.

Example 2.2 Consider the following differential constraint [44]:

$$\left(q_2^2 - q_1^2 - q_3 \right) \dot{q}_1 + \left(q_3 - q_2^2 - q_1 q_2 \right) \dot{q}_2 + q_1 \dot{q}_3 = 0. \tag{2.20}$$

When written in matrix form, the integrability conditions are

$$\begin{bmatrix} q_1 q_2 + q_2^2 - q_3 - q_1^2 + q_2^2 - q_3 & 0 & 3 q_2 \\ -q_1 & 0 & -q_1^2 + q_2^2 - q_3 & -2 \\ 0 & -q_1 & -q_1 q_2 - q_2^2 + q_3 & 1 \end{bmatrix} \begin{bmatrix} \frac{\partial \beta}{\partial q_1} \\ \frac{\partial \beta}{\partial q_2} \\ \frac{\partial \beta}{\partial q_3} \\ \beta \end{bmatrix} = 0.$$

After Gaussian elimination it is found that

$$\begin{bmatrix} q_1 q_2 + q_2^2 - q_3 - q_1^2 + q_2^2 - q_3 & 0 & 3 q_2 \\ 0 & -\frac{q_1 \left(q_1^2 - q_2^2 + q_3 \right)}{q_1 q_2 + q_2^2 - q_3} & -q_1^2 + q_2^2 - q_3 & \frac{q_1 q_2 - 2 q_2^2 + 2 q_3}{q_1 q_2 + q_2^2 - q_3} \\ 0 & 0 & 0 & \frac{q_1^2 - q_1 q_2 + q_2^2 - q_3}{q_1^2 - q_2^2 + q_3} \end{bmatrix} \begin{bmatrix} \frac{\partial \beta}{\partial q_1} \\ \frac{\partial \beta}{\partial q_2} \\ \frac{\partial \beta}{\partial q_3} \\ \beta \end{bmatrix} = 0.$$

The fourth row shows that β is free when

$$q_1^2 - q_1 q_2 + q_2^2 - q_3 = 0, \tag{2.21}$$

which is the holonomic equivalent to (2.20). This can be checked by substituting (2.21) into (2.20).

Example 2.3 Consider the classical example of a disc that rolls vertically in the plane without slipping [41].

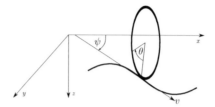

The coordinates used to describe the motion of the disc are the x and y coordinates of the disc's centre, the angular position θ of the disc, and its orientation ψ relative to the x-axis. To ensure pure rolling one must enforce the (rolling) constraint

$$v = r\dot{\theta}, \tag{2.22}$$

where r is the disc's radius. Projecting the velocity vector onto the x and y axes gives

$$\dot{x} = r\dot{\theta}\cos\psi \tag{2.23}$$
$$\dot{y} = r\dot{\theta}\sin\psi. \tag{2.24}$$

One can establish that these constraints are nonholonomic, by showing that the disc can be rolled from an arbitrary initial configuration $x_i, y_i, \psi_i, \theta_i$ to an arbitrary final configuration $x_f, y_f, \psi_f, \theta_f$ as follows:

(1) Place the disc at $x_i, y_i, \psi_i, \theta_i$;
(2) roll the disc from $x_i, y_i, \psi_i, \theta_i$ to x_f, y_f along any path of length $r(\theta_f - \theta_i + 2k\pi)$;
(3) rotate the disc around the vertical axis to ψ_f.

This confirms that the two constraints do not impose a constraint on the generalized coordinates.

Nonholonomy can also be established using (2.19). Since the problem has four generalized coordinates, there are six conditions associated with each of (2.23) and (2.24). Relevant to the present argument are these:

$$\frac{\partial\beta_1}{\partial\psi} = 0 \tag{2.25}$$

$$-\frac{\partial\beta_1}{\partial\psi}r\cos\psi + \beta_1 r\sin\psi = 0 \tag{2.26}$$

$$\frac{\partial\beta_2}{\partial\psi} = 0 \tag{2.27}$$

$$-\frac{\partial\beta_2}{\partial\psi}r\sin\psi - \beta_2 r\cos\psi = 0, \tag{2.28}$$

in which β_1 and β_2 are the integrating factors related to (2.23) and (2.24) respectively. It follows that $\beta_1 r\sin\psi = 0$ and $\beta_2 r\cos\psi = 0$, which means that $\beta_1 \equiv 0$ and $\beta_2 \equiv 0$, indicating that (2.23) and (2.24) are nonholonomic. In the special case that $\psi = k\pi$ (k integer), β_1 is arbitrary indicating that (2.23) is holonomic and can be integrated to give $x = r\theta + $ a constant. In this case the disc is rolling parallel to the x-axis and $\dot{y} = 0$; (2.24) plays no further part. A similar argument can be developed using (2.24) when $\psi = \pi/2 + k\pi$. In each of these special cases only two variables are involved and differential relationships between two variables are always integrable [44].

2.2.6   Kinetic energy

Leibniz (1646–1716) noticed that in many mechanical systems the quantity $\sum_i m_i v_i^T v_i$ was conserved. He called this the *vis viva* of the system. This quantity (apart from a factor of two), is what we now call kinetic energy, which for rigid bodies generally takes the form

$$T = \frac{m}{2} v_G^T v_G + \frac{1}{2} \omega^T J_G \omega, \tag{2.29}$$

where m is the body's mass, v_G the velocity of its mass centre, ω is the body's angular velocity, and J_G is the body's inertia tensor with respect to its mass centre (see Section 2.2.2). The first term accounts for the body's translation energy, while the second accounts for its rotation energy. Alternatively, the kinetic energy can be written as

$$T = \frac{1}{2} \omega^T J_O \omega, \tag{2.30}$$

in which the inertia J_O is computed with respect to a fixed point O in the body, which is instantaneously absolutely stationary.

Example 2.4 Consider the non-slipping rolling wheel in the figure below:

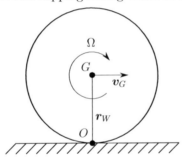

The wheel's kinetic energy is given by

$$T = \frac{m}{2} v_G^T v_G + \frac{1}{2} \Omega^T J_G \Omega,$$

where m is the wheel's mass and J_G is its spin moment of inertia. If r_W is a vector pointing from O to G, then for non-slip rolling

$$v_G = \Omega \times r_W$$

and

$$\begin{aligned} T &= \frac{m}{2} (\Omega \times r_W)^T (\Omega \times r_W) + \frac{1}{2} \Omega^T J_G \Omega \\ &= \frac{1}{2} \Omega^T \left(m r_W^2 + J_G \right) \Omega \\ &= \frac{1}{2} \Omega^T J_O \Omega \end{aligned}$$

by the parallel axis theorem. Point O is fixed in the wheel and momentarily absolutely stationary.

2.2.7 Potential energy

Consider a mechanical system that is subject to a force $\boldsymbol{F} = [F_x, F_y, F_z]^T$ and an infinitesimal displacement $\boldsymbol{ds} = [dx, dy, dz]^T$ of the force's point of application. The work done is

$$dW = \boldsymbol{F} \cdot \boldsymbol{ds} = F_x dx + F_y dy + F_z dz, \tag{2.31}$$

which can be rewritten in terms of the generalized coordinates that describe the system configuration as

$$dW = \sum_{i=1}^{n} Q_i dq_i, \tag{2.32}$$

where the Q_is are the components of the *generalized force* $\boldsymbol{Q}$.

Of central importance in analytical mechanics are forces that are derivable from a scalar *work function* $V(q_1, q_2, \cdots, q_n, t)$,[2] for which

$$Q_i = -\frac{\partial V}{\partial q_i}. \tag{2.33}$$

In the case that V is not explicitly time dependent, the related forces are *conservative*, which implies that the work done depends on the initial and final states only, and not on the path taken.[3] In this case the work function can be interpreted as a *potential energy function* $V = V(\boldsymbol{q})$. The work done by the force between an initial state $\boldsymbol{q}_i$ and a final states $\boldsymbol{q}_f$ is

$$W = V(\boldsymbol{q}_i) - V(\boldsymbol{q}_f). \tag{2.34}$$

Typical potential energy expressions are those related to the energy stored in a spring with stiffness k and deflection l

$$V = k\frac{l^2}{2} \Rightarrow Q_k = -kl \tag{2.35}$$

and to gravitational acceleration g of a body with mass M and centre of mass at $\boldsymbol{r}_G$

$$V = -M\boldsymbol{r}_G \cdot \boldsymbol{g} \Rightarrow Q_g = Mg; \tag{2.36}$$

the potential is negative when the vertical axis is pointing downwards.

---

[2] In the most general case the work function may be of the form $V = V(\boldsymbol{q}, \dot{\boldsymbol{q}}, t)$ [44].

[3] Consider the closed contour C in the figure below, and suppose that the work done going from A to B is independent of the path taken.

This means that $\int_C \boldsymbol{F} \cdot \boldsymbol{ds} = 0$, since one could choose the forward path from A to B to be the return path taken in reverse. It then follows from Stoke's theorem that $\int_S \nabla \times \boldsymbol{F} \cdot \boldsymbol{dS} = 0$ for any surface bounded by C. Since this is true for any contour C containing A and B it follows that $\nabla \times \boldsymbol{F} = 0$; conservative force fields are irrotational. We therefore conclude that $\boldsymbol{F} = \nabla V$ for some scalar potential function V.

Forces derivable from work functions are called *monogenic*, while forces such as friction, that are not derivable from work functions, are called *polygenic* [44].

2.2.8   Lagrange's equations

Lagrange's equations are now introduced. We will assume temporarily that the system is not subject to constraints and that all the external forces are monogenic. Hamilton's principle, or the *principle of stationary action*, states that the motion of an arbitrary mechanical system evolves in such a way that a certain definite integral, called the *action integral*, is stationary for all possible variations of the system configuration. The action integral S is given by

$$S = \int_{t_1}^{t_2} \mathcal{L}(q_1, \cdots q_n, \dot{q}_1, \cdots \dot{q}_n, t)dt \qquad \mathcal{L} = T - V, \tag{2.37}$$

where $\mathcal{L}$ is the Lagrangian, T is the kinetic energy, and V is the potential energy of the system. If $\boldsymbol{q}$ is the vector of generalized coordinates, (2.37) can be written more compactly as

$$S = \int_{t_1}^{t_2} \mathcal{L}(\boldsymbol{q}, \dot{\boldsymbol{q}}, t)dt. \tag{2.38}$$

The principle of stationary action requires that the first variation in the action integral

$$\delta S = \delta \int_{t_1}^{t_2} \mathcal{L}(\boldsymbol{q}, \dot{\boldsymbol{q}}, t)dt \tag{2.39}$$

vanishes. Standard variational arguments [44] show that condition (2.39) can only be satisfied if the system Lagrangian satisfies the n Euler–Lagrange equations [43]

$$\frac{d}{dt}\left(\frac{\partial \mathcal{L}}{\partial \dot{\boldsymbol{q}}}\right) - \frac{\partial \mathcal{L}}{\partial \boldsymbol{q}} = 0. \tag{2.40}$$

If holonomic constraints are present, the generalized coordinates $q_1, \cdots, q_n$ are not independent, but are restricted by constraints of the form

$$f_j(\boldsymbol{q}, t) = 0, \qquad (j = 1, \cdots, m), \tag{2.41}$$

which can be written in vector form as $\boldsymbol{f}(\boldsymbol{q}, t) = 0$. These equations allow m of the q_i s to be expressed in terms of the remaining $n - m$ variables; the problem is thus reduced to one having $n - m$ degrees of freedom. We can proceed in two ways. Either we eliminate $n - m$ generalized coordinates 'up front', thus transforming the constrained problem in n variables into an unconstrained problem in $n - m$ variables, or we append the constraint conditions to the integrand of the action integral using *Lagrange multipliers*. In the Lagrange multiplier method the integrand in (2.37) is modified to

$$\hat{\mathcal{L}} = \mathcal{L} - \boldsymbol{\lambda}^T \boldsymbol{f}, \tag{2.42}$$

in which $\boldsymbol{\lambda}$ and $\boldsymbol{f}$ are vectors of Lagrange multipliers and constraints respectively. Following the introduction of the Lagrange multipliers, (2.40) can be expressed as

$$\frac{d}{dt}\left(\frac{\partial\hat{\mathcal{L}}}{\partial\dot{\boldsymbol{q}}}\right) - \frac{\partial\hat{\mathcal{L}}}{\partial\boldsymbol{q}} = 0, \tag{2.43}$$

which highlights the fact that holonomic constraints can be treated within a variational framework. If one considers the constraints to be part of the potential energy function

$$\hat{V} = V + \boldsymbol{\lambda}^T\boldsymbol{f}, \tag{2.44}$$

$\hat{V}$ will generate the conservative forces, as well as the forces of constraint. Equation (2.43) can also be expressed as

$$\frac{d}{dt}\left(\frac{\partial\mathcal{L}}{\partial\dot{\boldsymbol{q}}}\right) - \frac{\partial\mathcal{L}}{\partial\boldsymbol{q}} = \boldsymbol{k}, \tag{2.45}$$

in which the generalized reaction force $\boldsymbol{k}$ is given by

$$\boldsymbol{k} = -A(\boldsymbol{q})^T\boldsymbol{\lambda}; \tag{2.46}$$

$\boldsymbol{k}$ is an n-vector, $\boldsymbol{\lambda}$ is an m-vector and $A(\boldsymbol{q})$ is an $m \times n$ matrix, whose entries are given by

$$A_{ji}(\boldsymbol{q}) = \left\{\frac{\partial f_j}{\partial q_i}\right\}.$$

The Lagrange multipliers provide the monogenic forces of reaction that maintain the kinematic constraints.[4]

Nonholonomic systems (coined by Hertz in 1894) are typified by velocity-dependent constraints that are not derivable from position constraints. As a result of the velocity constraints, nonholonomic systems are not variational, but the basic mechanics are still governed by Newton's second law. There are some fascinating differences between nonholonomic systems and classical Lagrangian systems. First, the momenta conjugate to cyclic coordinates (Section 2.3.2) may not be conserved as they would be in a holonomic system [41]; further details follow. Second, even in the absence of external forces and dissipation, nonholonomic systems may possess asymptotically stable relative equilibria despite being (energy) conservative. This property will be illustrated in the context of Čhaplygin's Sleigh (Section 2.8), the rolling disc (Section 2.10), and the Whipple bicycle (Chapter 5).

Since nonholonomic constraints place no restrictions on the generalized coordinates, initial values on the q_i s can be assigned arbitrarily. The velocities, however, are constrained by

$$F_{j1}\dot{q}_1 + \cdots + F_{jn}\dot{q}_n + F_{jt} = 0; \qquad (j = 1, \cdots, m), \tag{2.47}$$

in which the F_{ij} s are given functions of the q_i s. In this case we can assign arbitrarily n initial coordinates and $n - m$ initial velocities; (2.47) serves to eliminate the other

---

[4] Note that no $\partial\boldsymbol{f}/\partial t$ term appears in (2.46); virtual displacements only consider displacements of the coordinates. Under these conditions the forces of constraint remain perpendicular to the virtual displacements (even if the constraint is time-varying). In the case of time-varying constraints, however, the real work done by the constraint force may be non-zero.

m initial velocities. In the case of nonholonomic constraints the forces of constraint are not drivable from a work function, but are given by

$$k = -A(q)^T \lambda \qquad (2.48)$$

where the matrix $A(q)$ is given by

$$A_{ji}(q) = \{F_{ji}\} \qquad A(q)\dot{q} = -F_t,$$

which is similar in form to (2.46). Once more these forces act on the system so as to enforce the given nonholonomic constraints; (2.48) does not include the terms F_{jt} even when the constraints are rheonomic.

Finally, one can characterize external polygenic forces (e.g. aerodynamic forces and friction) in terms of their virtual work [44]. Let that work be

$$\delta W_j = Q_{j1}\delta q_1 + \cdots + Q_{jn}\delta q_n \qquad (j = 1, \cdots, m), \qquad (2.49)$$

where Q_{ji} are the components of the generalized forces and δq_i are the components of the virtual displacement. These (polygenic) forces again produce a right-hand-side term in the Lagrange equations

$$k_i = \sum_{j=1}^{m} Q_{ji} \qquad (i = 1, \cdots, n). \qquad (2.50)$$

We conclude this subsection with a number of examples that either illustrate key points of the theory, or else uncover issues we have not discussed in our brief overview of classical mechanics.

Example 2.5 Consider a particle whose position is (x, y) in a rotating coordinate system x–y, which rotates relative to an inertial frame X–Y through some time-varying angle $\theta(t)$.

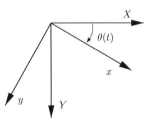

The coordinates of the particle in the inertial frame are given by

$$\begin{bmatrix} X \\ Y \end{bmatrix} = \begin{bmatrix} \cos\theta & -\sin\theta \\ \sin\theta & \cos\theta \end{bmatrix} \begin{bmatrix} x \\ y \end{bmatrix}. \qquad (2.51)$$

The particle's velocity comes from differentiating (2.51) with respect to time

$$\begin{bmatrix} \dot{X} \\ \dot{Y} \end{bmatrix} = \begin{bmatrix} \cos\theta & -\sin\theta \\ \sin\theta & \cos\theta \end{bmatrix} \left\{ \begin{bmatrix} \dot{x} \\ \dot{y} \end{bmatrix} - \dot{\theta} \begin{bmatrix} y \\ -x \end{bmatrix} \right\}. \qquad (2.52)$$

The particle's acceleration is thus

$$
\begin{bmatrix} \ddot{X} \\ \ddot{Y} \end{bmatrix} = \begin{bmatrix} \cos\theta & -\sin\theta \\ \sin\theta & \cos\theta \end{bmatrix} \left\{ \begin{bmatrix} \ddot{x} \\ \ddot{y} \end{bmatrix} - \dot{\theta}^2 \begin{bmatrix} x \\ y \end{bmatrix} - \ddot{\theta} \begin{bmatrix} y \\ -x \end{bmatrix} - 2\dot{\theta} \begin{bmatrix} \dot{y} \\ -\dot{x} \end{bmatrix} \right\}.
$$
(2.53)

The first terms on the right-hand side of (2.53) are rotating-frame acceleration terms, the second terms are the *centrifugal acceleration* terms, the third terms are the *Euler acceleration* terms that derive from the angular acceleration of the moving frame (relative to the inertial frame), while the last terms are the *Coriolis acceleration* terms. These accelerations (forces) are sometimes referred to as *fictitious, apparent,* or *non-inertial* accelerations (forces), because they result from the use of non-inertial frames of reference.

Example 2.6 Consider the system of interconnected masses illustrated in Figure 2.2.

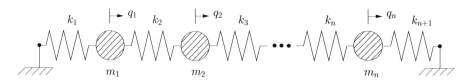

Figure 2.2: A string of interconnected springs and masses with colinear translational freedoms. The masses have mass m_i, the springs have stiffness k_i, and the displacement of the masses from their rest positions is denoted q_i.

A direct application of Newton's second law gives

$$
M\ddot{q} + Kq = 0
$$
(2.54)

in which the mass matrix is

$$
M = \begin{bmatrix} m_1 & 0 & \cdots & 0 & 0 \\ 0 & m_2 & 0 & \cdots & 0 \\ \vdots & \vdots & \vdots & & \vdots \\ 0 & 0 & \cdots & 0 & m_n \end{bmatrix}
$$
(2.55)

and the stiffness matrix is

$$
K = \begin{bmatrix} k_1 + k_2 & -k_2 & 0 & \cdots & & 0 \\ -k_2 & k_2 + k_3 & -k_3 & \ddots & & \vdots \\ 0 & \vdots & \vdots & \vdots & & 0 \\ \vdots & \ddots & k_{n-1} & k_{n-1} + k_n & -k_n \\ 0 & \cdots & 0 & -k_n & k_n + k_{n+1} \end{bmatrix}.
$$
(2.56)

By introducing the vector of intermediate variables p we can write these equations as

$$\begin{bmatrix} 0 & I \\ M & 0 \end{bmatrix} \begin{bmatrix} \dot{p} \\ \dot{q} \end{bmatrix} = \begin{bmatrix} I & 0 \\ 0 & -K \end{bmatrix} \begin{bmatrix} p \\ q \end{bmatrix}, \tag{2.57}$$

or equivalently

$$\begin{bmatrix} \dot{p} \\ \dot{q} \end{bmatrix} = \begin{bmatrix} 0 & -M^{-1}K \\ I & 0 \end{bmatrix} \begin{bmatrix} p \\ q \end{bmatrix}. \tag{2.58}$$

If a matrix H satisfies $JH = (JH)^T$, in which

$$J = \begin{bmatrix} 0 & I_n \\ -I_n & 0 \end{bmatrix},$$

then H is called Hamiltonian [55]. The eigenvalues of Hamiltonian matrices have well-known symmetry properties—if λ_i is an eigenvalue of H, then so is $-\lambda_i$. If the entries of H are real, and if λ_i is an eigenvalue, then so is $\bar{\lambda}_i$ (the bar denotes complex conjugate). Since the mass and stiffness matrices are symmetric, it is easy to show that the $2n \times 2n$ matrix in (2.58) is Hamiltonian. The $2n$ eigenvalues of H are given by $\pm i\omega_i$, where the ω_is are the system's n modal frequencies.

The kinetic energy of each mass is $\frac{1}{2}m_i\dot{q}_i^2$, and so the system's total kinetic energy is given by the quadratic form $T(\dot{q}) = \frac{1}{2}(\dot{q}^T M \dot{q})$. From this it follows that the inertial force term in (2.54) is given by the vector $\frac{d}{dt}(\partial T/\partial \dot{q}) = M\ddot{q}$. It is also clear that the strain energy in the central springs is given by $\frac{1}{2}k_i(q_i - q_{i-1})^2$, while that in the two end springs is given by $\frac{1}{2}k_1 q_1^2$ and $\frac{1}{2}k_{n+1}q_n^2$; we make no distinction between the energy associated with compression and that associated with spring extensions. This means that the total strain (potential) energy is given by the quadratic form $V(q) = \frac{1}{2}(q^T K q)$ and the stiffness force term in (2.54) is given by the vector $(\partial V/\partial q) = Kq$.

Example 2.7 When some coordinate systems are used the mass matrix becomes a function of the displacement coordinates. To see how this comes about we consider briefly the spring pendulum in Figure 2.3. Suppose the origin of an inertial polar coordinate system is located at the pendulum pivot. In this coordinate system the mass centre of the pendulum bob is located at

$$l = (r_0 + r)e^{i\theta}. \tag{2.59}$$

The complex quantity in (2.59) can be thought of as a radial vector that points away from the pendulum pivot at an angle of θ radians to the vertical; see Figure 2.3. The acceleration of the bob's mass centre can be found by differentiating the position vector twice

$$\ddot{l} = (\ddot{r} - (r_0 + r)\dot{\theta}^2)e^{i\theta} + i(2\dot{\theta}\dot{r} + (r_0 + r)\ddot{\theta})e^{i\theta}; \tag{2.60}$$

the second term in (2.60) represents a radial vector at an angle of $\theta + \frac{\pi}{2}$ to the vertical. Summing forces in the radial direction gives

$$m\ddot{r} = mg\cos\theta - kr + m(r_0 + r)\dot{\theta}^2 \tag{2.61}$$

in which k is the spring stiffness. Summing forces in the tangential direction gives

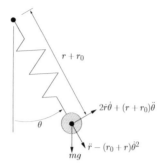

Figure 2.3: Spring pendulum system. The pendulum angle is θ, the pendulum's relaxed length is r_0, the pendulum's extended length is $r_0 + r$, and the bob mass is m.

$$(r_0 + r)\ddot{\theta} + 2\dot{\theta}\dot{r} = -g\sin\theta. \tag{2.62}$$

Equations (2.61) and (2.62) can be combined as

$$\begin{bmatrix} m & 0 \\ 0 & r_0 + r \end{bmatrix} \begin{bmatrix} \ddot{r} \\ \ddot{\theta} \end{bmatrix} = \begin{bmatrix} mg\cos\theta - kr + m(r_0 + r)\dot{\theta}^2 \\ -2\dot{\theta}\dot{r} - g\sin\theta \end{bmatrix} \tag{2.63}$$

which is of the general form

$$M(\boldsymbol{q})\ddot{\boldsymbol{q}} = F(\boldsymbol{q}, \dot{\boldsymbol{q}})$$

in which the mass matrix is position dependent. It is clear that the swing mode is coupled to the heave mode through a Coriolis acceleration term, while the heave mode is coupled to the swing mode via both gravitational and centripetal acceleration terms.[5]

2.2.9 Lagrange's equations in quasi-velocities

The Lagrange equations, as discussed in Section 2.2.8, are not always an efficient way of deriving a compact set of equations of motion. It is difficult, for example, to study the motion of tops and gyroscopes using the Euler–Lagrange equations in generalized coordinates and generalized velocities. As an alternative, Euler showed that expressing the equations of motion of moving bodies in terms of angular velocity components, relative to body-fixed axes, was effective in producing compact descriptions of the system dynamics. These angular velocity components are examples of *quasi-velocities* that may be used instead of conventional generalized velocities. Another example of the application of quasi-velocities is in vehicle dynamics, where velocities in vehicle-fixed axes gives a compact description of the vehicle's motion.

---

[5] For those already familiar with Lagrange mechanics, it should be easy to check that the pendulum system's kinetic energy is given by $T(\dot{\boldsymbol{q}}) = \frac{m}{2}(\dot{r}^2 + (r_0 + r)^2\dot{\theta}^2)$; that the potential energy is given by $V(\boldsymbol{q}) = \frac{1}{2}kr^2 - mg(r_0 + r)\cos\theta$; and that the Lagrangian is given by $\mathcal{L} = T - V$. Direct substitution of this Lagrangian into (2.40) gives (2.63). Is there anything you can say about $T(\dot{\boldsymbol{q}}) + V(\boldsymbol{q})$?

In the classical Lagrangian formalism nonholonomic constraints are appended to Lagrange's equations using Lagrange multipliers; see (2.45). The subsequent elimination of these multipliers using (2.47) can be a cumbersome process. In the approach to be described here, a judicious choice of quasi-velocities allows for the inclusion of nonholonomic constraints without invoking Lagrange multipliers. Procedurally, the unconstrained equations are derived first, with the constraints then enforced by setting the associated quasi-velocities to zero. The equations of motion relating to the nonholonomic constraints can then be used to find the forces of constraint; an example of the application of this process will be given in Section 2.8.

In Section 2.2.8 it was shown that the Lagrange's equations of motion take the form

$$\frac{d}{dt}\left(\frac{\partial \mathcal{L}}{\partial \dot{q}}\right) - \frac{\partial \mathcal{L}}{\partial q} = k \tag{2.64}$$

where $\mathcal{L}$ is given by (2.37), q is the vector of generalized variables, $\dot{q}$ is the vector of generalized velocities, and k is the vector of generalized forces.

As explained above, it is often convenient to replace $\dot{q}$ with the vector of quasi-velocities $v = [v_1, \cdots, v_n]^T$. Following [56], we assume that the quasi-velocities are a linear function of the the the generalized velocities as follows:

$$v_i = \alpha_{1i}\dot{q}_1 + \alpha_{2i}\dot{q}_2 + \cdots + \alpha_{ni}\dot{q}_n = \sum_{j=1}^{n} \alpha_{ji}\dot{q}_j, \tag{2.65}$$

in which the α_{ji}s are functions of the generalized coordinates q. The corresponding set of differentials ds_i are given by

$$ds_i = \sum_{j=1}^{n} \alpha_{ji}dq_j, \qquad i = 1, 2, \cdots, n. \tag{2.66}$$

It follows from Schwarz theorem that (2.66) cannot be integrated to obtain the s_is unless $\frac{\partial \alpha_{ji}}{\partial q_k} = \frac{\partial \alpha_{ki}}{\partial q_j}$. This is why the v_is are called *quasi-velocities* [56–58]. In matrix form (2.65) becomes

$$v = \alpha^T \dot{q}; \tag{2.67}$$

the reason for introducing the transpose will soon become apparent. We will assume that α is non-singular and so $\dot{q} = \alpha^{-T}v$; quasi-velocities v_i corresponding to nonholonomic constraints will necessarily be zero.

Our aim is now to rewrite (2.64) in terms of q and v instead of q and $\dot{q}$. Starting with the first term of (2.64), in which $\mathcal{L}(q, \dot{q}) = \bar{\mathcal{L}}(q, v)$, there holds

$$\frac{\partial \mathcal{L}}{\partial \dot{q}_i} = \sum_{j=1}^{n} \frac{\partial \bar{\mathcal{L}}}{\partial v_j}\frac{\partial v_j}{\partial \dot{q}_i} = \sum_{j=1}^{n} \alpha_{ij}\frac{\partial \bar{\mathcal{L}}}{\partial v_j}. \tag{2.68}$$

Using matrix notation, (2.68) can be used to show that

$$\frac{d}{dt}\left(\frac{\partial \mathcal{L}}{\partial \dot{q}}\right) = \frac{d}{dt}\left(\alpha\frac{\partial \bar{\mathcal{L}}}{\partial v}\right) = \alpha\frac{d}{dt}\left(\frac{\partial \bar{\mathcal{L}}}{\partial v}\right) + \dot{\alpha}\frac{\partial \bar{\mathcal{L}}}{\partial v}, \tag{2.69}$$

where

$$\dot{\alpha}_{ij} = \sum_{k=1}^{n} \frac{\partial \alpha_{ij}}{\partial q_k} \dot{q}_k = \dot{\boldsymbol{q}}^T \frac{\partial \alpha_{ij}}{\partial \boldsymbol{q}} = \boldsymbol{v}^T \alpha^{-1} \frac{\partial \alpha_{ij}}{\partial \boldsymbol{q}}. \tag{2.70}$$

The second term of (2.64) can be rewritten as

$$\frac{\partial \mathcal{L}}{\partial q_i} = \frac{\partial \bar{\mathcal{L}}}{\partial q_i} + \left(\frac{\partial \boldsymbol{v}}{\partial q_i}\right)^T \frac{\partial \bar{\mathcal{L}}}{\partial \boldsymbol{v}} = \frac{\partial \bar{\mathcal{L}}}{\partial q_i} + \dot{\boldsymbol{q}}^T \frac{\partial \alpha}{\partial q_i} \frac{\partial \bar{\mathcal{L}}}{\partial \boldsymbol{v}} = \frac{\partial \bar{\mathcal{L}}}{\partial q_i} + \boldsymbol{v}^T \alpha^{-1} \frac{\partial \alpha}{\partial q_i} \frac{\partial \bar{\mathcal{L}}}{\partial \boldsymbol{v}}. \tag{2.71}$$

The Lagrange equations of motion in terms of the quasi-coordinates $\boldsymbol{q}$ and $\boldsymbol{v}$ can now be obtained from (2.69)–(2.71) after premultiplication by α^{-1}:

$$\frac{d}{dt}\left(\frac{\partial L(\boldsymbol{q}, \boldsymbol{v})}{\partial \boldsymbol{v}}\right) + \alpha^{-1}\gamma \frac{\partial L(\boldsymbol{q}, \boldsymbol{v})}{\partial \boldsymbol{v}} - \alpha^{-1}\frac{\partial L(\boldsymbol{q}, \boldsymbol{v})}{\partial \boldsymbol{q}} = \alpha^{-1}\boldsymbol{k}, \tag{2.72}$$

where the bar in $\bar{L}$ has been dropped, and where the $n \times n$ matrix γ is given by

$$\gamma = \begin{bmatrix} \ddots & \vdots & \cdot \cdot \\ \cdots & \boldsymbol{v}^T \alpha^{-1} \frac{\partial \alpha_{ij}}{\partial \boldsymbol{q}} & \cdots \\ \cdot \cdot & \vdots & \ddots \end{bmatrix} - \begin{bmatrix} \vdots \\ \boldsymbol{v}^T \alpha^{-1} \frac{\partial \alpha}{\partial q_i} \\ \vdots \end{bmatrix}. \tag{2.73}$$

It is worth emphasizing that each term in the first part of (2.73) is a scalar quantity, while those in the second part are $1 \times n$ row vectors. The right-hand-side term $\alpha^{-1}\boldsymbol{k}$ in (2.72) is the projection of the generalized forces onto the quasi-velocity system.

Example 2.8 We will derive the Lagrange equations in quasi-velocities for a body moving in a horizontal x–y plane. The body has translational velocities u, v along an axis system attached to its mass centre, and yaw ω. The absolute position of the body is given by x, y while the orientation is given by the angle ψ.
The quasi-velocity vector $\boldsymbol{v} = [u, v, \omega]^T$ replaces $\dot{\boldsymbol{q}} = [\dot{x}, \dot{y}, \dot{\psi}]^T$, with the corresponding Lagrangian given by $\mathcal{L} = T(\boldsymbol{v})$, where T is the kinetic energy. The relationship between $\boldsymbol{v}$ and $\dot{\boldsymbol{q}}$ is obtained by trigonometry as

$$\alpha^T = \begin{bmatrix} \cos \psi & \sin \psi & 0 \\ -\sin \psi & \cos \psi & 0 \\ 0 & 0 & 1 \end{bmatrix}, \tag{2.74}$$

which is introduced into (2.73) to give

$$\alpha^{-T}\gamma = \begin{bmatrix} 0 & -\omega & 0 \\ \omega & 0 & 0 \\ -v & u & 0 \end{bmatrix}. \tag{2.75}$$

The resulting equations of motion are

$$\frac{d}{dt}\frac{\partial T}{\partial u} - \omega \frac{\partial T}{\partial v} = k_u \tag{2.76}$$

$$\frac{d}{dt}\frac{\partial T}{\partial v} + \omega \frac{\partial T}{\partial u} = k_v \tag{2.77}$$

$$\frac{d}{dt}\frac{\partial T}{\partial \omega} - v\frac{\partial T}{\partial u} + u\frac{\partial T}{\partial v} = k_\omega \tag{2.78}$$

which corresponds to those reported in, for example, [8]. In Section 4.2 the same equations will be derived with the alternative Newton–Euler approach, in the context of modelling a car moving on a flat road.

Example 2.9 We will now derive Lagrange's equations in quasi-velocities for a body moving in space. The body has translational velocities u, v, w and rotational velocities $\omega_x, \omega_y, \omega_z$ along a body-fixed axis system attached to its mass centre. The absolute position of the body is given by x, y, z, while the orientation is given by three angles θ, μ, ϕ.
The quasi-velocity vector $\boldsymbol{v} = [u, v, w, \omega_x, \omega_y, \omega_z]^T$ is used in place of $\dot{\boldsymbol{q}} = [\dot{x}, \dot{y}, \dot{z}, \dot{\theta}, \dot{\mu}, \dot{\phi}]$.
The Lagrangian is $\mathcal{L}(\boldsymbol{v}, \boldsymbol{q}) = T(\boldsymbol{v}) - V(\boldsymbol{q})$, where T is the kinetic energy and V is the potential energy. The relationship between $\boldsymbol{v}$ and $\dot{\boldsymbol{q}}$ is given by[6]

$$
\alpha^T =
\begin{bmatrix}
c_\theta c_\mu & s_\theta c_\mu & -s_\mu & 0 & 0 & 0 \\
-s_\theta c_\phi + c_\theta s_\mu s_\phi & c_\theta c_\phi + s_\theta s_\mu s_\phi & c_\mu s_\phi & 0 & 0 & 0 \\
s_\theta s_\phi + c_\theta s_\mu c_\phi & -c_\theta s_\phi + s_\theta s_\mu c_\phi & c_\mu c_\phi & 0 & 0 & 0 \\
0 & 0 & 0 & -s_\mu & 0 & 1 \\
0 & 0 & 0 & c_\mu s_\phi & c_\phi & 0 \\
0 & 0 & 0 & c_\mu c_\phi & -s_\phi & 0
\end{bmatrix}
$$

which is introduced into (2.73) to give

$$
\alpha^{-T}\gamma =
\begin{bmatrix}
0 & -\omega_z & \omega_y & 0 & 0 & 0 \\
\omega_z & 0 & -\omega_x & 0 & 0 & 0 \\
-\omega_y & \omega_x & 0 & 0 & 0 & 0 \\
0 & -w & v & 0 & -\omega_z & \omega_y \\
w & 0 & -u & \omega_z & 0 & -\omega_x \\
-v & u & 0 & -\omega_y & \omega_x & 0
\end{bmatrix}.
\tag{2.79}
$$

The resulting equations of motion are

---

[6] A general method for deriving the relationships between quantities defined in different axis systems will be explained in Section 2.5. In this example the orientation is obtained according to the yaw–pitch–roll $(\theta{-}\mu{-}\phi)$ convention. The 3×3 diagonal blocks in (2.79) are skew-symmetric matrix representations of $\boldsymbol{\omega} \times \cdot$, where $\boldsymbol{\omega}$ is the angular velocity of the moving body with respect to a stationary axis system.

$$\frac{d}{dt}\frac{\partial \mathcal{L}}{\partial u} - \omega_z \frac{\partial \mathcal{L}}{\partial v} + \omega_y \frac{\partial \mathcal{L}}{\partial w} + \sin \mu \frac{\partial \mathcal{L}}{\partial z} = k_u \tag{2.80}$$

$$\frac{d}{dt}\frac{\partial \mathcal{L}}{\partial v} + \omega_z \frac{\partial \mathcal{L}}{\partial u} - \omega_x \frac{\partial \mathcal{L}}{\partial w} - \cos \mu \sin \phi \frac{\partial \mathcal{L}}{\partial z} = k_v \tag{2.81}$$

$$\frac{d}{dt}\frac{\partial \mathcal{L}}{\partial w} - \omega_y \frac{\partial \mathcal{L}}{\partial u} + \omega_x \frac{\partial T}{\partial v} - \cos \mu \cos \phi \frac{\partial \mathcal{L}}{\partial z} = k_w \tag{2.82}$$

$$\frac{d}{dt}\frac{\partial \mathcal{L}}{\partial \omega_x} - w\frac{\partial \mathcal{L}}{\partial v} + v\frac{\partial \mathcal{L}}{\partial w} - \omega_z \frac{\partial \mathcal{L}}{\partial \omega_y} + \omega_y \frac{\partial \mathcal{L}}{\partial \omega_z} = k_{\omega_x} \tag{2.83}$$

$$\frac{d}{dt}\frac{\partial \mathcal{L}}{\partial \omega_y} + w\frac{\partial \mathcal{L}}{\partial u} - u\frac{\partial \mathcal{L}}{\partial w} + \omega_z \frac{\partial \mathcal{L}}{\partial \omega_x} - \omega_x \frac{\partial \mathcal{L}}{\partial \omega_z} = k_{\omega_y} \tag{2.84}$$

$$\frac{d}{dt}\frac{\partial \mathcal{L}}{\partial \omega_z} - v\frac{\partial \mathcal{L}}{\partial u} + u\frac{\partial \mathcal{L}}{\partial v} - \omega_y \frac{\partial \mathcal{L}}{\partial \omega_x} + \omega_x \frac{\partial \mathcal{L}}{\partial \omega_y} = k_{\omega_z} . \tag{2.85}$$

In Section 7.5 the same equations will be derived using a Newton–Euler formulation for a car moving on a three-dimensional road.

2.3 Conservation laws

The dynamic properties of a mechanical system are described in terms of the temporal evolution of its generalized coordinates from a given initial condition. While the generalized coordinates are almost always time-varying, there are certain quantities that remain constant during the motion and depend only on the initial conditions. These quantities are called *integrals of the motion* [43], some of which are of great importance and derive from the homogeneity (translation invariance) and isotropy (rotation invariance) of space and time. These constants of the motion are referred to as *conserved quantities*.[7]

2.3.1 Energy

We consider first the conservation law that derives from the *homogeneity of time*. If all the impressed forces can be derived from a potential, and the resulting system Lagrangian is time-shift invariant (scleronomic), there holds

$$\frac{d\mathcal{L}}{dt} = \frac{\partial \mathcal{L}}{\partial q} \cdot \dot{q} + \frac{\partial \mathcal{L}}{\partial \dot{q}} \cdot \ddot{q}; \tag{2.86}$$

in which the $\cdot$ denotes a dot product. There is no $\frac{\partial \mathcal{L}}{\partial t}$ term, since $\mathcal{L}$ is not an explicit function of time (by assumption). In the case of unconstrained systems, the $\frac{\partial \mathcal{L}}{\partial q}$ term can be replaced by $\frac{d}{dt}(\frac{\partial \mathcal{L}}{\partial \dot{q}})$ using (2.40) so that

---

[7] In some important work E. Noether (1918) showed that every symmetry of the action of a physical system has a corresponding conservation law. By 'symmetry' we mean any transformation of the generalized coordinates, the associated generalized velocities, and time that leaves the value of the Lagrangian unaffected.

$$\frac{d\mathcal{L}}{dt} = \frac{d}{dt}\left(\frac{\partial\mathcal{L}}{\partial\dot{\boldsymbol{q}}}\right)\cdot\dot{\boldsymbol{q}} + \frac{\partial\mathcal{L}}{\partial\dot{\boldsymbol{q}}}\cdot\ddot{\boldsymbol{q}}$$

$$= \frac{d}{dt}\left(\frac{\partial\mathcal{L}}{\partial\dot{\boldsymbol{q}}}\cdot\dot{\boldsymbol{q}}\right)$$

or

$$\frac{d}{dt}\left(\frac{\partial\mathcal{L}}{\partial\dot{\boldsymbol{q}}}\cdot\dot{\boldsymbol{q}} - \mathcal{L}\right) = 0.$$

The quantity

$$E = \frac{\partial\mathcal{L}}{\partial\dot{\boldsymbol{q}}}\cdot\dot{\boldsymbol{q}} - \mathcal{L} \tag{2.87}$$

is therefore constant as a result of temporal homogeneity. In the case that time-invariant holonomic constraints are included, (2.87) still holds with $\mathcal{L}$ replaced by $\hat{\mathcal{L}}$ given by (2.42).

Homogeneity in time results in energy conservation in the case of (scleronomic) non-holonomic systems too. To see this observe that the $\frac{\partial\mathcal{L}}{\partial\boldsymbol{q}}$ term in (2.86) can be replaced by $\frac{d}{dt}(\frac{\partial\mathcal{L}}{\partial\dot{\boldsymbol{q}}}) + A(\boldsymbol{q})^T\boldsymbol{\lambda}$ using (2.45) and (2.46). Since the nonholonomic constraints are satisfied during the motion, $A(\boldsymbol{q})^T\boldsymbol{\lambda}\cdot\dot{\boldsymbol{q}} = \boldsymbol{\lambda}^T A(\boldsymbol{q})\dot{\boldsymbol{q}} = 0$, and the arguments relating to holonomic systems hold good.

In sum, the requirements for the conservation of energy are that the impressed forces be derivable from a potential function and that the Lagrangian and constraints are not explicitly dependent on time.

If the Lagrangian of a mechanical system is of the form $\mathcal{L} = T(\boldsymbol{q},\dot{\boldsymbol{q}}) - V(\boldsymbol{q})$, in which T is quadratic in the velocities, there holds

$$\frac{\partial\mathcal{L}}{\partial\dot{\boldsymbol{q}}}\cdot\dot{\boldsymbol{q}} = \frac{\partial T}{\partial\dot{\boldsymbol{q}}}\cdot\dot{\boldsymbol{q}} = 2T$$

and (2.87) can be rewritten as

$$E = T(\boldsymbol{q},\dot{\boldsymbol{q}}) + V(\boldsymbol{q}), \tag{2.88}$$

which is the sum of the system's kinetic and potential energy; time homogeneity leads to the *law of conservation of energy*.

Example 2.10 Consider the one-dimensional harmonic oscillator shown here:

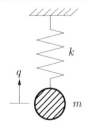

Ignoring gravity, the system's kinetic and potential energies are given by

$$T = \frac{1}{2}m\dot{q}^2 \qquad \text{and} \qquad V = \frac{1}{2}kq^2$$

respectively. The Lagrangian is given by

$$\mathcal{L} = \frac{1}{2}m\dot{q}^2 - \left(\frac{1}{2}kq^2\right).$$

Solving (2.40) gives

$$\ddot{q} + \omega^2 q = 0,$$

where

$$\omega = \sqrt{\frac{k}{m}}.$$

It is easy to establish that $q(t) = A\sin\omega t$ and that $E = T + V = A^2 k/2$. In this case the Lagrangian is time-shift invariant, and system's total internal energy is constant and a function only of the initial conditions.

2.3.2 Linear momentum

Another conservation law derives from homogeneity, that is, translation invariance. In this case we derive the system properties that remain unchanged under an infinitesimal translational generalized displacement of the entire system. Let us consider, therefore, a small (constant) displacement δq in the generalized coordinates, with the generalized velocities unchanged. That is

$$\delta\mathcal{L} = \frac{\partial\mathcal{L}}{\partial q}\delta q. \tag{2.89}$$

Since $\delta\mathcal{L} = 0$ by assumption, and δq is arbitrary, there holds

$$\frac{\partial\mathcal{L}}{\partial q} = 0. \tag{2.90}$$

In the case of an unconstrained system (2.40) gives

$$\frac{d}{dt}\left(\frac{\partial\mathcal{L}}{\partial\dot{q}}\right) = 0$$

and so the generalized momenta

$$\boldsymbol{p} = \frac{\partial\mathcal{L}}{\partial\dot{q}} \tag{2.91}$$

are constant (conserved). In the case of holonomic constraints, the same arguments hold with $\mathcal{L}$ replaced by $\hat{\mathcal{L}}$ in (2.42). In the case of nonholonomic systems, (2.45) and (2.46) give

$$\boldsymbol{p} = -\int_0^t A(\boldsymbol{q})^T \boldsymbol{\lambda} d\tau, \tag{2.92}$$

which may be time-varying.

Generalized coordinates that do not appear in the Lagrangian, while their derivatives do, are called *cyclic*, *ignorable*, or *kinosthenic* variables [41, 44]. These variables satisfy (2.90) and so the related momenta are conserved if the Lagrange equations have no right-hand-side terms. Typical examples are the position and yaw of a vehicle on a flat road and the spin angle of a homogeneous wheel.

Example 2.11 Consider a particle of mass m in a horizontal Cartesian inertial reference frame. In this case the Lagrangian is

$$\mathcal{L} = \frac{m}{2}(\dot{x}^2 + \dot{y}^2).$$

Since the Lagrangian is not explicitly time dependent, energy is conserved. It also follows from (2.40) that $m\dot{x}$ and $m\dot{y}$ are constant. This is an elementary illustration of the conservation of linear momentum, which derives from a Lagrangian that is not a function of position.

Example 2.12 Consider in the following figure a simple planar pendulum suspended from pivot point P, which is free to move along the x-axis.

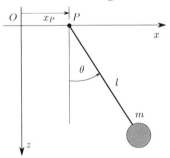

The pendulum rod is massless and has length l; the bob mass is m. The pendulum's angle of rotation is θ. The system's kinetic energy is computed using (2.29)

$$T = \frac{m}{2}\left(\dot{x}_P^2 + 2l\dot{x}_P\dot{\theta}\cos\theta + l^2\dot{\theta}^2\right),\tag{2.93}$$

and the system's potential energy is given by (2.36)

$$V = -mgl\cos\theta.\tag{2.94}$$

The Lagrangian is

$$\mathcal{L} = \frac{m}{2}\left(\dot{x}_P^2 + 2l\dot{x}_P\dot{\theta}\cos\theta + l^2\dot{\theta}^2\right) + mgl\cos\theta.\tag{2.95}$$

Since the Lagrangian is not an explicit function of time, the system is conservative.
The equations of motion are derived using (2.40). The θ-related equation is

$$\frac{d}{dt}\left(\frac{\partial\mathcal{L}}{\partial\dot{\theta}}\right) - \frac{\partial\mathcal{L}}{\partial\theta} = 0,\tag{2.96}$$

which gives

$$l\ddot{\theta} = -\ddot{x}_P\cos\theta - g\sin\theta,\tag{2.97}$$

which is the familiar pendulum equation when x_P is fixed (anywhere along the x-axis).

Since the Lagrangian is neither an explicit function of time nor a function of x_P, the system is conservative with the linear momentum

$$\frac{\partial \mathcal{L}}{\partial \dot{x}_P} = m(\dot{x}_P + l\dot{\theta}\cos\theta) \tag{2.98}$$

conserved.

Now suppose that x_P is subject to the more general constraint

$$f(x_P, t) = x_P - u(t) = 0. \tag{2.99}$$

The augmented Lagrangian (2.42) is given by

$$\hat{\mathcal{L}} = \frac{m}{2}\left(\dot{x}_P^2 + 2l\dot{x}_P\dot{\theta}\cos\theta + l^2\dot{\theta}^2\right) + mgl\cos\theta + \lambda(x_P - u(t)), \tag{2.100}$$

and the swing equation becomes

$$l\ddot{\theta} = -\left(\ddot{u}(t)\cos\theta + g\sin\theta\right). \tag{2.101}$$

The linear displacement equation is

$$\frac{d}{dt}\left(\frac{\partial \mathcal{L}}{\partial \dot{x}_P}\right) - \frac{\partial \mathcal{L}}{\partial x_P} = -A(x_P)\lambda, \tag{2.102}$$

where

$$A(x_P) = \frac{\partial f(x_P, t)}{\partial x_P} = 1. \tag{2.103}$$

Therefore

$$\lambda = m\left(l\dot{\theta}^2\sin\theta - \ddot{u}(t) - l\ddot{\theta}\cos\theta\right) \tag{2.104}$$

which gives the force of constraint

$$k_{x_P} = -A(x_P)\lambda = m\left(\ddot{u}(t) + l\ddot{\theta}\cos\theta - l\dot{\theta}^2\sin\theta\right). \tag{2.105}$$

There are three cases to consider. If $u = x_P = 0$ (2.97) becomes

$$l\ddot{\theta} = -g\sin\theta \tag{2.106}$$

and the system is conservative. In the case that $\dot{u} = \dot{x}_P$ is constant, one can replace the original coordinate system with one in which x_P is stationary. This new axis system is *inertial*, (2.97) applies, and the system is again conservative. If $\ddot{u} = \ddot{x}_P(t) \neq 0$, the swing equation becomes (2.101), (real) work is done by the constraint force and the system is no longer conservative.

2.3.3 Angular momentum

Spatial isotropy is related to the mechanical property of systems that remain invariant under spatial rotations. In this case we consider next an infinitesimal rotation of the entire system and then seek the condition under which the Lagrangian remains unchanged. As shown in the diagram, the vector $\delta\phi$ has a prescribed axis of rotation and magnitude ϕ.

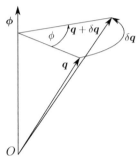

The change in the generalized coordinates δq resulting from the infinitesimal but fixed rotation ϕ is given by

$$\delta q = \delta\phi \times q,$$

while the change in the generalized velocities is given by

$$\delta\dot{q} = \delta\phi \times \dot{q}.$$

If these expressions are substituted into the condition for the variation in the Lagrangian

$$\delta\mathcal{L} = \frac{\partial\mathcal{L}}{\partial q}\cdot\delta q + \frac{\partial\mathcal{L}}{\partial\dot{q}}\cdot\delta\dot{q} \tag{2.107}$$

to remain unchanged, we obtain

$$0 = \frac{\partial\mathcal{L}}{\partial q}\cdot\delta\phi\times q + \frac{\partial\mathcal{L}}{\partial\dot{q}}\cdot\delta\phi\times\dot{q}$$
$$= \frac{d}{dt}\left(\frac{\partial\mathcal{L}}{\partial\dot{q}}\right)\cdot\delta\phi\times q + \frac{\partial\mathcal{L}}{\partial\dot{q}}\cdot\delta\phi\times\dot{q}$$

in the case of a holonomic system. Using a standard triple product identity one obtains

$$0 = \delta\phi\left(q\times\frac{d}{dt}\left(\frac{\partial\mathcal{L}}{\partial\dot{q}}\right) + \dot{q}\times\frac{\partial\mathcal{L}}{\partial\dot{q}}\right)$$
$$= \delta\phi\left(\frac{d}{dt}\left(q\times p\right)\right),$$

in which $p = \frac{\partial\mathcal{L}}{\partial\dot{q}}$ is the generalized momentum. Since $\delta\phi$ is arbitrary, it follows that $(d/dt)(q\times p) = 0$ and so

$$H = q \times p \tag{2.108}$$

is constant; H is the *angular momentum*, or the *moment of momentum*.

Example 2.13 Consider once more a free particle of mass m in a polar inertial reference frame. As shown in the following sketch,

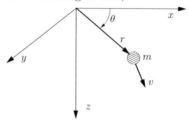

the particle is free to move in the x–y plane, with the z-axis orthogonal to the horizontal plane. The Lagrangian is again the kinetic energy of the particle and is given by

$$\mathcal{L} = \frac{m}{2}(\dot{r}^2 + r^2\dot{\theta}^2)$$

in this case. The Euler–Lagrange equations in r and θ are given by

$$\frac{d}{dt}\left(\frac{\partial \mathcal{L}}{\partial \dot{r}}\right) - \frac{\partial \mathcal{L}}{\partial r} = 0 \tag{2.109}$$

and

$$\frac{d}{dt}\left(\frac{\partial \mathcal{L}}{\partial \dot{\theta}}\right) = 0 \tag{2.110}$$

respectively. Equation (2.109) gives

$$\ddot{r} = r\dot{\theta}^2 \tag{2.111}$$

that is the centrifugal acceleration usually associated with circular motion. Equation (2.110) shows that

$$\frac{\partial \mathcal{L}}{\partial \dot{\theta}} = mr^2\dot{\theta},$$

is constant, which is the magnitude of the angular momentum. If one recognizes the vectorial implications of this result, we see that the angular momentum (as a vector), $\boldsymbol{H}$, is given by (Section 2.2.2)

$$\boldsymbol{H} = r\hat{e}_r \times m(\dot{r}\hat{e}_r + r\dot{\theta}\hat{e}_\theta),$$

in which $\hat{e}_r$ is a unit vector in the radial direction in the x–y plane, while $\hat{e}_\theta$ is also in the x–y plane and orthogonal to $\hat{e}_r$. By simplifying the cross product, one sees that the angular momentum is in the direction of $\hat{e}_z$ along the z-axis:

$$\boldsymbol{H} = mr^2\dot{\theta}\hat{e}_z.$$

Solving (2.110) gives

$$\ddot{\theta} = -\frac{2\dot{r}\dot{\theta}}{r}. \tag{2.112}$$

It is easy to show that (2.111) and (2.112) have solutions

$$r \cos \theta = at + b \tag{2.113}$$
$$r \sin \theta = ct + d \tag{2.114}$$

with the constants a, b, c, and d determined by the initial conditions.[8] As one would expect, this solution represents rectilinear motion in polar coordinates.

2.4 Hamilton's equations

In the Lagrangian framework the state of the system is described in terms of the n generalized coordinates q_i, their derivatives $\dot{q}_i$, and the time t. In an alternative framework, due to *Hamilton*, the system state is described in terms of the generalized coordinates q_i, the generalized momenta p_i, and t. Hamilton's canonical equations replace the original n (second-order) Lagrangian differential equations with $2n$ first-order equations. As we will now show Lagranges equations are converted into the Hamiltonian formalism using the *Legendre Transformation*. The discovery of the canonical equations led to a new era in theoretical mechanics.

2.4.1 Legendre transform

The Legendre transformation connects two alternative descriptions of the same physics through functions of related ('conjugate') variables. Suppose a function of n variables is given

$$F = F(u_1, \cdots, u_n) \tag{2.115}$$

and that we introduce a new set of variables by means of the transformation

$$\boldsymbol{v} = \frac{\partial F}{\partial \boldsymbol{u}}, \tag{2.116}$$

where $\boldsymbol{u} = [u_1, \cdots, u_n]^T$. If

$$det \left[\frac{\partial^2 F}{\partial u_i \partial u_j} \right] \neq 0 \tag{2.117}$$

the v_i s are independent and (2.116) can be solved for the u_i s in terms of the v_i s. In mechanics, the regularity condition is satisfied if $\mathcal{L}$ is given as the kinetic energy T with a positive definite generalized mass matrix minus the potential energy V. We will not consider degenerate systems in which the u_i s are not independent.

A new function G can now be defined as follows

$$G = \boldsymbol{u}^T \boldsymbol{v} - F. \tag{2.118}$$

By solving (2.116) for the u_i s, we can express G in terms of the v_i s alone

$$G = G(v_1, \cdots, v_n). \tag{2.119}$$

Taking variations in G, one obtains

---

[8] Differentiate (2.113) and (2.114) twice, and eliminate $\ddot{r}$ and $\ddot{\theta}$ using (2.111) and (2.112).

$$\delta G = \sum_{i=1}^{n} \frac{\partial G}{\partial v_i} \delta v_i \tag{2.120}$$

$$= \sum_{i=1}^{n} \left(u_i \delta v_i + \left(v_i - \frac{\partial F}{\partial u_i} \right) \delta u_i \right) \tag{2.121}$$

$$= \sum_{i=1}^{n} u_i \delta v_i. \tag{2.122}$$

using (2.116). This means that

$$\frac{\partial G}{\partial v_i} \delta v_i = \sum_{i=1}^{n} u_i \delta v_i,$$

and so

$$\boldsymbol{u} = \frac{\partial G}{\partial \boldsymbol{v}}, \tag{2.123}$$

which establishes the duality of the Legendre transform. The new variables are the partial derivatives of the old functions with respect to the old variables and the old variables are the partial derivatives of the new functions with respect to the new variables. In sum:

Old System	*New System*
variables: $u_1, \cdots, u_n$;	variables: $v_1, \cdots, v_n$;
Function $F = F(u_1, \cdots, u_n)$;	Function $G = G(v_1, \cdots, v_n)$;

System Transformations

$\boldsymbol{v} = \dfrac{\partial F}{\partial \boldsymbol{u}}$;	$\boldsymbol{u} = \dfrac{\partial G}{\partial \boldsymbol{v}}$;
$G = \boldsymbol{u}^T \boldsymbol{v} - F$;	$F = \boldsymbol{u}^T \boldsymbol{v} - G$;
$G = G(v_1, \cdots, v_n)$.	$F = F(u_1, \cdots, u_n)$.

In the case that F is a function of two independent sets of variables

$$F(w_1, \cdots, w_m; u_1, \cdots, u_m), \tag{2.124}$$

where the w_i s play no part in the transformation. A new function G is again defined as

$$G = \boldsymbol{u}^T \boldsymbol{v} - F, \tag{2.125}$$

in which the w_i s are passive variables. If we return to (2.121), and compute the complete variation of G by allowing the u_i s and w_i s to vary arbitrarily, we obtain

$$\frac{\partial F}{\partial w} = -\frac{\partial G}{\partial w} \qquad (2.126)$$

and (2.123) using (2.116) as before.

2.4.2 Canonical equations

As shown in (2.37), the Lagrangian function $\mathcal{L}$ is a function of n position coordinates q_i, n velocities $\dot{q}_i$, and possibly the time t. We can now apply Legendre's transformation to $\mathcal{L}$, with the $\dot{q}_i$ the n active variables, and the q_i s and t the $n+1$ passive variables. In this scheme

$$\boldsymbol{p} = \frac{\partial \mathcal{L}}{\partial \dot{\boldsymbol{q}}} \qquad (2.127)$$

are the n *generalized momenta*. The new function is

$$\mathcal{H} = \boldsymbol{p}^T \dot{\boldsymbol{q}} - \mathcal{L}, \qquad (2.128)$$

which we recognize from (2.87) as the total system energy. Solving (2.127) for the $\dot{q}_i$ s, and substituting them into (2.128), gives

$$\mathcal{H} = \mathcal{H}(q_1, \cdots, q_n; p_1, \cdots, p_n, t). \qquad (2.129)$$

The basic features of the original and transformed systems are:

Old System	*New System*
Function: Lagrangian function $\mathcal{L}$;	Function: Hamiltonian function $\mathcal{H}$;
Variables: generalized velocities $\dot{q}_i$;	Variables: generalized momenta p_i;

The passive variables are the generalized positions and time. The two systems are therefore described by

$$\boldsymbol{p} = \frac{\partial \mathcal{L}}{\partial \dot{\boldsymbol{q}}}; \qquad\qquad \dot{\boldsymbol{q}} = \frac{\partial \mathcal{H}}{\partial \boldsymbol{p}};$$

$$\mathcal{H} = \boldsymbol{p}^T \dot{\boldsymbol{q}} - \mathcal{L}; \qquad\qquad \mathcal{L} = \boldsymbol{p}^T \dot{\boldsymbol{q}} - \mathcal{H};$$

$$\mathcal{H} = \mathcal{H}(q_1, \cdots, q_n; p_1, \cdots, p_n; t). \qquad \mathcal{L} = \mathcal{L}(q_1, \cdots, q_n; \dot{q}_1, \cdots, \dot{q}_n; t).$$

The passive variables transform according to (2.126)

$$\frac{\partial \mathcal{L}}{\partial \boldsymbol{q}} = -\frac{\partial \mathcal{H}}{\partial \boldsymbol{q}} \qquad (2.130)$$

$$\frac{\partial \mathcal{L}}{\partial t} = -\frac{\partial \mathcal{H}}{\partial t}. \qquad (2.131)$$

If we substitute (2.40) into (2.130), we obtain

$$\dot{\boldsymbol{p}} = -\frac{\partial \mathcal{H}}{\partial \boldsymbol{q}}. \qquad (2.132)$$

The Lagrangian equations of motion have thus been replaced by a new set of equations called the *canonical equations of Hamilton*:

$$\dot{q} = \frac{\partial \mathcal{H}}{\partial p} \qquad \dot{p} = -\frac{\partial \mathcal{H}}{\partial q}. \tag{2.133}$$

These equations are equivalent to the original Lagrangian equations; they are just in a new form.

In Section 2.2.8 we showed that constrained mechanical systems can be described by Euler–Lagrange equations with extra terms corresponding to the constraint forces. These equations take the form

$$\frac{d}{dt}\left(\frac{\partial \mathcal{L}}{\partial \dot{q}}\right) - \frac{\partial \mathcal{L}}{\partial q} = -A^T(q)\lambda, \tag{2.134}$$

in which

$$A(q)\dot{q} = -F_t \tag{2.135}$$

describes the constraints. In the case of constrained problems, we can substitute (2.134) into (2.130) to obtain

$$\dot{p} = -\frac{\partial \mathcal{H}}{\partial q} - A^T(q)\lambda. \tag{2.136}$$

Further details relating to the incorporation of nonholonomic constraints into mechanical system modelling in a Hamiltonian framework can be found in [59].

Example 2.14 Consider once more the one-dimensional harmonic oscillator of Example 2.10. The Hamiltonian is given by (2.128)

$$\mathcal{H}(p, q) = \frac{1}{2m}p^2 + \frac{k}{2}q^2,$$

where $p = m\dot{q}$. Therefore

$$\dot{q} = \frac{\partial \mathcal{H}}{\partial p} = \frac{p}{m} \qquad \dot{p} = -\frac{\partial \mathcal{H}}{\partial q} = -kq.$$

These two Hamiltonian equations are clearly equivalent to the second-order Lagrange equation obtained previously.

2.4.3   Poisson brackets

Werner Heisenberg formulated a theory of quantum mechanics, known as matrix mechanics, in which an object called the *commutator* $[A, B] = AB - BA$ plays a central role; A and B are matrices. Matrices A and B are said to commute under multiplication if $AB - BA = 0$. It is self-evident that the commutator is *antisymmetric*:

$$[A, B] = -[B, A].$$

Linearity is easily checked:

$$[\alpha A + \beta B, \ C] = (\alpha A + \beta B)C - C(\alpha A + \beta B)$$
$$= \alpha(AC - CA) + \beta(BC - CB)$$
$$= \alpha[A, \ C] + \beta[B, \ C].$$

The commutator also satisfies a *product rule*:

$$[AB, \ C] = ABC - CAB$$
$$= ABC - CAB - ACB + ACB$$
$$= A[B, \ C] + [A, \ C]B.$$

Finally, the commutator satisfies the *Jacobi identity*:

$$0 = A(BC - CB) - (BC - CB)A + B(CA - AC) - (CA - AC)B$$
$$+ C(AB - BA) - (AB - BA)C$$
$$= [A, \ [B, \ C]] + [B, \ [C, \ A]] + [C, \ [A, \ B]].$$

A function theoretic equivalent of the commutator is another bracket, called the *Poisson bracket*, which has an analogous definition. In the context of mechanics, if we suppose that $F(\boldsymbol{q}, \boldsymbol{p})$ and $G(\boldsymbol{q}, \boldsymbol{p})$ are functions of the generalized position and momentum, then the Poisson bracket is defined as

$$\{F, G\} = \sum_{i=1}^{n} \frac{\partial F}{\partial q_i} \frac{\partial G}{\partial p_i} - \frac{\partial F}{\partial p_i} \frac{\partial G}{\partial q_i}. \tag{2.137}$$

The Poisson bracket also satisfies the *antisymmetry, lineariy, product rule*, and *Jacobi identity* properties:

$$\{F, G\} = -\{G, F\};$$
$$\{\alpha F + \beta G, H\} = \alpha\{F, H\} + \beta\{G, H\};$$
$$\{FG, H\} = F\{G, H\} + \{F, H\}G;$$
$$0 = \{F, \{G, H\}\} + \{G, \{H, F\}\} + \{H, \{F, G\}\}. \tag{2.138}$$

Functions that commute with the Hamiltonian are of great importance in mechanics. For any function $F(\boldsymbol{q}, \boldsymbol{p}, t)$, where $\boldsymbol{p}$ and $\boldsymbol{q}$ are governed by $\mathcal{H}$, there holds

$$\frac{dF}{dt} = \frac{\partial F}{\partial \boldsymbol{q}} \cdot \dot{\boldsymbol{q}} + \frac{\partial F}{\partial \boldsymbol{p}} \cdot \dot{\boldsymbol{p}} + \frac{\partial F}{\partial t}$$
$$= \{F, \mathcal{H}\} + \frac{\partial F}{\partial t}. \tag{2.139}$$

Functions of the dynamical variables that remain constant during the motion of the system are the *integrals of the motion*. It follows from (2.139) that F is an integral of the motion if

$$\{F, \mathcal{H}\} + \frac{\partial F}{\partial t} = 0.$$

If the integral of the motion is not explicitly dependent on time, then

$$\{F, \mathcal{H}\} = 0.$$

If $\mathcal{H}$ is not explicitly time dependent,

$$\{\mathcal{H}, \mathcal{H}\} = 0$$

shows that $\frac{d\mathcal{H}}{dt} = 0$ and therefore that the total system energy is conserved.

It is immediate from (2.137) that the canonical equations (2.133) can be written as

$$\{p_j, \mathcal{H}\} = -\frac{\partial \mathcal{H}}{\partial q_j} \quad \text{and} \quad \{q_j, \mathcal{H}\} = \frac{\partial \mathcal{H}}{\partial p_j}. \tag{2.140}$$

The first equation shows that if p_j commutes with $\mathcal{H}$, $\mathcal{H}$ is independent of q_j, which is therefore a conserved quantity. This establishes an important relationship between ignorable coordinates and conserved quantities in the language of Poisson brackets.

An important property of the Poisson bracket is Poisson's theorem, which says that if F and G commute with $\mathcal{H}$, then so does $\{F, G\}$. This is easy to show if F and G do not depend explicitly on time using the Jacobi identity:

$$0 = \{F, \{G, \mathcal{H}\}\} + \{G, \{\mathcal{H}, F\}\} + \{\mathcal{H}, \{F, G\}\},$$

then $\{F, \mathcal{H}\} = 0$ and $\{G, \mathcal{H}\} = 0$ implies that $\{\mathcal{H}, \{F, G\}\} = 0$. As shown in [43] the extension to the case that F and G are explicitly time dependent is only slightly more complicated to prove.

Poisson's theorem, in principle, allows one to find a third integral of the motion given two others. Not all the integrals of the motion computed in this way will be 'new'; some may be old integrals, or constants that may be zero. If a body is in free motion in space, then p_x, p_y, and p_z are all constant. Yet knowing that $\dot{p}_x = 0$ and $\dot{p}_y = 0$ does not allow one to deduce that $\dot{p}_z = 0$, since $\{p_x, p_y\} = 0$ provides no new information.

Example 2.15 The *Kepler problem* is one of the fundamental problems in classical mechanics and is as ubiquitous as the harmonic oscillator in the mechanics literature.[9] The Kepler problem is also a simple and informative illustration of the use of Poisson brackets. Goldstein [41] dedicates a whole chapter to two-body central-force problems and so a far more detailed treatment of this problem can be found there. In essence, the Kepler problem studies the dynamics of a unit point mass in an inverse-square law central force field:

$$\ddot{r} + \frac{\kappa r}{r^3} = 0, \tag{2.141}$$

where r is a position (radius) vector, κ is a constant, and $r = |r|$. The central force motion of two bodies about their mass centre can always be reduced to an equivalent one-body problem [41].

---

[9]   The Kepler laws of planetary motion have been of interest ever since the appearance of Newton's *Principia*. In 1911 Rutherford proposed a planetary model for the hydrogen atom, whereby an electron rotates on a planetary orbit about a charged atomic core. This model is unstable (and therefore wrong), because the rotating electron would radiate away its energy until the atom collapsed.

The first thing to note is that the motion is planar. If we cross (2.141) with $\boldsymbol{r}$ we obtain

$$\boldsymbol{r} \times \ddot{\boldsymbol{r}} = 0$$

and so the angular momentum vector

$$\boldsymbol{H} = \boldsymbol{r} \times \dot{\boldsymbol{r}}, \qquad (2.142)$$

is constant. Since $\boldsymbol{H}$ is perpendicular to both $\boldsymbol{r}$ and $\dot{\boldsymbol{r}}$, the constancy of $\boldsymbol{H}$ implies that the orbit is confined to a plane.

We can therefore describe the problem in terms of cylindrical (rather than spherical) coordinates with Lagrangian

$$\mathcal{L} = \frac{1}{2}\left(\dot{r}^2 + (r\dot{\theta})^2\right) + \frac{\kappa}{r},$$

and Hamiltonian

$$\mathcal{H} = \frac{1}{2}\left(p_r^2 + \frac{p_\theta^2}{r^2}\right) - \frac{\kappa}{r},$$

where

$$p_r = \dot{r} \quad \text{and} \quad p_\theta = r^2\dot{\theta}.$$

Unit vectors in the $\boldsymbol{r}$, $\boldsymbol{\theta}$, and $\boldsymbol{z}$ directions are denoted $\hat{\boldsymbol{r}}$, $\hat{\boldsymbol{\theta}}$, and $\hat{\boldsymbol{z}}$ respectively. Since these vectors are orthogonal, the cross product of any two provides the third; $\hat{\boldsymbol{z}}$ is chosen perpendicular to the motion.

Using (2.140) one can see that

$$\dot{p}_\theta = \{p_\theta, \mathcal{H}\} = 0,$$

which shows that p_θ is a conserved quantity. Since $\mathcal{H}$ is not explicitly time-varying, the system energy is also conserved.

The motion-specifying equation (2.141) can be written as

$$0 = \ddot{\boldsymbol{r}} + \frac{\kappa\boldsymbol{r}}{r^3} = \ddot{\boldsymbol{r}} + \frac{\kappa}{H}\dot{\theta}\hat{\boldsymbol{r}} = \frac{d}{dt}\left(\dot{\boldsymbol{r}} - \frac{\kappa}{H}\hat{\boldsymbol{\theta}}\right),$$

since $\dot{\hat{\boldsymbol{\theta}}} = -\dot{\theta}\hat{\boldsymbol{r}}$. The *Hamiltonian vector*

$$\boldsymbol{L} = \dot{\boldsymbol{r}} - \frac{\kappa}{H}\hat{\boldsymbol{\theta}}$$

is thus also a conserved quantity. Since the cross product of two conserved quantities is also conserved,

$$\boldsymbol{K} = \boldsymbol{L} \times \boldsymbol{H} = \dot{\boldsymbol{r}} \times \boldsymbol{H} - \kappa\hat{\boldsymbol{r}}$$

which is the *Laplace–Runge–Lenz* vector, is also conserved [41]. The vectors $\boldsymbol{H}$, $\boldsymbol{L}$, and $\boldsymbol{K}$ are mutually orthogonal, with $\boldsymbol{H}$ normal to the orbital plane.

Carrying out the requisite differentiations gives [41, 60]:

$$\{\boldsymbol{L}, \mathcal{H}\} = 0$$
$$\{\boldsymbol{K}, \mathcal{H}\} = 0$$
$$\{H_i, H_j\} = \epsilon_{ijk} H_k$$
$$\{H_i, K_j\} = \epsilon_{ijk} K_k$$
$$\{K_i, K_j\} = -2\mathcal{H}\epsilon_{ijk} H_k,$$

where ϵ_{ijk} is the Levi–Civita symbol. These identities highlight the conserved quantities in the Kepler problem in terms of Poisson brackets and Poisson's theorem.

2.5 Frames, velocity, and acceleration

In vehicle dynamics it is common to use at least two orthonormal coordinate systems to describe the motion: a reference (inertial) coordinate system and a moving coordinate system attached to the vehicle. The most general displacement of a rigid body, and thus also of an orthonormal reference frame, is a translation plus a rotation; this is *Chasles' theorem.*[10] The orientation of one frame relative to another can be described by a maximum of three successive rotations about the coordinate axes (*Euler's first theorem*), or by a single rotation about a specific axis (*Euler's second theorem*). Therefore the relative motion between two reference frames in a three-dimensional space is usually described by three translations and three rotations. We begin this section by deriving the vectorial relationships relating position, velocity, and acceleration in an inertial frame and in a moving frame. We then introduce the matrix notation used to deal with the modelling of a generic three-dimensional motion and the related position, velocity, and acceleration relationships.

A moving reference frame is characterized by the position of its origin $\boldsymbol{r}_{OC}$ and its angular velocity $\boldsymbol{\omega}$ with respect to the inertial frame.

The position $\boldsymbol{r}_{OP}$ of a point P in the inertial frame can be described by

$$\boldsymbol{r}_{OP} = \boldsymbol{r}_{OC} + \boldsymbol{r}_{CP}, \tag{2.143}$$

where $\boldsymbol{r}_{OC}$ is the position of the origin of the moving frame and $\boldsymbol{r}_{CP}$ is the position of P in the moving frame.

The velocity $\boldsymbol{v}_P$ of the point P can be obtained by differentiating (2.143) with respect to time

$$\boldsymbol{v}_P = \frac{d\boldsymbol{r}_{OP}}{dt} = \dot{\boldsymbol{r}}_{OC} + \boldsymbol{\omega} \times \boldsymbol{r}_{CP} + \dot{\boldsymbol{r}}_{CP}, \tag{2.144}$$

where $\dot{\boldsymbol{r}}_{OC}$ is the velocity of the origin of the moving frame, $\boldsymbol{\omega} \times \boldsymbol{r}_{CP}$ is the transferred velocity, and $\dot{\boldsymbol{r}}_{CP}$ is the velocity of P relative to the moving frame; the 'dot' denotes

---

[10] There is a stronger form of the theorem stating that the most general displacement is a translation plus a rotation along a given axis, the so-called *screw axis*.

component-wise differentiation with respect to time. In the case of a moving frame of reference, the complete time derivation has a 'dot term' and a $\boldsymbol{\omega}\times$ angular velocity term.[11]

The acceleration $\boldsymbol{a}_{OP}$ of the point P is obtained by differentiating (2.144) with respect to time

$$\boldsymbol{a}_P = \frac{d^2\boldsymbol{r}_{OP}}{dt^2} = \ddot{\boldsymbol{r}}_{OC} + \dot{\boldsymbol{\omega}}\times\boldsymbol{r}_{CP} + \boldsymbol{\omega}\times(\boldsymbol{\omega}\times\boldsymbol{r}_{CP}) + 2\boldsymbol{\omega}\times\dot{\boldsymbol{r}}_{CP} + \ddot{\boldsymbol{r}}_{CP}, \qquad (2.145)$$

where $\ddot{\boldsymbol{r}}_{OC}$ is the acceleration of the origin of the moving frame, $\dot{\boldsymbol{\omega}}\times\boldsymbol{r}_{CP}$ is the Euler acceleration term, $\boldsymbol{\omega}\times(\boldsymbol{\omega}\times\boldsymbol{r}_{CP})$ is the centrifugal term, $2\boldsymbol{\omega}\times\dot{\boldsymbol{r}}_{CP}$ is the Coriolis term, and $\ddot{\boldsymbol{r}}_{CP}$ is the acceleration of P relative to the moving frame.

It goes without saying that the vector quantities in (2.143), (2.144), and (2.145) must be expressed in the same reference frame prior to any vector addition calculations. The vector $\boldsymbol{a}_P$ represents the absolute acceleration of P, which may be expressed in the coordinates of the moving reference frame if so desired.

In the case that P is stationary in the moving frame, the velocity and acceleration expressions simplify to

$$\boldsymbol{v}_P = \dot{\boldsymbol{r}}_{OC} + \boldsymbol{\omega}\times\boldsymbol{r}_{CP} \qquad (2.146)$$

$$\boldsymbol{a}_P = \ddot{\boldsymbol{r}}_{OC} + \dot{\boldsymbol{\omega}}\times\boldsymbol{r}_{CP} + \boldsymbol{\omega}\times(\boldsymbol{\omega}\times\boldsymbol{r}_{CP}). \qquad (2.147)$$

It is often convenient to use transformation matrices to change the coordinates of a point P expressed in frame $\mathcal{C}$ (for example the moving frame attached to the vehicle) $\boldsymbol{r}_{CP}^{\mathcal{C}}$ to a different frame $\mathcal{O}$ (for example the inertial ground frame) $\boldsymbol{r}_{OP}^{\mathcal{O}}$. The superscript is used to indicate the coordinate system in which the coordinates are expressed. Equation (2.143) can be written in the following matrix form

$$\begin{bmatrix} \boldsymbol{r}_{OP}^{\mathcal{O}} \\ 1 \end{bmatrix} = \mathcal{T}_{OC}\begin{bmatrix} \boldsymbol{r}_{CP}^{\mathcal{C}} \\ 1 \end{bmatrix} \qquad (2.148)$$

where

$$\mathcal{T}_{OC} = \begin{bmatrix} \mathcal{R}_{OC} & \boldsymbol{r}_{OC}^{\mathcal{O}} \\ 0 & 1 \end{bmatrix} \qquad (2.149)$$

is the 4×4 transformation matrix, which consists of a 3×3 rotation matrix $\mathcal{R}_{OC}$ and a 3×1 translation vector $\boldsymbol{r}_{OC}^{\mathcal{O}}$. The columns of $\mathcal{R}_{OC}$ are the components of the unit vectors of the frame $\mathcal{C}$ axes expressed in the coordinates of the frame $\mathcal{O}$, while the $\boldsymbol{r}_{OC}^{\mathcal{O}}$ is the position of the origin of the frame $\mathcal{C}$ expressed in the coordinates of frame $\mathcal{O}$. General rotation matrices satisfy the conditions $\mathcal{R}\mathcal{R}^T = I$ with the added constraint

---

[11] The relationship between the time derivative in an inertial frame $\mathcal{O}$ and the time derivative in a moving frame $\mathcal{C}$ moving with angular velocity $\boldsymbol{\omega}$ is (Chapter IV in [42])

$$\left.\frac{d\cdot}{dt}\right|_{\mathcal{O}} = \left.\frac{d\cdot}{dt}\right|_{\mathcal{C}} + \boldsymbol{\omega}\times\cdot.$$

$\det(\mathcal{R}) = 1$; rotation matrices form a subset of the orthogonal matrices [41]. In some references one sees three-dimensional rotations referred to as members of SO(3), which is the (three-dimensional) special orthogonal group [53].

The inverse of the transformation matrix $\mathcal{T}_{OC}$ is

$$\mathcal{T}_{OC}^{-1} = \begin{bmatrix} \mathcal{R}_{OC}^T & -\mathcal{R}_{OC}^T r_{OC}^O \\ 0 & 1 \end{bmatrix}. \tag{2.150}$$

The coordinates of P expressed in frame $\mathcal{C}$ can be obtained from those given in frame $\mathcal{O}$ by inversion of (2.148)

$$\begin{bmatrix} r_{CP}^C \\ 1 \end{bmatrix} = \mathcal{T}_{OC}^{-1} \begin{bmatrix} r_{OP}^O \\ 1 \end{bmatrix}. \tag{2.151}$$

The transformations defined above for points can be applied to vectors by replacing the '1' on the right-hand side of (2.148) with a '0', which suppresses the translational component of the transformation:

$$\begin{bmatrix} r_{CP}^O \\ 0 \end{bmatrix} = \mathcal{T}_{OC} \begin{bmatrix} r_{CP}^C \\ 0 \end{bmatrix}. \tag{2.152}$$

The vectors on both sides of (2.152) have subscript 'CP', while in the case of points r_{OP} and r_{CP} are used; see (2.148).

While the translation component of the transformation is straightforward, the rotational element component requires further discussion. Every rotation matrix can be expressed in terms of an axis-of-rotation unit-vector n and a rotation angle φ [41,53]. For any rotation matrix $\mathcal{R}$ there exists a unit vector n and an angle φ such that $\mathcal{R} = \mathcal{R}(n, \varphi)$, where

$$\mathcal{R}(n, \varphi) = I + (1 - \cos(\varphi))S^2(n) + \sin(\varphi)S(n) \tag{2.153}$$

is *Rodrigues' rotation formula*, and $S(n)$ is a skew-symmetric matrix

$$S(n) = \begin{bmatrix} 0 & -n_3 & n_2 \\ n_3 & 0 & -n_1 \\ -n_2 & n_1 & 0 \end{bmatrix}, \tag{2.154}$$

in which n_1, n_2, and n_3 are the x-, y-, and z-axis components of n.

The matrix $\mathcal{R}_{123}$ related to the orientation of a body in space can be defined by three successive rotations about given axes performed in a specific sequence.[12] The combined rotation is given by

$$\mathcal{R}_{123} = \mathcal{R}_1 \mathcal{R}_2 \mathcal{R}_3. \tag{2.155}$$

In principle, there are $3^3 = 27$ possible combinations of the three basic rotations, but only 12 of them can be used to represent arbitrary three-dimensional rotations.

---

[12] The order does not matter for 'small' rotations.

These 12 combinations avoid degenerate consecutive rotations around the same axis (such as x–x–y) that would reduce the degrees of freedom that can be represented. The most widespread sequences are z–x–z (usually referred to as Euler 'x-convention'), z–y–z (Euler 'y-convention', popular in quantum and nuclear mechanics), x–y–z (usually referred to as Tait–Bryan), z–y–x (heading–attitude–bank, popular in aerospace) and z–x–y (yaw–roll–pitch, popular in vehicle dynamics).

The rotated frame x–y–z may be imagined to be initially aligned with X–Y–Z, before undergoing the three elemental rotations. Its successive orientations may be described as:

1. X–Y–Z (initial configuration),
2. X'–Y'–Z' (first rotation),
3. X''–Y''–Z'' (second rotation),
4. x–y–z (final configuration).

Suppose that the sequence z–x–y is chosen

$$\mathcal{R}_1 = \mathcal{R}(\mathbf{e}_z, \psi) = \begin{bmatrix} c_\psi & -s_\psi & 0 \\ s_\psi & c_\psi & 0 \\ 0 & 0 & 1 \end{bmatrix} \tag{2.156}$$

$$\mathcal{R}_2 = \mathcal{R}(\mathbf{e}_x, \phi) = \begin{bmatrix} 1 & 0 & 0 \\ 0 & c_\phi & -s_\phi \\ 0 & s_\phi & c_\phi \end{bmatrix} \tag{2.157}$$

$$\mathcal{R}_3 = \mathcal{R}(\mathbf{e}_y, \theta) = \begin{bmatrix} c_\theta & 0 & s_\theta \\ 0 & 1 & 0 \\ -s_\theta & 0 & c_\theta \end{bmatrix}, \tag{2.158}$$

in which $\mathcal{R}(\cdot, \cdot)$ is defined in (2.153), with $\mathbf{e}_x$, $\mathbf{e}_y$ and $\mathbf{e}_z$ unit vectors in the x-, y-, and z-axis directions, respectively, in the appropriate reference frame. We have used the shorthand $s_{(\cdot)}$ and $c_{(\cdot)}$ to denote the sine and cosine of angle $(\cdot)$. The resulting yaw–roll–pitch rotation matrix is

$$\mathcal{R} = \mathcal{R}_1\mathcal{R}_2\mathcal{R}_3 = \begin{bmatrix} -s_\psi s_\phi s_\theta + c_\psi c_\theta & -s_\psi c_\phi & s_\psi s_\phi c_\theta + c_\psi s_\theta \\ c_\psi s_\phi s_\theta + s_\psi c_\theta & c_\psi c_\phi & -c_\psi s_\phi c_\theta + s_\psi s_\theta \\ -c_\phi s_\theta & s_\phi & c_\phi c_\theta \end{bmatrix} \tag{2.159}$$

The three-rotations matrix in (2.159) can be alternatively expressed in terms of the single-rotation matrix $\mathcal{R}(\mathbf{n}, \varphi)$ in (2.153). The axis of rotation $\mathbf{n}$ is the eigenvector corresponding to the unity eigenvalue of (2.159)

$$\mathcal{R}\mathbf{n} = \mathbf{n}, \tag{2.160}$$

since the rotation axis remains the same in the initial and final reference frames. It follows from (2.153) that the angle φ can be obtained from

$$\cos\varphi = \frac{Tr(\mathcal{R}) - 1}{2} \qquad \sin\varphi = \frac{\mathcal{R}(kj) - \mathcal{R}(jk)}{2n_i} \qquad (i, j, k = 1, 2, 3 \text{ cyclic}), \tag{2.161}$$

where $Tr(\mathcal{R})$ is the trace of $\mathcal{R}$, n_i is the i-th entry of $\mathbf{n}$, and $\mathcal{R}(kj)$ is the kj-th entry of $\mathcal{R}$.

Our next goal is to link rotation matrices to the vector cross product defined by

$$\mathbf{a} \times \mathbf{b} = \begin{bmatrix} a_2 b_3 - a_3 b_2 \\ a_3 b_1 - a_1 b_3 \\ a_1 b_2 - a_2 b_1 \end{bmatrix}$$

in which $\mathbf{a}$ and $\mathbf{b}$ are vectors. Since the cross product (by $\mathbf{a}$) is a linear operation, it has a matrix representation $S(\mathbf{a})$ and so

$$\mathbf{a} \times \mathbf{b} = S(\mathbf{a})\mathbf{b}, \tag{2.162}$$

where $S(\mathbf{a})$ is the skew symmetric matrix defined in (2.154). It is important to note that rotations commute with the cross product. That is $\mathcal{R}(\mathbf{a} \times \mathbf{b}) = (\mathcal{R}\mathbf{a}) \times (\mathcal{R}\mathbf{b})$.[13] It follows that $\mathcal{R}(\mathbf{a} \times \mathbf{b}) = \mathcal{R}S(\mathbf{a})(\mathcal{R}^T \mathcal{R})\mathbf{b}$ and so

$$\mathcal{R}S(\mathbf{a})\mathcal{R}^T = S(\mathcal{R}\mathbf{a}). \tag{2.163}$$

We will use the properties of cross products to derive a number of standard formulae relating to motions in rotating reference frames. Since $I = \mathcal{R}\mathcal{R}^T$, $d(\mathcal{R}\mathcal{R}^T)/dt = \dot{\mathcal{R}}\mathcal{R}^T + (\dot{\mathcal{R}}\mathcal{R}^T)^T = 0$, and we see that $\dot{\mathcal{R}}\mathcal{R}^T = -(\dot{\mathcal{R}}\mathcal{R}^T)^T$, which shows that $\dot{\mathcal{R}}\mathcal{R}^T$ is skew-symmetric. Parallel arguments using $d(\mathcal{R}^T\mathcal{R})/dt = 0$ show that $\mathcal{R}^T\dot{\mathcal{R}}$ is skew-symmetric too.

We can now derive expressions for the velocities and accelerations using matrix notation. The absolute velocity of a point P, expressed in the frame $\mathcal{O}$, is obtained by differentiating (2.148) with respect to time

$$v_P^{\mathcal{O}} = \dot{\mathcal{T}}_{OC} \begin{bmatrix} r_{CP}^{\mathcal{C}} \\ 1 \end{bmatrix} + \mathcal{T}_{OC} \begin{bmatrix} \dot{r}_{CP}^{\mathcal{C}} \\ 0 \end{bmatrix} \tag{2.164}$$

$$= \dot{r}_{OC}^{\mathcal{O}} + \dot{\mathcal{R}}_{OC} r_{CP}^{\mathcal{C}} + \mathcal{R}_{OC} \dot{r}_{CP}^{\mathcal{C}}; \tag{2.165}$$

the first term is the translational velocity of the origin of the moving frame, the second is the transferred velocity, and the third is the relative velocity. The second term in (2.165) can be rewritten as $(\dot{\mathcal{R}}_{OC}\mathcal{R}_{OC}^T)r_{CP}^{\mathcal{O}}$ and the third as $\dot{r}_{CP}^{\mathcal{O}}$. Comparison with (2.144) shows that the skew-symmetric matrix for the angular velocity of the frame $\mathcal{C}$ expressed in the frame $\mathcal{O}$ is

$$S(\omega_{OC}^{\mathcal{O}}) = \dot{\mathcal{R}}_{OC}\mathcal{R}_{OC}^T. \tag{2.166}$$

It is often convenient to express the velocity of the moving frame in its own coordinate system. Premultiplied by the inverse transformation matrix, (2.164) and (2.165) become

$$v_P^{\mathcal{C}} = (\mathcal{T}_{OC}^{-1}\dot{\mathcal{T}}_{OC}) \begin{bmatrix} r_{CP}^{\mathcal{C}} \\ 1 \end{bmatrix} + \begin{bmatrix} \dot{r}_{CP}^{\mathcal{C}} \\ 0 \end{bmatrix} \tag{2.167}$$

$$= \mathcal{R}_{OC}^T \dot{r}_{OC}^{\mathcal{O}} + (\mathcal{R}_{OC}^T \dot{\mathcal{R}}_{OC}) r_{CP}^{\mathcal{C}} + \dot{r}_{CP}^{\mathcal{C}}. \tag{2.168}$$

---

[13] This can be established as follows: if $\mathbf{a} \times \mathbf{b} = \mathbf{c}$, then $\mathcal{R}(\mathbf{a} \times \mathbf{b}) = \mathcal{R}\mathbf{c}$. If $\mathbf{a}' = \mathcal{R}\mathbf{a}$, $\mathbf{b}' = \mathcal{R}\mathbf{b}$, and $\mathbf{c}' = \mathcal{R}\mathbf{c}$, then $\mathbf{a}' \times \mathbf{b}' = \mathbf{c}'$, since the relative orientations and lengths of these vectors have not changed. Consequently $(\mathcal{R}\mathbf{a}) \times (\mathcal{R}\mathbf{b}) = \mathcal{R}\mathbf{c} = \mathcal{R}(\mathbf{a} \times \mathbf{b})$ as required.

Comparison with (2.144) shows that the angular velocity of frame $\mathcal{C}$ expressed in its own coordinate system is given by

$$S(\omega_{OC}^C) = \mathcal{R}_{OC}^T \dot{\mathcal{R}}_{OC}. \tag{2.169}$$

The acceleration vector can be computed by differentiating the velocity in (2.164) with respect to time:

$$
\begin{aligned}
a_P^O &= \frac{d}{dt}\left(\dot{\mathcal{T}}_{OC}\mathcal{T}_{OC}^{-1}\begin{bmatrix} r_{OP}^O \\ 1 \end{bmatrix} + \mathcal{T}_{OC}\begin{bmatrix} \dot{r}_{CP}^C \\ 0 \end{bmatrix} \right) \\
&= \frac{d}{dt}\left(\dot{\mathcal{T}}_{OC}\mathcal{T}_{OC}^{-1} \right)\begin{bmatrix} r_{OP}^O \\ 1 \end{bmatrix} + \dot{\mathcal{T}}_{OC}\mathcal{T}_{OC}^{-1}\left(\dot{\mathcal{T}}_{OC}\begin{bmatrix} r_{CP}^C \\ 1 \end{bmatrix} \right. \\
&\quad \left. + \mathcal{T}_{OC}\begin{bmatrix} \dot{r}_{CP}^C \\ 0 \end{bmatrix} \right) + \dot{\mathcal{T}}_{OC}\begin{bmatrix} \dot{r}_{CP}^C \\ 0 \end{bmatrix} + \mathcal{T}_{OC}\begin{bmatrix} \ddot{r}_{CP}^C \\ 0 \end{bmatrix} \\
&= \frac{d}{dt}\left(\dot{\mathcal{T}}_{OC}\mathcal{T}_{OC}^{-1} \right)\begin{bmatrix} r_{OP}^O \\ 1 \end{bmatrix} + (\dot{\mathcal{T}}_{OC}\mathcal{T}_{OC}^{-1})(\dot{\mathcal{T}}_{OC}\mathcal{T}_{OC}^{-1})\begin{bmatrix} r_{OP}^O \\ 1 \end{bmatrix} \\
&\quad + 2(\dot{\mathcal{T}}_{OC}\mathcal{T}_{OC}^{-1})\begin{bmatrix} \dot{r}_{CP}^C \\ 0 \end{bmatrix} + \mathcal{T}_{OC}\begin{bmatrix} \ddot{r}_{CP}^C \\ 0 \end{bmatrix} \tag{2.170} \\
&= \ddot{r}_{OC}^O + \frac{d}{dt}(\dot{\mathcal{R}}_{OC}\mathcal{R}_{OC}^T)r_{CP}^O + (\dot{\mathcal{R}}_{OC}\mathcal{R}_{OC}^T)(\dot{\mathcal{R}}_{OC}\mathcal{R}_{OC}^T)r_{CP}^O \\
&\quad + 2(\dot{\mathcal{R}}_{OC}\mathcal{R}_{OC}^T)\dot{r}_{CP}^O + \mathcal{R}_{OC}\ddot{r}_{CP}^C. \tag{2.171}
\end{aligned}
$$

Again, comparison with (2.145) allows one to recognize the acceleration of the origin of the frame, the Euler acceleration, the centrifugal acceleration, the Coriolis acceleration, and the relative acceleration. The absolute acceleration in (2.170) and (2.171) can be expressed in the moving frame by premultiplication by $\mathcal{T}^{-1}$ and $\mathcal{R}^T$ respectively.

We conclude the section by noting that the summation rule holds for the angular velocity. Assuming a combination of the three rotation $\mathcal{R}_1$, $\mathcal{R}_2$, and $\mathcal{R}_3$, there holds $\omega = \omega_1 + \omega_2 + \omega_3$. If we use the three rotations in (2.156)–(2.158) we find that the angular velocities of the frames in their own coordinate systems are

$$S(\omega_1^1) = \mathcal{R}_1^T \dot{\mathcal{R}}_1 = \begin{bmatrix} 0 & -\dot{\psi} & 0 \\ \dot{\psi} & 0 & 0 \\ 0 & 0 & 0 \end{bmatrix} \rightarrow \omega_1^1 = \begin{bmatrix} 0 \\ 0 \\ \dot{\psi} \end{bmatrix} \tag{2.172}$$

$$S(\omega_2^2) = \mathcal{R}_2^T \dot{\mathcal{R}}_2 = \begin{bmatrix} 0 & 0 & 0 \\ 0 & 0 & -\dot{\phi} \\ 0 & \dot{\phi} & 0 \end{bmatrix} \rightarrow \omega_2^2 = \begin{bmatrix} \dot{\phi} \\ 0 \\ 0 \end{bmatrix} \tag{2.173}$$

$$S(\omega_3^3) = \mathcal{R}_3^T \dot{\mathcal{R}}_3 = \begin{bmatrix} 0 & 0 & \dot{\theta} \\ 0 & 0 & 0 \\ -\dot{\theta} & 0 & 0 \end{bmatrix} \rightarrow \omega_3^3 = \begin{bmatrix} 0 \\ \dot{\theta} \\ 0 \end{bmatrix} \tag{2.174}$$

and the resulting angular velocity is

$$\omega_{123}^3 = \omega_1^3 + \omega_2^3 + \omega_3^3 = \begin{bmatrix} -s_\theta \dot{\psi} c_\phi + c_\theta \dot{\theta} \\ s_\phi \dot{\psi} + \dot{\theta} \\ c_\theta \dot{\psi} c_\phi + s_\theta \dot{\theta} \end{bmatrix} \tag{2.175}$$

which is expressed in the coordinates of the final frame.

2.6 Equilibria, stability, and linearization

We will make extensive use of linear models, because linear equations are easy to solve and their stability properties are readily assessed using eigenvalues. Also, in many nonlinear problems in mechanics, linearized models provide a satisfactory approximate solution. In cases when linear models are not enough, a study of the local dynamics is often a useful initial step. When studying linear behaviour one examines small perturbations around an equilibrium (trim) condition. If

$$\dot{x}(t) = f(x(t)),$$

then x_0 is a trim state if $f(x_0) = 0$. In the case that $T = \frac{1}{2} p^T M(q)^{-1} p$, $V = V(q)$, and $\mathcal{H} = T + V$, and there are no cyclic coordinates, we have:

Theorem 2.1 *The point $q = q_0$, $p = p_0$ is an equilibrium position if and only if $p_0 = 0$ and q_0 is a stationary point of the potential energy:*

$$\left. \frac{\partial V}{\partial q} \right|_{q_0} = 0. \tag{2.176}$$

Proof Under the assumptions relating to the energy functions, Hamilton's equations take the form:

$$\begin{bmatrix} \dot{p} \\ \dot{q} \end{bmatrix} = \begin{bmatrix} -\frac{\partial \mathcal{H}}{\partial q} \\ \frac{\partial \mathcal{H}}{\partial p} \end{bmatrix} = \begin{bmatrix} -\frac{\partial T}{\partial q} - \frac{\partial V}{\partial q} \\ M^{-1}(q) p \end{bmatrix}. \tag{2.177}$$

Suppose the system is in equilibrium. The second row of (2.177) shows that $\dot{q} = 0$ implies that $p_0 = 0$, and the first row establishes that $\dot{p} = 0$ implies (2.176), since $\frac{\partial T}{\partial q} = -\frac{1}{2} p^T M^{-1} \frac{\partial M}{\partial q} M^{-1} p$ vanishes when $p_0 = 0$. If (2.176) is satisfied and $p_0 = 0$, the first row of (2.177) establishes that $\dot{p} = 0$, while the second row shows that $\dot{q} = 0$ and so the system is in equilibrium. □

The next result shows that trim conditions corresponding to minima in the potential energy function are stable in the sense of Lyapunov [45].

Theorem 2.2 *If the point q_0 is a local minimum in the potential energy function, the equilibrium $q = q_0$ is stable in the sense of Lyapunov.*

Proof Suppose $V(q_0) = h$. For sufficiently small $\epsilon > 0$, the solution set $q : V(q) \le h + \epsilon$, containing q_0, will be in a small neighbourhood of q_0. Furthermore, the corresponding region of the state space $\{p, q : E(q, p) \le h + \epsilon\}$ will be in a small neighbourhood of $p = 0$ and q_0; see Figure 2.4. Since the energy remains constant throughout the flow, $E(q_0, p_0) \leq h + \epsilon$, implies that $E(q, p) \leq h + \epsilon$. Thus for any initial condition (q_0, p_0) close to $(q_0, 0)$, the corresponding flow $(q(t), p(t))$ also remains close to $(q_0, 0)$. □

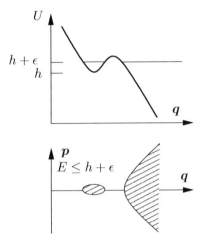

Figure 2.4: Stable equilibrium associated with $V(q_0) = h$; the shaded areas correspond to $E(p, q) \leq h + \epsilon$.

The final result of this section shows that one can derive the linear equations of motion by considering only the quadratic parts of the energy functions. We can assume without loss of generality that the coordinate system has been chosen so that the equilibrium position is $q_0 = 0$.

Theorem 2.3 *In order to linearize the Lagrangian system*

$$\frac{d}{dt} \frac{\partial \mathcal{L}}{\partial \dot{q}} = \frac{\partial \mathcal{L}}{\partial q}$$

in the neighbourhood of $q = 0$, one may replace the kinetic energy by $T_2 = \frac{1}{2} \dot{q}^T M(0) \dot{q}$ and the potential energy by its quadratic part

$$V_2 = \frac{1}{2} q^T \left(\frac{\partial^2 V}{\partial q_i \partial q_j} \Big|_{q=0} \right) q.$$

Proof Since the inertial force terms are determined by partial derivatives with respect to the $\dot{q}_i$ s, the linear inertial forces are determined by the quadratic terms in T, which are precisely those in T_2. The imposed forces derive from partial derivatives with respect to the q_i s, and the linear imposed force terms come from the quadratic terms in V, which are those in V_2. □

2.7 Time-reversal symmetry and dissipation

Time reversibility is a topic that lies on the border of physics and philosophy and has caused more than its fair share of controversy. In one oft-quoted reference [61], the idea that 'all known laws of physics are invariant under time reversal' is studied; alternative discussions on time reversibility can be found in the excellent texts [62,63]. In the case of classical mechanics it can be argued that Newton's first and third laws make no reference to the direction of time and would therefore have identical forms in a time-reversed universe. In the case of Newton's second law one might consider a particle of mass m at position $\boldsymbol{r}$ in a force field of the form $\boldsymbol{F}(\boldsymbol{r})$ and so

$$m\ddot{\boldsymbol{r}} = \boldsymbol{F}(\boldsymbol{r}). \tag{2.178}$$

This equation can be solved subject to boundary conditions on $\boldsymbol{r}(t_0)$ and $\dot{\boldsymbol{r}}(t_0)$. If $m\ddot{\boldsymbol{\gamma}}(t) = \boldsymbol{F}(\boldsymbol{\gamma}(t))$, then $\boldsymbol{\gamma}(t)$ is a solution to (2.178). If t is replaced by $\tau = -t$ one obtains

$$m\ddot{\boldsymbol{\gamma}}(\tau) = \boldsymbol{F}(\boldsymbol{\gamma}(\tau)) \tag{2.179}$$

and so $\boldsymbol{\gamma}(\tau)$ is a possible trajectory if $\boldsymbol{\gamma}(t)$ is; recall $\frac{d^2}{dt^2} = \frac{d^2}{d\tau^2}$. Under the given assumptions on the force, Newton's second law is *time reversal invariant*. The time reversal leaves $\boldsymbol{r}$ unchanged, but reverses the sign of the velocities $d\boldsymbol{r}/dt = -(d\boldsymbol{r}/d\tau)$. The same arguments show that the system has time-reversal symmetry even with a more general force of the form $\boldsymbol{F}(\boldsymbol{r}, \dot{\boldsymbol{r}}, t)$ that satisfies

$$\boldsymbol{F}(\boldsymbol{r}, \dot{\boldsymbol{r}}, t) = \boldsymbol{F}(\boldsymbol{r}, -\dot{\boldsymbol{r}}, -t) \tag{2.180}$$

under the mapping $t \to -t$, $\boldsymbol{r} \to \boldsymbol{r}$, $\dot{\boldsymbol{r}} \to -\dot{\boldsymbol{r}}$. These arguments can be developed in a similar way in a Lagrangian framework [64].

Example 2.16 Consider an ideal lossless simple pendulum described by

$$\ddot{\theta} l + g\theta = 0,$$

which has solution

$$\theta(t) = \frac{\dot{\theta}_0}{\omega} \sin(\omega t) + \theta_0 \cos(\omega t)$$

$$\dot{\theta}(t) = \dot{\theta}_0 \cos(\omega t) - \theta_0 \omega \sin(\omega t),$$

where $\omega = \sqrt{g/l}$, and θ_0 and $\dot{\theta}_0$ are initial conditions on the pendulum's position and velocity respectively. If we now reverse the sign of t and $\dot{\theta}_0$, there holds

$$\theta(-t) = -\frac{\dot{\theta}_0}{\omega} \sin(-\omega t) + \theta_0 \cos(-\omega t) = \theta(t)$$

$$\dot{\theta}(-t) = -\dot{\theta}_0 \cos(-\omega t) - \theta_0 \omega \sin(-\omega t) = -\dot{\theta}(t).$$

This means that if we were to film a few cycles of the pendulum's motion it would be impossible to tell if one was watching the film being played forwards or backwards. The backward time motion satisfies the same laws of physics as the forward motion and in this case (2.180) holds. If we now consider a more realistic pendulum with losses, it will be obvious if the observed motion is running forwards or backwards, because the forward motion will show decay, while the reversed observation will show growth. Linear damping forces take the form $-c\dot{\theta}$, for c constant, which do not satisfy (2.180).

The previous example suggests an association between conservative systems and time-reversibility. The next two examples show that conservative systems are not necessarily time reversible, nor are time-reversible systems necessarily conservative! The first is adapted from [65] and shows that conservative systems need not be time reversible. The second example shows that time-reversible systems need not be conservative [66].

Example 2.17 Consider a unit mass with one translational degree of freedom in a well-like potential field described by $V(x) = (x^4 - x^6)/2$. The kinetic energy of the unit mass is $T(\dot{x}) = \dot{x}^2/2$. The initial position of the mass is at $x = +1$ m; see Figure 2.5. A direct application of Lagrange's equation shows that the mass's equation of motion is $\ddot{x} = 3x^5 - 2x^3$. Solving this equation for $x(0) = 1$ and $\dot{x}(0) = 0$ gives $x(t) = (1-t^2)^{-1/2}$ and so $\dot{x}(t) = t(1-t^2)^{-3/2}$. This means that the kinetic energy of the mass, as a function of time, is given by $T(t) = t^2(1-t^2)^{-3}/2$, while $T(t) + V(t) = 0$. It is now clear that the mass's position, velocity, and kinetic energy all approach infinity in the limit as $t \to 1$. At the same time the kinetic energy approaches infinity, the potential energy approaches minus infinity, thereby keeping the total system energy constant (and in fact zero). For times greater than unity the system comprises a force field alone, since the mass has left the universe being considered. This system cannot be

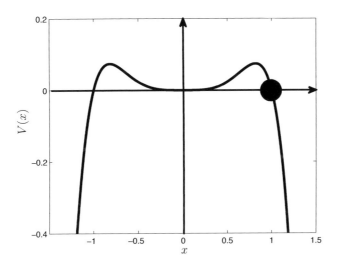

Figure 2.5: Unit mass in a potential force field.

time reversed despite being conservative. As a result of the singularity at $t = 1$, we cannot deduce the system's reversibility from the equations of motion alone and time reversibility may only be a 'local' property.

Example 2.18 If $x(t)$ and $y(t)$ are solutions to the equations

$$\dot{x} = -2\cos x - \cos y \qquad x(t_0) = x_0$$
$$\dot{y} = -2\cos y - \cos x \qquad y(t_0) = y_0, \qquad (2.181)$$

then so are $-x(-t)$ and $-y(-t)$. This means that this system is time reversible. The fixed points satisfy $\dot{x} = \dot{y} = 0$. Solving the corresponding algebraic equations gives $\cos x^* = 0$ and $\cos y^* = 0$, or $(x^*, y^*) = (\pm\frac{\pi}{2}, \pm\frac{\pi}{2})$. Linearization around (x^*, y^*) gives

$$\frac{\partial f}{\partial x} = \begin{bmatrix} 2\sin x^* & \sin y^* \\ \sin x^* & 2\sin y^* \end{bmatrix},$$

which has eigenvalues $(1, 3)$ at $(\frac{\pi}{2}, \frac{\pi}{2})$, $\pm\sqrt{3}$ at $(\pm\frac{\pi}{2}, \mp\frac{\pi}{2})$, and $(-1, -3)$ at $(-\frac{\pi}{2}, -\frac{\pi}{2})$. As shown in the phase portrait in Figure 2.6, the fixed points are respectively a repeller, two saddle points, and an attractor. A conservative system must be free of attractors, because the system energy cannot be both time-invariant and non-constant throughout the basin of attraction.

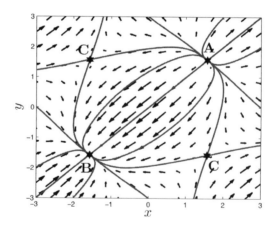

Figure 2.6: Phase portrait of the system given in (2.181). Point A is a repeller, point B an attractor, and the points C are saddles.

Engineers are always making use of time-reversibility-destroying effects such as friction, damping, hysteresis, and Ohmic losses, even though physicists might take the view that these loss-creating influences are 'non-physical' contrivances that are not part of the fundamental make-up of nature. It might be suggested that one has to consider the more basic properties of nature's micro-structure, where forces are conservative and relate to such things as atomic interactions. The damped-pendulum example poses a deep theoretical problem that is encapsulated in Loschmidt's objection [61], also known as the irreversibility paradox, which asks: 'how can a macroscopic system be irreversible (like the damped pendulum), while its microscopic constituents are conservative?' This puts the time-reversal symmetry of (almost) all known low-level fundamental physical processes at odds with any attempt to infer from them the second law of thermodynamics that describes the behaviour of macroscopic systems. In an attempt to explain this paradox, one could argues that the motion of the damped pendulum is slowed by the transference of kinetic energy from the pendulum to the surrounding medium atoms in the form of heat [61]. Since the laws of physics governing atomic interactions are reversible, each collision must be reversed, causing a cooperative transfer of energy back to the bob, which will then be consequently accelerated returning it to its original position (if one waits long enough). The *Poincaré recurrence theorem* states that conservative systems will, after a sufficiently long but finite time (the *Poincaré recurrence time*), return to a state 'close' to the initial state.

During modelling exercises one must be mindful of the difference between the true properties of nature and the characteristics of the model being used to describe it. As we have shown, if one assumes that air resistance can be described by a velocity-dependent damping force (linear damping), the *predicted* motion of the pendulum will be irreversible. If one believes Newton's laws, there is nothing wrong with the mechanics, but the model being used to describe nature may be brought into question. Classical mechanics (as a theory) is neutral on the subject of time reversibility and is equally compatible with forces that produce reversible behaviours, and phenomenological or empirical forces such as friction, which cause irreversible motions.

Bearing Example 2.16 in mind, one might postulate that conservative (position-dependent) forces result in reversible motions, while time- and velocity-dependent forces such as friction do not. The next two examples show that the relationship between forces that result in reversible motions and ones that do not is more subtle than this. Indeed, conservative systems may 'appear' dissipative if one does not look too far into the future (much less than the Poincaré recurrence time). The first example is taken from statistical mechanics, while the second is taken from engineering.

Example 2.19 In physics there are many systems that are describable by equations of motion of the form

$$m\ddot{x} + c\dot{x} + \frac{\partial V(x)}{\partial x} = F \tag{2.182}$$

in which $c\dot{x}$ is a damping term and F is a driving force. A simple example is the *Brownian motion* of colloidal particles floating in a liquid medium. In this case m is the mass of the particle, c is a damping coefficient, $V(x)$ is the potential acting on the particle, and F is a stochastic force, which causes the particles to undergo irregular 'jiggle' motions. This force is describable in terms of its mean and variance:

$$\mathcal{E}\{F(t)\} = 0,$$
$$\mathcal{E}\{F(t)F(t')\} = 2mk_BTc\delta(t-t'), \tag{2.183}$$

where $\mathcal{E}\{\cdot\}$ represents the expected value, or statistical average, over the ensemble of realizations of the force $F(t)$ [67]. Boltzmann's constant is given by k_B, and T is the

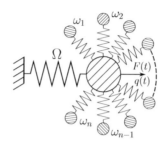

Figure 2.7: Harmonic oscillator in a thermal bath.

temperature. The appearance of the damping constant c in the variance means that the damping force and the random force are related. This relationship between the deterministic damping and the random forces is the subject of the *fluctuation–dissipation theorem*, which states that the random force must have a power spectrum determined by the damping [67]. In sum, (2.182) is a second-order stochastic differential equation driven by white noise and is known in the physics literature as the *Langevin equation*.

In classical studies of Brownian motion, the so-called *Caldeira–Leggett system-plus-reservoir model* is used [68–70], which is illustrated in Figure 2.7. Since this system is conservative, it follows from the Poincaré recurrence theorem that this system cannot display damping effects, because this would preclude a return to the system's initial state.

The Lagrangian associated with this system is

$$\mathcal{L} = \frac{1}{2}\dot{q}^2 + \sum_{i=1}^{n} \frac{m_i}{2}\dot{q}_i^2 - \frac{1}{2}\Omega^2 q^2 - \sum_{i=1}^{n} \frac{k_i}{2}(q_i - q)^2, \tag{2.184}$$

in which the particle mass is assumed to be unity. The primary oscillator frequency is Ω, the heat bath spring stiffnesses are $k_i = \gamma^2$, the heat bath masses are $m_i = \gamma^2 \left(\frac{n}{i\omega_c}\right)^2$, with the equispaced bath frequencies given by $\omega_i = i\omega_c/n$, and ω_c is some cut-off frequency. Direct calculation using Lagrange's equations (2.40) gives

$$\ddot{q} = -\Omega^2 q + \sum_{i=1}^{n} k_i (q_i - q) \tag{2.185}$$

$$\ddot{q}_i = -\omega_i^2 (q_i - q). \tag{2.186}$$

The Laplace transform of (2.185) is

$$q(s)\left(s^2 + \Omega^2 + \sum_{i=1}^{n} k_i\right) = \sum_{i=1}^{n} k_i q_i(s) + sq(0^+) + \dot{q}(0^+) \tag{2.187}$$

while the Laplace transform of (2.186) is

$$q_i(s) = \frac{\omega_i^2 q(s)}{s^2 + \omega_i^2} + \frac{sq_i(0^+) + \dot{q}_i(0^+)}{s^2 + \omega_i^2}. \tag{2.188}$$

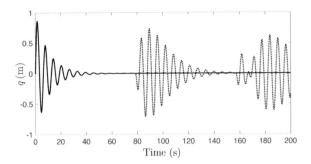

Figure 2.8: Impulse response of a harmonic oscillator in a thermal bath; $\Omega = 1$, $\gamma = 0.1$, $n = 250$ and $\omega_c = 20$. The dotted curve is the response of the full model (2.189), while the solid curve is the response of the second-order damped model given in (2.190).

Substituting (2.188) into (2.187), assuming that $q(0^+) = 0$ and $\dot{q}(0^+) = 0$, gives

$$q(s)\left(s^2 + \Omega^2 + s\frac{n\pi\gamma^2}{2\omega_c}\left(\frac{2\omega_c}{\pi n}\sum_{i=1}^{n}\frac{s}{s^2 + (\frac{i\omega_c}{n})^2}\right)\right) = \overbrace{\sum_{i=1}^{n}\left(\frac{sq_i(0^+) + \dot{q}_i(0^+)}{s^2 + (\frac{i\omega_c}{n})^2}\right)}^{F(s)}.$$

(2.189)

The right-hand side of (2.189) can be interpreted as random forcing that comes from the initial positions and momenta of the bath masses. Since

$$\lim_{n\to\infty}\left(\frac{2\omega_c}{\pi n}\sum_{i=1}^{n}\frac{s}{s^2 + (\frac{i\omega_c}{n})^2}\right) = 1,$$

when n is large, one might expect (2.189) to behave like

$$\ddot{q} + \frac{n\pi\gamma^2}{2\omega_c}\dot{q} + \Omega^2 q = F(t),$$

(2.190)

which is in the form of the Langevin equation (2.182), which predicts damping.

In problems of this type, high-order conservative descriptions are often replaced by simple low-order models with dissipation. Figure 2.8 shows the impulse responses of the full-order system (2.189) and the second-order system (2.190). It is clear that on some finite horizon, which may be large (but smaller than the Poincaré recurrence time), the high-order conservative model is almost indistinguishable from a low-order dissipative description.

Example 2.20 We now return to Example 2.6, and consider again the one-dimensional mass–spring transmission line illustrated in Figure 2.9. The input of this system is the force F, while the output is the velocity $\dot{q}$ of the left-hand mass m. A force balance on the first mass gives

$$m\ddot{q} = F + k(q_1 - q).$$

(2.191)

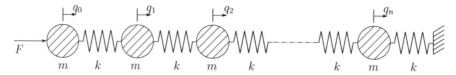

Figure 2.9: Transmission line lumped approximation.

Forces balances on the interior masses yield

$$m\ddot{q}_i = k(q_{i-1} - 2q_i + q_{i+1}) \qquad i = 1, \cdots, n-1, \qquad (2.192)$$

while a force balance on the last mass is given by

$$m\ddot{q}_n = k(q_{n-1} - 2q_n). \qquad (2.193)$$

The step response of this system is given in Figure 2.10 and it can be seen that the mechanical admittance[14] is $F/\dot{q} = \pm 80$, which corresponds to pure (positive and negative) damping, with the sign switching every 50 s. As one would expect, there is

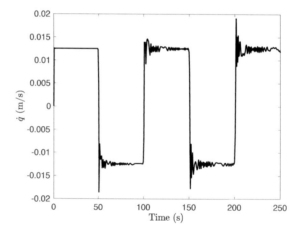

Figure 2.10: Step response $\dot{q}$ of the lumped transmission line model as a function of time; $k = 1600$, $m = 4$ and $n = 500$.

a travelling wave in the lumped-mass approximation of the transmission system. This is best understood by considering the limiting case when $n \to \infty$.

Consider the transmission element illustrated in Figure 2.11, which is at a distance x along the line. It is immediate from that figure that the propagated force is given

---

[14] By exploiting the mechanical-electrical analogy, where current $\leftrightarrow$ force and voltage $\leftrightarrow$ velocity, it follows that the admittance is given by either current/voltage or force/velocity [71].

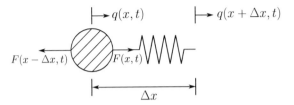

Figure 2.11: Small element of a continuous transmission line.

by

$$F(x, t) = k(q(x + \Delta x, t) - q(x, t))$$
$$= K \frac{q(x + \Delta x, t) - q(x, t)}{\Delta x}, \qquad (2.194)$$

in which $K = k\Delta x$.[15] Taking limits in (2.194) and differentiating with respect to time gives

$$\frac{\partial F(x, t)}{\partial t} = K \frac{\partial \dot{q}(x, t)}{\partial x}. \qquad (2.195)$$

A force balance on the mass element gives

$$\Delta x \rho \frac{\partial \dot{q}(x, t)}{\partial t} = F(x + \Delta x, t) - F(x, t)$$

in which $m = \rho \Delta x$, where ρ is the lengthwise mass density. Taking limits gives

$$\rho \frac{\partial \dot{q}(x, t)}{\partial t} = \frac{\partial F(x, t)}{\partial x}. \qquad (2.196)$$

If we now differentiate (2.195) with respect to time, and (2.196) with respect to position, and eliminate the mixed partial derivatives, we obtain

$$\frac{\partial^2 F(x, t)}{\partial t^2} = \frac{K}{\rho} \frac{\partial^2 F(x, t)}{\partial x^2}, \qquad (2.197)$$

which is the well-known wave equation in which $v = \sqrt{\frac{K}{\rho}}$ is the wave-propagation velocity. In the same way one can also obtain

$$\frac{\partial^2 \dot{q}(x, t)}{\partial t^2} = \frac{K}{\rho} \frac{\partial^2 \dot{q}(x, t)}{\partial x^2}; \qquad (2.198)$$

the force and velocity waves satisfy the same wave equation.

---

[15] In the limit as the spring length goes to zero $k \to \infty$. For that reason we introduce K, which is well defined in this limit.

There is an important relationship between the force and velocity waves. Using (2.195) we obtain

$$\frac{\partial F(x,t)}{\partial t} = K \frac{\partial \dot{q}(x,t)}{\partial t} \frac{\partial t}{\partial x}$$
$$= \pm \frac{K}{v} \frac{\partial \dot{q}(x,t)}{\partial t}$$
$$= \pm \sqrt{K\rho} \frac{\partial \dot{q}(x,t)}{\partial t}.$$

It now follows by integration that

$$F(x,t) = D_O \dot{q}(x,t), \tag{2.199}$$

in which $D_O = \pm \sqrt{K\rho}$ is the characteristic damping of the transmission line. In the case of the forward wave D_O is positive, while D_O is negative for the backward wave.

It can be shown by direct calculation that

$$f(x,t) = F_+ \overrightarrow{f}\left(t - \frac{x}{v}\right) + F_- \overleftarrow{f}\left(t + \frac{x}{v}\right) \tag{2.200}$$

is a solution to (2.197) (and by analogy to (2.198)) in which v is the wave-propagation velocity. The first term in (2.200) represents a forward-travelling wave with amplitude F_+, while the second term corresponds to a backward-propagating wave with amplitude F_-. Since the wave equation is linear, these terms can be considered separately. Suppose a pulse at $t = 0$ is given by $\overrightarrow{f}(x,0)$; see Figure 2.12. At a later moment in time, Δt say, the difference $\Delta t - x/v$ will have the same value as at $t = 0$ if we consider position $x + \Delta t v$. This means that the pulse $\overrightarrow{f}(x,0)$ will move unaltered in shape to position $x + \Delta t v$. Since Δt is arbitrary, the pulse will move continuously and unaltered in shape to the right with velocity v. Parallel arguments show that the second term in (2.200) represents a wave propagating in the negative x-direction.

This means that the transmission line, which is made up of only lossless components, is time reversible (reverse the sign of t and v in (2.200)), and yet the driving-point admittance $F(x,t)/\dot{q}(x,t) = \pm D_O$ is a damper. In our example the propagation velocity is $v = \sqrt{K/\rho} = \sqrt{1600/4} = 20$ and the length is 500. This means that the round-trip time is 50 as shown. It is now clear that, in principle, the line's driving-point admittance can be 'designed' to look like a (positive) damper for arbitrarily long periods.

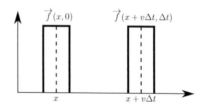

Figure 2.12: A wave moving along a lossless transmission line, with velocity v, that is unaltered in shape.

2.8 Čhaplygin's sleigh

The most widely studied class of mechanical systems involves conservative forces constrained by holonomic constraints. Systems of this type cannot have exponential stability in any of their configuration variables. It is therefore paradoxical that the most widely studied of all mechanical systems cannot have the most fundamental of all engineering requirements—stability! In contrast, the rolling disc (Section 2.10), or axisymmetric tops spinning on a level surface, are conservative nonholonomic systems that can have exponential stability in some of their configuration variables. In order to examine these issues in more detail we study here a well-known example that demonstrates that exponential stability in conservative systems can be a direct consequence of nonholonomy [72].

Čhaplygin's sleigh is a nonholonomic system comprising a rigid body that is free to slide over a frictionless horizontal plane; see Figure 2.13. The rigid body has mass m with centre of mass G and polar moment of inertia J. There is a skate C at a distance l from G. The skate and the body-fixed unit vector $\hat{e}_x$ are aligned with CG. The body-fixed unit vector $\hat{e}_y$ lies on the plane and is normal to $\hat{e}_x$. The unit vector $\hat{e}_z = \hat{e}_x \times \hat{e}_y$ is normal to the plane. The unit vector $\hat{e}_x$ has yaw angle ψ relative to the positive inertial x-axis. The skate constraint ensures that the velocity of point C on the sleigh is $v_C = u\hat{e}_x$ and that the reactive force on the sleigh is thus $F_r\hat{e}_y$. The absolute position of G is $r_G = r_C + l\hat{e}_x$. Since $\dot{\hat{e}}_x = \omega\hat{e}_y$, where $\omega = \dot{\psi}$ is the sleigh's angular velocity, and $\dot{\hat{e}}_y = -\omega\hat{e}_x$, the velocity and acceleration of the mass centre are given by

$$v_G = u\hat{e}_x + \omega l\hat{e}_y \quad \text{and} \quad a_G = \dot{u}\hat{e}_x + u\omega\hat{e}_y + \dot{\omega}l\hat{e}_y - \omega^2 l\hat{e}_x \quad (2.201)$$

respectively.

We can now derive the equations of motion by taking moments about G using (2.1) and (2.11):

$$F = ma_G \quad \text{and} \quad (r_C - r_G) \times F = J\dot{\omega}\hat{e}_z. \quad (2.202)$$

If we 'dot' the first of these equations with $\hat{e}_x$, while making use of (2.201), we get

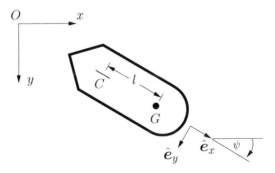

Figure 2.13: Čhaplygin's sleigh is a rigid body that is constrained to move on a frictionless plane. The origin of the body-fixed axis system is at C.

$$\dot{u} = \omega^2 l. \tag{2.203}$$

Since the sleigh is constrained in the $\hat{e}_y$ direction, the reaction force must be in this direction, and it cannot do work. If we 'dot' the first equation in (2.202) with $\hat{e}_y$, we get the reaction force as

$$\boldsymbol{F}_r = m(u\omega + l\dot{\omega})\hat{e}_y. \tag{2.204}$$

If we substitute $\boldsymbol{r}_G - \boldsymbol{r}_C = l\hat{e}_x$ in the second equation in (2.202) one obtains

$$\dot{\omega} = -\frac{lF_r}{J} \tag{2.205}$$

$$= -\frac{ml}{J + ml^2}u\omega. \tag{2.206}$$

The dynamics of the sleigh are fully determined by the solution of (2.203) and (2.206), which are quadratically coupled first-order nonlinear differential equations. The sleigh's initial speed $u_0 = u(0)$ and its initial angular velocity $\omega_0 = \omega(0)$ are assumed specified. When solving these equations we will specify the initial conditions in terms of the sleigh's initial speed u_0, and its total system energy. In order to keep track of the yaw angle and mass centre of the sleigh (in inertial coordinates) we need to solve

$$\begin{bmatrix} \dot{x}_G \\ \dot{y}_G \\ \dot{\psi} \end{bmatrix} = \begin{bmatrix} c_\psi & s_\psi & 0 \\ -s_\psi & c_\psi & 0 \\ 0 & 0 & 1 \end{bmatrix} \begin{bmatrix} u \\ \omega l \\ \omega \end{bmatrix} \tag{2.207}$$

in which the initial conditions $x_G(0)$, $y_G(0)$, and $\psi(0)$ are assumed specified. This means that the sleigh's dynamics are described by five first-order differential equations.

The sleigh's equations of motion can also be obtained using Lagrange's equation. To do this we will assume that the velocity component of the mass centre in the (inertial) x-direction is given by $\dot{x}_G$, while that in the y-direction is given by $\dot{y}_G$. The system Lagrangian (and the system internal energy in this case) is given by

$$\mathcal{L} = \frac{m}{2}\left(\dot{x}_G^2 + \dot{y}_G^2\right) + \frac{J}{2}\omega^2; \tag{2.208}$$

note that all the position variables are cyclic. Referring to Figure 2.13, we see that

$$0 = \omega l + \dot{x}_G \sin\psi - \dot{y}_G \cos\psi \tag{2.209}$$

ensures that the sleigh is absolutely stationary in the $\hat{e}_y$ direction; this is a nonholonomic constraint that must be appended to (2.208). Using (2.45) and (2.48) gives

$$m\ddot{x}_G = -\lambda \sin\psi \tag{2.210}$$
$$m\ddot{y}_G = \lambda \cos\psi \tag{2.211}$$
$$J\dot{\omega} = -\lambda l. \tag{2.212}$$

In order to reconcile (2.203), (2.204), and (2.206) with (2.210), (2.211), and (2.212), we observe that (2.211)$\cos\psi$ − (2.210)$\sin\psi$ gives

$$m\left(\ddot{y}_G \cos\psi - \ddot{x}_G \sin\psi\right) = \lambda. \tag{2.213}$$

The expression $\ddot{y}_G \cos\psi - \ddot{x}_G \sin\psi$ is the acceleration of G in the $\hat{e}_y$ direction; as shown in (2.201), this is just $(\dot{\omega}l + u\omega)\hat{e}_y$. Hence

$$m(u\omega + \dot{\omega}l) = \lambda = F_r$$

by (2.204); (2.205) and (2.212) are thus equivalent. Next, (2.210)$\cos\psi$ + (2.211)$\sin\psi$ gives

$$0 = \ddot{x}_G \cos\psi + \ddot{y}_G \sin\psi. \tag{2.214}$$

We now use (2.214) and $\boldsymbol{v}_G = (\dot{u} - l\omega^2)\hat{e}_x$, which comes from differentiating $\boldsymbol{r}_G = \boldsymbol{r}_C + l\hat{e}_x$ twice with respect to time, to establish that

$$0 = \ddot{x}_G \cos\psi + \ddot{y}_G \sin\psi$$
$$= \dot{u} - l\omega^2,$$

which is (2.203). It is clear that (2.205) and (2.212) are the same.

Finally, we derive the equations of motion of the sleigh using the Lagrangian approach in quasi-velocities; see Section 2.2.9. Initially we will neglect the nonholonomic constraint associated with the skate. Accordingly, the unconstrained velocity of the skate is $\boldsymbol{v}_C = u\hat{e}_x + v\hat{e}_y$, with the unconstrained kinetic energy given by $T = \frac{m}{2}(u^2 + (v+\omega l)^2) + \frac{J}{2}\omega^2$. The vector of quasi-velocities $[u, v, \omega]^T$ is used in place of $\dot{\boldsymbol{q}} = [\dot{x}, \dot{y}, \dot{\psi}]^T$; u is the longitudinal velocity of the skate, v is its lateral velocity, and ω its yaw rate. The relationship between the quasi-velocities and the generalized velocities is given by (2.65), which takes the form of (2.74). In anticipation of the introduction of the nonholonomic constraint, the generalized force is $\boldsymbol{k} = [0, F_r, 0]^T$. The unconstrained equations of motion are obtained from (2.76)–(2.78), and are given by

$$m(\dot{u} - \omega^2 l - \omega v) = 0 \tag{2.215}$$
$$m(u\omega + l\dot{\omega} + \dot{v}) = F_r \tag{2.216}$$
$$(J + ml^2)\dot{\omega} + m(lu\omega + uv - v\dot{u} + l\dot{v}) = 0. \tag{2.217}$$

In order to enforce the nonholonomic constraint we set $v = \dot{v} = 0$ in (2.215)–(2.217), in which case (2.215) becomes (2.203), and (2.217) becomes (2.206). Under the assumed velocity constraint, F_r in (2.216) is the constraint force and corresponds to (2.204).

We will now focus on the dynamics of the sleigh. In order to solve analytically (2.203) and (2.206) we begin by eliminating ω. From (2.203) we see that $\dot{\omega} = \ddot{u}/(2l\omega)$, which can be substituted into (2.206) to obtain

$$\ddot{u} = -\frac{2ml}{J+ml^2}(u\dot{u}) = -\frac{2ml}{J+ml^2}\left(\frac{1}{2}\frac{d}{dt}(u^2)\right). \tag{2.218}$$

Integrating (2.218) gives the Riccati equation

$$\dot{u} = -\frac{ml}{J+ml^2}u^2 + \frac{2lE}{J+ml^2}, \tag{2.219}$$

in which the second term is a constant of integration. If we now define $U = \sqrt{\frac{2E}{m}}$ and $r = (l\sqrt{2mE})/(J + ml^2)$, it is straightforward to verify that the Riccati equation (2.219) has solution[16]

$$u = U \frac{u_0 + U \tanh(rt)}{U + u_0 \tanh(rt)}, \qquad (2.220)$$

which is a function of U and the initial speed u_0. A closed-form expression for the yaw rate is obtained by substituting (2.220) into (2.203) and (2.219). This gives

$$\omega = \sqrt{\frac{2E(U^2 - u_0^2)}{J + ml^2}} \frac{1}{U \cosh(rt) + u_0 \sinh(rt)}. \qquad (2.221)$$

Substituting (2.203) into (2.219) gives

$$\left(\frac{J}{m} + l^2\right) \omega^2 + u^2 = U^2, \qquad (2.222)$$

which is the equation of an ellipse that characterize the motion (flow) of the system. Referring to (2.208), we see that the system's total internal energy is given by

$$E = \frac{1}{2}\left(mu^2 + (ml^2 + J)\omega^2\right), \qquad (2.223)$$

since $|\boldsymbol{v}_G|^2 = u^2 + l^2\omega^2$, and

$$U^2 = \frac{2E}{m} = u^2 + \left(\frac{J}{m} + l^2\right)\omega^2.$$

In the case that $\omega = 0$, $u = U$ and all the system energy is translational; it therefore follows that $|u_0| \leq U$. This means that $|u_0 \tanh(\beta t)| \leq U$, establishing that (2.219) has no finite escape times. Also, $\lim_{t\to\infty} u(t) = U$ for all non-equilibrium initial conditions and the flow is attracted to the positive u-axis along constant energy trajectories as shown in Figure 2.14.

By exploiting the relationship between the speed, the energy, and the yaw rate given in (2.223), we can express ω in terms of the simpler expression

$$\omega = \frac{U\omega_0}{U \cosh(rt) + u_0 \sinh(rt)}, \qquad (2.224)$$

which admits both positive and negative values of ω_0 facilitating trajectories around the right- and left-hand sides of Figure 2.14. Equation (2.224) can be integrated to give the yaw angle

$$\psi = \frac{2U\omega_0}{r\sqrt{U^2 - u_0^2}}\left(\tan^{-1}\left\{\frac{U\tanh(\frac{rt}{2}) + u_0}{\sqrt{U^2 - u_0^2}}\right\} - \tan^{-1}\left\{\frac{u_0}{\sqrt{U^2 - u_0^2}}\right\}\right) + \psi_0, \quad (2.225)$$

with ψ_0 the initial yaw angle.

---

[16] The Riccati equation $\dot{u} = au^2 + b$ can be solved using the substitution $u = -\dot{w}/(aw)$, which results in the linear second-order equation $\ddot{w} + abw = 0$.

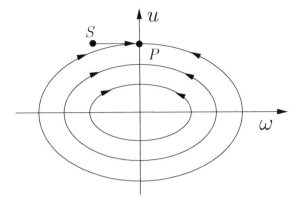

Figure 2.14: Flows of the Čhaplygin sleigh take the form of constant-energy ellipses. The only stable equilibrium points are along the positive u-axis; P is one such point. The point S is an exemplar initial condition for the linearized model, which then flows to P.

The sleigh's reaction force can be found by substituting (2.220) and (2.221) into (2.204), while making use of (2.206). This gives

$$F_r = 2F_r^{max} \sqrt{1 - \frac{u_0}{U} \left(\frac{(\frac{u_0}{U} \cosh(rt) + \sinh(rt))}{(\frac{u_0}{U} \sinh(rt) + \cosh(rt))^2} \right)} \qquad (2.226)$$

with

$$|F_r| \le F_r^{max} = \frac{E(1 - k^2)}{lk^3}, \qquad (2.227)$$

where $k^2 = 1 + \frac{J}{ml^2}$. This result appears to have been found for the first time by Carathéodory [73].

If we linearize equations (2.203) and (2.206) around the equilibrium point $u = u_* \ge 0$ and $\omega = 0$, we see that these equations become decoupled, with solutions $\dot{u} = 0$ and $\omega = \delta\omega_0 e^{(-mlu^*t)/(J+ml^2)}$, in which $\delta\omega_0$ is an arbitrary small perturbation in ω. Thus if the system is perturbed to point S, say, in Figure 2.14, it will decay back to P along a horizontal constant-velocity trajectory. It is important to observe that since the exponential decay in ω is not accompanied by a change in the speed of the sleigh, energy is necessarily dissipated. Equation (2.223) becomes

$$E = \frac{1}{2} \left(mu_*^2 + (ml^2 + J)\delta\omega_0^2 e^{(-2mlu_*t)/(J+ml^2)} \right),$$

which represents a dissipative motion since $\dot{E} < 0$ for any $u_* > 0$. The energy conserving properties of the nonlinear system come from the quadratic coupling terms in (2.203) and (2.206), which are destroyed by linearization.

Example 2.21 We consider the case $l = 0.1\,\mathrm{m}$, $m = 1.0\,\mathrm{kg}$, $J = 0.1\,\mathrm{kg\,m^2}$, $u_0 = 1.99\,\mathrm{m/s}$, and $E = 2\,\mathrm{J}$ (which results in $\omega(0) = 0.6\,\mathrm{rad/s}$ and $U = 2\,\mathrm{m/s}$), with $x_G(0) = 0$, $y_G(0) = 0$, and $\psi(0) = 0$. As shown in Figure 2.15 (a), the sleigh's mass

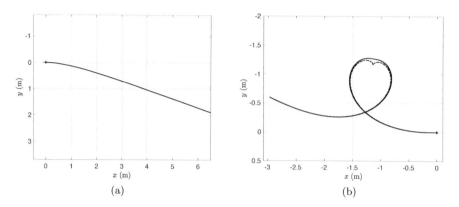

Figure 2.15: Trajectory of the mass centre and skate of the Čhaplygin sleigh: (a) positive initial velocity ($u_0 = 1.99$ m/s); (b) negative initial velocity ($u_0 = -1.99$ m/s). The mass centre is shown as the solid curve, while the skate position is the dot-dash curve; the system was simulated for 3.5 s.

centre moves to the right along the positive x-axis. Since $u_0 < U$, some of the initial stored energy is in the form of rotational kinetic energy. It follows from (2.220) that this energy is soon transferred into translational kinetic energy causing the sleigh to complete its turn, and then move away in a straight line.

   In the case that $u_0 = -1.99$ m/s ($\omega(0) = 0.6$ rad/s), when the skate starts out in front of the mass centre, the motion is more complex and consequently more interesting. First, it follows from (2.220), that $\lim_{t\to\infty} u(t) = 2$ m/s. As shown in Figure 2.16 (a), the forward speed starts at $u_0 = -1.99$ m/s, passes through zero at 1.65 s, and then approaches 2 m/s. Since the system energy remains constant, energy is transferred from translational kinetic energy into rotational kinetic energy and then back again. During the period of high angular velocity, the sleigh 'flips around' so that the skate follows the mass centre in a stable configuration. The ground-plane trajectory is shown in Figure 2.15 (b). In this case the sleigh's mass centre moves off to the left along the negative x-axis following the skate. It then turns gently to the left and starts to slow down. At a relatively low speed it enters into the neighbourhood of the cusp, 'flips', and then moves away with the skate following the mass centre. The sleigh then speeds up and moves away along a rectilinear path. Figure 2.17 shows the reaction force at the sleigh's skate and F_r^{max} during the manoeuvre. The small negative initial reaction force cause an increase in the sleigh's angular velocity causing it to turn to the left; see Figure 2.15 (a). At approximately 1.2 s the reaction force reaches its minimum and begins to increase. When the reaction force changes sign (at approximately 1.7 s), the angular velocity is at its peak, the translational speed is zero, and the sleigh is in mid 'flip'. Once the sleigh has 'flipped', it continues to turn left despite a change in sign in the reaction force, but this comes from the fact that the body-fixed vector $\hat{e}_y$ has flipped too. The reaction force then decays away as the sleigh transitions into its rectilinear terminal motion.

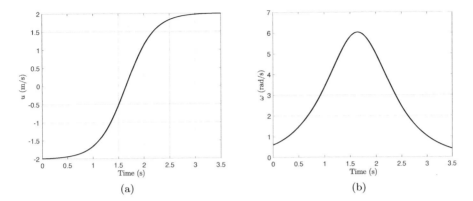

Figure 2.16: Speed (a) and yaw rate (b) of the Čhaplygin sleigh.

2.8.1 T-symmetry of the Čhaplygin sleigh

As we will now demonstrate, it is possible to establish the time reversibility of the Čhaplygin sleigh by studying the solutions of the describing equations. To begin, (2.220) and the odd symmetry of the hyperbolic tangent function established that $u(-t) = -u(t)$ when u_0 is replaced by $-u_0$. In Equation (2.224) the symmetry of the hyperbolic cosine function and the odd symmetry of the hyperbolic sine function show that $\omega(-t) = -\omega(t)$ when u_0 is replaced by $-u_0$, and ω_0 is replaced by $-\omega_0$. In the same way (2.225) shows that $\psi(t) = \psi(-t)$ when $u_0 \to -u_0$, $\omega_0 \to -\omega_0$, and ψ_0 remains unchanged. It now follows from (2.207) that $\dot{x}_G(t) = -\dot{x}_G(-t)$ and $\dot{y}_G(t) = -\dot{y}_G(-t)$. When these equations are integrated with respect to time, we find that $x_G(t) = x_G(-t)$ and $y_G(t) = y_G(-t)$, which proves that the sleigh follows the same trajectory under time reversals. Finally, (2.204) can be written as

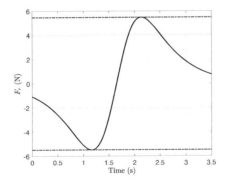

Figure 2.17: Sleigh reaction force. The dot-dash lines are given by $\pm F_r^{max}$ in equation (2.227).

$$\boldsymbol{F}_r = \frac{mJu\omega}{J + ml^2}\hat{\boldsymbol{e}}_y \tag{2.228}$$

using (2.206), which shows that the skate reaction force is invariant under time reversals satisfying (2.180).

2.9 Dynamics of a rolling ball

Rolling contacts are a key feature of all road-going vehicles. While modern modelling techniques involve the concept of 'slip' (Chapter 3), rolling contacts were modelled classically using nonholonomic constraints, which is a reasonable approximation for tyres rolling at very low speeds. In order to introduce this aspect of vehicle dynamics we will study the motion of a ball rolling on a rough surface, which is a well-known class of nonholonomic systems. When solving this problem one thinks of a point of contact P, which is a material point on the surface of the ball, that is in instantaneous contact with a corresponding material point P′ on the underlying rough surface. In the case of an ideal non-slipping contact one assumes that the points P and P′ are relatively stationary, that is, they have zero relative velocity. This modelling assumption imposes a kinematic constraint that forces the velocities of P and P′ to be equal in inertial space. In addition, the vertical components of the positions of P and P′ must remain equal so that the ball remains in contact with the surface below it. In the vernacular of nonholonomic systems the rolling contact is characterized by a holonomic constraint in the direction normal to the plane, and two nonholonomic constraints in the plane of the underlying rough surface. Rolling-ball problems are simplified by the fact that homogeneous axisymmetric spheres have uniform inertia properties characterized by the fact that any axis passing through the sphere's mass centre is a principal axis with a common moment of inertia. There is therefore no need for the use of Euler angles, or other similar device, to characterize the ball's orientation in space.

We begin with a study of a ball rolling down an inclined surface, which is an entry-level problem that is usually solved using an elementary two-dimensional analysis. The solution to this problem is followed by the analysis of a ball rolling on the surface of an inclined turntable. The analyses presented in this section follow for the most part the excellent treatment given to these problems in [46].

2.9.1 Ball on an incline

The study of the dynamics of a ball rolling down an inclined surface is essentially a planar problem, which we have nonetheless chosen to solve using three-dimensional vector mechanics. At first sight this might seem unnecessarily complicated, but this approach will soon pay dividends.

In this section we study the dynamics of a ball rolling down a non-rotating inclined turntable; see Figure 2.18 where $\Omega = 0$. In the next section we will consider the effect of $\Omega \neq 0$. The homogeneous and axisymmetric ball is assumed free to roll on the inclined surface without slipping. The ball has radius a, mass m, and moment of inertia μma^2 (for any axis passing through the ball's mass centre). In the case of a uniform sphere $\mu = \frac{2}{5}$, while in the case of a spherical 'thin' shell $\mu = \frac{2}{3}$.

The origin of an inertial coordinate system is located at the central point O on the surface of the inclined turntable. A unit vector **k** points down the turntable's spindle.

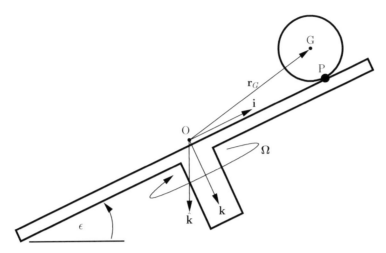

Figure 2.18: Rolling ball on an inclined turntable. The diagram shows a ball rolling on the surface of a turntable that is inclined at an angle ϵ to the horizontal. The turntable rotates at angular velocity Ω. The ball's contact point is located at P, while the ball's mass centre is at G. The unit vectors $\mathbf{i}$ and $\mathbf{k}$ form part of an orthogonal coordinate system that is completed by $\mathbf{j}$, which points out of the page. The unit vector $\hat{\mathbf{k}}$ points in the direction of gravity.

The unit vector $\mathbf{i}$, which lies in the page, points away from O in a radial 'uphill' direction. This right-handed orthogonal axis system is completed with the unit vector $\mathbf{j} = \mathbf{k} \times \mathbf{i}$ that points out of the page. The unit vector $\hat{\mathbf{k}}$ points away from O in the direction of gravity. The ball's mass centre is at G and is located by a vector $\mathbf{r}_G$. The radius vector $a\mathbf{k}$ points from G to P. The force of reaction at the turntable contact point is $\mathbf{F}_r$, while the gravitational force is $mg\hat{\mathbf{k}}$.

We will derive the equations of motion using Newton's laws as given in (2.1) and (2.11). Summing forces on the ball gives

$$m\ddot{\mathbf{r}}_G = \mathbf{F}_r + mg\hat{\mathbf{k}}. \tag{2.229}$$

If the ball's angular velocity is $\boldsymbol{\omega}$, then balancing moments at the ball's mass centre gives

$$\mu m a^2 \dot{\boldsymbol{\omega}} = a\mathbf{k} \times \mathbf{F}_r. \tag{2.230}$$

Eliminating the reaction force $\mathbf{F}_r$ from (2.229) and (2.230) yields

$$\mu m a^2 \dot{\boldsymbol{\omega}} = am\mathbf{k} \times (\ddot{\mathbf{r}}_G - g\hat{\mathbf{k}}). \tag{2.231}$$

Vector multiplication by $\mathbf{k}$ gives:

$$\mu a \dot{\boldsymbol{\omega}} \times \mathbf{k} = \left(\mathbf{k} \times (\ddot{\mathbf{r}}_G - g\hat{\mathbf{k}})\right) \times \mathbf{k}. \tag{2.232}$$

Since the centre of mass of the ball must move in a plane parallel to the incline, $\ddot{\mathbf{r}}_G \perp \mathbf{k}$, and it follows that $(\mathbf{k} \times \ddot{\mathbf{r}}_G) \times \mathbf{k} = \ddot{\mathbf{r}}_G$. It is also evident that $(\mathbf{k} \times \hat{\mathbf{k}}) \times \mathbf{k} = -\sin(\epsilon)\mathbf{i}$. Therefore (2.232) becomes

$$\mu a \dot{\boldsymbol{\omega}} \times \mathbf{k} = \ddot{\mathbf{r}}_G + g \sin(\epsilon)\mathbf{i}. \tag{2.233}$$

Under the 'no-slip' assumption, the contact point P that is regarded as a material point on the surface of the ball must be absolutely stationary. This means that its velocity (2.144) and acceleration (2.145) are

$$\dot{\mathbf{r}}_G + \boldsymbol{\omega} \times a\mathbf{k} = 0, \tag{2.234}$$

$$\ddot{\mathbf{r}}_G + \dot{\boldsymbol{\omega}} \times a\mathbf{k} = 0. \tag{2.235}$$

It therefore follows from (2.233) and (2.235) that

$$(\mu + 1)\ddot{\mathbf{r}}_G = -g \sin(\epsilon)\mathbf{i}. \tag{2.236}$$

As anticipated, the dynamics of the ball are governed by the three equations in (2.236) that are independent of position variables. Integrating this equation gives

$$\dot{\mathbf{r}}_G = -\frac{g}{1 + \mu} \sin(\epsilon)t\mathbf{i} + \dot{\mathbf{r}}_G(0) \tag{2.237}$$

in which $\dot{\mathbf{r}}_G(0)$ is a vector-valued constant of integration that represents the initial velocity of the ball's mass centre, and t is the time.

In order to understand the spinning motion of the ball we return to (2.234), which we multiply vectorially by $\mathbf{k}$ to obtain

$$\begin{aligned}
0 &= \mathbf{k} \times (\dot{\mathbf{r}}_G + a\boldsymbol{\omega} \times \mathbf{k}) \\
&= \mathbf{k} \times \dot{\mathbf{r}}_G + a(\boldsymbol{\omega} - \mathbf{k}(\boldsymbol{\omega} \cdot \mathbf{k})) \\
&= \mathbf{k} \times \dot{\mathbf{r}}_G + a(\boldsymbol{\omega} - \omega_k \mathbf{k}),
\end{aligned} \tag{2.238}$$

where $\omega_k = \boldsymbol{\omega} \cdot \mathbf{k}$ is the polar component of the angular velocity. In the second line we employ the useful vector identity $\mathbf{a} \times (\mathbf{b} \times \mathbf{c}) = \mathbf{b}(\mathbf{a} \cdot \mathbf{c}) - \mathbf{c}(\mathbf{a} \cdot \mathbf{b})$. Scalar multiplication of (2.231) by $\mathbf{k}$ gives $\dot{\boldsymbol{\omega}} \cdot \mathbf{k} = \dot{\omega}_k = 0$ and so ω_k is constant. We therefore have from (2.238)

$$\boldsymbol{\omega} = \omega_k \mathbf{k} + \frac{1}{a}\dot{\mathbf{r}}_G \times \mathbf{k}, \tag{2.239}$$

or equivalently substituting (2.237) into (2.239)

$$\boldsymbol{\omega} = \omega_k \mathbf{k} + \frac{g \sin(\epsilon)t}{a(1 + \mu)}\mathbf{j} + \frac{1}{a}\dot{\mathbf{r}}_G(0) \times \mathbf{k}. \tag{2.240}$$

The first term is the constant polar spin term, the second term corresponds to the angular acceleration produced by the in-plane component of the reaction force $\mathbf{F}_r$, while the final term is due to the ball's initial translational velocity.

As an alternative, the balance of moments can be computed with respect to the material contact point P, which is stationary. In this case the angular momentum (2.8) is $\mathbf{H}_P = J_P \boldsymbol{\omega} = (\mu m a^2 + m a^2)\boldsymbol{\omega}$, where the parallel axis theorem (2.13) has been used to computed J_P, and the equation of motion (2.11) is $m a^2(\mu + 1)\dot{\boldsymbol{\omega}} = -a\mathbf{k} \times mg\hat{\mathbf{k}}$ which is identical to (2.230) when (2.235) is introduced.

2.9.2 Ball on an inclined turntable

A more advanced problem that illustrates the analysis of systems containing non-holonomic constraints and rotating bodies is the ball and turntable illustrated in Figure 2.18. This examples builds upon the results obtained in the previous section by adding the effect of a constant turntable angular velocity Ω.

We start from the balance of moments (2.233), which remains good even when the turntable rotation Ω is included. The effect of rotation enters the problem through the rolling constraint. Since the ball rolls without slipping, it follows that any material contact point P on the surface of the ball must have the same instantaneous velocity as the corresponding material contact point P' on the surface of the turntable. Unlike the $\Omega = 0$ case, the point P' on the turntable's surface has a non-zero velocity $\Omega \times (\mathbf{r}_G + a\mathbf{k})$, in which $\mathbf{r}_G + a\mathbf{k}$ is a vector pointing from O to P. It thus follows that

$$\dot{\mathbf{r}}_G + a\boldsymbol{\omega} \times \mathbf{k} - \Omega \times (\mathbf{r}_G + a\mathbf{k}) = 0. \tag{2.241}$$

Since Ω is parallel with $\mathbf{k}$, the last term in (2.241) is zero. Differenting this equation gives

$$\ddot{\mathbf{r}}_G + a\dot{\boldsymbol{\omega}} \times \mathbf{k} = \Omega \times \dot{\mathbf{r}}_G. \tag{2.242}$$

Substituting (2.242) into (2.233) yields:

$$\ddot{\mathbf{r}}_G + g\sin(\epsilon)\mathbf{i} = \mu\left(\Omega \times \dot{\mathbf{r}}_G - \ddot{\mathbf{r}}_G\right), \tag{2.243}$$

or what is the same

$$\ddot{\mathbf{r}}_G = \frac{\mu}{1+\mu}\Omega \times \dot{\mathbf{r}}_G - \frac{g}{1+\mu}\sin(\epsilon)\mathbf{i}. \tag{2.244}$$

Again, the dynamics of the ball are governed by the three equations in (2.244) and no position variables appear. Integrating (2.244) (with respect to time t) gives:

$$\dot{\mathbf{r}}_G = \frac{\mu}{1+\mu}\Omega \times (\mathbf{r}_G - \mathbf{c}_0) - \frac{g}{1+\mu}\sin(\epsilon)t\mathbf{i} \tag{2.245}$$

in which $\mathbf{c}_0$ is a vector-valued constant of integration. The first term describes circular motion about an initial centre of rotation $\mathbf{c}_0$, with angular speed $\frac{\mu|\Omega|}{1+\mu}$ in the plane of the turntable. The second term is an inclination-related drift term. If $\epsilon = 0$, the ball will move in a circle with centre $\mathbf{c}_0$ and period

$$T = \frac{2\pi(1+\mu)}{\mu|\Omega|}. \tag{2.246}$$

The solution of (2.245) is thus the superposition of an orbital motion and a drift term.

If the initial position $\mathbf{r}_G(0)$ is known, it follows from (2.245) that the initial velocity $\dot{\mathbf{r}}_G(0)$ and the initial centre of rotation $\mathbf{c}_0$ are related. It is immediate from (2.245) that

$$\dot{\mathbf{r}}_G(0) = \frac{\mu}{1+\mu}\Omega \times (\mathbf{r}_G(0) - \mathbf{c}_0).$$

Vector multiplication by $\mathbf{k}$ gives

$$\mathbf{c}_0 = \mathbf{r}_G(0) - \left(\frac{1+\mu}{\mu|\mathbf{\Omega}|}\right)\dot{\mathbf{r}}_G(0) \times \mathbf{k}. \tag{2.247}$$

This equation shows that for a given initial position $\mathbf{r}_G(0)$, $\dot{\mathbf{r}}_G(0)$ and $\mathbf{c}_0$ are uniquely related.

Example 2.22 Figure 2.19 shows the trajectory of a ball rolling without slipping on a horizontal turntable that is rotating at $|\mathbf{\Omega}| = 1\,\mathrm{rad/s}$. The radius of the ball is $a = 2.5\,\mathrm{cm}$ and it is assumed to have a spherical shell ($\mu = \frac{2}{3}$). It follows from (2.246) that the period of rotation is $T = 5\pi\,\mathrm{s}$. The ball's initial contact position is $\mathbf{i}$ with its initial mass centre velocity $\dot{\mathbf{r}}_G(0) = \mathbf{j}$. These initial conditions imply that the centre of rotation is at $\mathbf{c}_0 = -1.5\mathbf{i} - a\mathbf{k}$, and that a circular orbital radius of 2.5 m results.

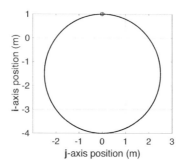

Figure 2.19: Trajectory of the mass centre of a hollow ball rolling on a level turntable. The simulation is run for $25\pi\,\mathrm{s}$, which represents precisely 5 orbits of period $5\pi\,\mathrm{s}$.

If $\mathbf{\Omega} = \mathbf{0}$, (2.245) shows that the ball will simply roll 'down the hill' according to (2.237). If both terms are present, a drifting circular motion results. At first sight it would appear that (2.245) suggests a cyclic motion that drifts 'downhill'. This is not true however, because the gravitational term interacts with the time-varying radius $\mathbf{r}_G$ to produce a 'sideways' epicycloidal motion. One can see this by rewriting (2.245) as

$$\dot{\mathbf{r}}_G = \frac{\mu}{1+\mu}\mathbf{\Omega} \times \left(\mathbf{r}_G - \mathbf{c}_0 + \frac{g\sin(\epsilon)t}{\mu|\mathbf{\Omega}|}\mathbf{j}\right). \tag{2.248}$$

In the case $|\mathbf{\Omega}| = 0$, this becomes, by substitution of (2.247)

$$\dot{\mathbf{r}}_G = -\frac{g}{1+\mu}\sin(\epsilon)t\mathbf{i} + \dot{\mathbf{r}}_G(0), \tag{2.249}$$

which is the same as (2.237); as expected we recover the stationary-incline solution. Equation (2.248) represents epicyclic motion about a centre of rotation $\mathbf{c}_0 - \frac{g\sin(\epsilon)t}{\mu|\mathbf{\Omega}|}\mathbf{j}$ that drifts in the negative $\mathbf{j}$ direction.

Example 2.23 Figure 2.20 shows the uniform sideways drifting of the ball's mass centre predicted by (2.248), when the turntable is inclined at $0.005\,\mathrm{rad}$. The other parameters are the same as those in the previous example.

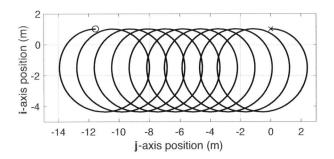

Figure 2.20: Trajectory of the mass centre of a hollow ball rolling on an inclined turntable. The turntable is rotating at $1\,\mathrm{rad/s}$ and is inclined at $0.005\,\mathrm{rad}$. The initial position of the ball's mass centre is $\mathbf{r}_G(0) = \mathbf{i} - a\mathbf{k}$, while its initial velocity is $\dot{\mathbf{r}}_G(0) = \mathbf{j}$. The simulation is run for $50\pi\,\mathrm{s}$.

We conclude this section with a brief analysis of the ball's rotational behaviour. It follows from (2.229) and (2.230) that

$$\mu a\dot{\boldsymbol{\omega}} = \mathbf{k} \times \ddot{\mathbf{r}}_G + g\sin(\epsilon)\mathbf{j}. \qquad (2.250)$$

Scalar multiplication of this equation by $\mathbf{k}$ gives $\dot{\boldsymbol{\omega}} \cdot \mathbf{k} = 0$ and so the polar spin of the ball ω_k is again an arbitrary constant. Vector multiplication of the rolling constraint equation (2.241) by $\mathbf{k}$ gives

$$\boldsymbol{\omega} = \mathbf{k}\omega_k - \frac{|\boldsymbol{\Omega}|}{a}(\mathbf{r}_G + a\mathbf{k}) - \frac{1}{a}\mathbf{k} \times \dot{\mathbf{r}}_G. \qquad (2.251)$$

In order to determine the ball's angular velocity in terms of its position alone, we substitute (2.245) into (2.251) to obtain

$$\boldsymbol{\omega} = \mathbf{k}\omega_k - \boldsymbol{\Omega} - \frac{|\boldsymbol{\Omega}|}{a(1+\mu)}\mathbf{r}_G - \frac{\mu|\boldsymbol{\Omega}|}{a(1+\mu)}\mathbf{c}_0 + \frac{g\sin(\epsilon)t}{a(1+\mu)}\mathbf{j}. \qquad (2.252)$$

The ball's initial angular velocity can be expressed in terms of its initial position and centre of rotation by

$$\boldsymbol{\omega}(0) = \mathbf{k}\omega_k - \boldsymbol{\Omega} - \frac{|\boldsymbol{\Omega}|}{a(1+\mu)}\mathbf{r}_G(0) - \frac{\mu|\boldsymbol{\Omega}|}{a(1+\mu)}\mathbf{c}_0. \qquad (2.253)$$

Force-generating contact.   Road vehicle tyres cannot realistically be represented by nonholonomic rolling constraints. Instead, they are best modelled as force producers, with the force proportional to some form of slip quantity. To see how this might come about we begin by representing the constraint equation (2.241) with an equivalent reaction or friction force. Substituting (2.244) into (2.229) yields

$$\mathbf{F}_r = \frac{m\mu}{1+\mu}\left(\boldsymbol{\Omega} \times \dot{\mathbf{r}}_G + g\sin(\epsilon)\mathbf{i}\right) - mg\cos(\epsilon)\mathbf{k}. \qquad (2.254)$$

This friction force is clearly dynamically equivalent to the no-slipping constraint (2.241) and will therefore produce identical motions.

To obtain a first insight into tyre force-generating mechanisms, we define the slip vector

$$s = \frac{(\dot{\mathbf{r}}_G + a\boldsymbol{\omega} \times \mathbf{k} - \boldsymbol{\Omega} \times \mathbf{r}_G)}{|\dot{\mathbf{r}}_G - \boldsymbol{\Omega} \times \mathbf{r}_G|}, \tag{2.255}$$

which is defined whenever $|\dot{\mathbf{r}}_G - \boldsymbol{\Omega} \times \mathbf{r}_G| \neq 0$. The numerator in this expression represents the departure, or slippage, from perfect rolling, while the denominator represents the speed of the ball's mass centre relative to the moving surface below it. We now suppose that whenever $s \neq 0$ a reaction force is developed that acts to negate the slip. One possibility is simply to set

$$\mathbf{F}_r = -C_s s - mg\cos(\epsilon)\mathbf{k} \tag{2.256}$$

in which C_s is a friction-related stiffness constant (longitudinal slip stiffness; see Chapter 3). Substituting (2.256) into (2.229) we see that

$$s = -(m\ddot{\mathbf{r}}_G - mg\hat{\mathbf{k}})/C_s,$$

which shows that in the limit as $C_s \to \infty$ (2.241) is enforced.

Example 2.24 The simulation shown in Figure 2.21(a) shows that a friction force of the form (2.256), with $C_s = 200$, produces a motion that is almost identical to the non-slipping case given in Figure 2.19. If the friction coefficient is reduced to $C_s = 10$, the trajectory begins to deviate from that in Figure 2.19 as is shown in Figure 2.21(b). If the turntable is inclined at 0.005 rad, with $C_s = 10$, one obtains Figure 2.22, which should be compared with Figure 2.20 in order to appreciate the influence of contact slippage. The parameters and initial conditions are the same as those in the previous examples, with $\boldsymbol{\omega}(0) = \mathbf{0}$ and $m = 0.05\,\text{kg}$.

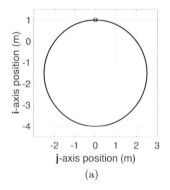

(a)

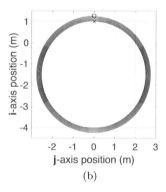

(b)

Figure 2.21: Trajectory of the mass centre of a hollow slipping ball rolling on a level turntable. The horizontal turntable is rotating at $1\,\text{rad/s}$ and the slipping stiffness is $C_s = 200$ (a) and $C_s = 10$ (b). The simulation is run for $25\pi\,\text{s}$, which represents 5 orbits of period $5\pi\,\text{s}$.

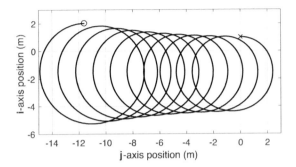

Figure 2.22: Trajectory of the mass centre of a hollow slipping ball rolling on an inclined turntable with $C_s = 10$. The turntable is rotating at $1\,\mathrm{rad/s}$ and is inclined at $0.005\,\mathrm{rad}$. The simulation is run for $50\pi\,\mathrm{s}$.

2.10   Rolling disc

The rolling disc is a prototypical nonholonomic system comprising a thin, uniform disc that rolls without slipping on a horizontal surface. This famous problem has been studied by many authors including for example Goldstein [41], Neĭmark and Fufaev [48], O'Reilly [74], Pars [75], Routh [76], Synge and Griffith [40], and Bloch and his colleagues [77, 78]. This example is preparatory in the context of a study of road vehicles for several reasons:

1. It is relatively simple, but involves a rigid body with large rotations in three dimensions. This type of rigid-body behaviour occurs in both bicycles and motor-cycles;
2. It involves two nonholonomic constraints that are associated with pure non-slip rolling, which are typical in the early modelling of bicycle wheels;
3. In common with bicycles and motorcycles, it has speed-dependent stability prop-erties.

   The rolling-disc problem involves six degrees of freedom related to the position and orientation of the body. There are one holonomic and two nonholonomic constraints relating to the rolling contact that constrain the disc's motion. The holonomic con-straint ensures that the disc and plane remain in contact, while the nonholonomic constraints ensure pure rolling. As we will show, the disc's translational position, its yaw angle, and its rotational position are all cyclic variables that do not enter the equations of motion; see Section 2.3.2.

2.10.1   Introduction

The rotation of the disc will be represented by three Euler angles ψ, ϕ, and θ, which denote respectively the yaw, the roll, and the spin of the disc as it rolls across a smooth surface. We will take the reference configuration to be one in which the disc lies flat on a horizontal plane Π; see Figure 2.23. If the mass of the disc is m, with its radius R, then the inertia of the disc about its mass centre, and in its reference position, is:

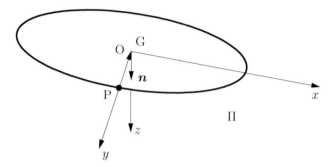

Figure 2.23: Disc in its reference position on a horizontal plane Π. The mass centre of the disc is at G, which in the reference position is coincident with the origin O of the inertial axis system $Oxyz$. The vector n is normal to the plane of the disc at G. A vector r_{PG}^{ref} points from the point P on the edge of the disc towards G.

$$
J_G^{ref} = \begin{bmatrix} \frac{mR^2}{4} & 0 & 0 \\ 0 & \frac{mR^2}{4} & 0 \\ 0 & 0 & \frac{mR^2}{2} \end{bmatrix}. \tag{2.257}
$$

Next, we rotate the disc through an angle ψ about e_z, which is a unit vector in the positive z-axis direction. This is followed by a rotation about the now-rotated e_x vector by an angle ϕ, which points in the positive x-axis direction. Lastly, we rotate the disc through an angle θ about the now-twice-rotated e_z vector. To see how these calculations work out in detail we refer to the *Rodrigues formula* (2.153). Defining $R_1 = R(e_z, \psi)$, $\bar{R}_2 = R(R_1 e_x, \phi)$, and $\bar{R}_3 = R(\bar{R}_2 R_1 e_z, \theta)$, the final rotation matrix is given by $R_I = \bar{R}_3 \bar{R}_2 R_1$, where

$$
R_I = \begin{bmatrix} c_\theta c_\psi - s_\theta c_\phi s_\psi & -c_\theta s_\psi c_\phi - c_\psi s_\theta & s_\phi s_\psi \\ s_\theta c_\phi c_\psi + c_\theta s_\psi & c_\psi c_\theta c_\phi - s_\psi s_\theta & -s_\phi c_\psi \\ s_\theta s_\phi & c_\theta s_\phi & c_\phi \end{bmatrix}. \tag{2.258}
$$

Observe that this is the same as $R_I = R_1 R_2 R_3$, where $R_2 = R(e_x, \phi)$ and $R_3 = R(e_z, \theta)$, and that each of the Euler angles is time-varying.

In order to find the disc's (mass-centred) moment of inertia in an arbitrary position, expressed in the inertial coordinate system, we use:

$$
J_G = R_I J_G^{ref} R_I^T = \frac{mR^2}{4} \begin{bmatrix} s_\psi^2 s_\phi^2 + 1 & -c_\psi s_\psi s_\phi^2 & s_\phi s_\psi c_\phi \\ -c_\psi s_\psi s_\phi^2 & 1 + c_\psi^2 s_\phi^2 & -s_\phi c_\psi c_\phi \\ s_\phi s_\psi c_\phi & -s_\phi c_\psi c_\phi & 1 + c_\phi^2 \end{bmatrix}. \tag{2.259}
$$

The disc is seen in its general position in Figure 2.24.

2.10.2   Rolling constraints

If we were to consider the motion of the disc in free space, neglecting temporarily any interactions with the plane Π in Figure 2.23, it would have six degrees of freedom.

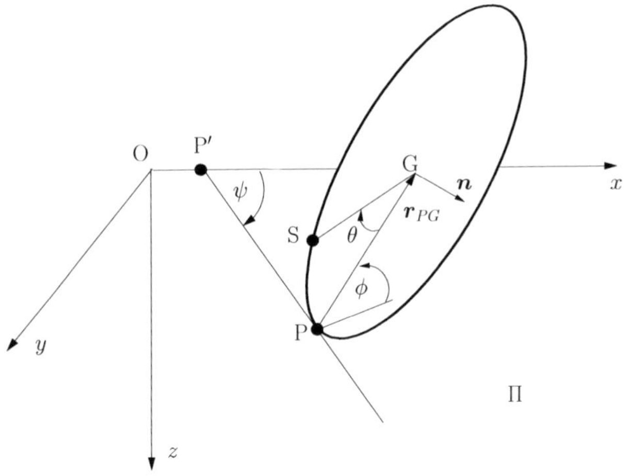

Figure 2.24: Disc rolling on a horizontal plane Π. The yaw angle is denoted by ψ, the roll angle by ϕ, and the spin angle by θ. The mass centre of the disc is at G, while the disc contacts the plane Π at P; the material point S is coincident with the (unspun) ground contact point P once every revolution of the disc. The vector n is normal to the plane of the disc at G. The axis system $Oxyz$ is inertial.

These freedoms would be the three translations $x_G(t)$, $y_G(t)$, and $z_G(t)$ of the disc's mass centre, and the three Euler angles $\psi(t)$, $\phi(t)$, and $\theta(t)$. When non-slip rolling (on the plane Π) is considered, these freedoms are constrained. In order to find the constraint equations we begin by considering the vector r_{PG}^{ref}, which points from the point P on the periphery of the disc to the mass centre G in Figure 2.23. If this vector is allowed to yaw and then roll we obtain:

$$
r_{PG} = \bar{\mathcal{R}}_2 \mathcal{R}_1 \begin{bmatrix} 0 \\ -R \\ 0 \end{bmatrix} = -R \begin{bmatrix} -s_\psi c_\phi \\ c_\psi c_\phi \\ s_\phi \end{bmatrix}, \tag{2.260}
$$

which is a vector pointing from the centre of the disc to the point of contact with the plane Π. Note that this vector is *not* fixed to the disc (body-fixed), but is 'unspun'; it has not been subjected to the θ rotation $\bar{\mathcal{R}}_3$. The vector r_{PG} can be seen in Figure 2.24.

This vector can also be found by first considering the vector

$$
\mathbf{r} = \mathbf{n} \times \mathbf{e}_z,
$$

which lies along the line of intersection $P'P$ between the plane of the disc and the ground plane Π. The vector r_{PG} can then be found using the alternative expression

$$
r_{PG} = -R \frac{\mathbf{r}}{\|\mathbf{r}\|} \times \mathbf{n}, \tag{2.261}
$$

which is the more common approach to this calculation.

The velocity of the disc's mass centre can be found using:

$$\mathbf{v}_G = \mathbf{v}_P + \boldsymbol{\omega} \times \mathbf{r}_{PG}. \tag{2.262}$$

The disc's angular velocity $\boldsymbol{\omega}$ is

$$\boldsymbol{\omega} = \begin{bmatrix} \dot{\theta}s_\phi s_\psi + \dot{\phi}c_\psi \\ \dot{\phi}s_\psi - \dot{\theta}s_\phi c_\psi \\ \dot{\psi} + \dot{\theta}c_\phi \end{bmatrix}, \tag{2.263}$$

which can be computed using (2.166).

Since the point P, when regarded as a material point on the periphery of the disc in instantaneous contact with the ground, must be stationary, we substitute $\mathbf{v}_P = 0$ into (2.262) to obtain:

$$\mathbf{v}_G = \boldsymbol{\omega} \times \mathbf{r}_{PG}$$
$$= R \begin{bmatrix} \dot{\psi}c_\psi c_\phi - \dot{\phi}s_\psi s_\phi + \dot{\theta}c_\psi \\ \dot{\psi}s_\psi c_\phi + \dot{\phi}c_\psi s_\phi + \dot{\theta}s_\psi \\ -\dot{\phi}c_\phi \end{bmatrix}. \tag{2.264}$$

The third entry in $\mathbf{v}_G$ can be integrated to give the holonomic constraint

$$z_G(t) = -Rs_\phi. \tag{2.265}$$

The remaining two entries give the nonholonomic constraints:

$$\dot{x}_G(t) = R(\dot{\psi}c_\psi c_\phi - \dot{\phi}s_\psi s_\phi + \dot{\theta}c_\psi) \tag{2.266}$$
$$\dot{y}_G(t) = R(\dot{\psi}s_\psi c_\phi + \dot{\phi}c_\psi s_\phi + \dot{\theta}s_\psi). \tag{2.267}$$

2.10.3 Angular momentum balance

The disc's six degrees of freedom are the three translations $x_G(t)$, $y_G(t)$, and $z_G(t)$ of the disc's mass centre, and the three Euler angles $\psi(t)$, $\phi(t)$, and $\theta(t)$. The force acting at the disc's contact point P is $\mathbf{F}_P$; the direction of this force is unknown (at this stage). A linear momentum balance gives:

$$\mathbf{F}_P = m\mathbf{a}_G - mg\mathbf{e}_z; \tag{2.268}$$

note that the positive $\mathbf{e}_z$ direction points 'downwards'. An angular momentum balance about the mass centre G gives:

$$0 = \mathbf{r}_{PG} \times \mathbf{F}_P + J_G\dot{\boldsymbol{\omega}} + \boldsymbol{\omega} \times J_G\boldsymbol{\omega}$$
$$= m\mathbf{r}_{PG} \times (\mathbf{a}_G - g\mathbf{e}_z) + J_G\dot{\boldsymbol{\omega}} + \boldsymbol{\omega} \times J_G\boldsymbol{\omega}$$
$$= m\mathbf{r}_{PG} \times (\dot{\boldsymbol{\omega}} \times \mathbf{r}_{PG} + \boldsymbol{\omega} \times \dot{\mathbf{r}}_{PG} - g\mathbf{e}_z) + J_G\dot{\boldsymbol{\omega}} + \boldsymbol{\omega} \times J_G\boldsymbol{\omega}, \tag{2.269}$$

where (2.262) was used to eliminate the disc's three translational freedoms.[17]

---

[17] In (2.269) we used the relationship

Direct calculation using (2.269) gives the disc's equations of motion as follows:

$$0 = \frac{4}{mR^2} \begin{bmatrix} -s_\psi c_\phi & c_\psi c_\phi & s_\phi \\ c_\psi & s_\psi & 0 \\ \frac{1}{2}s_\phi s_\psi & -\frac{1}{2}s_\phi c_\psi & \frac{1}{2}c_\phi \end{bmatrix} \begin{bmatrix} s_\phi\ddot{\psi} - 2\dot{\theta}\dot{\phi} \\ \frac{4c_\phi g}{R} + 6s_\phi\dot{\theta}\dot{\psi} + 5\ddot{\phi} + 5s_\phi c_\phi\dot{\psi}^2 \\ 3c_\phi\ddot{\psi} + 3\ddot{\theta} - 5s_\phi\dot{\phi}\dot{\psi} \end{bmatrix}. \tag{2.271}$$

Since

$$\det\left(\begin{bmatrix} -s_\psi c_\phi & c_\psi c_\phi & s_\phi \\ c_\psi & s_\psi & 0 \\ \frac{1}{2}s_\phi s_\psi & -\frac{1}{2}s_\phi c_\psi & \frac{1}{2}c_\phi \end{bmatrix}\right) = -\frac{1}{2} \tag{2.272}$$

the disc's motion is fully determined by the Euler angles, which must satisfy:

$$0 = s_\phi\ddot{\psi} - 2\dot{\theta}\dot{\phi} \tag{2.273}$$

$$0 = \frac{4c_\phi g}{R} + 6s_\phi\dot{\theta}\dot{\psi} + 5\ddot{\phi} + 5s_\phi c_\phi\dot{\psi}^2 \tag{2.274}$$

$$0 = 3c_\phi\ddot{\psi} + 3\ddot{\theta} - 5s_\phi\dot{\phi}\dot{\psi}. \tag{2.275}$$

$$M = \frac{d\mathbf{H}}{dt} = \frac{\partial\mathbf{H}}{\partial t} + \boldsymbol{\omega} \times \mathbf{H} \tag{2.270}$$

in which M is the external moment and $\mathbf{H}$ is the rigid body's angular momentum. The apparent rate of change of $\mathbf{H}$ in a body-fixed moving frame of reference is $\frac{\partial\mathbf{H}}{\partial t} = J_G\dot{\boldsymbol{\omega}}$, since J_G is constant in this frame. The second term $\boldsymbol{\omega} \times \mathbf{H} = \boldsymbol{\omega} \times J_G\boldsymbol{\omega}$ comes from (2.147). A physical interpretation of the $\boldsymbol{\omega} \times J_G\boldsymbol{\omega}$ term may be of interest:

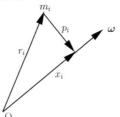

In the figure we consider a rigid body with instantaneous angular velocity vector $\boldsymbol{\omega}$. Let O be the body's mass centre and m_i the mass of a constituent mass element that is located at r_i relative to the mass centre. Let p_i represent a vector from m_i to $\boldsymbol{\omega}$ and perpendicular to it. Let x_i be the projection of r_i on to $\boldsymbol{\omega}$. Then:

$$p_i = x_i - r_i = (r_i \cdot \boldsymbol{\omega})\frac{\boldsymbol{\omega}}{\omega^2} - r_i \qquad \text{where} \qquad \omega = |\boldsymbol{\omega}|$$

and so

$$\omega^2 r_i \times p_i = (r_i.\boldsymbol{\omega})(r_i \times \boldsymbol{\omega}).$$

Now the angular momentum of the body is:

$$\mathbf{H} = \sum m_i r_i \times (\boldsymbol{\omega} \times r_i) = \sum m_i(\boldsymbol{\omega}(r_i \cdot r_i) - r_i(\boldsymbol{\omega} \cdot r_i))$$

so that

$$\boldsymbol{\omega} \times \mathbf{H} = \sum m_i(\boldsymbol{\omega} \cdot r_i)(r_i \times \boldsymbol{\omega}) = \sum m_i\omega^2(r_i \times p_i).$$

Hence the term $\boldsymbol{\omega} \times \mathbf{H}$ is the moment about the body's mass centre of the centripetal forces acting on the constituent particles making up the body, as it spins around the instantaneous angular velocity vector $\boldsymbol{\omega}$.

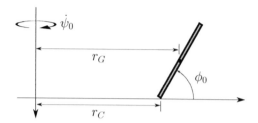

Figure 2.25: Rolling disc in steady motion.

2.10.4   Equilibrium solutions

The equations describing the disc's dynamics have a number of dynamic equilibria that we will consider briefly. To begin, we find the velocity of the disc's mass centre at a constant roll angle, and constant spin and yaw rates. It follows from (2.266) and (2.267) that for constant yaw and spin rates $\dot{\psi}_0$ and $\dot{\theta}_0$, respectively, and a constant roll angle ϕ_0, the velocity is given by

$$\boldsymbol{v}_G^0 = R(\dot{\psi}_0 c_{\phi 0} + \dot{\theta}_0) \begin{bmatrix} c_{\psi 0} \\ s_{\psi 0} \\ 0 \end{bmatrix}, \tag{2.276}$$

which has constant magnitude $|\boldsymbol{v}_G^0| = v_G^0 = R(\dot{\psi}_0 c_{\phi 0} + \dot{\theta}_0)$, and lies in the horizontal plane. It is clear from (2.276) and Figure 2.25, that under steady-state conditions, the disc's mass centre undergoes circular motion around a stationary vertical axis. The radius of rotation of the contact point is r_C, while that of the mass centre is r_G. Since $v_G^0 = \dot{\psi}_0 r_G$, there holds

$$r_G \dot{\psi}_0 = R(\dot{\psi}_0 c_{\phi 0} + \dot{\theta}_0), \tag{2.277}$$

with

$$r_C = r_G - R c_{\phi 0}. \tag{2.278}$$

Thus, for steady motion the 'spin' angular velocity $\dot{\theta}_0$ is related to the 'precession' angular velocity $\dot{\psi}_0$ by

$$\dot{\theta}_0 = \frac{r_C}{R} \dot{\psi}_0. \tag{2.279}$$

Eliminating $\dot{\theta}_0$ from (2.274) gives:

$$\dot{\psi}_0^2 = \frac{-4g c_{\phi 0}}{s_{\phi 0}(6 r_C + 5 R c_{\phi 0})} \tag{2.280}$$

under steady-state conditions. In the case that $0 < \phi_0 < \pi/2$, the denominator in (2.280) is positive, since r_C is non-negative (by definition), while the numerator is negative. This leads to a non-physical imaginary precession angular velocity. We conclude that steady motion is not possible for this range of roll angles, which accords with the intuitive physical notion that the disc must 'lean' into circular motion.

In the case that $\pi/2 < \phi_0 < \pi$, the numerator in (2.280) is positive with the denominator also positive provided $r_C > -5Rc_{\phi_0}/6$, or what is the same

$$r_G > r_G^{min} = \frac{Rc_{\phi_0}}{6}. \tag{2.281}$$

In addition to the commonly observed $r_G > 0$ case, steady motion is still possible with small negative values of r_G. Once $\dot{\psi}_0$ has been found using (2.280), $\dot{\theta}_0$ follows from (2.279).

In the case that $r_G = 0$, $r_C = -Rc_{\phi_0}$, and $\dot{\theta}_0 = -\dot{\psi}_0 c_{\phi_0}$, there hold

$$\dot{\psi}_0 = \sqrt{\frac{4g}{Rs_{\phi_0}}} \tag{2.282}$$

and

$$\dot{\theta}_0 = c_{\phi_0} \sqrt{\frac{4g}{Rs_{\phi_0}}}. \tag{2.283}$$

Equations (2.282) and (2.283) correspond to the whirling motion associated with a spinning disc that has its centre of mass stationary. This is the motion associated with the toy known as Euler's disc [79].

In the case that $\phi_0 = \pi/2$, a solution is possible if $\dot{\psi}_0 = 0$ and $\dot{\theta}_0 > 0$ is arbitrary. This corresponds to the disc rolling in a straight line; if this motion is to be stable $\dot{\theta}_0$ must be 'large enough'. Alternatively, one could have $r_G = r_C = 0$, $\dot{\theta}_0 = 0$ and $\dot{\psi}_0$ arbitrary. This corresponds to the disc spinning on its side in a fixed vertical position (with arbitrary yaw velocity).

The reaction force given in (2.268) is easily evaluated under steady-state conditions by differentiating (2.276) with respect to time. This gives

$$\dot{v}_G^0 = R\dot{\psi}_0(\dot{\psi}_0 c_{\phi_0} + \dot{\theta}_0) \begin{bmatrix} -s_{\psi_0} \\ c_{\psi_0} \\ 0 \end{bmatrix}, \tag{2.284}$$

which has magnitude

$$|\dot{v}_G^0| = |a_G^0| = a_G^0 = \dot{\psi}_0 v_G^0,$$

and points towards the centre of rotation. This means that the ground-contact force is

$$F_P = ma_G^0 = m(v_G^0)^2/r_G,$$

which points towards the centre of rotation too.

Example 2.25 Consider a disc of radius $R = 0.025\,\text{m}$ under the rolling conditions given in Table 2.1. In the first case (2.281) has been violated and so steady motion is not possible. The second case corresponds to the common case of a disc rolling in a circle. In the third case the disc's mass centre is on the vertical axis of rotation as shown in Figure 2.25 and a 'whirling' motion results. When $r_G = 0$, (2.280) simplifies to (2.282) and it is clear that as $\phi_0 \to \pi$, the precession angular velocity $\dot{\psi}_0$ increases without bound. The fourth case illustrates this phenomenon.

Table 2.1 Rolling conditions

Case	ϕ_0	r_C	r_G	r_G^{min}	$\dot{\psi}_0$	$\dot{\theta}_0$
1	$3\pi/4$	0	-0.0177	-0.0029	—	—
2	$3\pi/4$	0.5	0.482	-0.0029	3.671	73.422
3	$3\pi/4$	0.0177	0.0	-0.0029	47.11	33.31
4	$179\pi/180$	0.025	0.0	-0.0042	299.89	299.85

2.10.5 Lagrange's equation

As is well known, it is also possible to derive the equations of motion using energy-based procedures. The disc's kinetic energy is:

$$T = \frac{m}{2}(\dot{x}_G^2(t) + \dot{y}_G^2(t) + \dot{z}_G^2(t)) + \frac{1}{2}\boldsymbol{\omega}_I^T I_{cm}\boldsymbol{\omega}_I, \tag{2.285}$$

while its potential energy is

$$V = -mgz_G(t). \tag{2.286}$$

Combining these, we obtain the Lagranian

$$\mathcal{L} = T - V. \tag{2.287}$$

We can immediately reduce the number of degrees of freedom (generalized coordinates) from six to five by eliminating $z_G(t)$ using the holonomic constraint (2.265); it is also possible to incorporate this holonomic constraint in its differential form. In order to incorporate the holonomic constraint (2.265), and the nonholonomic conditions (2.266) and (2.267) into the problem formulation, we will append these constraints to the Lagrangian using Lagrange multipliers. The equations of motion then follow from

$$\frac{d}{dt}\left(\frac{\partial \mathcal{L}}{\partial \dot{q}_j}\right) - \left(\frac{\partial \mathcal{L}}{\partial q_j}\right) = -\sum_{k=1}^{3}\lambda_k a_{kj} \qquad \text{for} \quad j = 1, 2, \cdots 6 \tag{2.288}$$

in which (2.266) and (2.267) (and optionally) (2.265) have been written as

$$0 = \sum_{j=1}^{6} a_{ij}\dot{q}_j + a_i \qquad \text{for} \quad i = 1, \cdots, 3. \tag{2.289}$$

The (nine) equations in (2.288) and (2.289) give nine unknown quantities; six second derivatives and three Lagrange multipliers. The Lagrange multipliers specify the force of constraint $\mathbf{F}_P$. By direct calculation one obtains:

$$\lambda_1 = m\frac{d^2}{dt^2}x_G(t) \tag{2.290}$$

$$\lambda_2 = m\frac{d^2}{dt^2}y_G(t) \tag{2.291}$$

$$\lambda_3 = m\frac{d^2}{dt^2}z_G(t) - mg, \tag{2.292}$$

which we recognize as the three components of the contact-point reaction force. Eliminating the Lagrange multipliers and the translational freedoms again gives:

$$0 = s_\phi \ddot{\psi} - 2\dot{\theta}\dot{\phi}$$

$$0 = \frac{4c_\phi g}{R} + 6s_\phi \dot{\theta}\dot{\psi} + 5\ddot{\phi} + 5s_\phi c_\phi \dot{\psi}^2$$

$$0 = 3c_\phi \ddot{\psi} + 3\ddot{\theta} - 5s_\phi \dot{\phi}\dot{\psi}.$$

2.10.6 Rolling stability

The stability of the rolling disc has been investigated by a number of authors including for example Routh [76], Synge and Griffith [40], and Neĭmark and Fufaev [48]. If one rolls a compact disc over table top it will roll stably provided the initial speed is 'high enough'. In this case the disc will even jump over small objects and then continue on its way; try it with a compact disc! Once the speed starts to reduce, the disc will begin to wobble and then it will tilt to one side and fall over—the roll dynamics of the disc become unstable. As we will show later, this type of behaviour is found in bicycle dynamics too.

Suppose the disc is rolling along a circular trajectory at constant speed on a horizontal plane Π. Under these conditions we have

$$\dot{\psi} = \dot{\psi}_0, \quad \phi = \phi_0, \quad \text{and} \quad \dot{\theta} = \dot{\theta}_0$$

in which $(\cdot)_0$ denotes a constant steady value. Expressions for the equilibrium values of the yaw and roll rates are given in (2.280) and (2.279) respectively.

Equations (2.273), (2.274), and (2.275) can be linearized for small perturbations from the previously specified steady-state cornering trim condition to obtain

$$0 = s_{\phi_0}\ddot{\psi} - 2\dot{\theta}_0\dot{\phi} \tag{2.293}$$

$$0 = 5\ddot{\phi} - \frac{4g}{R}s_{\phi_0}\phi + 6s_{\phi_0}\dot{\theta}_0\dot{\psi} + 6s_{\phi_0}\dot{\theta}\dot{\psi}_0 + 6c_{\phi_0}\dot{\theta}_0\dot{\psi}_0\phi$$

$$+5\dot{\psi}_0^2\phi(c_{\phi_0}^2 - s_{\phi_0}^2) + 10c_{\phi_0}s_{\phi_0}\dot{\psi}_0\dot{\psi} \tag{2.294}$$

$$0 = 3c_{\phi_0}\ddot{\psi} + 3\ddot{\theta} - 5s_{\phi_0}\dot{\psi}_0\dot{\phi} \tag{2.295}$$

in which ψ, ϕ, and θ are small perturbations. Integrating (2.293) and (2.295) gives respectively

$$\dot{\psi} = (C_1 + 2\dot{\theta}_0\phi)/s_{\phi_0} \tag{2.296}$$

$$\dot{\theta} = \frac{5}{3}s_{\phi_0}\dot{\psi}_0\phi - c_{\phi_0}(C_1 + 2\dot{\theta}_0\phi)/s_{\phi_0} + C_2, \tag{2.297}$$

in which C_1 and C_2 are constants of integration. Substituting (2.296) and (2.297) into (2.294) gives the following second-order linear differential equation for the roll angle:

$$\ddot{\phi} + \Xi\phi = \Delta \tag{2.298}$$

in which

$$\Xi = \frac{12}{5}\dot{\theta}_0^2 + \frac{14}{5}c_{\phi_0}\dot{\theta}_0\dot{\psi}_0 + \dot{\psi}_0^2 - \frac{4g}{5R}s_{\phi_0} \tag{2.299}$$

with Δ a constant that is irrelevant for present purposes. In the case that $\Xi < 0$ divergent roll behaviour will result. If $\Xi > 0$, bounded, but undamped roll oscillations will occur. The case $\Xi = 0$ requires a more sophisticaed (nonlinear) analysis as was pointed out in [48]. If one forms the equation

$$M(s) \begin{bmatrix} \phi(s) \\ \psi(s) \\ \theta(s) \end{bmatrix} = 0,$$

by taking Laplace transforms of (2.293) to (2.295), then the corresponding sixth-order characteristic polynomial is given by $\det(M(s))$. In the case that $\Xi > 0$ the roots of $\det(M(s))$ are of the form $\{i\lambda, -i\lambda, 0, 0, 0, 0\}$ in which $\lambda > 0$ for some real λ. If $\Xi < 0$ the roots of $\det(M(s))$ are of the form $\{\lambda, -\lambda, 0, 0, 0, 0\}$ in which $\lambda > 0$ is real. If $\Xi = 0$ the roots of $\det(M(s))$ are all at the origin. In the special case of $\dot{\psi}_0 = 0$ and $\phi_0 = \pi/2$, we obtain

$$\Xi = \frac{12}{5}\dot{\theta}_0^2 - \frac{4g}{5R}. \tag{2.300}$$

The condition $\Xi > 0$, or what is the same, $\dot{\theta}_0 > \sqrt{g/(3R)}$, is the condition for straight-running stability given in Routh [76].

3

Tyres

3.1 Background

Road–tyre interactions have been modelled classically using two nonholonomic constraints (such as those discussed in Sections 2.9 and 2.10), which is a reasonable approximation at very low speeds. At the end of the nineteenth century this model came into question when it was recognized (using tests with rubber models) that there were distinct adhesion and sliding regions in the tyre's contact patch. This discovery led to the concept of (longitudinal) 'tyre slip' [80]. The same ideas were later applied to lateral slip, which was recognized as being responsible for the directional stability of guided wheels and 'shimmy' [81].[1] It was again shown by measurement that the lateral force and the lateral slip were proportional to each other for small slips, and the concept of cornering stiffness (originally named the 'side-slip' factor) was introduced. The first investigation into shimmy was even earlier [82], when a two-degree-of-freedom (tramping/roll and shimmy/yaw) model of the front axle of an automobile was studied. The tyres do not play a central roll in this phenomenon, because the two motions are coupled through a gyroscopic moment, hence the term gyroscopic shimmy. More detailed studies of the mechanics of tyres started in the 1940s when the seminal work of von Schlippe and Fromm introduced the basics of the brush and string models, which are related to the concept 'tyre relaxation' (or tyre lag) [83]. The role of tyres (and lateral suspension compliance) in shimmy was of great concern, particularly in the context of aircraft landing gear. Many more studies on the challenging topic of tyre dynamics have been published since. A comprehensive treatment of tyres can be found in [8].

In the twenty-first century shimmy phenomena are still reported in motorcycles [84] and bicycles [85, 86], while much attention is devoted to the design of shimmy-free landing gear [87, 88].

This chapter aims to explain the mechanisms and modelling issues related to the generation of tyre forces. Physical models such as the brush and string models are used to clarify the basic concepts. Building upon these findings, the empirical models widespread in vehicle dynamics analyses are discussed. Finally, an overview of some of the advanced models currently used in industry is given.

---

[1] Shimmy is an oscillation of a wheel around its swivel/steering axis. It is usually called wobble in the context of motorcycle and bicycle dynamics.

Dynamics and Optimal Control of Road Vehicles. D. J. N. Limebeer and M. Massaro.
© D. J. N. Limebeer and M. Massaro 2018. Published in 2018 by Oxford University Press.
DOI: 10.1093/oso/9780198825715.001.0001

3.2   Tyre forces and slips

The tyre makes contact with the road in a contact patch. The interaction between the tyre and the road is described by a force acting on the centre C of the contact patch and a moment; see Figure 3.1. This force and moment are usually described in terms of their (three) components. The tyre force components are the longitudinal force F_x, the cornering (side) force F_y, and the normal force (load) F_z. The three moment components are the overturning couple M_x, the rolling resistance M_y, and the yawing moment M_z.[2] These forces and moments depend on the tyre's radial deflection δr, its angular velocity Ω, the longitudinal slip κ, the lateral slip α, and the camber angle γ, and the turn slip φ_t (that combine to give the spin slip φ_s).

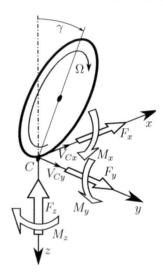

Figure 3.1: Tyre forces and moments.

Since tyres are flexible bodies they have loaded and unloaded radii given by r and r_0 respectively, with the difference given by:

$$\delta r = r_0 - r. \tag{3.1}$$

For a straight-running wheel in free rolling the *effective rolling radius* r_e is defined as:

$$r_e = \frac{V_x}{\Omega_0}, \tag{3.2}$$

where V_x is the velocity of the wheel centre in the x-direction; the wheel's free-rolling angular velocity is Ω_0. Under these circumstances $V_x = V_{Cx}$; see Figures 3.1 and 3.2.

---

[2]   It is convenient to define the whole yawing moment as $M_z = M_z' + M_z^r$, where M_z' is the component related to the lateral distortion of the tyre contact patch (the only component of a zero-width tyre), while M_z^r is the component related to the longitudinal distortion of a finite-width tyre.

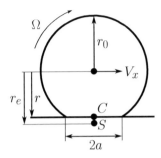

Figure 3.2: Unloaded radius r_0, loaded radius r, effective radius r_e, point C (centre of the contact patch which has length $2a$), and slip point S.

The wheel is said to be in *free rolling* if it is rolling without the application of a driving or braking torque. In free rolling a small longitudinal (braking) force F_x arises as a result of the rolling resistance moment M_y. The wheel is said to be in a state of *pure rolling* when a driving torque is applied to the wheel that overcomes the rolling resistance moment. As one will appreciate, the unloaded, effective, and loaded radii are related by $r_0 \geq r_e \geq r$; see Figure 3.2. Relationship (3.2) can be used to determine the effective rolling radius by experiment.

In order to define the notion of 'longitudinal slip', which is denoted κ, we need to introduce the slip point S. This point is attached to the wheel body at distance r_e from the wheel spindle and forms the centre of rotation when the wheel is in pure rolling; see Figure 3.2. If the wheel is driven, or under braking, the tyre tread begins to 'slip' relative to the road with slip velocity:

$$V_{Sx} = V_{Cx} - \Omega r_e, \tag{3.3}$$

where V_{Cx} is the longitudinal velocity of the ground contact point; see Figure 3.1. The longitudinal slip velocity V_{Sx} is positive under braking. The longitudinal slip (or slip ratio) κ is defined as:

$$\kappa = -\frac{V_{Sx}}{V_{Cx}}, \tag{3.4}$$

where κ is negative under braking and $\kappa = -1$ when the wheel is locked. Alternatively,

$$\kappa = \frac{\Omega r_e - V_{Cx}}{V_{Cx}} = \frac{V_r - V_{Cx}}{V_{Cx}} = \frac{V_r}{V_{Cx}} - 1, \tag{3.5}$$

where the rolling velocity V_r and the forward velocity V_{Cx} are related by

$$V_r = \Omega r_e = V_{Cx} - V_{Sx} = V_{Cx}(1 + \kappa). \tag{3.6}$$

The lateral slip $\tan \alpha$ is defined as the ratio of the lateral velocity of point S and the forward velocity of point C:

$$\tan \alpha = -\frac{V_{Sy}}{V_{Cx}}. \tag{3.7}$$

It is often assumed $V_{Sy} \approx V_{Cy}$. In this case the side-slip angle is the angle between the velocity of point C and the wheel's plane of symmetry. The negative sign is introduced so that positive lateral slips produce positive lateral forces.

The slip velocity components V_{Sx} an V_{Sy} can be combined into a slip vector $\boldsymbol{V}_S$ given by

$$\boldsymbol{V}_S = \begin{bmatrix} V_{Sx} \\ V_{Sy} \end{bmatrix}. \tag{3.8}$$

The wheel camber γ is defined as the angle between the wheel diameter through C and a vector normal to the road; see Figure 3.1. It is anticipated that for a car, camber-related lateral forces and moments are much smaller than those related to the lateral slip, since the wheel camber is seldom more than a few degrees. In the case of a motorcycle, however, when the camber can reach values of 50–60 deg, the bulk of the tyre's lateral force is camber related.

The turn slip φ_t is defined as the ratio of the yaw velocity $\dot{\psi}$ and the forward velocity of the contact centre C:

$$\varphi_t = -\frac{\dot{\psi}}{V_{Cx}}. \tag{3.9}$$

This reduces to the path curvature when the lateral slip is zero, since in this case $V_{Cx} = R\dot{\psi}$, where R is the turn radius of curvature. Again, the minus sign is introduced to produce a positive lateral force with a positive turn slip. Turn-slip-related lateral forces and moments are usually only significant for small cornering radii—in parking manoeuvres for example, or when steering from standstill. Turn slip and camber are components of the spin slip φ_s, which is the quantity that is actually related to the force- and moment-generation mechanisms.

The spin slip is defined as the negative ratio of the vertical component of the absolute angular velocity of the wheel ω_z to the forward velocity of the contact centre C:

$$\varphi_s = -\frac{\omega_z}{V_{Cx}}. \tag{3.10}$$

As usual, the minus produces a positive lateral force with a positive spin slip. The angular velocity is related to the yaw rate $\dot{\psi}$, the speed of revolution Ω, and the camber γ via:

$$\omega_z = \dot{\psi} - \Omega \sin\gamma. \tag{3.11}$$

A wheel that experiences zero side slip but both components of spin slip is illustrated in Figure 3.3. Equation (3.10) is thus the sum of two terms, with the first related to the turn slip and the second to the camber:

$$\varphi_s = -\frac{\dot{\psi}}{V_{Cx}} + \frac{\Omega \sin\gamma}{V_{Cx}} = \varphi_t + \frac{\Omega \sin\gamma}{V_{Cx}}. \tag{3.12}$$

In the case that $\alpha = 0$ and $\kappa = 0$ this becomes

$$\varphi_s = \varphi_t|_{\alpha=0} + \left.\frac{\Omega \sin\gamma}{V_{Cx}}\right|_{\kappa=0} = -\frac{1}{R} + \frac{1}{R_\gamma}, \tag{3.13}$$

where the equivalent camber curvature is given by

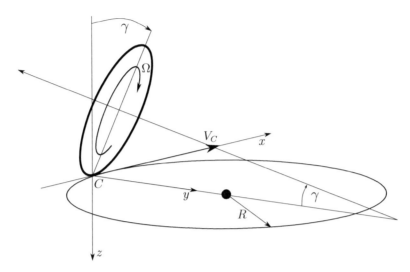

Figure 3.3: Turn-related slip associated with a cambered turning wheel. In this diagram zero lateral slip ($\alpha = 0$) is assumed.

$$\left.\frac{1}{R_\gamma}\right|_{\kappa=0} = \left.\frac{\Omega \sin \gamma}{V_{Cx}}\right|_{\kappa=0} = \frac{1}{r_e} \sin \gamma. \tag{3.14}$$

It follows from (3.13) that the spin slip is zero when $R = R_\gamma$. This occurs when the path curvature matches the curvature of the path taken by the cambered tyre's contact patch through the contact region.

In order to correct for the camber-related curvature arising from the flattening of the tyre contact patch, we introduce the load-induced deflection factor ϵ_γ as follows:

$$\varphi_s = -\frac{\dot\psi}{V_{Cx}} + (1 - \epsilon_\gamma)\frac{\Omega \sin \gamma}{V_{Cx}}. \tag{3.15}$$

3.3 Sliding velocities

As the tyre tread rolls into the contact area it will adhere to the road, distort, and generate friction-related ground-contact forces in the plane of the road. When the tread deflection exceeds a limit value it will begin to slide. The contact area is therefore made up typically of adhesion and sliding regions.

In this section we derive expressions for the sliding velocity of an arbitrary tread-base material point in the contact area. These sliding velocities will be used to compute the tyre deflections resulting from slip. The resulting tyre forces and moments can then be found by integration over the contact patch.

The tyre tread behaviour will be analysed with the help of Figure 3.4. The axis system $OXYZ$ is fixed in the road and will be treated as inertial. The Z-axis (not shown) points into the road. The $Cxyz$ coordinate system moves with the wheel and has its origin, C, at the tyre's contact centre. The x–y plane is in the road plane with the z-axis again pointing downwards. The y-axis is the projection of the wheel spindle in the ground plane.

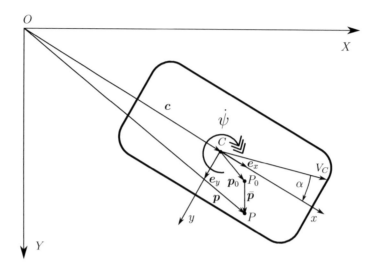

Figure 3.4: Plan view of the tyre contact patch. The vector $\boldsymbol{c}$ represents the tyre contact centre. The vector $\boldsymbol{p}_0$ represents the position of an arbitrary unstressed material point P_0 in the tread base. The friction-related deformation is given by the vector $\bar{\boldsymbol{p}}$; $\boldsymbol{p} = \boldsymbol{c} + \boldsymbol{p}_o + \bar{\boldsymbol{p}}$.

The sliding velocity $\boldsymbol{V}_C$ will be expressed in the $Cxyz$ coordinate frame and has a longitudinal component V_{Cx} and lateral component V_{Cy} in the $\boldsymbol{e}_x$ and $\boldsymbol{e}_y$ directions respectively; see Figure 3.4. The coordinates of the arbitrary stressed material point P (expressed on the $Cxyz$ frame) are $x = x_0 + \bar{x}$ and $y = y_0 + \bar{y}$, where $\boldsymbol{p}_0 = (x_0, y_0, 0)$ represents the position of an arbitrary unstressed material point, possibly including deformations related to the normal load. The additional deformation $\bar{\boldsymbol{p}} = (\bar{x}, \bar{y}, 0)$ is induced by tyre–road friction. The absolute Eulerian velocity $\boldsymbol{V}_P$ of the material point P, expressed in the moving reference frame, is:

$$\boldsymbol{V}_P = \boldsymbol{V}_C + \dot{\psi}\boldsymbol{e}_z \times (\boldsymbol{p}_0 + \bar{\boldsymbol{p}}) + (\dot{\boldsymbol{p}}_0 + \dot{\bar{\boldsymbol{p}}}) \tag{3.16}$$

or in coordinate form

$$\begin{aligned}
V_{Px} &= V_{Cx} - \dot{\psi}\,(y_0 + \bar{y}) + (\dot{x}_0 + \dot{\bar{x}}) \\
V_{Py} &= V_{Cy} + \dot{\psi}\,(x_0 + \bar{x}) + (\dot{y}_0 + \dot{\bar{y}}),
\end{aligned} \tag{3.17}$$

where V_{Px} is in the $\boldsymbol{e}_x$ direction and V_{Py} is in the $\boldsymbol{e}_y$ direction. The first term is related to the translation of the origin of the moving frame, the second term is related to the rotation of the moving frame, and the third term is related to the velocity of the material point within the moving frame.

We will now relate the velocity of P_0 (in the moving reference frame) to the angular velocity of a cambered wheel. Figure 3.5 shows a general point P_0 in the tyre contact area along with the velocity of that point in the moving wheel frame. The kinematic geometry indicates that

$$\begin{aligned}
\dot{x}_0 &= -\Omega(r_e - y_0 \sin\gamma) \\
\dot{y}_0 &= \quad -\Omega x_0 \sin\gamma,
\end{aligned} \tag{3.18}$$

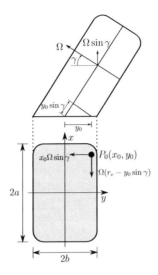

Figure 3.5: Velocity of a general point P_0 in the tyre contact patch.

where Ω is the wheel-spin velocity and r_e the effective rolling radius; $\Omega \sin \gamma$ is a yaw velocity term. In order to accommodate carcass distortion, (3.18) is modified to include a load-induced deflection factor $(1 - \epsilon_\gamma)$ in the camber term $\sin \gamma$.[3] In addition x_0, y_0 will be approximated by x, y, since the friction deflections $\bar{x}, \bar{y}$ are assumed 'small':

$$
\begin{aligned}
\dot{x}_0 &= -\Omega \left(r_e - y \left(1 - \epsilon_\gamma \right) \sin \gamma \right) \\
\dot{y}_0 &= -\Omega x \left(1 - \epsilon_\gamma \right) \sin \gamma.
\end{aligned}
\tag{3.19}
$$

When a cambered car tyre makes contact with a frictionless (icy) road, load-induced distortion results in an almost straight contact trajectory, while for a motorcycle tyre the contact trajectory is curved. When cambered tyres roll on real roads (i.e. with friction), the line of contact is almost straight in both cases. In the case of car tyres (almost) no camber-related friction forces arise, since the contact trajectory is approximately straight. With motorcycle tyres, however, camber-related friction forces straighten the contact trajectory.

We will now consider the influence of friction on an arbitrary material point on the tread surface as it rolls into and out of the contact region. The friction-related deformations are given by $\bar{x}$ and $\bar{y}$ in Figure 3.4, and are functions of x_0, y_0 and time. It therefore follows that

$$
\begin{aligned}
\dot{\bar{x}} &= \frac{d\bar{x}}{dt} = \frac{\partial \bar{x}}{\partial x_0} \dot{x}_0 + \frac{\partial \bar{x}}{\partial y_0} \dot{y}_0 + \frac{\partial \bar{x}}{\partial t} \\
\dot{\bar{y}} &= \frac{d\bar{y}}{dt} = \frac{\partial \bar{y}}{\partial x_0} \dot{x}_0 + \frac{\partial \bar{y}}{\partial y_0} \dot{y}_0 + \frac{\partial \bar{y}}{\partial t}.
\end{aligned}
\tag{3.20}
$$

---

[3] It might be expected that the effect of ϵ_γ on $\dot{y}_0$ has a significant influence on the tyre's ability to generate camber-related lateral forces. In the case of a motorcycle tyre $\epsilon_\gamma \approx 0$ (negligible correction), while for car tyres $\epsilon_\gamma \approx 0.5$–$0.7$ (large correction) resulting in small $\dot{y}_0$. These differences relate mainly to the different geometry and structure of motorcycle and car tyres. A large ϵ_γ value is related to the 'straightening' of the contact line as a result of the flattening of the loaded tyre.

The second terms in (3.20) are usually much smaller than the other terms and $\partial/\partial x_0$ can be replaced by $\partial/\partial x$ since $\bar{x}$ is small, thus the following approximation will be used from now on:

$$\dot{\bar{x}} \approx \frac{\partial \bar{x}}{\partial x}\dot{x}_0 + \frac{\partial \bar{x}}{\partial t}$$
$$\dot{\bar{y}} \approx \frac{\partial \bar{y}}{\partial x}\dot{x}_0 + \frac{\partial \bar{y}}{\partial t} \qquad (3.21)$$

Substituting (3.19) and (3.21) into (3.17), together with $V_{Cx} = V_{Sx} + V_r$, $V_r = r_e\Omega = V_{Cx}(1+\kappa)$, and the expressions for $\kappa, \alpha, \varphi_s$ given in (3.4), (3.7), (3.15), gives the following expressions for the sliding velocities of a material point in the contact patch that is located at (x, y):

$$V_{Px} = V_{Cx} - \dot{\psi}y + \dot{x}_0\left(1 + \frac{\partial \bar{x}}{\partial x}\right) + \frac{\partial \bar{x}}{\partial t}$$

$$= V_{Sx} + V_r - y\dot{\psi} + (\Omega y(1-\epsilon_\gamma)\sin\gamma - V_r)\left(1 + \frac{\partial \bar{x}}{\partial x}\right) + \frac{\partial \bar{x}}{\partial t}$$

$$= \frac{\partial \bar{x}}{\partial t} + V_{Sx} + V_{Cx}y\varphi_s - V_r\frac{\partial \bar{x}}{\partial x}$$

$$= \frac{\partial \bar{x}}{\partial t} - V_{Cx}\kappa + V_{Cx}y\varphi_s - V_{Cx}(1+\kappa)\frac{\partial \bar{x}}{\partial x}, \qquad (3.22)$$

$$V_{Py} = V_{Cy} + \dot{\psi}x + \dot{y}_0 + \frac{\partial \bar{y}}{\partial x}\dot{x}_0 + \frac{\partial \bar{y}}{\partial t}$$

$$= V_{Sy} + x\dot{\psi} - \Omega x(1-\epsilon_\gamma)\sin\gamma + \frac{\partial \bar{y}}{\partial x}(\Omega y(1-\epsilon_\gamma)\sin\gamma - V_r) + \frac{\partial \bar{y}}{\partial t}$$

$$= \frac{\partial \bar{y}}{\partial t} + V_{Sy} - V_{Cx}x\varphi_s - V_r\frac{\partial \bar{y}}{\partial x}$$

$$= \frac{\partial \bar{y}}{\partial t} - V_{Cx}\tan\alpha - V_{Cx}x\varphi_s - V_{Cx}(1+\kappa)\frac{\partial \bar{y}}{\partial x}, \qquad (3.23)$$

where the terms $\partial \bar{x}/\partial x\,\Omega y(1-\epsilon_\gamma)\sin\gamma$ and $\partial \bar{y}/\partial x\,\Omega y(1-\epsilon_\gamma)\sin\gamma$ have been neglected, because y, $\partial \bar{x}/\partial x$, $\partial \bar{y}/\partial x$ are assumed 'small'.

If we now recognize the distance travelled s as the independent variable (rather than time), one obtains

$$s = V_C t \approx V_{Cx}t \xrightarrow{V_{Cx}=const} ds = V_{Cx}dt. \qquad (3.24)$$

The final expressions for the sliding velocity of a point located at (x, y) in the contact patch, which will be used in the development of the string model in Section 3.7, are

$$V_{Px} = \left(\frac{\partial \bar{x}}{\partial s} - \kappa - \left(\frac{\partial \bar{x}}{\partial x}\right)(1+\kappa) + y\varphi_s\right)V_{Cx}$$
$$V_{Py} = \left(\frac{\partial \bar{y}}{\partial s} - \tan\alpha - \left(\frac{\partial \bar{y}}{\partial x}\right)(1+\kappa) - x\varphi_s\right)V_{Cx} \qquad (3.25)$$

The steady-state sliding velocities in the contact patch, which will be used in the brush model in Section 3.4, are obtained by assuming that $\partial \bar{x}/\partial s = \partial \bar{y}/\partial s = 0$:

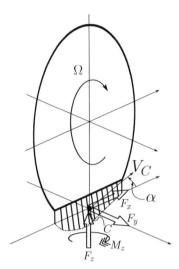

Figure 3.6: Brush model (under pure side slip). The leading and trailing bristles are undeflected; the dotted line represents the deflected contact patch.

$$V_{Px} = \left(-\kappa - \left(\tfrac{\partial \bar{x}}{\partial x}\right)(1+\kappa) + y\varphi_s\right) V_{Cx}$$
$$V_{Py} = \left(-\tan\alpha - \left(\tfrac{\partial \bar{y}}{\partial x}\right)(1+\kappa) - x\varphi_s\right) V_{Cx}. \tag{3.26}$$

Equations (3.26) show that the velocity field in the contact patch is fully determined by the slips and the position: $(\kappa, \alpha, \varphi_s, x, y) \mapsto (V_{Px}, V_{Py})$. Since the sliding velocities determine the tyre deflections, which result in the tyre forces and moments, κ, α, and φ_s serve as the slip quantities related to the force- and moment-generating mechanisms.

It is noted that the longitudinal slip κ is related to the longitudinal deflection $\bar{x}$, the lateral slip $\tan\alpha$ is related to the lateral deflection $\bar{y}$, while both the longitudinal and lateral deflections are related to the spin slip φ_s.

3.4 Steady-state behaviour: the brush model

In this section the 'brush model' is used to explain the essentials of steady-state tyre force generation. As shown in Figure 3.6, the brush model consists of a single row of elastic bristles distributed circumferentially around the periphery of a thin disc. As these bristles enter the contact region they are deflected, thereby producing tyre forces and moments; these deflections only occur within the contact patch. The contact patch is assumed to extend between $-a \leq x \leq a$; the rolling resistance is disregarded. When the tyre rolls freely, the tread elements ('bristles') remain vertical and move from the leading edge of the contact region to the trailing edge without deflection and without developing any forces. The basic brush model is based on the theory given in [89], with carcass compliance effects included thereafter [90,91]. The steady-state characteristics of the brush model are analysed in [92]. Asymmetric contact pressure distributions are considered in [93], while more general distributions are reported in [94,95]. In most cases a parabolic pressure distribution is assumed due to its mathematical simplicity.

3.4.1 Pure side slip

The steady-state deflection in the adhesion region in the case of pure side slip is obtained from (3.26) with $V_{Py} = 0$ (adhesion), and with $\kappa = \varphi_s = 0$ (pure lateral slip)

$$\frac{\partial \bar{y}}{\partial x} = -\tan \alpha \Rightarrow \bar{y} = -x \tan \alpha + c_{\bar{y}}, \tag{3.27}$$

where $c_{\bar{y}}$ is the constant of integration. The boundary condition $\bar{y}(a) = 0$ yields

$$\bar{y}(a) = -a \tan \alpha + c_{\bar{y}} = 0 \Rightarrow c_{\bar{y}} = a \tan \alpha. \tag{3.28}$$

The lateral deflection on the contact patch is therefore given by

$$\bar{y} = (a - x) \tan \alpha = (a - x) \sigma_y, \tag{3.29}$$

where

$$\sigma_y = \tan \alpha. \tag{3.30}$$

Equation (3.29) implies that a tyre material point in full adhesion moves in a straight line once it has entered the contact patch. Note that $\sigma_y = \tan \alpha / (1+\kappa)$ in the combined slip context—when longitudinal and lateral slipping are present simultaneously—see Section 3.4.4.

The lateral force per unit length of contact patch is related to the lateral stiffness per unit length K_{Py} and the lateral deflection of the tyre, and is given by

$$q_y = K_{Py}\bar{y}. \tag{3.31}$$

The following quadratic force distribution is assumed across the contact region:

$$q_z = \frac{3F_z}{4a}\left\{1 - \left(\frac{x}{a}\right)^2\right\}; \tag{3.32}$$

the normal load is thus given by $F_z = \int_{-a}^{+a} q_z dx$.

The maximum possible local side force is related to the lateral friction coefficient μ_y and local normal load q_z

$$q_y^{\max} = \mu_y q_z, \tag{3.33}$$

while the maximum lateral deflection is

$$\bar{y}_{\max} = \frac{q_y^{\max}}{K_{Py}}. \tag{3.34}$$

As the tread base material enters the contact patch, the strain in the tread material builds up until it reaches a transition point x_t, where adhesion between the rubber and road transitions into sliding. At the transition point x_t between adhesion and sliding the lateral force is at its maximum, which is given by

$$K_{Py}(a - x_t)|\sigma_y| = \mu_y \frac{3F_z}{4a^3}(a - x_t)(a + x_t). \tag{3.35}$$

Equation (3.35) gives

$$\frac{x_t}{2a} = \frac{2K_{Py}a^2}{3\mu_y F_z}|\sigma_y| - \frac{1}{2} = \theta_{Py}|\sigma_y| - \frac{1}{2}, \tag{3.36}$$

in which the nondimensional tyre parameter θ_{Py} is given by

$$\theta_{Py} = \frac{2K_{Py}a^2}{3\mu_y F_z}. \tag{3.37}$$

Since $x_t \leq a$, it is immediate from (3.36) that

$$|\sigma_y| \leq \frac{1}{\theta_{Py}}. \tag{3.38}$$

This implies that $|\sigma_y| = 1/\theta_{Py}$ corresponds to full sliding.

The tyre side force and aligning moment are now computed by integration across the contact region. The lateral force is split into its adhesion and sliding components, so that $F_y = F_{ya} + F_{ys}$. The adhesion component is given by

$$F_{ya} = \int_{x_t}^{a} K_{Py}\bar{y}dx = \int_{x_t}^{a} K_{Py}(a - x)\sigma_y dx$$

$$= \frac{K_{Py}\sigma_y}{2}(a - x_t)^2 = 3\mu_y F_z \theta_{Py}\sigma_y (1 - \theta_{Py}|\sigma_y|)^2, \tag{3.39}$$

while the sliding component is

$$F_{ys} = \int_{-a}^{x_t} \mu_y q_z dx = \int_{-a}^{x_t} \mu_y \frac{3F_z}{4a^3}(a^2 - x^2)dx = \frac{3\mu_y F_z}{4a^3}\left(a^2 x_t + \frac{2a^3}{3} - \frac{x_t^3}{3}\right)$$

$$= 3\mu_y F_z \theta_{Py}\sigma_y \left(\theta_{Py}|\sigma_y| - \frac{2}{3}(\theta_{Py}\sigma_y)^2\right). \tag{3.40}$$

The full lateral force is obtained by adding (3.39) to (3.40)

$$F_y = F_{ya} + F_{ys} = 3\mu_y F_z \theta_{Py}\sigma_y \left(1 - \theta_{Py}|\sigma_y| + \frac{1}{3}(\theta_{Py}\sigma_y)^2\right)$$

$$= \mu_y F_z \left(\left(\frac{a + x_t}{2a} - 1\right)^3 + 1\right) \text{sgn}(\sigma_y). \tag{3.41}$$

In the case of full sliding $(x_t = a)$ (3.41) simplifies to[4]

---

[4] Equation (3.41) and (3.45) can only be used for $|\sigma_y| \leq 1/\theta_{Py}$. In the case that $|\sigma_y| > 1/\theta_{Py}$, $F_y = \mu_y F_z \text{sgn}(\sigma_y)$ and $M_z = 0$.

$$F_y = \mu_y F_z \text{sgn}(\sigma_y).$$
(3.42)

The yawing moment is computed in terms of its adhesion and sliding components and so $M'_z = M'_{za} + M'_{zs}$, where the adhesion moment is given by

$$M'_{za} = \int_{x_t}^{a} K_{Py} \bar{y} x dx = \int_{x_t}^{a} K_{Py}(a - x)\sigma_y x dx$$
$$= -\mu_y F_z \theta_{Py} \sigma_y a \left(1 - 6\theta_{Py}|\sigma_y| + 9(\theta_{Py}\sigma_y)^2 - 4(\theta_{Py}|\sigma_y|)^3\right),$$
(3.43)

while the sliding component is given by

$$M'_{zs} = \int_{-a}^{x_t} \mu_y q_z x dx = \int_{-a}^{x_t} \mu_y \frac{3F_z}{4a^3}(a^2 - x^2)x dx$$
$$= -\mu_y F_z \theta_{Py} \sigma_y a \left(3\theta_{Py}|\sigma_y| - 6(\theta_{Py}\sigma_y)^2 + 3(\theta_{Py}|\sigma_y|)^3\right).$$
(3.44)

The expression for the full moment is obtained by adding (3.43) to (3.44) to obtain:

$$M'_z = -\mu_y F_z \theta_{Py} \sigma_y a \left(1 - 3\theta_{Py}|\sigma_y| + 3(\theta_{Py}\sigma_y)^2 - (\theta_{Py}|\sigma_y|)^3\right)$$
$$= -\mu_y F_z a \left(\frac{a + x_t}{2a}\right)\left(1 - \frac{a + x_t}{2a}\right)^3 \text{sgn}(\sigma_y).$$
(3.45)

The negative sign ensures that the moment opposes the direction of turning, that is, it is an *aligning* moment. In other words, the point at which the lateral force acts is behind the centre of the contact patch. Setting $\frac{\partial M'_z}{\partial x_t} = 0$ gives $x_t = -a/2$, or what is the same $\sigma_y = 1/4\theta_{Py}$ and $|M'_z| = 27/256 = 0.105$.

The pneumatic trail is the distance behind the contact centre at which the resultant side force F_y acts and is computed as

$$t = -\frac{M'_z}{F_y} = a\frac{1 - 3\theta_{Py}|\sigma_y| + 3(\theta_{Py}\sigma_y)^2 - (\theta_{Py}|\sigma_y|)^3}{3 - 3\theta_{Py}|\sigma_y| + (\theta_{Py}\sigma_y)^2}$$
$$= a\left(\frac{\left(1 - \left(\frac{x_t + a}{2a}\right)\right)^3}{3 - 3\left(\frac{x_t + a}{2a}\right) + \left(\frac{x_t + a}{2a}\right)^2}\right).$$
(3.46)

In the case of full sliding $x_t = a$ and so $t = M'_z = 0$. It follows by direct substitution that at peak aligning moment, that is, when $x_t = -a/2$, there obtains $t = 27a/148 = 0.182a$, while at $|\sigma_y| \ll 1$ (where $x_t \approx -a$) it is $t = a/3$.

Figure 3.7 shows the general form of the lateral tyre force, aligning moment, and pneumatic trail generated by the brush model as given in equations (3.41), (3.45), and (3.46) respectively.

The cornering stiffness, aligning moment stiffness, and trail at vanishing slip are computed from (3.41), (3.45), and (3.46) for $\sigma_y \approx \alpha \approx 0$:

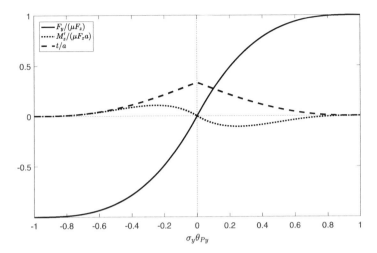

Figure 3.7: Normalized lateral tyre force, aligning moment, and pneumatic trail derived using (3.41), (3.45), and (3.46) respectively.

$$C_\alpha^{F_y} = \left.\frac{\partial F_y}{\partial \alpha}\right|_{\alpha=0} = 2K_{Py}a^2 \tag{3.47}$$

$$C_\alpha^{M_z'} = -\left.\frac{\partial M_z'}{\partial \alpha}\right|_{\alpha=0} = \frac{2}{3}K_{Py}a^3 \tag{3.48}$$

$$t_0 = \frac{a}{3}. \tag{3.49}$$

It can be reasonably assumed that $a \propto F_z^{1/2}$, resulting in $F_y \propto F_z$ and $M_z' \propto F_z^{3/2}$. In reality, F_y is smaller than $\propto F_z$, because K_{Py} reduces with F_z, and in most cases F_y/F_z and $C_\alpha^{F_y}$ have finite maxima; see Section 3.5. Finally, $C_\alpha^{F_y}$ and $C_\alpha^{M_z'}$ are larger than there measured values, while t_0 is smaller, because the elasticity of the tyre's carcass acts in series with the elasticity of the tread.

3.4.2 Pure longitudinal slip

The steady-state deflection, in the adhesion region in the case of pure longitudinal slip, is obtained from (3.26) with $V_{Px} = 0$ (adhesion assumed), and with $\tan \alpha = \varphi_s = 0$ (pure longitudinal slip assumed),

$$\frac{\partial \bar{x}}{\partial x} = -\frac{\kappa}{1+\kappa} \Rightarrow \bar{x} = -\frac{\kappa}{1+\kappa}x + c_{\bar{x}}, \tag{3.50}$$

where $c_{\bar{x}}$ is a constant of integration. This constant can be computed from the boundary condition $\bar{x}(a) = 0$:

$$\bar{x}(a) = -\frac{\kappa}{1+\kappa}a + c_{\bar{x}} = 0 \Rightarrow c_{\bar{x}} = \frac{\kappa}{1+\kappa}a. \tag{3.51}$$

Therefore the longitudinal deflection on the contact patch is:

$$\bar{x} = (a - x)\frac{\kappa}{1 + \kappa} = (a - x)\sigma_x \tag{3.52}$$

where

$$\sigma_x = \frac{\kappa}{1 + \kappa}. \tag{3.53}$$

All the relationships obtained for the lateral force F_y generated by σ_y, hold valid in the case of the longitudinal force F_x generated from σ_x, mutatis mutandis:

$$F_x = 3\mu_x F_z \theta_{Px}\sigma_x \left(1 - \theta_{Px}|\sigma_x| + \frac{1}{3}(\theta_{Px}\sigma_x)^2\right)$$

$$= \mu_x F_z \left(\left(\frac{x_t + a}{2a} - 1\right)^3 + 1\right) \mathrm{sgn}(\sigma_x); \tag{3.54}$$

where

$$\theta_{Px} = \frac{2K_{Px}a^2}{3\mu_x F_z}. \tag{3.55}$$

The maximum force $F_x = \mu_x F_z$ is again attained in total sliding,[5] that is, when $|\sigma_x|$ achieves equality in

$$|\sigma_x| \leq \frac{1}{\theta_{Px}}. \tag{3.56}$$

The longitudinal force slip stiffness has a form similar to (3.47), where the longitudinal stiffness per unit length K_{Px} replaces the lateral stiffness K_{Py}:

$$C_\kappa^{F_x} = \frac{\partial F_x}{\partial \kappa}\bigg|_{\kappa=0} = 2K_{Px}a^2. \tag{3.57}$$

The F_x vs $\sigma_x\theta_{Px}$ characteristic is essentially the same as that in Figure 3.7, when replacing $F_y, \sigma_y, \theta_{Py}$ with $F_x, \sigma_x, \theta_{Px}$.

3.4.3 Pure spin slip

The steady-state deflection in the adhesion region in the case of pure spin slip is obtained from (3.26) with $V_{Px} = V_{Py} = 0$, and with $\tan\alpha = \kappa = 0$

$$\frac{\partial\bar{x}}{\partial x} = y\varphi_s \Rightarrow \bar{x} = yx\varphi_s + c_{\bar{x}} \tag{3.58}$$

$$\frac{\partial\bar{y}}{\partial x} = -x\varphi_s \Rightarrow \bar{y} = -\frac{x^2}{2}\varphi_s + c_{\bar{y}} \tag{3.59}$$

where $c_{\bar{x}}$ and $c_{\bar{y}}$ are constants of integration related to the boundary conditions $\bar{x}(a) = 0$ and $\bar{y}(a) = 0$

---

[5] Equation (3.54) can only be used for $|\sigma_x| \leq 1/\theta_{Px}$. In the case that $|\sigma_x| > 1/\theta_{Px}$, $F_x = \mu_x F_z \mathrm{sgn}(\sigma_x)$.

$$\bar{x}(a) = ya\varphi_s + c_{\bar{x}} = 0 \Rightarrow c_{\bar{x}} = -ya\varphi_s \tag{3.60}$$

$$\bar{y}(a) = -\frac{a^2}{2}\varphi_s + c_{\bar{y}} = 0 \Rightarrow c_{\bar{y}} = \frac{a^2}{2}\varphi_s. \tag{3.61}$$

Therefore the longitudinal and lateral deflections in the contact patch are

$$\bar{x} = -y(a - x)\varphi_s \tag{3.62}$$

$$\bar{y} = \frac{1}{2}(a^2 - x^2)\varphi_s. \tag{3.63}$$

Equation (3.63) implies that a tyre material point in adhesion moves along a circle of radius R in pure turn slip, and along a straight line in pure camber.[6]
In pure camber

$$\bar{y} = -y_{\gamma 0} = \frac{1}{2}(a^2 - x^2)\frac{(1 - \epsilon_\gamma)}{r_e}\sin\gamma, \tag{3.64}$$

which is the friction-related deflection in the adhesion region, or what is the same, the bristle contact line for a cambered tyre on a frictionless surface.

The force and moment resulting from φ_s will again be computed by integration over the adhesion and sliding regions. In the case that both adhesion and sliding occur, the analysis is more complicated. We will therefore consider the case of pure adhesion first in order to gain insight into the basic characteristics of the camber and turn slip force- and moment-generation mechanisms. While this is clearly an approximation, we make a 'thin tyre' approximation in the more general case in which both adhesion and sliding occur.

---

[6] (a) In pure turn slip $\varphi_s = -1/R$ and the distance between a circle of radius R and the straight-line segment of length $2a$ at $Y = R\sqrt{1 - \left(\frac{a}{R}\right)}$, the line of contact on a frictionless surface, is:

$$R\sqrt{1 - \left(\frac{a}{R}\right)^2} - R\sqrt{1 - \left(\frac{x}{R}\right)^2} \approx R\left(1 - \frac{1}{2}\left(\frac{a}{R}\right)^2\right) - R\left(1 - \frac{1}{2}\left(\frac{x}{R}\right)^2\right) = \frac{1}{2}(a^2 - x^2)\frac{-1}{R}.$$

(a)

(b)

(b) In pure camber $\varphi_s = \frac{(1 - \epsilon_\gamma)}{r_e}\sin\gamma$ and the distance between the line of contact of a straight-running cambered tyre on a frictionless surface and the straight-line segment of length $2a$ is found by integration of $\dot{y}_0$ in (3.19) using the transformation $\dot{y}_0 = \frac{dy_0}{dx}\frac{dx}{dt} = \frac{dy_0}{dx}(-\Omega r_e)$:

$$\int_{-a}^{x} \frac{1}{-\Omega r_e}(-\Omega\hat{x}(1 - \epsilon_\gamma)\sin\gamma)d\hat{x} = \frac{1}{2}(x^2 - a^2)\frac{(1 - \epsilon_\gamma)}{r_e}\sin\gamma = -\frac{1}{2}(a^2 - x^2)\varphi_s = y_{\gamma 0}.$$

In the case of full adhesion, the lateral deflection $\bar{y}$ given in (3.63) is symmetrical around the centre of the contact region and will thus not contribute to the yawing moment. The lateral force is obtained by integrating over the contact patch (of length $2a$ and width $2b$); only the lateral deflections contribute to the lateral force

$$F_y = \int_{-a}^{+a} \int_{-b}^{+b} \frac{K_{Py}}{2b} \bar{y} dy dx = \iint \frac{K_{Py}}{2b} \frac{1}{2}(a^2 - x^2)\varphi_s dy dx$$
$$= \frac{2}{3} K_{Py} a^3 \varphi_s. \tag{3.65}$$

The z-axis moment is also obtained by integration over the contact patch. As one would expect by symmetry, only the longitudinal deflections contribute to the moment

$$M_z^r = -\int_{-a}^{+a} \int_{-b}^{+b} \frac{K_{Px}}{2b} \bar{x} y dy dx = \iint \frac{K_{Px}}{2b} y^2 (a - x)\varphi_s dy dx$$
$$= \frac{2}{3} K_{Px} a^2 b^2 \varphi_s. \tag{3.66}$$

Despite their simplicity, there are a number of points that are worth making in connection with (3.65) and (3.66). Since the spin slip has a positive camber-related term and a negative turn-slip-related term, F_y and M_z^r can change direction according to which of these terms is dominant. If the turn-slip-related term is dominant, the side force F_y in (3.65) is negative, and thus directed away from the centre of path curvature; see Figure 3.3. This force therefore tends to 'straighten up' the tyre's ground-plane trajectory. A turn-slip-dominant moment in (3.66) is negative and will thus also act to 'straighten up' the tyre's ground-plane trajectory. In the camber-dominant case, the spin slip will increase the tyre's path curvature (misaligning moment). In the pure adhesion case the spin-slip-related force is proportional to φ_s, while the spin-slip-related moment vanishes in the limit $b \to 0$. Note that sometimes (3.65) and (3.66) are derived using the longitudinal and lateral stiffness per unit area K'_{Px} and K'_{Py}, instead of the longitudinal and lateral stiffnesses per unit length $K_{Px} = K'_{Px} 2b$ and $K_{Py} = K'_{Py} 2b$ used herein.

The spin-slip-force stiffness and moment stiffness are:

$$C_{\varphi_s}^{F_y} = \left.\frac{\partial F_y}{\partial \varphi_s}\right|_{\varphi_s=0} = \frac{2}{3} K_{Py} a^3 = C_\alpha^{M'_z} \tag{3.67}$$

$$C_{\varphi_s}^{M_z^r} = \left.\frac{\partial M_z^r}{\partial \varphi_s}\right|_{\varphi_s=0} = \frac{2}{3} K_{Px} a^2 b^2. \tag{3.68}$$

It is noted that the spin-slip lateral force stiffness $C_{\varphi_s}^{F_y}$ in (3.67) is the same as the aligning moment lateral slip stiffness $C_\alpha^{M'_z}$ in (3.48). This reciprocity property is lost when spin-induced lateral slip resulting from the yaw compliance of the tyre is considered [96]; see also Section 3.4.6 for a discussion on induced/apparent slips.

The force and moment can be rewritten to separate explicitly the contributions of turn slip φ_t (or curvature $1/R$) and camber γ. Using (3.15) gives

$$F_y = C_{\varphi_s}^{F_y}\left(\varphi_t + \frac{1-\epsilon_\gamma}{r_e}\sin\gamma\right)\Big|_{\alpha=0} = -C_{\varphi_s}^{F_y}\left(\frac{1}{R} - \frac{1-\epsilon_\gamma}{r_e}\sin\gamma\right)$$

$$M_z^r = C_{\varphi_s}^{M_z^r}\left(\varphi_t + \frac{1-\epsilon_\gamma}{r_e}\sin\gamma\right)\Big|_{\alpha=0} = -C_{\varphi_s}^{M_z^r}\left(\frac{1}{R} - \frac{1-\epsilon_\gamma}{r_e}\sin\gamma\right).$$

(3.69)

Experimental tests [96] suggest that $\epsilon_\gamma \approx 0.5-0.7$ for radial ply car tyres. This has the effect of introducing a factor of 2 to 3 between the road curvature- and camber-related terms in (3.69);[7] this observation is supportive of the discussion following (3.19).

The influence of spin-related sliding is now considered. If we again invoke the quadratic pressure distribution (3.32), the lateral stiffness (3.31), and the lateral strain displacement (3.63), there holds

$$K_{Py}\frac{1}{2}(a^2 - x^2)\varphi_s^{sl} = \mu_y \frac{3F_z}{4a^3}(a^2 - x^2)$$

and so the spin slip on the limit of sliding is given by

$$\varphi_s^{sl} = \frac{3\mu_y F_z}{2K_{Py}a^3} = \frac{1}{a\theta_{Py}}.$$

(3.70)

If one substitutes φ_s^{sl} into (3.65) there obtains $F_y = \mu_y F_z$. Note that (3.70) resembles (3.38) and (3.56) for lateral and longitudinal slips under conditions of full sliding. However, in this case it is anticipated that for $|\varphi_s| > \varphi_s^{sl}$ only the front half of the patch and the trailing edge are sliding, while full sliding occurs only when $|\varphi_s| \to \infty$ (a tyre steered from standstill); see the remark following (3.74).

The analysis of spin-slip sliding in the case of a two-dimensional tyre tread base model is complicated [8]. For that reason we will now analyse the side-force- and moment-generation process related to spin slip using the 'thin' brush-type tyre model. In this model the side force is generated by the lateral deflections $\bar{y}$ given in (3.31) and the parabolic pressure distribution (3.32). Since both the lateral deflection (3.63) and the pressure distribution are parabolic, no sliding will occur for $\varphi_s \le \varphi_s^{sl}$, with the adhesion limit then reached throughout the contact patch.

To analyse the sliding case we make use of Figure 3.8. For $\varphi_s > \varphi_s^{sl}$ all material points on the front half of the contact patch $(0 < x < a)$ slide simultaneously. The contact line is dictated by the maximum lateral deflection at the limit of adhesion. It follows from (3.32), (3.33), and (3.34) that

$$\bar{y}^{max} = \frac{q_y^{max}}{K_{Py}} = \frac{3}{4}\mu_y F_z \frac{a^2 - x^2}{a^3}\frac{1}{K_{Py}} = \frac{a^2 - x^2}{2a\theta_{Py}}$$

(3.71)

with θ_{Py} defined in (3.37).

In the case of pure turn slip (Figure 3.8 (a)) the bristles move away from $y = 0$, which would be their position on a frictionless surface, to $y = \bar{y}^{max}$. In the case of pure

---

[7] The parameter ϵ_γ can be computed from $(1 - \epsilon_\gamma)/r_e = C_\gamma^{F_y}/C_{\varphi_t}^{F_y}$, since $C_\gamma^{F_y} = C_{\varphi_s}^{F_y}(1 - \epsilon_\gamma)/r_e$ and $C_{\varphi_t}^{F_y} = C_{\varphi_s}^{F_y}$.

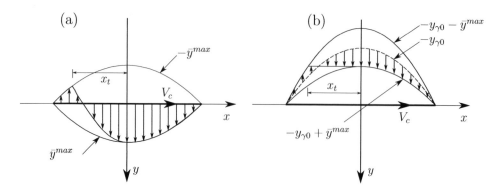

Figure 3.8: Tread deformation for a 'thin' brush model under pure turn slip (a), or pure camber (b) [8]. The curves $y_{0\gamma}$ and $\bar{y}_{max}$ are given in (3.64) and (3.71) respectively.

camber (Figure 3.8 (b)) the bristles move away from $y = -y_{\gamma 0}$, which would be their position on a frictionless surface, to $y = -y_{\gamma 0} + \bar{y}^{\max}$. Once the contact line reaches the rear half of the contact region, sliding ceases and adhesion resumes. The reason for this is that friction forces constrain the tread base material to follow a curved path in the case of pure turn slip, and a straight line in the case of pure camber whenever the friction limit is not exceeded; see comment following (3.63). In the rear half of the contact region the bristle deflection $\bar{y}$ reduces from its maximum value of $\bar{y}^{\max} = a^2/2a\theta_{Py} = \frac{1}{2}a^2\varphi_s^{sl}$ at the centre of the contact patch, and adhesion is re-established; see Figure 3.8. Therefore, in pure turn slip, the tread material switches from the $\bar{y}^{\max}$ boundary to the $-\bar{y}^{\max}$ boundary (with the curvature dictated by φ_t), where the friction limit is once again reached. In the case of pure camber, the tread base material moves to the $-y_{\gamma 0} - y^{\bar{m}ax}$ boundary, where the friction limit is again reached. In both cases the trailing edge is sliding in the opposite direction (as compared with the sliding in the front half of the contact patch). It is clear that for $\varphi > \varphi_{sl}$ a moment M'_z must be generated, because the deflection $\bar{y}$ is no longer symmetric and F_y reduces, because of the sign change in the trailing edge deflection.

We now need to find the transition point x_t at which sliding is re-established. To do this we must respect the maximal deflection constraint (3.71) at $x = 0$ and allow the sliding region to cross from the $+\bar{y}^{\max}$ boundary to the $-\bar{y}^{\max}$ boundary (pure turn slip), or from $-y_{\gamma 0} + \bar{y}^{\max}$ to $-y_{\gamma 0} - \bar{y}^{\max}$ (pure camber), whenever $\varphi_s > \varphi_s^{\max}$. Following (3.63) and (3.71)

$$\bar{y}_{x_t \leq x \leq 0} = \frac{1}{2}a^2\varphi_s^{sl} - \frac{1}{2}x^2\varphi_s \tag{3.72}$$

which gives (using (3.71))

$$\frac{1}{2}a^2\varphi_s^{sl} - \frac{1}{2}x_t^2|\varphi_s| = -\frac{a^2 - x_t^2}{2a\theta_{Py}} \tag{3.73}$$

and so

$$\frac{x_t}{2a} = \frac{-1}{\sqrt{2(1 + a\theta_{Py}|\varphi_s|)}}. \tag{3.74}$$

The negative root has been selected, because the transition point is on the rear half of the contact patch. Note that when $|\varphi_s| = \varphi_s^{sl}$ $x_t = -a$, and when $|\varphi_s| \to \infty$ $x_t = 0$.

The contact force is now computed by integration over the three sections of the contact patch: sliding trailing edge (I), adhesion rear half (II), and sliding front half (III)

$$F_y^I = -\int_{-a}^{x_t} K_{Py}(\bar{y}^{\max})dx = -K_{Py}\int_{-a}^{x_t} \frac{a^2 - x^2}{2a\theta_{Py}}dx = -\frac{K_{Py}}{6a\theta_{Py}}\left(2a^3 + 3a^2x_t - x_t^3\right)$$

$$F_y^{II} = \int_{x_t}^{0} K_{Py}\frac{1}{2}\left(a^2\varphi_s^{sl} - x^2\varphi_s\right)dx = \frac{K_{Py}x_t}{6}\left(x_t^2\varphi_s - \frac{3a}{\theta_{Py}}\right)$$

$$F_y^{III} = \int_{0}^{a} K_{Py}\bar{y}^{\max}dx = \frac{1}{2}\mu_y F_z.$$

The whole lateral force is obtained by adding the three components above and introducing θ_{Py} and x_t from (3.37) and (3.74):

$$F_y = F_y^I + F_y^{II} + F_y^{III} = \mu_y F_z\sqrt{2}\frac{1}{\sqrt{1 + a\theta_{Py}|\varphi_s|}}. \tag{3.75}$$

Similarly, the moment is obtained by integration on the same three sections

$$M_z^I = -\int_{-a}^{x_t} K_{Py}(\bar{y}^{\max})xdx = \frac{K_{Py}}{8a\theta_y}\left(a^2 - x_t^2\right)^2$$

$$M_z^{II} = \int_{x_t}^{0} K_{Py}\frac{1}{2}\left(a^2\varphi_s^{sl} - x^2\varphi_s\right)xdx = -\frac{K_{Py}x_t^2}{4\theta_{Py}}\left(a - \frac{1}{2}\theta_{Py}x_t^2\varphi_s\right)$$

$$M_z^{III} = \int_{0}^{a} K_{Py}\bar{y}^{\max}xdx = \frac{3}{16}a\mu_y F_z.$$

The full moment is obtained by adding the three components above and introducing θ_{Py} and x_t from (3.37) and (3.74):

$$M_z' = M_z^I + M_z^{II} + M_z^{III} = \frac{3}{8}\mu_y F_z a\frac{a\theta_{Py}|\varphi_s| - 1}{a\theta_{Py}|\varphi_s| + 1}\text{sgn}(\varphi_s). \tag{3.76}$$

The general form of (3.75) and (3.76) is shown in Figure 3.9.[8] It is clear that the side force grows linearly in the adhesion region, reaching a peak at the limit of sliding; after that it decays away as a result of opposing force contributions either side of the transition point x_t. As predicted in (3.66), M_z' is zero in the adhesion region, with the moment growing after that as a result of the unbalanced forces distribution across the contact patch under sliding. The force and moment as the spin approaches infinity can be determined by allowing $|\varphi_s| \to \infty$ in (3.75) and (3.76):

$$F_y = 0 \qquad M_z^{\max} = \frac{3}{8}\mu_y F_z a. \tag{3.77}$$

In this case $x_t = 0$ which results in the rear half of the contact patch sliding in the opposite direction to the front half; see Figure 3.8.

---

[8] Equations (3.75) and (3.76) can only be used for $|\varphi_s| \geq \varphi_s^{sl}$, while at lower values of $|\varphi_s|$ (3.65) and (3.66) hold.

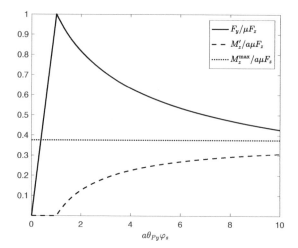

Figure 3.9: Normalized spin-slip-related force and moment for a 'thin' brush model, derived using (3.65), (3.66), (3.75), and (3.76).

It is interesting to note that experimental studies suggested a similar behaviour [91]

$$M_z^{\max} = \frac{3}{8}\mu F_z \left(a + \frac{2}{3}b\right), \tag{3.78}$$

where the additional term is related to the contact patch width; recall that the derivation of (3.77) was carried out using a 'thin' brush model. In general, the accurate modelling of the behaviour at standstill is not trivial.

In sum, for spin slip up to φ_s^{sl}, the moment is related only to the longitudinal deflection $\bar{x}$, $M_z = M_z^r$, which is misaligning in the case of pure camber and aligning in the case of pure turn slip. Above the sliding limit φ_s^{sl} an additional moment M_z' is generated that arises from the lateral deflection $\bar{y}$ (again misaligning in the case of pure camber and aligning in the case of pure spin slip). The force F_y is related to the lateral deflection $\bar{y}$ only and increases with the spin slip up to φ_s^{sl}. It then reduces and is zero at large spin slip, where the moment attains its maximum value.

3.4.4 Combined lateral and longitudinal slip

We will assume for simplicity that $K_P = K_{Px} = K_{Py}$, $\mu = \mu_x = \mu_y$, and $\theta_p = \theta_{Px} = \theta_{Py}$. The velocities of tread-base material points in the the contact region are given by $V_r = \Omega r_e$. Equations (3.52) and (3.29) give the relationship between the longitudinal tread deflection $\bar{x}$ and the lateral tread deflection $\bar{y}$, with the theoretical slips σ_x and σ_y given by

$$\begin{bmatrix} \bar{x} \\ \bar{y} \end{bmatrix} = (a - x) \begin{bmatrix} \sigma_x \\ \sigma_y \end{bmatrix}, \tag{3.79}$$

where the relationships between the theoretical and practical slip quantities are

$$\sigma_x = \frac{\kappa}{1+\kappa} \qquad \sigma_y = \frac{\tan\alpha}{1+\kappa} \qquad (3.80)$$

according to (3.53) and (3.30)[9] respectively.

Equation (3.79) highlights the idea that the tyre deflections are governed independently by σ_x and σ_y (and not by κ and $\tan\alpha$).

The transition point can be computed in the same way it was in the case of pure lateral force by introducing σ in place of σ_y (compare with (3.35))

$$K_P(a - x_t)\sigma = \mu \frac{3F_z}{4a^3}(a - x_t)(a + x_t) \qquad (3.81)$$

where

$$\sigma = \sqrt{\sigma_x^2 + \sigma_y^2} \qquad (3.82)$$

to obtain

$$\frac{x_t}{2a} = \theta_P \sigma - \frac{1}{2}, \qquad (3.83)$$

which can be compared with (3.36). Again full sliding ($x_t = +a$) occurs for $\sigma \geq 1/\theta_P$. The expressions for the force, aligning moment, and pneumatic trail are found in the same way:

$$\begin{bmatrix} F_x \\ F_y \end{bmatrix} = \frac{F}{\sigma}\begin{bmatrix} \sigma_x \\ \sigma_y \end{bmatrix} \qquad (3.84)$$

where

$$F = 3\mu F_z \theta_P \sigma \left(1 - \theta_P \sigma + \frac{1}{3}(\theta_P \sigma)^2\right) \qquad \text{if } \sigma \leq 1/\theta_P \qquad (3.85)$$

$$= \mu F_z \qquad \text{if } \sigma > 1/\theta_P \qquad (3.86)$$

$$M'_z = -tF_y \qquad (3.87)$$

$$t = a\frac{1 - 3\theta_P\sigma + 3(\theta_P\sigma)^2 - (\theta_P\sigma)^3}{3 - 3\theta_P\sigma + (\theta_P\sigma)^2}. \qquad (3.88)$$

Note that the isotropic stiffness and friction assumptions mean that the deflection is in the direction opposite to the slip vector $\boldsymbol{V}_s$.

Figure 3.10 illustrates a key feature of the coupling between the longitudinal and lateral forces, which is that the generation of one comes at the expense of the other. Figure 3.10 (a) shows how the lateral force F_y, at a fixed side-slip angle α, drops away when the tyre is asked simultaneously to generate braking or driving forces ($|F_y| \to 0$ when $\kappa \to \infty$). Figure 3.10 (b) shows how the longitudinal force F_x, at a fixed side-slip angle κ, drops away when the tyre is asked simultaneously to generate a side

---

[9] It can be shown that a theoretical slip of $\sigma_y = \frac{\tan\alpha}{1+\kappa}$ is obtained in place of $\sigma_y = \tan\alpha$ when integrating (3.26) under the assumption of constant κ rather than $\kappa = 0$.

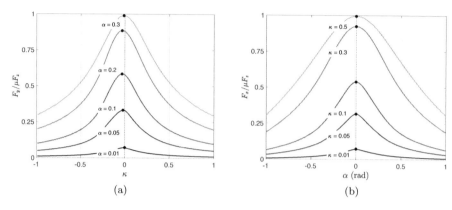

(a)   (b)

Figure 3.10: The influence of combined slip in the simple brush model. Graph (a) shows the normalized lateral force, for various fixed values of side-slip angle α, as a function of the longitudinal slip κ. Graph (b) shows the normalized longitudinal force, for various fixed values of longitudinal slip κ, as a function of α.

force ($|F_x| \to 0$ when $|\tan\alpha| \to \infty$, that is, $|\alpha| \to \pi/2$). It will be observed that the peak of the side force characteristic in Figure 3.10 (a) is biased towards the braking abscissa.[10] Experimental evidence shows that often the shape of these curves is even more asymmetric than predicted by the brush model. Possible reasons are increased contact length during braking and brake force-induced slip angle.

The spin-slip effect can be included in the present formulation by replacing the side slip $\tan\alpha$ in (3.80) with $\tan\alpha^*$, where

$$\alpha^* = \alpha + \frac{C_{\varphi_s}^{F_y}}{C_{\alpha}^{F_y}}\varphi_s \tag{3.89}$$

has the actual side-slip angle and an equivalent side-slip angle representing the effect of spin slip. The residual moment $M_z^r = C_{\varphi_s}^{M_z^r}\varphi_s$ can also be added to the aligning moment M_z' in (3.87) to give the total yawing moment

$$M_z = M_z' + M_z^r. \tag{3.90}$$

3.4.5 Combined lateral, longitudinal, and spin slip

As with all analytic physics-based models, at some point they become too complex to be practically useful. In particular, the brush model discussed cannot realistically be used to treat combined lateral, longitudinal, and spin slip, especially if one wishes also

---

[10] Figure 3.10 (a) is asymmetric, because σ_y varies asymmetrically in κ (for fixed α); $\sigma_y \propto 1/(1+\kappa)$ which increases under braking and decreases under driving. In addition, σ_x is asymmetric in κ. Figure 3.10 (b) is symmetric; because σ_x is constant for constant κ, variations in α only affect σ_y, and $\sigma_y(\alpha) = -\sigma_y(-\alpha)$.

to accommodate arbitrary pressure distributions, non-isotropic stiffness and friction properties, and finite tread widths. One may, however, consider using numerical models such as *TreadSim* [8] in more complex situations.

In practice, vehicle simulation models usually use an empirical tyre model such as those reported in Section 3.5. The parameters that populate these models are derived from experimental tests using various tyre-testing machines such as test trailers (e.g. TNO Tyre Test Trailer), flat-track test machines (e.g. MTS Flat Trac, Calspan TIRF), flat-plank testers (e.g. [97]), flat-disk machines (e.g. [98]), and internal or external drum test machines (e.g. [99]).

3.4.6 Summary of analytical models

The role of physics-based models, such as the brush model described earlier in this chapter, is to uncover the basic tyre force- and moment-generating mechanisms. Both quantities arise as a result of friction-induced tread material distortion and sliding, as quantified by the longitudinal slip κ, the lateral slip $\tan\alpha$, and the spin slip φ_s. For small values of slip, the forces and moments are proportional to slip. The constants of proportionality are the slip stiffnesses $C_\kappa^{F_x}$, $C_\alpha^{F_y}$, $C_{\varphi_s}^{F_y}$, $C_\alpha^{M_z'}$, and $C_{\varphi_s}^{M_z^r}$, which can be either computed or measured. Under small-slip conditions the majority of the contact patch is in adhesion, with tread material sliding only a minor influence. As the slip increases, so does sliding. Under increased slip conditions the relationships with the resulting forces and moments become increasingly nonlinear until saturation occurs. In the above analysis we have seen that full-patch sliding can occur in pure side slip when $F_y = \mu_y F_z \text{sgn}(\sigma_y)$; that full-patch sliding can occur in pure longitudinal slip when $F_x = \mu_x F_z \text{sgn}(\sigma_x)$; and that full-patch sliding can occur in pure spin slip when $F_y = 0$ and $M_z = \frac{3}{8}\mu_y F_z a$.

Numerical simulations can be found in [8] which go well beyond the capabilities of the brush models discussed here. These include such things as a sliding-velocity-dependent friction coefficient, a flexible tyre carcass, a finite tread width ($b > 0$), and combined lateral and spin slip. In the case of a friction coefficient that decreases with velocity, the tyre forces exhibit a peak that is found in most experimental testing. If the tyre has a flexible carcass, a reduction in the slip stiffness is expected and observed, which is the result of the tyre's reduced structural stiffness. The flexible carcass also causes an increase in the pneumatic trail. When spin slip is present, the yaw compliance of the tyre combined with the finite width of the contact patch leads to a force-induced side slip which increases $C_{\varphi_s}^{F_y}$ (and reduces $C_{\varphi_s}^{M_z}$). This effect is also related to the loss of the reciprocity property ($C_{\varphi_s}^{F_y} = C_\alpha^{M_z}$), which is experimentally observed. Finally, the yaw compliance causes an apparent side slip which increases/decreases the camber-related side force when braking/accelerating.

3.5 Steady-state behaviour: the 'magic formula'

The dominant tyre model currently in use is the so called magic formula model. This is an empirical parametric model that has its parameter values optimized to fit experimentally measured steady-state tyre force and moment data. magic formula models

provide an accurate representation of the experimental data, but with a model complexity much lower than that required of physics-based models of similar accuracy.
The cost of this modelling approach is the loss of physical insight combined with non-
physical parameters. The model formulae are usually functions of the normal load and
more recently the inflation pressure. A number of scaling factors are used to extrapo-
late the model's behaviour to conditions away from those of the original testing.

The development of the magic formula tyre model began in the late 1980s [100–102].
The physics-based approach to combined slip conditions is used in this work. In later
versions of the magic formulae, the friction circle is replaced by an empirical formula
introduced by Michelin [103]. The effect of inflation pressure was introduced explicitly
into the formula in [104]. There are still some important influences, such as temperature, which are not currently represented. That said, some of the manufacturers have
developed temperature-dependent magic formulae. The aim of this section is discuss
the key features of these models, without going into all the ins and outs of the various different versions of the magic formula models. Once an overview understanding
is obtained, it is relatively easy to move between different versions and upgrades;
widely used versions are reported in [105] and [106], with a motorcycle version available in [107]. The latest version is reported in [8], which is essentially the same as that
reported in [104], which can be downloaded from the Internet.[11]

The brush model analysis given in Section 3.4 showed that the tyre force and moment depend predominantly on the longitudinal slip κ, the lateral slip $\tan\alpha$, the spin
slip φ_s, and vertical load F_z. The magic formulae will therefore have these quantities
as inputs. Under most circumstances, only the camber component of the spin slip is
important and so the most widespread versions of the magic formula use the camber
γ instead of the more general spin slip φ_s.

3.5.1 Pure slip

The basic magic formula for the pure longitudinal force is:

$$F_x = D_x \sin(C_x \arctan(B_x\kappa - E_x(B_x\kappa - \arctan(B_x\kappa)))) \tag{3.91}$$

in which B_x is the stiffness factor, C_x is the shape factor, and E_x is the curvature
factor. If $C_x > 1$, $D_x = \mu_x F_z$ gives the peak value of F_x; μ_x is the friction coefficient for
longitudinal forces. If $E_x < 1$, $F_x > 0$ when $\kappa \to +\infty$. The curve given in (3.91) passes
through the origin $(F_x(0) = 0)$ and has slope $B_x C_x D_x$ there, which corresponds to
the longitudinal slip stiffness $C_\kappa^{F_x}$ estimated in (3.57). The peak force is D_x (assuming
$C_x > 1$) and then saturates at:

$$\lim_{\kappa=\infty} F_x = \begin{cases} D_x \sin(C_x\pi/2)\mathrm{sgn}(1 - E)\mathrm{sgn}(B) & E \neq 1 \\ D_x \sin(C_x \arctan(\pi/2))\mathrm{sgn}(B) & E = 1. \end{cases} \tag{3.92}$$

If B_x, C_x, D_x, and E_x are constant, the magic formula (3.91) generates an antisymmetric curve. It then reaches a maximum/minimum and subsequently tends towards horizontal asymptotes. One may wish to introduce offsets to this curve in order

---

[11] See http://www.mate.tue.nl/mate/pdfs/11281.pdf.

to accommodate such things as rolling resistance ($F_x \neq 0$ when $\kappa = 0$—see the free-rolling convention in Section 3.2). This can be achieved by replacing κ with $\kappa + S_H$ and/or F_x with $F_x + S_V$ for example. Setting $S_H = -M_y/(rB_xC_xD_x)$, which gives $F_x(0) = -M_y/r$, where M_y is the rolling resistance moment. Alternatively, if $S_H = 0$, $S_V = -M_y/r$ also gives $F_x(0) = -M_y/r$.

In the standard case that $C_x > 1$ and $E_x < 1$ the parameters can be determined as follows: D_x is obtained from the peak value $D_x = F_x^{\max}$; C_x is obtained from the asymptote $F_x^\infty = \lim_{\kappa \to \infty} F_x$, since $C_x = 2\arcsin(F_x^\infty/D_x)/\pi$—see (3.92); B_x is obtained from the slope at the origin, since $B_x = \partial F_x/\partial\kappa|_{\kappa=0}/C_xD_x$ (hence stiffness parameter); and finally E_x is obtained from $E_x = (B_x\kappa_m - \tan(\pi/2C_x))/(B_x\kappa_m - \arctan(B_x\kappa_m))$ where κ_m corresponds to the position of the peak. In practice the parameters are determined using nonlinear fitting algorithms. When the coefficients B_x, C_x, E_x are held constant, and $D_x = \mu_x F_z$, the longitudinal force is proportional to F_z and nonlinearly dependent on the slip.

In the case of the lateral force F_y, there are essentially two formulae depending on its intended use; for car, truck, or motorcycle tyres. In the case of car tyres, the formula resembles (3.91) and becomes

$$F_y = D_y \sin(C_y \arctan(B_y \tan\alpha - E_y(B_y \tan\alpha - \arctan(B_y \tan\alpha)))), \qquad (3.93)$$

in which $\tan\alpha$ replaces κ. In the case of motorcycle tyres

$$\begin{aligned} F_y = D_y \sin(C_y \arctan(B_y \tan\alpha - E_y(B_y \tan\alpha - \arctan(B_y \tan\alpha))) \\ +C_\gamma \arctan(B_\gamma\gamma - E_\gamma(B_\gamma\gamma - \arctan(B_\gamma\gamma)))), \qquad (3.94) \end{aligned}$$

where the effect of camber γ is explicitly accomodated; $C_y + C_\gamma < 2$ to prevent the side force from becoming negative at large slip and large camber angles. In the car version of the formula (3.93), the influence of camber is accounted for with a camber-dependent vertical shift. This approach is only appropriate when the camber force is small and thus cannot be used for motorcycle tyres. The cornering stiffness is $C_\alpha^{F_y} = B_yC_yD_y$ (recall (3.47)), while the camber stiffness is $C_\gamma^{F_y} = B_\gamma C_\gamma D_\gamma$ (recall footnote 7 and (3.67)).

In the case of F_y horizontal and vertical shifts can be used. For example, one may wish to recognize asymmetries in the tyre's construction that produce ply-steer, or pseudo-side-slip, which require a horizontal shift in the lateral slip plus a vertical shift in the force. Conicity, or pseudo-camber, can be accounted for with an horizontal shift in the camber and a vertical shift in the force.

While the magic formula parameters are typically load dependent, there are three whose load dependence is particularly important: the cornering stiffness $B_yC_yD_y$, the lateral peak force D_y, and the longitudinal peak force D_x. When considering these effects, the forces become nonlinearly dependent on the load. The cornering stiffness usually has a maximum, p_1, at a certain load, $F_z = p_2$, and is thus modelled as

$$B_yC_yD_y = p_1 \sin(2\arctan(F_z/p_2)). \qquad (3.95)$$

This effect has significant implications for vehicle handling, including its over- or under-steering behaviour; see Section 4.2.1.

The friction coefficients usually decrease as the load increases, which translates into the nonlinear load dependency of the peak force factors D_x and D_y, which are modelled as:

$$D_y = \mu_y F_z = (p_{y3} + p_{y4} df_z) F_z \tag{3.96}$$
$$D_x = \mu_x F_z = (p_{x3} + p_{x4} df_z) F_z \tag{3.97}$$

in which $p_{y4} < 0$ and $p_{x4} < 0$ and $df_z = (F_z - F_{z0})/F_{z0}$ where F_{z0} is some reference load. This effect has implications for load transfer effects; see Section 7.2.3.

The formulae used for longitudinal and lateral force can also be used for the lateral-slip-related aligning moment. However, a slightly modified formula, the 'cos-magic formula', is usually preferred for the pneumatic trail that is used to generate the aligning moment:

$$t = D_t \cos(C_t \arctan(B_t \tan \alpha - E_t(B_t \tan \alpha - \arctan B_t \tan \alpha))). \tag{3.98}$$

The aligning moment is then computed as $M_z' = -t F_y|_{\gamma=0}$, where $F_y|_{\gamma=0}$ is the lateral force related to the lateral slip $\tan \alpha$. The aligning moment stiffness is $C_\alpha^{M_z'} = D_t B_y C_y D_y$ (recall (3.48)). The reason the camber-related lateral force is not taken into account is because its associated misaligning moment is not generated by the resultant of the lateral force moving behind the centre of the contact patch. Indeed, the camber-related resultant force passes through the centre of the contact patch to the limit of sliding, and then moves in front of it under sliding; see Section 3.4.3 and particularly Figure 3.8. As a final remark, the right-hand side of (3.98) is sometimes multiplied by $\cos \alpha = V_{Cx}/V_C$ to handle properly the case of large slip angles (where $t \approx 0$, and thus $M_z' \approx 0$) and possibly also reverse running (when t changes sign).

The misaligning component of M_z, which is mainly related to the camber and conicity, is called the residual torque M_z^r and is fitted with the formula

$$M_z^r = D_r \cos(\arctan B_r \tan \alpha), \tag{3.99}$$

where D_r is the peak value, which is attained when the lateral slip $\tan \alpha$ is zero. The term D_r contains the effect of camber, which is the dominant cause of misaligning moment (as shown in Section 3.4.3), since:

$$D_r = F_z(q_{r1}\gamma + q_{r2}|\gamma|\gamma). \tag{3.100}$$

The misaligning moment stiffness is $C_\gamma^{M_z^r} = F_z q_{r1}$ (recall footnote 7 and (3.68)). The whole moment results from the aligning moment and the residual moment:

$$M_z = M_z' + M_z^r. \tag{3.101}$$

The rolling resistance (not considered in the brush model in Section 3.4) is usually modelled as:

$$M_y = f_s F_z \tag{3.102}$$
$$f_s = (q_{s1} + q_{s2}V_r + q_{s3}V_r^4)r_0, \tag{3.103}$$

in which f_s is the rolling resistance coefficient, q_{s1} governs the initial (low speed) resistance, q_{s2} provides a small speed-related increase, and q_{s3} provides a further sharp

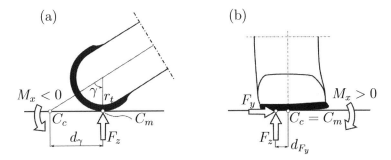

Figure 3.11: Overturning moment M_x and definitions of possible tyre-force points of application C. (a): the contribution related to camber γ is $M_x^{C_c} = -F_z d_\gamma$ (with $d_\gamma = r_t \tan\gamma$) and $M_x^{C_m} = 0$. (b): the contribution related to lateral force is $M_x^{C_c} = M_x^{C_m} = F_z d_{F_y}$ (shown positive in the figure, but negative values are possible).

rise in the rolling resistance that occurs when the critical velocity of the tyre is exceeded (standing waves produce large deformation and energy loss that eventually lead to tyre failure). The rolling resistance is caused by the effective point of application of F_z moving to the front half of the contact patch (due to hysteresis of the tread rubber).

The last of the three moment components is the overturning moment M_x. It is clear that this moment will depend on, amongst other things, the application point of the tyre-force system; see C in Figure 3.11. In car tyres it is common to use $C = C_c$, which is on the line of intersection between the wheel plane with the road surface. With motorcycle tyres it is more usual to define $C = C_m$, which is at the centre of the contact patch, as in the brush model described in Section 3.4.

In the case that C_c is used, the overturning moment has a camber-related contribution and a contribution related to the tyre lateral force F_y. The camber-related contribution is negative for $\gamma > 0$, because the force application point moves in the inclination direction (see Figure 3.11 (a)). Intuition might suggest that a positive lateral force $F_y > 0$ results in a negative moment (F_z moves in the direction of the lateral force), as a result of carcass lateral distortion. This is, however, not always the case, because an asymmetric distribution of the vertical loading may occur, resulting in a positive moment (F_z moves in a direction opposite to the F_y direction) (see Figure 3.11 (b)).[12]

When the C_m convention is used, the overturning moment derives from the lateral force alone, because the camber-related contribution is automatically included by geometry. This convention is particularly convenient for motorcycle tyres, where M_x is camber dominated. Another advantage of the convention will become apparent when dealing with coupling conditions, and in particular with the M_z arising from F_x when camber is present; see the discussion following (3.113).

---

[12] This phenomenon has been observed in car/truck tyres at low vertical loads, with a possible sign reversal (from $M_x > 0$ to $M_x < 0$) when some vertical load is exceeded [8].

In sum, in the case that the tyre forces are applied in C_c, the moment is modelled as:

$$M_x^{C_c} = \left(q_{x1} - q_{x2}\gamma + q_{x3}\frac{F_y}{F_z}\right)F_z, \tag{3.104}$$

while in the case that the forces system is applied at C_m the moment is:

$$M_x^{C_m} = \left(q_{x1} + q_{x3}\frac{F_y}{F_z}\right)F_z. \tag{3.105}$$

In these equations q_{x1} is an offset term, which can often be neglected; $q_{x2} > 0$ is a factor related to camber, with $d_\gamma = q_{x2}\gamma r_0$, producing a good fit to experimental data for small γ; q_{x3} is related to the effect of lateral force, with $d_{F_y} = q_{x3}r_0 F_y/F_z$, which can be either positive or negative.[13]

The normal load is usually modelled as

$$F_z = p_z \max(\delta r, 0)\left(1 + p_{z2}|\Omega|\right) \tag{3.106}$$

where p_z is the static vertical stiffness of the tyre. The p_{z2} term governs the tyre growth resulting from the centrifugal force acting on the belt (which leads to an increase in the tyre radius with speed).

3.5.2 Combined slip

Equation (3.84) showed that the generation of F_x is compromised by the simultaneous generation of F_y and vice versa. In the magic formula world, cosinusoidal weighting functions G are used to represent this combined-force compromise. These weights reflect the effects of $\tan\alpha$ on F_x and κ on F_y as follows:

$$G_{x\alpha} = \cos(C_{x\alpha}\arctan(B_{x\alpha}\tan\alpha)) \tag{3.107}$$
$$G_{y\kappa} = \cos(C_{y\kappa}\arctan(B_{y\kappa}\kappa)), \tag{3.108}$$

where the Bs influence the shape of G, while the Cs are related to the function's support. The combined-slip forces are:

$$F_{xc} = F_x G_{x\alpha} \tag{3.109}$$
$$F_{yc} = F_y G_{y\kappa} \tag{3.110}$$

while the moment is:

$$M_{zc} = M_z'(\sigma_y^{eq}) + M_z^r(\sigma_y^{eq}) + s(\gamma, F_z)F_x. \tag{3.111}$$

The effective lateral slip is given by σ_y^{eq} and accounts for combined-slip influences and thus depends on κ, while s is the lever arm for the longitudinal force:

---

[13] When C_m is a force application point that recognizes the lateral deformation of the carcass, (3.105) only has the q_{x1} term, which can usually be neglected and so $M_x = 0$ in this case.

$$\sigma_y^{eq} = \sqrt{(\tan\alpha)^2 + \left(\frac{B_x C_x D_x}{B_y C_y D_y}\right)^2 \kappa^2 \mathrm{sgn}(\alpha)} \tag{3.112}$$

$$s = \left(s_{z1} + s_{z2}\frac{F_y}{F_{zo}} + s_{z3}\gamma\right). \tag{3.113}$$

It is worth noting that when employing the C_m convention (see Figure 3.11), $s_{z3} = 0$, because the camber-related component is automatically included in the setup geometry.[14] A shift $\kappa \to \kappa + S_{Hy\kappa}$ may be introduced in (3.108) to account for the (small) increase in lateral force F_y experienced at moderate braking before the peak in $G_{y\kappa}$ is reached and F_y begins to decay (this effect is explained at the end of Section 3.4.4). If this shift is used, (3.108) must be divided by $\cos(C_{y\kappa}\arctan B_{y\kappa} S_{Hy\kappa})$ to ensure that $G_{y\kappa}(0) = 1$. In the same way, a κ-dependent vertical shift $S_{Vy\kappa}$ can be introduced in (3.110), which recognizes any κ-induced ply-steer ; the apparent lateral slip associated with longitudinal-force induces deformation of the tread. Note that F_z and M_y are not affected by combined-slip effects.

Combined-slip effects can also be modelled using a more physical approach using concepts derived from the brush model in Section 3.4.4. The advantage of this approach is the retention of physical ideas and the further use of the pure slip curves—the empirical-coupling G functions are not required (or the combined slip measurement data used to parametrize them). The drawback is that the fit to experimental data is not as good as that achieved with the empirical approach.

The tyre forces are computed from the theoretical slips introduced above (the practical slips κ and $\tan\alpha$ can also be used), with the camber transformed into an equivalent slip angle:

$$\sigma_x = \frac{\kappa}{1+\kappa} \qquad \sigma_y = \frac{\tan\alpha^*}{1+\kappa} \qquad \sigma = \sqrt{\sigma_x^2 + \sigma_y^2} \tag{3.114}$$

where

$$\alpha^* = \alpha + \frac{B_\gamma C_\gamma D_\gamma}{B_y C_y D_y}\gamma. \tag{3.115}$$

The forces are now computed as:

$$F_{xc} = \frac{\sigma_x}{\sigma} F_x(\sigma) \tag{3.116}$$

$$F_{yc} = \frac{\sigma_y}{\sigma} F_y(\sigma). \tag{3.117}$$

In the same way the moment is:

$$M_z = \frac{\sigma_y}{\sigma} M_z'(\sigma) + q_{r1} F_z\gamma, \tag{3.118}$$

in which the first term is the aligning moment and the second term the residual moment (related to camber). The residual moment is sometimes multiplied by a factor that

---

[14] When C_m recognizes the lateral deformation of the carcass, (3.113) only has the s_{z1} term, which can often be neglected and so $s(\gamma, F_x) = 0$.

makes the moment zero at large slip angles, say $1/(1+100\alpha^2)$. When comparing (3.118) with (3.111) it is clear that the F_x term is missing, since it is assumed that γ is small, and thus that this term is negligible, or the C_m convention is used (see Figure 3.11). Note that (3.116), (3.117), and (3.118) require no additional empirical coefficients.

In summary, the physical coupling approach requires: the pure longitudinal force curve (F_x vs. κ); the pure lateral force curve (F_y vs. $\tan\alpha$); the pneumatic trail curve (t vs. $\tan\alpha$); the lateral force camber stiffness ($\partial F_y/\partial\gamma = B_\gamma C_\gamma D_\gamma$); and the yawing moment camber stiffness ($\partial M_z^r/\partial\gamma = q_{r1}F_z$).

3.6 Behaviour in non-reference conditions

The characterization of tyre properties in terms of forces and torques, as functions of slip, is now a standardized procedure amongst car, motorcycle, and tyre manufacturers. It is again worth remembering that these measurements do not provide 'absolute' tyre properties, but the response of the tyre under specific test conditions (normal load, inflation pressure, test surface, and temperature).

For this reason it is useful to introduce a number of scaling factors in the magic formulae that were obtained under reference test conditions in order to predict (qualitatively) tyre behaviour under different (non-test) conditions. The idea is to exploit the fact that the pure slip curves remain broadly similar in shape when the tyre runs under conditions that are different from the reference conditions. Typical test conditions include the tyre's rated (nominal) load (F_{z0}), rated inflation pressure (p_0), given road surface (μ_0), and temperature (T_p). These ideas build on the normalization theory of [90] as introduced in [108, 109]. Alternatively, the scaling factors of a specific tyre can be identified by testing under non-reference conditions [110].

3.6.1 Effect of normal load

A particular instance of a 'variable parameter' relates to changes in the normal load away from reference test conditions F_{z0}. The magic formulae can be extended to include changes in various coefficients through the normalized change in vertical load df_z

$$df_z = \frac{F_z - F_{z0}}{F_{z0}}. \tag{3.119}$$

These changes are usually accommodated as either linear or quadratic variations of the form [8]

$$A = A_0 + A_1 df_z + A_2 df_z^2 \tag{3.120}$$

in which A_0 is the reference value of A, and A_1 and A_2 represent linear and quadratic variations respectively in df_z.

The qualitative effect of an increase in normal load is as follows: an increase in the size of the contact patch, an increase in rolling resistance, an increase in the yawing moment, a change in the cornering stiffness (that depends on normal load and the tyre characteristics; see (3.95)). As already pointed out in (3.96)–(3.97), there is usually a reduction in the maximum achievable tyre friction as a consequence of an increase in the normal load. A large contact patch is beneficial on dry surfaces, while a small one can be beneficial in wet conditions; see Section 3.6.4.

Basic tyre models include only a linear dependence of tyre force and moment on the normal load F_z. This is achieved by keeping constant the following coefficients introduced in Section 3.5: $B_{x,y,\gamma}$, $C_{x,y,\gamma}$, $E_{x,y,\gamma}$, $\mu_{x,y}$, D_t, q_{r1}, q_{r2}. Note that this results in constant slip stiffnesses per unit load: $C_\kappa^{F_x}/F_z = B_x C_x \mu_x$, $C_\alpha^{F_y}/F_z = B_y C_y \mu_y$, $C_\gamma^{F_y}/F_z = B_\gamma C_\gamma \mu_y$, $C_\alpha^{M_z}/F_z = D_t B_y C_y \mu_y$, $C_\gamma^{M_z}/F_z = q_{r1}$.

Good results are normally obtained when the variations of $C_\alpha^{F_y}/F_z$ and $\mu_{x,y}$ only depend on the normal load. This results in a nonlinear dependence of the tyre force and moment on F_z.

3.6.2 Effect of inflation pressure

The effect of pressure and pressure variations is a recent development of the magic formulae [104]. The changes are treated in much the same way as the changes in the normal load with a normalized change dp_i:

$$dp_i = \frac{p_i - p_0}{p_0}, \tag{3.121}$$

where p_0 is the reference inflation pressure.

The qualitative effect of an increase in inflation pressure is as follows: a reduction in the size of the contact patch, a reduction of rolling resistance, a reduction in the yawing moment, a change in the cornering stiffness that depends on the normal load and tyre characteristics [104, 111, 112]. These effects are usually neglected in simple models.

3.6.3 Effect of road surface

Tyre models are really tyre–road interaction models that can be significantly affected by changes in the road conditions. There are, however, some general trends that hold good [113, 114]. For example, when moving to wet (non hydroplaning) conditions, the friction peak is reduced, but there are generally only small changes in the cornering stiffness. In the same way, operation on different road surfaces may produce changes in the friction peak coefficient, but the cornering stiffness remains virtually unchanged. More generally, provided the road surface is 'rigid' relative to the tyre carcass, the cornering stiffness remains essentially the same (as predicted by the theoretical brush model).[15] However, when considering off-road surfaces such as gravel, a reduction in the cornering stiffness is to be expected [115, 116]. In these cases the peak in the lateral force characteristic moves towards higher slips and may even disappear completely. On very soft surfaces a bulldozing effect akin to 'ploughing' may occur [117]. Similar changes to the longitudinal force characteristics may also be expected [116].

---

[15] Note that changing the friction coefficients $\mu_{x,y}$ only, results in a unwanted change in the slip stiffnesses, unless one correspondingly changes $B_{x,y,\gamma}$ in order to keep the factor $B_{x,y,\gamma} C_{x,y,\gamma} \mu_{x,y}$ constant.

As common sense would suggest, different road conditions are likely to call for changes in driving strategy [118,119]. For example, on a wet or poor road surface, one would expect to brake earlier on the entrance to corners, accelerate more gently on the exit from corners, and possibly modify the line taken through bends [120].

3.6.4 Hydroplaning

Hydroplaning or *aquaplaning* is the partial or total lost of contact between the tyre and the road as a consequence of the hydrodynamic pressure generated by the tyre rolling over a layer of water [121]; this phenomenon is usually associated with driving/riding through puddled water at elevated speeds. In the case of partial hydroplaning, the effect is localized around the leading edge of the contact patch. In extreme cases one experiences a sustained loss of traction and the vehicle becomes non-responsive to steering inputs. For motorcycles, hydroplaning is particularly serious, because the vehicle becomes unstable as well as uncontrollable when the tyre force system is compromised.

The phenomenon was first observed in the late 1950s, whilst its basics were reported in the early 1960s [122], together with a simple formula for the prediction of the hydroplaning critical speed. Under hydroplaning conditions the mean pressure on the contact patch F_z/A is equal to the mean hydrodynamic pressure p_h given by

$$p_h = \frac{1}{2}\rho C_L(V_C^h)^2 \qquad \Rightarrow \qquad V_C^h = \sqrt{\frac{2}{\rho C_L}\frac{F_z}{A}} = \sqrt{\frac{p_h}{K}}, \qquad (3.122)$$

where ρ is the water density, $A = 4ab$ is the area of the contact patch, V_C^h is the tyre hydroplaning critical speed, and $K = 322$ in the original study, where it was further assumed that the inflation pressure was the mean pressure over the contact patch.[16] Equation (3.122) gives critical speeds of 25 m/s (90 km/h), and 50 m/s with inflation pressures of 2 bar (typical of car tyres) and 8 bar (typical of truck tyres) respectively, and is still in use to determine the hydroplaning potential of road surfaces [123]. When the vehicle speed exceeds the critical hydroplaning speed given by (3.122) by more than 8 km/h, and the water film thickness exceeds 1.5 mm, the hydroplaning risk is deemed to be 'high'. If the water film thickness is less than 0.5 mm, conditons are deemed to be safe from a hydroplaning perspective [123]. Depending on the tread pattern, and the shape and size of the contact patch, K varies over the range 250 (high-performance grooved tyres) to 500 (slick tyre).

The road macro-texture and tyre tread depth influence the onset of dynamic hydroplaning in two ways. First, they have a direct effect on the critical hydroplaning speed, because they provide a pathway for water to escape from the road–tyre interface. Second, they have an indirect effect on the critical hydroplaning speed, since the larger the macro-texture, the deeper the water must be to cause hydroplaning. However, the road surface must also have the proper micro-texture to develop adequate friction.

---

[16] The two pressures would indeed be exactly the same if the carcass bending stiffness was zero (membrane assumption). In practice the inflation pressure is somewhat lower (e.g. 2 bars vs 3 bars).

Another factor influencing hydroplaning potential is the shape of the contact patch. Rounded contact patches (motorcycle tyres) result in higher critical speeds as compared with rectangular contact patches (car tyres). If β is the angle between the normal to the leading edge and the direction of travel, the velocity component contributing to the dynamic pressure is $V \cos \beta$; aircraft wings are swept back for the same reason. Detailed numerical analyses of the tyre hydroplaning behaviour based on coupled CFD (computational fluid dynamics) and FEM (finite element method) were reported in the late 1990s [124].

3.6.5 Thermal effects

While temperature has a significant effect on tyre behaviour, thermal influences are not currently included in magic formula tyre models. The effect of temperature is particularly important in the racing context, where high-performance tyre compounds provide maximal tyre–road friction over a limited range of temperatures [125]. Variations in the track surface roughness and the ambient temperature have led tyre manufacturers to develop a range of compounds, each with its own friction and durability characteristics.

The relationships between friction generation and the compound's viscoelastic properties have been examined by several authors [126–128]. These studies are based on the seminal work of Grosch [129], who demonstrated that the adhesive and hysteretic components of friction can be characterized by a spacial-frequency-dependent master curve. The frequency–temperature equivalence described in [130] uses a scaling factor to correct these master curves for temperature variations.

Friction is an energy-dissipation mechanism that results in surface degradation and wear. With this in mind, Moore [131] examines the relationship between friction, fatigue, abrasion, and wear mechanisms for road car tyre rubber compounds. In the case of slick racing tyres, in combination with a rough, dry track, high friction coefficients and abrasive wear and *graining* are dominant phenomena [132]. A qualitative assessment of these processes is made in [133], which highlights the importance of the frequency of road asperities. The large body of experimental work, which is summarized in [134–136], shows that temperature, slip velocity, normal load, and the topology of abrasive surfaces are all key friction-influencing features.

Figure 3.12 shows how the tyre friction coefficient might vary with tyre tread temperature; the friction factor λ_μ is the ratio between the friction at the optimal reference temperature and the friction available at other temperatures T_p. Peak friction is achieved between the temperatures T_{p1} and T_{p2}; one might expect the midpoint between T_{p1} and T_{p2} to be of the order $100°$ C with $\lambda_\mu > 0.95$ over the tyre's operating temperature range. The longitudinal and lateral tyre forces, as modelled by magic formulae, are scaled by λ_μ; these details are often commercially sensitive.

In order to exploit tyre thermal sensitivities in simulation, or optimal control studies, it is necessary to track the tyre tread temperature variations. This can be done using a simple lumped-parameter model that describes the heat flows into the tread mass; high-frequency temperature variations associated with the rotation of the wheels are not considered. The dominant heat flows considered in the analysis are shown schematically in Figure 3.13 and consist of the following:

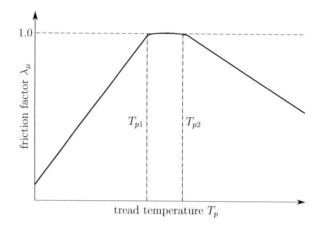

Figure 3.12: Friction factor λ_μ against the tyre tread temperature T_p.

(a) Q_1—heat generation in the sliding region of the contact patch
(b) Q_2—heat generation due to tyre carcass deflection
(c) Q_3—convective cooling through ambient air surrounding tyre
(d) Q_4—conductive cooling in the non-sliding region of the contact patch

The tyre tread surface temperature T_p is described by:

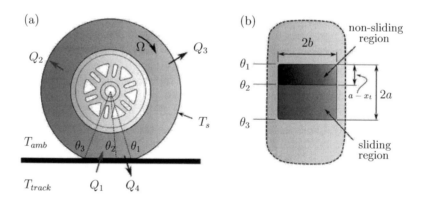

Figure 3.13: (a) Dominant heat flows that determine the tyre tread temperature T_p. The angle θ_1 defines the angle at which the leading edge of the contact patch is located. The angle θ_2 denotes the transition between the adhesion and the sliding regions, while θ_3 marks the end of the trailing edge. (b) View from the bottom of the tyre detailing the contact-patch geometry.

$$m_t \, c_t \, \frac{d}{dt} T_p = Q_1 + Q_2 - Q_3 - Q_4, \tag{3.123}$$

where c_t is the specific heat capacity of the tyre tread and m_t its mass. The heat generated from the sliding region of the contact patch is calculated through summation of longitudinal and lateral slip powers:

$$Q_1 = p_1 \, V_{Cx} \, (|F_x| \kappa + |F_y| \tan \alpha), \tag{3.124}$$

where the slip quantities α, κ, and the tyre forces F_x, F_y, are derived from the tyre model. The parameter p_1 accounts for the proportion of frictional power lost to the track in the sliding region of the contact patch. Heat generation due to bulk tyre deflection is treated in a manner similar to that described in [126], with the parameters p_2 through p_4 describing efficiency terms. These relate to the magnitude of heat generated as a result of longitudinal, lateral, and normal tyre forces

$$Q_2 = V_{Cx} \, (p_2 \, |F_x| + p_3 \, |F_y| + p_4 \, |F_z|). \tag{3.125}$$

The convective cooling of each tyre through the ambient air temperature T_{amb} is dealt with using the cooling law:

$$Q_3 = p_5 \, V_{Cx}^{p_6} \, (T_p - T_{amb}), \tag{3.126}$$

where the heat transfer coefficient $V_{Cx}^{p_6}$ is a nonlinear function of the vehicle speed [137] and parameters p_5 and p_6 are used to control the variations in airflow associated with each corner of the vehicle. The conductive cooling through the non-sliding region of the contact patch is determined by:

$$Q_4 = h_t \, A_{ns} \, (T_p - T_{track}), \tag{3.127}$$

where h_t is the heat transfer coefficient between the track and tyre, A_{ns} is the non-sliding area of the contact patch, and T_{track} the track temperature. The width of the contact patch $2b$ is assumed constant, with the length $2a$ defined by a fractional power β of the normal load:

$$2a = a_{cp} \, F_z^{\beta}. \tag{3.128}$$

The constant a_{cp} reflects contact-patch length deformation. Following the Brush model convention (see Section 3.4), the transition point between the sliding and non-sliding region is x_t and is given by

$$\frac{x_t}{2a} = \hat{\theta}_P |\sigma| - \frac{1}{2}, \tag{3.129}$$

where $\hat{\theta}_P$ is an estimate of θ_P; see (3.83). The resulting non-sliding contact-patch area is calculated using:

$$A_{ns} = 2b \, (a - x_t). \tag{3.130}$$

The six parameters p_1 through p_6 can be determined using nonlinear least-squares fitting against measured data [125]. Figure 3.14 shows the measured and predicted temperature variations for a Formula One car on one lap of a closed-circuit track. Significant are the facts that the temperature profiles of the four tyres are significantly different and that the tyre tread surface temperature can vary by almost $100°$ C within a single lap of approximately 4.6 km.

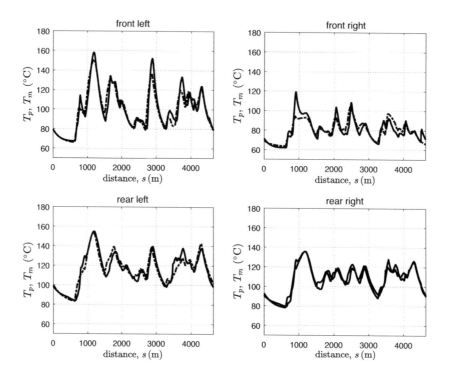

Figure 3.14: Tyre tread surface temperature measurements T_m (shown solid) and model values T_p (shown dot-dashed) [125].

3.7 Unsteady behaviour: string model

In this section the 'string model', first introduced in [138], is used to explain the essentials of transient tyre force generation. Later developments include [139] (derivation of frequency response), [92] (string with tread element), and [140] (computation of exact model response). More recently [87] compared several earlier models. In this development rolling resistance is neglected, complete adhesion is assumed, and all deformations are treated as small. As a consequence the resulting model is linear and holds only for small perturbations around straight-running trim conditions. The discussion is restricted to lateral motion (lateral slip α and spin slip φ_s) and the resulting lateral force F_y and moment M_z' (note that M_z' is used in place of $M_z = M_z' + M_z^r$, because the tyre is considered thin).

As shown in Figure 3.15, the tyre is treated as a loaded string in equilibrium. In the radial direction the string is held taught by a distributed radial force associated with the tyre's inflation. In the axial direction the string is supported elastically relative to the wheel-centre plane; no circumferential freedom being allowed. The string contacts a smooth horizontal road over the length of the contact patch, but extends over the tyre's perimeter. The string offers no resistance to bending, and so the only internal force is the tension F that acts in a direction tangent to the deflection curve. The

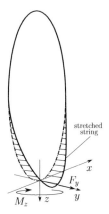

Figure 3.15: Tyre tread modelled as a stretched string.

ordinate of the deflection is given by $\bar{y}(x)$, in the $(Cxyz)$ coordinate system that moves with the wheel; see Figure 3.4. The angle between the x-axis and the tangent to the deflection curve is θ. The string load per unit length in the y-axis direction is $p(x)$. Figure 3.16 shows a plan view of the taught string, which has been deflected away from the wheel centre line by the lateral loading $p(x)$, which is given by

$$
p(x) = \begin{cases} -K_{Py}\bar{y} & |x| > a \\ q_y - K_{Py}\bar{y} & |x| \le a, \end{cases} \tag{3.131}
$$

where q_y is the frictional shear force per unit length, K_{Py} is the tyre lateral stiffness per unit length (as in the brush model), and $\bar{y}$ is the lateral deflection of the tread-base material. The equilibrium equations for an element between x and $x + dx$ are

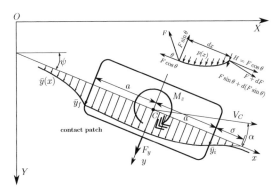

Figure 3.16: Plan view of the single-string model and its position with respect to the fixed reference frame. The insert shows the balance of forces on a string element of length dx.

$$\frac{d}{dx}(F\cos\theta) = 0 \qquad\qquad (3.132)$$

$$\frac{d}{dx}(F\sin\theta) = -p(x). \qquad\qquad (3.133)$$

Equation (3.132) shows that the horizontal component of the tension $H = F\cos\theta$ is constant. Eliminating F in (3.133) gives

$$H\frac{d}{dx}(\tan\theta) = -p(x), \qquad\qquad (3.134)$$

or since $\tan\theta = d\bar{y}/dx$

$$H\frac{d^2\bar{y}}{dx^2} = K_{Py}\bar{y} \qquad\qquad (3.135)$$

from (3.131) when $|x| > a$ and

$$H\frac{d^2\bar{y}}{dx^2} = K_{Py}\bar{y} - q_y \qquad\qquad (3.136)$$

when $|x| \le a$. The general form of the solution to (3.135) is easily checked to be

$$\bar{y} = c_i e^{-x/\sigma} + c_f e^{x/\sigma}, \qquad\qquad (3.137)$$

where

$$\sigma = \sqrt{\frac{H}{K_{Py}}} \qquad\qquad (3.138)$$

with c_i and c_f constants. The solution to (3.136) includes a particular integral

$$\bar{y} = c_i e^{-x/\sigma} + c_f e^{x/\sigma} + \frac{q_y}{K_{Py}}. \qquad\qquad (3.139)$$

Referring to Figure 3.15, it is physically reasonable to suppose also that $\lim_{x\to\infty} \bar{y}(x) = 0$, and so

$$\bar{y} = c_i e^{-x/\sigma} \qquad x > a. \qquad\qquad (3.140)$$

Similarly

$$\bar{y} = c_f e^{x/\sigma} \qquad\qquad (3.141)$$

when $x < -a$. This analysis assumes that the circumference of the tyre is much larger than the contact length. As a result the deflections for $x > a$ and $x < a$ become independent of one another.

Since $\bar{y}(a) = \bar{y}_i$ at the leading edge of the contact patch and $\bar{y}(-a) = \bar{y}_f$ at the trailing edge, there hold

$$\bar{y} = \bar{y}_i e^{(a-x)/\sigma} \qquad x > a \qquad\qquad (3.142)$$

$$\bar{y} = \bar{y}_f e^{(a+x)/\sigma} \qquad x < -a. \qquad\qquad (3.143)$$

Although the string is assumed to have zero bending stiffness, the physics of the rolling process suggests that the gradient $d\bar{y}/dx$ is continuous in $x = a$ (this is not necessarily true at $x = -a$). Since $\frac{d\bar{y}}{dx}\big|_{x=a} = -\bar{y}_i/\sigma = \tan\theta_i$, σ can be interpreted as a 'length' as shown in Figure 3.16, and is called the *relaxation length*.

3.7.1 Side slip and spin slip

In Section 3.3 we derived differential equations describing the sliding velocities of a rolling wheel. In the case of lateral slip α, and spin slip φ_s, with zero sliding velocities, it follows from (3.25) that

$$\frac{\partial \bar{y}}{\partial x} = \frac{\partial \bar{y}}{\partial s} - \alpha - x\varphi_s. \qquad (3.144)$$

We remind the reader that in this equation s denotes the distance travelled by the centre of the contact patch. The variables x and y denote the location of an arbitrary stressed material point with respect to the moving axes system $Cxyz$. This partial differential equation introduces 'dynamics' into the tyre model and will be solved using Laplace transforms; the travelled distance s (rather than time) is the independent variable in this case. The Laplace variable is denoted p and transformed variables are preceded by $\mathcal{L}$. The equation is now converted into an ordinary linear differential equation (in the independent variable x):

$$\frac{\partial \mathcal{L}\bar{y}}{\partial x} - p\mathcal{L}\bar{y} = -\mathcal{L}\alpha - x\mathcal{L}\varphi_s \rightarrow \mathcal{L}\bar{y} = c_{\bar{y}}e^{px} + \frac{1}{p}\left(\mathcal{L}\alpha + \left(\frac{1}{p} + x\right)\mathcal{L}\varphi_s\right). \qquad (3.145)$$

The constant $c_{\bar{y}}$ is obtained from the boundary condition on the leading edge of the contact patch:

$$\left.\frac{\partial \mathcal{L}\bar{y}}{\partial x}\right|_{x=a} = -\frac{\mathcal{L}\bar{y}_i}{\sigma} \rightarrow c_{\bar{y}} = -\frac{e^{-pa}}{p}\left(\frac{\mathcal{L}\bar{y}_i}{\sigma} + \frac{\mathcal{L}\varphi_s}{p}\right). \qquad (3.146)$$

Substituting (3.146) into (3.145) gives

$$\mathcal{L}\bar{y} = \frac{1}{p}\left(-\frac{\mathcal{L}\alpha + \left(\sigma + a + \frac{1}{p}\right)\mathcal{L}\varphi_s}{\sigma p + 1}e^{p(x-a)} + \mathcal{L}\alpha + \left(x + \frac{1}{p}\right)\mathcal{L}\varphi_s\right). \qquad (3.147)$$

The force F_y and the moment M_z' exerted on the tyre will now be computed with the aid of Figure 3.17. Summing forces in the y-axis direction gives

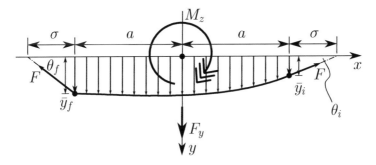

Figure 3.17: Free-body diagram of the taught string.

$$F_y = K_{Py} \int_{-a}^{+a} \bar{y} dx + F(\sin\theta_f + \sin\theta_i) = K_{Py} \int_{-a}^{+a} \bar{y} dx + H(\tan\theta_f + \tan\theta_i)$$

$$= K_{Py} \int_{-a}^{+a} \bar{y} dx + H\left(\frac{\bar{y}_i + \bar{y}_f}{\sigma}\right);$$
(3.148)

the lateral tyre force has to oppose the string tension and distort the tread base material. Taking moments around the origin in Figure 3.17 gives

$$M_z' = K_{Py} \int_{-a}^{+a} \bar{y} x dx + H(a\tan\theta_i + \bar{y}_i) - H(a\tan\theta_f + \bar{y}_f)$$

$$= K_{Py} \int_{-a}^{+a} \bar{y} x dx + H(\bar{y}_i - \bar{y}_f)\left(\frac{a+\sigma}{\sigma}\right).$$
(3.149)

It is now possible to find transfer functions between α and φ_s, and F_y and M_z', using Laplace-transformed versions of (3.148) and (3.149). To do this we will require (3.138), which is used to eliminate H, and (3.147), which is used to compute $\bar{y}_i$, $\bar{y}_f$, and $\bar{y}$. A direct, but tedious computation gives:

$$G_\alpha^{F_y} = \frac{K_{Py}}{p}\left[2(\sigma+a) - \frac{1}{p}\left(1 + \frac{\sigma p - 1}{\sigma p + 1}e^{-2pa}\right)\right]$$
(3.150)

$$G_{\varphi_s}^{F_y} = \frac{K_{Py}}{p^2}\left[2(\sigma+a) - \frac{1}{p}\left(1 + \frac{\sigma p - 1}{\sigma p + 1}e^{-2pa}\right)(p(\sigma+a)+1)\right]$$
(3.151)

$$G_\alpha^{M_z'} = -G_{\varphi_s}^{F_y}$$
(3.152)

$$G_{\varphi_s}^{M_z'} = \frac{K_{Py}}{p}\left[2a\left(\sigma(\sigma+a) + \frac{a^2}{3}\right) - \frac{1}{p}\left((\sigma+a)^2 - \frac{1}{p^2}\right)\right.$$

$$\left. + \frac{1}{p}\left(\sigma+a+\frac{1}{p}\right)^2 \frac{\sigma p - 1}{\sigma p + 1}e^{-2pa}\right],$$
(3.153)

where $G_\alpha^{F_y}$ is the transfer function mapping α to F_y and so on. As with the brush model, reciprocity between $G_\alpha^{M_z'}$ and $G_{\varphi_s}^{F_y}$ is found. As a result of linearity, the full lateral force and aligning moment responses come from superimposing (3.150) and (3.151), and (3.152) and (3.153) respectively.

In the context of the brush model, the cornering stiffness, the aligning moment stiffness, the spin-slip force stiffness, and the spin-slip moment stiffness were computed by considering limits in which $\alpha \to 0$ and $\varphi_s \to 0$. We can recover these relationships from (3.150), (3.151), (3.152), and (3.153) by invoking the final-value theorem associated with the Laplace transform. Formally

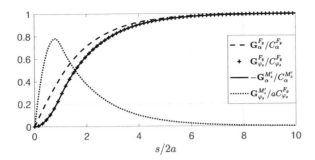

Figure 3.18: Normalized step responses for the transfer functions given in (3.150)–(3.153); $s/2a$ is the elapsed distance normalized to the length of the contact patch and $\sigma = 3a$.

$$C_\alpha^{F_y} = \lim_{p\to 0} \boldsymbol{G}_\alpha^{F_y} = 2K_{Py}(\sigma + a)^2 \tag{3.154}$$

$$C_{\varphi_s}^{F_y} = \lim_{p\to 0} \boldsymbol{G}_{\varphi_s}^{F_y} = 2K_{Py}a\left(\sigma(\sigma + a) + \frac{a^2}{3}\right) \tag{3.155}$$

$$C_\alpha^{M_z'} = \lim_{p\to 0} -\boldsymbol{G}_\alpha^{M_z'} = C_{\varphi_s}^{F_y} \tag{3.156}$$

$$C_{\varphi_s}^{M_z'} = \lim_{p\to 0} \boldsymbol{G}_\alpha^{M_z'} = 0 \tag{3.157}$$

with the pneumatic trail given by

$$t = \frac{C_\alpha^{M_z'}}{C_\alpha^{F_y}} = \frac{a\left(\sigma(\sigma + a) + \frac{a^2}{3}\right)}{(\sigma + a)^2}. \tag{3.158}$$

The brush-model slip stiffnesses in (3.47), (3.48), (3.67), and (3.68), and the pneumatic trail in (3.49), are recovered when $\sigma = 0$ is substituted into (3.154)–(3.158). By referring to (3.138), we see that $\sigma = 0$ corresponds to the removal of the string tension.

The (normalized) step responses for the four transfer functions (3.150)–(3.153) are given in Figure 3.18. As is self-evident, each response comprises the sum of a rational response term and a delayed term, the delay being associated with the time of travel though the contact patch.

When the low-frequency/large-wavelength behaviour is of interest, the delay terms in (3.150)–(3.153) can be approximated by truncated power series expansions around $p = 0$.[17] From (3.150) one obtains

$$(\sigma p + 1)\boldsymbol{G}_\alpha^{F_y} = K_{Py}\left(2p^2\sigma(a + \sigma) + p(2a + \sigma) - 1 + e^{-2ap}(1 - \sigma p)\right)/p^2$$
$$= C_\alpha^{F_y}(1 - (a - t)p) + O(p^2), \tag{3.159}$$

---

[17] Specifically $\dfrac{e^{-2ap}}{p} = \dfrac{1}{p} - 2a + 2pa^2 + O(p^2)$ and $\dfrac{e^{-2ap}}{p^2} = \dfrac{1}{p^2} - \dfrac{2a}{p} + 2a^2 - \frac{4}{3}a^2p + O(p^2)$.

which is equivalent to the following first-order linear differential equation description

$$\sigma\frac{dF_y}{ds} + F_y = C_\alpha^{F_y}(\alpha - (a-t)\frac{d\alpha}{ds}),\tag{3.160}$$

where the pneumatic trail t is given in (3.158). Similar (somewhat tedious) calculations give

$$\sigma\frac{dF_y}{ds} + F_y = C_{\varphi_s}^{F_y}(\varphi_s - a\frac{d\varphi_s}{ds});\tag{3.161}$$

$$\sigma\frac{dM_z'}{ds} + M_z' = -C_\alpha^{M_z'}(\alpha - a\frac{d\alpha}{ds});\tag{3.162}$$

$$\sigma\frac{dM_z'}{ds} + M_z' = a\left((\sigma + \frac{a}{3})C_\alpha^{M_z'} + \frac{a}{15}(C_\alpha^{M_z'} - \frac{a\sigma C_\alpha^{F_y}}{\sigma + a})\right)\frac{d\varphi_s}{ds}.\tag{3.163}$$

Note that the whole lateral force results from the summation of the component related to the side slip (3.160) and the component related to the spin slip (3.161). Similarly, the whole yawing moment results from (3.162) and (3.163). Closely related alternative reduced-order tyre models can be found by exploiting the quasi-steady approximation $C_\alpha^{F_y} d\alpha/ds = dF_y/ds$, $C_{\varphi_s}^{F_y} d\varphi_s/ds = dF_y/ds$, and $C_\alpha^{M_z'} d\alpha/ds = dM_z/ds$. Substituting the first of these relationships into (3.160), the second into (3.161), and the third into (3.162) gives

$$(\sigma + a - t)\frac{dF_y}{ds} + F_y = C_\alpha^{F_y}\alpha\tag{3.164}$$

$$(\sigma + a)\frac{dF_y}{ds} + F_y = C_{\varphi_s}^{F_y}\varphi_s\tag{3.165}$$

$$(\sigma + a)\frac{dM_z'}{ds} + M_z' = C_\alpha^{M_z'}\alpha.\tag{3.166}$$

Note that (3.163) has been left as is. The effect of these approximations can be seen in Figure 3.19. A notable practical difference between the exact model and the approximated model is that the step response of the former is exponential only after the tyre has travelled a distance $2a$ (when the last tread point that was making contact before the step took place leaves the contact patch). In contrast, the approximated model gives exponential responses. This suggests that the approximated model is acceptable when the path frequencies are such that the related wavelengths are much larger than the contact length, which is the case for most handling manoeuvres and low-frequency instabilities. At higher frequencies more elaborate models may be required.

In addition to its physical relationship with the slope of the contact line as shown in Figure 3.17, the relaxation length σ has a dynamic meaning relating to the temporal development of tyre forces and moments. The longer the relaxation length, the slower the dynamic response of the tyre force/moment to changes in the inputs α and/or φ_s; since σ is a spatial constant, and $ds = V_{Cx}dt$, σ/V_{Cx} is a time constant. In other words, when using the model in (3.160)–(3.163), σ is the distance travelled and σ/V_{Cx} is the time taken for the tyre to generate $\sim 63\%$ of its steady-state force/moment

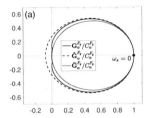
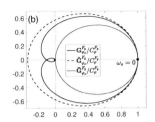
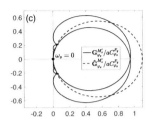

Figure 3.19: (a) shows normalized Nyquist diagrams of $G_\alpha^{F_y}$ given in (3.150), the rational approximation $\hat{G}_\alpha^{F_y}$ given in (3.160), and the rational approximation $\bar{G}_\alpha^{F_y}$ given in (3.164). (b) shows normalized Nyquist diagrams of $G_{\varphi_s}^{F_y}$ given in (3.151), the rational approximation $\hat{G}_{\varphi_s}^{F_y}$ given in (3.161), and the rational approximation $\bar{G}_{\varphi_s}^{F_y}$ given in (3.165). (c) shows normalized Nyquist diagrams of $G_{\varphi_s}^{M_z'}$ given in (3.153) and the rational approximation $\hat{G}_{\varphi_s}^{M_z'}$ given in (3.163); $\omega_s = 0$ represents zero spatial frequency.

after a step change in the (slip) input. When using (3.164)–(3.166) and (3.163), three different relaxation lengths are involved, namely $\sigma + a - t$, $\sigma + a$, and σ.

Various other approximate models have been developed over the years. We mention for example Von Schlippe's straight connection model, which assumes a straight contact line between the leading and trailing edges of the contact patch [138]; the straight tangent model, which assumes a straight contact line with the leading-edge gradient maintained across the contact patch; and the single-point contact model, which disregards the geometry of the contact patch. A comprehensive summary of this material can be found in [8].

3.7.2 Transfer function relationships

In this section we derive relationships between the transfer functions associated with the tyre slips α and φ_s, as well as those associated with the tyre lateral displacement y and orientation ψ. We will also show that the dynamic response of the string model reduces to its quasi-steady response when yaw oscillations take place around an axis located $\sigma + a$ ahead of the centre of the contact patch C.

In the case that $\gamma = 0$, and assuming $ds \approx V_{Cx}dt$, one obtains

$$\varphi_s = \varphi_t = -\frac{d\psi}{ds} \tag{3.167}$$

from (3.15) and so

$$\mathcal{L}\varphi_s = \mathcal{L}\varphi_t = -p\mathcal{L}\psi. \tag{3.168}$$

For a straight-running wheel subject to small yaw oscillations, $V_C \approx V_{Cx}$ and the wheel's lateral velocity is given by $V_{Cy} = V_{Sy} = \dot{y} - \psi V_{Cx}$. This gives

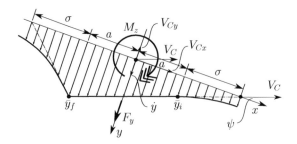

Figure 3.20: Wheel and tyre subject to small yaw oscillations around an axis located $\sigma + a$ ahead of the patch centre C.

$$\alpha = -\frac{V_{Sy}}{V_{Cx}} = \frac{V_{Cx}\psi - \dot{y}}{V_{Cx}} = \psi - \frac{1}{V_{Cx}}\frac{dy}{dt} = \psi - \frac{dy}{ds} \qquad (3.169)$$

and so

$$\mathcal{L}\alpha = \mathcal{L}\psi - p\mathcal{L}y. \qquad (3.170)$$

The relationships between the slip- and displacement-related transfer functions can now be found. From (3.168) and (3.170) there follows

$$\begin{aligned}
\mathcal{L}F_y &= \boldsymbol{G}_\alpha^{F_y}\mathcal{L}\alpha + \boldsymbol{G}_{\varphi_s}^{F_y}\mathcal{L}\varphi_s \\
&= \boldsymbol{G}_\alpha^{F_y}(\mathcal{L}\psi - p\mathcal{L}y) + \boldsymbol{G}_{\varphi_s}^{F_y}(-p\mathcal{L}\psi) \\
&= \underbrace{(\boldsymbol{G}_\alpha^{F_y} - p\boldsymbol{G}_{\varphi_s}^{F_y})}_{\boldsymbol{G}_\psi^{F_y}}\mathcal{L}\psi + \underbrace{(-p\boldsymbol{G}_\alpha^{F_y})}_{\boldsymbol{G}_y^{F_y}}\mathcal{L}y. \qquad (3.171)
\end{aligned}$$

In the same way the moments are described by

$$\begin{aligned}
\mathcal{L}M_z' &= \boldsymbol{G}_\alpha^{M_z'}\mathcal{L}\alpha + \boldsymbol{G}_{\varphi_s}^{M_z'}\mathcal{L}\varphi_s \\
&= \underbrace{(\boldsymbol{G}_\alpha^{M_z'} - p\boldsymbol{G}_{\varphi_s}^{M_z'})}_{\boldsymbol{G}_\psi^{M_z'}}\mathcal{L}\psi + \underbrace{(-p\boldsymbol{G}_\alpha^{M_z'})}_{\boldsymbol{G}_y^{M_z'}}\mathcal{L}y. \qquad (3.172)
\end{aligned}$$

It follows from (3.154) and (3.155), and (3.171) and (3.172), that

$$C_\psi^{F_y} = \lim_{p \to 0} \boldsymbol{G}_\psi^{F_y} = C_\alpha^{F_y}, \qquad (3.173)$$

$$C_\psi^{M_z'} = \lim_{p \to 0} -\boldsymbol{G}_\psi^{M_z'} = C_\alpha^{M_z'}. \qquad (3.174)$$

A further two relationships can be derived from an analysis of the response of the string model to the particular inputs contemplated in Figure 3.20. Since the lateral displacement of the contact patch centre is given by $\mathcal{L}y = -(a + \sigma)\mathcal{L}\psi$, there holds

$$\begin{aligned}
\mathcal{L}F_y &= \boldsymbol{G}_\psi^{F_y}\mathcal{L}\psi + \boldsymbol{G}_y^{F_y}\mathcal{L}y \\
&= \boldsymbol{G}_\psi^{F_y}\mathcal{L}\psi - (a + \sigma)\boldsymbol{G}_y^{F_y}\mathcal{L}\psi \\
&= \left(\boldsymbol{G}_\alpha^{F_y} - p\boldsymbol{G}_{\varphi_s}^{F_y} + (a + \sigma)p\boldsymbol{G}_\alpha^{F_y}\right)\mathcal{L}\psi. \qquad (3.175)
\end{aligned}$$

Direct calculation using (3.150), (3.151), and (3.154) gives

$$\mathcal{L}F_y = C_\alpha^{F_y}\mathcal{L}\psi. \tag{3.176}$$

The same reasoning leads to an equivalent relationship for the moment

$$\mathcal{L}M'_z = C_\alpha^{M'_z}\mathcal{L}\psi. \tag{3.177}$$

Equations (3.176) and (3.177) give

$$C_\alpha^{F_y} = G_\psi^{F_y} + p(a+\sigma)G_\alpha^{F_y} \tag{3.178}$$

$$C_\alpha^{M'_z} = G_\psi^{M'_z} + p(a+\sigma)G_\alpha^{M'_z}. \tag{3.179}$$

While (3.178) and (3.179) have been suggested by the analysis associated with Figure 3.20, they are true in general.

Six constraints on $\boldsymbol{G}_{\alpha,\varphi_s,y,\psi}^{F_y}$ and $\boldsymbol{G}_{\alpha,\varphi_s,y,\psi}^{M'_z}$ are thus given by (3.171), (3.172), and (3.178) and (3.179). The utility of these results comes from the fact that any one of $G_\alpha^{F_y}$, $G_{\varphi_s}^{F_y}$, $G_\psi^{F_y}$, and $G_y^{F_y}$ determines the other three; the same applies to the moment relationships $\boldsymbol{G}_{\alpha,\varphi_s,y,\psi}^{M'_z}$. If the tyre's frequency-response characteristics are determined experimentally using yaw oscillation testing, we can determine $G_\alpha^{F_y}$ using (3.173) and (3.178):

$$G_\alpha^{F_y} = \frac{C_\psi^{F_y} - G_\psi^{F_y}}{p(\sigma+a)}. \tag{3.180}$$

The transfer functions $\boldsymbol{G}_{\varphi_s}^{F_y}$ and $\boldsymbol{G}_y^{F_y}$ come from (3.171):

$$G_{\varphi_s}^{F_y} = \frac{G_\alpha^{F_y} - G_\psi^{F_y}}{p} \tag{3.181}$$

and

$$G_y^{F_y} = -pG_\alpha^{F_y}. \tag{3.182}$$

Parallel arguments hold for the moment quantities $\boldsymbol{G}_\alpha^{M'_z}$, $\boldsymbol{G}_{\varphi_s}^{M'_z}$, $\boldsymbol{G}_\psi^{M'_z}$, and $\boldsymbol{G}_y^{M'_z}$. In experiments the whole moment $M_z = M'_z + M_z^r$ is measured and so the M_z^r component must be subtracted before applying the formulas [8]. Another issue relates to the fact that these relationships only hold good within the context of the string model itself, and are only approximations of the physical reality.

3.8 Unsteady magic formulae

The string model analysis demonstrates that there is a lag, related to the relaxation length, in the force and moment responses to slip changes. The idea now is to extend the (steady-state) magic formula to unsteady conditions by using 'relaxed' slips as inputs to the magic formula. Under steady-state conditions this extended model reduces to the original magic formula model. If we restrict our modelling to low frequencies (up to

15 Hz) and large wavelengths (>1.5 m $\approx 24a$), which are the conditions usually found when simulating the handling of vehicles, the belt dynamics can be neglected and the force lag can be described using first-order equations [8, 105]. These equations are similar to those derived for the string model in (3.160), (3.161), and (3.162), or (3.164), (3.165), and (3.166), but in this case the relaxation-related quantities are determined experimentally. In general, the relaxation length(s) increase with increased normal load [141], reduce with increasing slip angle [142], reduce with increasing inflation pressure [112], and may increase at high speeds, because of the gyroscopic moments arising from the combination of the belt lateral distortion velocity and tyre rolling velocity [143].

Using the string model as a guide to their general form, the slip relaxation equations we will use are

$$\frac{\sigma_\kappa}{V_{Cx}} \frac{d\hat{\kappa}}{dt} + \hat{\kappa} = \kappa \qquad (3.183)$$

$$\frac{\sigma_\alpha}{V_{Cx}} \frac{d\hat{\alpha}}{dt} + \hat{\alpha} = \alpha \qquad (3.184)$$

$$\frac{\sigma_{\varphi_s}}{V_{Cx}} \frac{d\hat{\varphi}_s}{dt} + \hat{\varphi}_s = \varphi_s \qquad (3.185)$$

in which σ_κ, σ_α, and σ_{φ_s} are the three relaxation lengths in the model and $\hat{\kappa}$, $\hat{\alpha}$, and $\hat{\varphi}_s$ are the relaxed slips (or instantaneous slips), which are filtered versions of the slips κ, α, and φ_s. These relaxed slips are used as inputs for the (nonlinear) magic formulas of Section 3.5 to compute the tyre forces and moments

$$(F_x, F_y, M_x, M_y, M_z) = f(\hat{\kappa}, \hat{\alpha}, \hat{\varphi}_s, F_z). \qquad (3.186)$$

To summarize, the extended magic formula model consists of slip relaxation equations (3.183)–(3.185) and the magic formulae in (3.186). It is usually more convenient to use the magic formula defined in terms of γ, rather than φ_s, in which case the σ_{φ_s} and φ_s in (3.185) are replaced with σ_γ and γ. In vehicle dynamics applications the most important relaxation equation is (3.184), which is related to the lateral slip.

The relaxed model has a mechanical analogy that is described using Figure 3.21. A road-contact point $\hat{S}$ is introduced, which is distinct from the wheel slip point S defined in Section 3.2. It is sometimes assumed that points S and C are coincident, which corresponds to replacing the effective rolling radius r_e with the loaded radius r. The velocity difference between S and $\hat{S}$ causes the carcass springs to deflect. The two points are connected by a longitudinal spring $K_x^{F_x}$ and a lateral spring $K_y^{F_y}$, the carcass deformation being $\hat{x}$ in the longitudinal direction and $\hat{y}$ in the lateral direction. The tyre forces are applied to $\hat{S}$, and depend on the slips $\hat{\kappa}$, $\hat{\alpha}$ and the spin slip φ_s. The longitudinal and lateral slips associated with $\hat{S}$ are computed from their definitions in (3.4) and (3.7)

$$\hat{\kappa} = \kappa - \frac{\dot{\hat{x}}}{V_{Cx}} \qquad (3.187)$$

$$\hat{\alpha} = \alpha - \frac{\dot{\hat{y}}}{V_{Cx}}, \qquad (3.188)$$

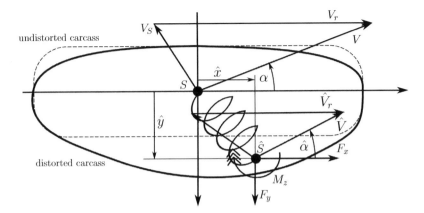

Figure 3.21: Plan view of a single-point tyre model showing longitudinal and lateral carcass deflections $\hat{x}$ and $\hat{y}$, respectively, which are modelled as linear springs.

where κ and α are the slips associated with S. At $\hat{S}$ there is balance between the forces arising from the spring and the forces arising from the slips

$$K_x^{F_x}\hat{x} = C_\kappa^{F_x}\hat{\kappa} \tag{3.189}$$

$$K_y^{F_y}\hat{y} = C_\alpha^{F_y}\hat{\alpha} + C_{\varphi_s}^{F_y}\varphi_s, \tag{3.190}$$

where it is noted that the spin-slip thrust $C_{\varphi_s}^{F_y}\varphi_s$ is assumed to develop instantaneously. However, as we will now show, the overall response of the tyre to variation in spin slip is not instantaneous, because of the coupling with the side-slip dynamics in (3.190). When introducing the expressions for slips given by (3.187) and (3.188) into (3.189) and (3.190)

$$K_x^{F_x}\hat{x} = C_\kappa^{F_x}\left(\kappa - \frac{\dot{\hat{x}}}{V_{Cx}}\right) \tag{3.191}$$

$$K_y^{F_y}\hat{y} = C_\alpha^{F_y}\left(\alpha - \frac{\dot{\hat{y}}}{V_{Cx}}\right) + C_{\varphi_s}^{F_y}\varphi_s, \tag{3.192}$$

where the lateral deformation will depend both on the contribution related to forces arising from α and forces arising from φ_s: $\hat{y} = \hat{y}_\alpha + \hat{y}_{\varphi_s}$. The side-slip and spin-slip contributions in (3.192) can now be separated by superposition so that

$$\frac{1}{V_{Cx}}\frac{C_\kappa^{F_x}}{K_x^{F_x}}\left(\frac{K_x^{F_x}}{C_\kappa^{F_x}}\dot{\hat{x}}\right) + \left(\frac{K_x^{F_x}}{C_\kappa^{F_x}}\hat{x}\right) = \kappa \tag{3.193}$$

$$\frac{1}{V_{Cx}}\frac{C_\alpha^{F_y}}{K_{\hat{y}}^{F_y}}\left(\frac{K_y^{F_y}}{C_\alpha^{F_y}}\dot{\hat{y}}_\alpha\right) + \left(\frac{K_y^{F_y}}{C_\alpha^{F_y}}\hat{y}_\alpha\right) = \alpha \tag{3.194}$$

$$\frac{1}{V_{Cx}}\frac{C_\alpha^{F_y}}{K_{\hat{y}}^{F_y}}\left(\frac{K_y^{F_y}}{C_{\varphi_s}^{F_y}}\dot{\hat{y}}_{\varphi_s}\right) + \left(\frac{K_y^{F_y}}{C_{\varphi_s}^{F_y}}\hat{y}_{\varphi_s}\right) = \varphi_s. \tag{3.195}$$

The conditions for equivalence between (3.193)–(3.195) and (3.183)–(3.185) are

$$\sigma_\kappa = \left(\frac{C_\kappa^{F_x}}{K_{\hat{x}}^{F_x}} \right) \tag{3.196}$$

$$\sigma_\alpha = \left(\frac{C_\alpha^{F_y}}{K_{\hat{y}}^{F_y}} \right) \tag{3.197}$$

$$\sigma_{\varphi_s} = \sigma_\alpha. \tag{3.198}$$

When side slip and spin slip occur simultaneously, the side slip should be computed using $\hat{\alpha} = \alpha - \hat{y}_\alpha/V_{Cx}$ rather than $\hat{\alpha} = \alpha - \hat{y}/V_{Cx}$ as in (3.188). The relaxation lengths given in (3.196)–(3.198) are the ratios of the tyre slip stiffnesses to the tyre structural stiffness. In this model the relaxation length related to spin slip (or camber) is clearly the same as the relaxation length related to lateral slip. In sum, equations (3.189) and (3.190) are employed when using the mechanical model, and the slips are computed as in (3.187) and (3.188).

In order for the transient model to automatically accommodate relaxation lengths that reduce with reduced normal loads or increasing slips, it is necessary to use the formulation based on the mechanical model in Figure 3.21, instead of that based on the relaxation equations (3.183)–(3.185). The two are equivalent for small slips when the springs are chosen such that (3.196)–(3.198) are satisfied, but for larger slips, the mechanical model has relaxation lengths that automatically reduce, because the numerators in (3.196)–(3.198) reduce, as the slips increase and/or the normal load reduces.

It has been noted experimentally that the lateral force related to camber is composed by a lagged (or relaxed) component (with $\sigma_{\varphi_s} \sim \sigma_\alpha$) and a non-lagged component [97,142]. This feature can be introduced into the model by replacing φ_s in (3.195) or (3.185) with $(1 - \epsilon_{NL})\varphi_s$, with ϵ_{NL} representing the non-lagged part of φ_s. The input of the magic formula in (3.186) is the sum of the lagged and unlagged components of spin slip. The lagged component is applied on the contact point $\hat{S}$ (as with the other forces), while the non-lagged component is applied directly to the wheel rim at S. The effect of ϵ_{NL} can be considered a secondary effect, which can be neglected in many vehicle dynamics simulations.

When a high-frequency tyre model is required (< 60–$100\,\mathrm{Hz}$), for ride comfort and vibration analysis for example, the dynamics of the carcass must be included in the model. In this case a rigid ring representation of the belt is usually employed; a flexible belt is necessary when even higher-frequency behaviours are required. The ring is suspended from the rim by (typically six) springs and dampers that represent the carcass compliance. The springs and dampers are then tuned to the frequencies and damping ratios of the lowest natural structural modes of the tyre. Other enhancements may include additional differential equations to model short-wavelength effects, such as overshoots [144], that clearly cannot be treated with a single first-order equation.

The MF-Swift [8, 145] is an example of a model implementing these advanced features.

3.8.1 Low-speed modelling

There are a number of precautions required if tyre models are to behave reliably at very low speeds. For example, V_{Cx} must be replaced by its modulus $|V_{Cx}|$ to allow for backwards running; the slip quantities (κ, α and φ_s) must be protected against zero-speed singularities (usually by adding a small ϵ term to the denominators of these expressions); all of the slip quantities must be limited to $|\kappa| < \kappa^{max}$, $|\alpha| < \alpha^{max}$, and $|\varphi_s| < \varphi_s^{max}$ to avoid unrealistic slip values at low speeds. At speeds below some minimum V^{low}, it may be necessary to enforce $\dot{\kappa} = 0$, $\dot{\alpha} = 0$, and $\dot{\varphi}_s = 0$ when the corresponding slip exceeds some maximum, which may be the slip corresponding to the peak of the associated tyre characteristic.

It can be seen from (3.191) that $\dot{\hat{x}} \rightarrow 0$ as $V_{C_x} \rightarrow 0$ since $\hat{x}$ must remain bounded. In other words, at very low speeds, the tyre behaves like a spring according to (3.189) and (3.190), and the carcass deflections are given by $\hat{x} = \sigma_\kappa \hat{\kappa}$, $\hat{y} = \sigma_\alpha \hat{\alpha}$ and $\hat{y}_{\varphi_s} = \sigma_\alpha \hat{\varphi}_s$. This low-speed pure-spring property means that the model equations are undamped (eigenvalue on the imaginary axis), and oscillatory behaviours will arise. Therefore, below some low-speed threshold, it may be necessary to add artificially a damper in parallel with the spring (when using the mechanical tyre model), or add a term proportional to V_{Sx} to the relaxed slips $\hat{\kappa}$, $\hat{\alpha}$, and $\hat{\varphi}_s$ (when the relaxation equations are used).

Most of the difficulties associated with low-speed tyre modelling are associated with the forces and moments associated with spin slip. The spring-like behaviour of the tyre at standstill means that the associated models generate an instantaneous spin-slip-related lateral force, whereas the lateral force should be zero (see Section 3.4.3). For this reason the lateral force should be artificially suppressed below a certain low speed V^{low}. It is known that special-purpose models are required for the accurate representation of the moment generated from spin slip at standstill [146].

3.9 Advanced models

There are several models available that are suitable for the simulation of tyre dynamics in a vehicle dynamics context. These models are generally coupled to multi-body software (MBS), which is used to simulate the dynamics of a car, truck, motorcycle, or aircraft.

Among the models with real-time capability, we mention: MF-Tyre, TMeasy, Tame-Tire, MuRiTyre, TRT/GreTa, and MF-Swift. Among the more complex models, which are still suitable for coupling with MBS but much slower than those already mentioned, are FTire and RMOD-K.

MF-Tyre is the official implementation of the magic formulae described above, which are combined with the relaxation equations. At the time of writing the current version is v6.2; however, older versions such as v5.2 (which achieved the status of an 'industry standard' for modelling passenger-car tyres in the late 1990s) are still widespread. MF-Tyre is generally used for handling studies with frequency up to 8–15 Hz.

TMeasy [147, 148] is a semi-physical model, which requires empirical slip-force curves (as in MF-Tyre), while the relaxation behaviour is derived from a (physical) carcass model, in much the same way as in the mechanical model described in Section

3.8. Coupling is accounted for using the concept of 'equivalent slip', which is similar to that given in Section 3.4.4. The identification of parameters requires only static and steady-state tests. The five parameters used to describe the slip vs force curves are identified for the nominal vertical load, and for twice the nominal vertical load in order to describe the digressive behaviour of the force-generating capability of the tyre due to increasing normal load (see comment around (3.95)–(3.96)). The model can deal with uneven roads as long as the contact patch is closed, that is, at long wavelengths.

TameTire [149, 150] (ThermAl and MEchanical TIRe Emulator) was developed by Michelin in order to include the effects of tyre temperature in vehicle handling simulations. In contrast with the previous models, it does not require the slip vs. force curves. These curves come from a physical model, which includes a flexible carcass described by three main stiffnesses (casing torsion, belt stiffness, and tread stiffness, all depending on load, inflation pressure, and tyre temperature). In the contact region the tread is described by brush elements, whose friction coefficient with the road is dependent on tyre pressure, tyre temperature, and sliding velocity. The brush model is important to determine the sliding region (where heat is generated) and the adhesion region (where the heat is transmitted to the track). The model predicts the average temperature, over a period of rotation on the tyre contact patch, in the belt and in the cavity. The inputs required are the ambient temperature, the road temperature, and the (three) initial tyre temperature(s).

MuRiTyre [151] is a physical discretized multi-rib tyre handling model, which can be parameterized without the requirement for traditional rolling-force and moment data. The number of longitudinal ribs can be increased or decreased in accordance with the processing power available, enabling real-time solutions. The force distribution along each rib is modelled, and used along with the deflected position of the rib, to calculate the ribs' contribution to the aligning, overturning, and rolling moments of the tyre. The tyre model has mechanical and thermal modes to predict tyre performance with variations in load and temperature. It also includes wear and degradation functions to allow the simulation of performance changes over time and use.

TRT/GrETA [152, 153] is a three-dimensional racing tyre model that takes into account the heat sources, flows, and dissipations in the tyre. The cooling influences of the track and the external air, as well as the heat flows inside the tyre, are modelled. The friction energy developed in the contact patch and the strain energy losses are modelled. The model generates the circumferential temperature distributions in the various tyre layers (surface, bulk, inner liner), as well as the internal heat flows.

MF-Swift [104, 145] (Short Wavelength Intermediate Frequency Tyre) is an extension of the MF-Tyre to higher frequencies (< 60–$100\,$Hz) and shorter wavelengths ($> 0.1 - 0.2m$). In this frequency range deformations of the tyre belt can be neglected and the tyre belt can be modelled as a rigid ring that is elastically suspended from the wheel rim. When short obstacles and rough roads are encountered, tyre geometry and elastic effects give rise to the nonlinear behaviours in the tyre forces, and to changes in the effective rolling radius. A 3D obstacle-enveloping model is used to calculate effective road inputs that simulate the tyre's behaviour as it moves over uneven road surfaces with the obstacle-enveloping behaviour of the tyre properly represented. With this approach a single-point-contact model can be used to simulate manoeuvres such as a tyre passing over a cleat, where there are three distinct areas of contact: in front of the cleat, on the cleat, and behind it.

FTire [154] (Flexible ring Tyre model) is a development of BRIT [155] (Brush and RIng Tyre model) that comprises structural and tread–road contact models. The structural model uses some 80–200 lumped-mass nodes to model the tyre's cord. These nodes are connected to the rim and to each other by several nonlinear inflation pressure-dependent stiffness, damping, and friction elements. The tread model comprises a number of massless friction elements that are placed between neighbouring belt segments to establish the road contact. These elements are sprung and deflect in accordance with the road profile and sliding velocities. The model also includes a thermal model, which is optionally activated and can compute the temperature of individual tread elements. The model can deal with dynamics up to about 150 Hz, while the computation time is 5–20 times real time. The model has an integrator for the calculation of the belt shape, which runs in parallel with the main integrator, but is synchronized to it.

The RMOD-K models are FEM-based tyre models with different levels of complexity [156, 157]. RMOD-K FEM is a complex nonlinear FEM model with different mesh densities in the structure and contact areas. RMOD-K FB (Flexible Belt) includes a flexible ring, which leads to a significant reduction in computational effort. RMOD-RB (Rigid Belt) is another rigid-ring model that uses an analytical description of the adhesion area, suitable for simulations with frequencies up to 100 Hz on smooth roads. It also includes a thermal model. Finally RMOD-K formula is an open, physical, steady-state tyre model.

Examples of tyre dataset can be found in [8] (car, motorcycle, and truck tyres), [98, 111, 112] (motorcycle tyres), [158] (aircraft tyres), and [159] (bicycle tyres).

4

Precursory Vehicle Modelling

4.1 Background

In Chapter 2 we studied several examples that introduce the underlying mechanics ideas required for road vehicle modelling. Predominant is the mechanics of the rolling contact that was studied in the context of the rolling ball (Section 2.9) and the rolling disc (Section 2.10). Our study of these examples made extensive use of (pure) non-slip rolling idealizations, in which any material point on the surface of the rolling ball, or the periphery of the rolling disc, which is in (instantaneous) contact with a stationary base surface is assumed stationary. These conditions are enforced by a pair of nonholonomic constraints—typically in two non-parallel ground-plane directions. The contact modelling is completed by a holonomic constraint in the vertical direction; see for example (2.234) and (2.235), and (2.264).

By the 1950s, the understanding of tyre and rolling-contact behaviour had improved substantially and it had become commonplace, although not universal, to regard the rolling contact as a force producer rather than a motion constraint [83]. In this abstraction the rolling-contact material at the ground plane 'slips' relative to the rolling surface and so has a non-zero absolute velocity. Under these conditions the linkage between the rolling body's rotational and translational velocities is lost. To model this slip-dependent behaviour it is necessary to introduce a slip-dependent force-generation mechanism, which might be an elementary description of the type given in (2.255) and (2.256), or a more sophisticated description of the type given in Chapter 3. These slip mechanisms can have a substantial impact on the dynamics of most vehicles. They can also produce new slip-related phenomena that can be highly undesirable. Obvious among these are skidding tyres; less obvious are various unstable oscillatory phenomena that we will study here, with more extensive treatments appearing in later chapters.

The focus of this chapter will be on precursory car and bicycle models, and an oscillatory phenomenon commonly referred to as 'shimmy'. In a subsequent chapters the single-track models introduced here will be developed into fully fledged bicycle and motorcycle models, while shimmy-related phenomena will reappear in the context of both bicycle and motorcycle dynamics.

4.2 Simple car model

Some of the fundamental properties of road cars can be developed with the help of simple single-track models of the type shown in Figure 4.1. In entry-level models of this type, many influences such as the suspension dynamics, chassis and tyre compliances, and lateral load transfer effects are ignored.

Dynamics and Optimal Control of Road Vehicles. D. J. N. Limebeer and M. Massaro.
© D. J. N. Limebeer and M. Massaro 2018. Published in 2018 by Oxford University Press.
DOI: 10.1093/oso/9780198825715.001.0001

In the figure the car's wheelbase is given by w; a and b locate the vehicle's mass centre relative to the wheel ground contacts; u and v ($v < 0$ in figure) are, respectively, the longitudinal and lateral components of the car's mass centre velocity $\boldsymbol{V}$; ν locates the vehicle's neutral steer point which is defined below; δ is the car's steering angle; α_f and α_r are, respectively, the front and rear tyre side-slip angles (which are defined in (3.7) and lie between the tyre centre velocity and the tyre longitudinal axis); β ($\beta < 0$ in the figure) is the vehicle side-slip angle (that lies between the vehicle mass centre velocity $\boldsymbol{V}$ and the vehicle's longitudinal axis); F_{fy} and F_{ry} are, respectively, the tyres' front and rear lateral forces; F_{fx} and F_{rx} are the front- and rear-tyre longitudinal drive or brake forces (not shown); w is the car's yaw velocity; a_y is the car mass centre's lateral acceleration; and $\mathcal{R}$ is the mass centre's radius of curvature, herein defined as the distance between the mass centre G and the instantaneous centre of rotation C.

The exact nature of the lateral acceleration a_y and the radius of curvature $\mathcal{R}$ require further analysis. The velocity of the car's mass centre (in body-fixed coordinates) is $\boldsymbol{V} = u\boldsymbol{i} + v\boldsymbol{j}$, and so the lateral acceleration is given by $d\boldsymbol{V}/dt = \dot{\boldsymbol{V}} + w\boldsymbol{k} \times \boldsymbol{V}$, where $\dot{\boldsymbol{V}}$ refers to the component-wise derivative. In particular $d\boldsymbol{V}/dt = \hat{a}_x\boldsymbol{i} + \hat{a}_y\boldsymbol{j}$, with $\hat{a}_x = \dot{u} - wv$ and $\hat{a}_y = \dot{v} + wu$; $\hat{a}_y$ is the lateral acceleration normal to the car's plane of symmetry . In contrast, the lateral acceleration a_y depicted in Figure 4.1 is normal to the velocity $\boldsymbol{V}$ and is given by $a_y = (w + \dot{\beta})V$. Under steady-state conditions the two

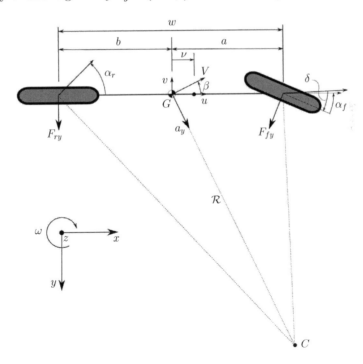

Figure 4.1: Kinematics of a single-track car model showing its basic geometric parameters. Also shown are the reference axes.

expressions become $\hat{a}_y^{ss} = \omega u$ and $a_y^{ss} = \omega V$. Under normal driving conditions, with a small side-slip angle, $\hat{a}_y \approx a_y$. The *velocity centre* or *instantaneous centre of rotation* C is obtained using (2.146) by setting $\boldsymbol{V} + \omega \boldsymbol{k} \times \boldsymbol{r}_C = \boldsymbol{0}$, which gives the location of point C as $\boldsymbol{r}_C = -\frac{v}{\omega}\boldsymbol{i} + \frac{u}{\omega}\boldsymbol{j}$. The distance between C and the symmetry plane of the car is $\hat{\mathcal{R}} = \frac{u}{\omega}$, which is in contrast to the distance $\mathcal{R} = V/(\omega + \dot{\beta})$ between C and G depicted in Figure 4.1. Again, under normal driving conditions, the two are very close and $\hat{\mathcal{R}} \approx \mathcal{R}$ for small angles. In general C does not coincide with the *acceleration centre*, which is the point with zero acceleration; the two centres do coincide under steady turning.

Balancing forces in the longitudinal and lateral directions, and balancing yaw moments around the car's mass centre, give the following well-known equations of motion (EOMs):[1]

$$m\dot{u} = m\omega v - F_{fy}\sin\delta + F_{fx}\cos\delta + F_{rx}, \qquad (4.1)$$

$$m\dot{v} = -m\omega u + F_{fy}\cos\delta + F_{fx}\sin\delta + F_{ry}, \qquad (4.2)$$

$$J\dot{\omega} = a(F_{fy}\cos\delta + F_{fx}\sin\delta) - bF_{ry}, \qquad (4.3)$$

in which m and J are, respectively, the car's mass and yaw moment of inertia; the model's state variables are u, v, and ω.

The three side-slip angles are computed using the kinematic relationships:

$$\beta = \arctan\left(\frac{v}{u}\right), \qquad (4.4)$$

$$\alpha_f = \delta - \arctan\left(\frac{\omega a + v}{u}\right), \qquad (4.5)$$

$$\alpha_r = \arctan\left(\frac{\omega b - v}{u}\right). \qquad (4.6)$$

The following small-angle approximations are often used:

$$\beta = \frac{v}{u}, \qquad (4.7)$$

$$\alpha_f = \delta - \frac{\omega a + v}{u}, \qquad (4.8)$$

$$\alpha_r = \frac{\omega b - v}{u}. \qquad (4.9)$$

Another classic assumption in the single-track car model is that the longitudinal velocity u is given and fixed. In this case (4.1) can be neglected and the system is described by v and ω only.[2] If it is further assumed that the vehicle is rear-wheel drive (RWD), and that the rolling resistance at the front wheel is negligible ($F_{fx} = 0$), the equations of motion simplify to the following classic form

---

[1]   These equations are easily derived using (2.76), (2.77), and (2.78), with $\mathcal{L} = \frac{m}{2}(u^2 + v^2) + \frac{J}{2}\omega^2$.

[2]   The term $m\omega v$ in (4.1) can also be neglected for small ω and v.

$$m\dot{v} = -m\omega u + F_{fy} + F_{ry} \tag{4.10}$$

$$J\dot{\omega} = aF_{fy} - bF_{ry}, \tag{4.11}$$

In the case of a front-wheel drive (FWD) or all-wheel drive (AWD), additional right-hand-side terms appear in (4.10) and (4.11); however, these additional terms are usually considered small and are thus neglected.[3]

When the vehicle accelerates, or brakes, there is a normal tyre load transfer between the front and rear axles. By balancing moments around the rear-wheel ground-contact point, and noting that the sum of the tyre down forces is equal to the weight of the vehicle, the front and rear normal loads (F_{fz} and F_{rz} respectively) can be computed as:

$$F_{rz} = mg\frac{a}{a+b} + ma_x\frac{h}{a+b} \tag{4.12}$$

$$F_{fz} = mg - F_{rz}, \tag{4.13}$$

where h is the mass centre's height and a_x is the longitudinal acceleration, for small angles $a_x \approx \hat{a}_x$.

We will model the tyre forces using a simple slip-based model of the type alluded to in the introduction to this section. The lateral forces generated by the tyres are assumed to depend linearly on the tyre normal loads and slip angles, and are computed using

$$F_{fy} = C_{f0}F_{fz}\left(\delta - \frac{\omega a + v}{u}\right), \tag{4.14}$$

$$F_{ry} = C_{r0}F_{rz}\left(\frac{\omega b - v}{u}\right), \tag{4.15}$$

where C_{f0} and C_{r0} are the normalized cornering stiffnesses (at the front and rear axles respectively), which can be estimated from the (front and rear) tyre properties using $C_\alpha^{F_y}/F_z$ (with $C_\alpha^{F_y}$ given by (3.47) under brush model assumptions and given by (3.154) under string model assumptions), or using $B_y C_y D_y/F_z$ (with B_y, C_y, D_y being the Pacejka's coefficients introduced in Section 3.5.1); the small angle approximations (4.8) and (4.9) are used.

The tyre model (4.14)–(4.15) is combined with the remaining EOMs (4.10)–(4.11), and the resulting linear EOMs are given by:

$$\dot{v} = -\frac{C}{mu}v + \left(\frac{-C\nu}{mu} - u\right)\omega + \frac{C_f}{m}\delta, \tag{4.16}$$

$$\dot{\omega} = \frac{-C\nu}{mk^2u}v - \frac{Cq^2}{mk^2u}\omega + \frac{C_f a}{mk^2}\delta, \tag{4.17}$$

in which

---

[3] In the case of FWD (4.1) gives (under small angle assumption) $F_{fx} = m(\dot{u} - \omega v) + F_{fy}\delta$, and the related $F_{fx}\delta$ and $aF_{fx}\delta$ terms in (4.10) and (4.11) are given in terms of this expression.

$$C_r = C_{r0}F_{rz} \tag{4.18}$$
$$C_f = C_{f0}F_{fz} \tag{4.19}$$
$$C\nu = C_f a - C_r b \tag{4.20}$$
$$C = C_f + C_r \tag{4.21}$$
$$Cq^2 = C_f a^2 + C_r b^2 \tag{4.22}$$
$$mk^2 = J. \tag{4.23}$$

This system represents a linear-time-invariant (LTI) model when the speed u is fixed, and a linear-time-varying (LTV) model in the case that u is varying. Following [8], the parameter ν is the distance between the centre of mass and the neutral steer point, q is the average moment arm, and k the radius of gyration. The neutral steer point, which is shown in Figure 4.1, has kinematic significance: a force applied through this point will cause the vehicle to move laterally without yawing.[4]

4.2.1 Constant speed case

When $u = u_0$, where u_0 is the steady-state velocity of the vehicle, (4.16) and (4.17) are time-invariant. In order to examine the car's behaviour, the response in the lateral acceleration, yaw velocity, and side-slip angle to perturbations in steering angle are determined. These quantities are given by the output response $\boldsymbol{y} = [\hat{a}_y, \omega, \beta]^T$, where the lateral acceleration is given by:

$$\hat{a}_y = \dot{v} + u_0\omega. \tag{4.24}$$

Here ω is a state, and β is given in (4.7).

The state-space model of the car from the steering angle δ to the output vector is given by[5]

$$A = \begin{bmatrix} -\frac{C}{mu_0} & \frac{-C\nu}{mu_0} - u_0 \\ \frac{-C\nu}{Ju_0} & -\frac{Cq^2}{Ju_0} \end{bmatrix} \quad B = \begin{bmatrix} \frac{C_f}{m} \\ \frac{aC_f}{J} \end{bmatrix} \quad \boldsymbol{x} = \begin{bmatrix} v \\ \omega \end{bmatrix}$$

$$C = \begin{bmatrix} -\frac{C}{mu_0} & -\frac{C\nu}{mu_0} \\ 0 & 1 \\ \frac{1}{u_0} & 0 \end{bmatrix} \quad D = \begin{bmatrix} \frac{C_f}{m} \\ 0 \\ 0 \end{bmatrix}. \tag{4.26}$$

---

[4] The expression of the neutral steer point is obtained adding an external lateral force F_e in (4.16) together with the related external moment νF_e in (4.17) and solving for F_e and ν under the assumption of $\dot{v} = \dot{\omega} = \omega = \delta = 0$. A closely related concept is the *static margin*, which is defined as $-\nu/l$.

[5] In the case that tyre relaxation is included ($\frac{\sigma_i}{u}\dot{F}_{iy} + F_{iy} = C_i\alpha_i$, where $i = f, r$, and σ is the relaxation length; see Chapter 3), the state and input matrices become

$$A = \begin{bmatrix} 0 & -u & \frac{1}{m} & \frac{1}{m} \\ 0 & 0 & \frac{a}{J} & -\frac{b}{J} \\ -\frac{C_f}{\sigma_f} & -\frac{C_f a}{\sigma_f} - \frac{u}{\sigma_f} & 0 \\ -\frac{C_r}{\sigma_r} & \frac{C_r b}{\sigma_r} & 0 & -\frac{u}{\sigma_r} \end{bmatrix} \quad B = \begin{bmatrix} 0 \\ 0 \\ \frac{C_f u}{\sigma_f} \\ 0 \end{bmatrix} \quad \boldsymbol{x} = \begin{bmatrix} v \\ \omega \\ F_{fy} \\ F_{ry} \end{bmatrix} \quad \boldsymbol{u} = [\delta]. \tag{4.25}$$

Steering implications. Under 'small angle' cornering, the yaw rate, the turn radius of curvature $\mathcal{C} = 1/\mathcal{R}$, and the vehicle's speed are related by $V\mathcal{C} = \omega$. This can be expressed in state-space form using $\mathcal{C} = C_\mathcal{C}\boldsymbol{x}$ with output matrix

$$C_\mathcal{C} = \begin{bmatrix} 0 & 1/V \end{bmatrix}. \tag{4.27}$$

It follows from (4.26) that the transfer function $g(s)$, with s the Laplace variable, between the steering angle and the path curvature is given by

$$g(s) = C_\mathcal{C}(sI - A)^{-1}B. \tag{4.28}$$

Under steady-state conditions the path curvature is given by

$$\begin{aligned}
\mathcal{C} &= -C_\mathcal{C}A^{-1}B\delta_{ss} \\
&= \frac{\delta_{ss}/w}{1 + \frac{V^2}{gw} \cdot \frac{(bC_r - aC_f)mg}{wC_rC_f}} \\
&= \frac{\delta_{ss}/w}{1 + \frac{V^2}{wg}\zeta}
\end{aligned} \tag{4.29}$$

in which

$$\zeta = \frac{(bC_r - aC_f)mg}{wC_rC_f} \tag{4.30}$$

is the *understeer coefficient* or *understeer gradient* [8].[6] Equation (4.29) can be solved for δ_{ss} to yield

$$\delta_{ss} = \frac{w}{\mathcal{R}} + \frac{a_y}{g}\zeta, \tag{4.31}$$

in which $\mathcal{R} = 1/\mathcal{C}$ is the path radius of curvature and $a_y = V^2\mathcal{C}$ is the car's lateral acceleration. It follows from (4.31) that when $\zeta = 0$, the steering angle is independent of speed; this case is usually referred to as *neutral* steering. When $\zeta > 0$ (i.e. $bC_r > aC_f$ and $\nu < 0$), the steering angle must increase in proportion to $\frac{a_y}{g}$ in order to maintain a constant path curvature. Vehicles with this property are referred to as *understeering*. In order to maintain a constant path curvature when $\zeta < 0$ (i.e. $bC_r < aC_f$ and $\nu > 0$), the steering angle must be decreased in proportion to $\frac{a_y}{g}$ and the vehicle is referred to as *oversteering*.

These concepts can also be interpreted in terms of the tyre slips (4.5) and (4.6):

$$\alpha_f - \alpha_r = \delta_{ss} - \frac{\omega(a+b)}{u} \rightarrow \delta_{ss} = \frac{w}{\mathcal{R}} + \alpha_f - \alpha_r \tag{4.32}$$

in which we suppose that $u \approx V$ and $\omega/V = 1/\mathcal{R}$. It follows that in a neutral vehicle $\alpha_f = \alpha_r$, in an understeering vehicle $\alpha_f > \alpha_r$, and in an oversteering vehicle $\alpha_f < \alpha_r$.

---

[6] The understeer gradient can be written in terms of the tyre cornering stiffness per unit load: $\zeta = \frac{mgb}{w}\frac{1}{C_f} - \frac{mga}{w}\frac{1}{C_r} = \frac{F_{zf0}}{C_f} - \frac{F_{zr0}}{C_r} = \frac{1}{C_{f0}} - \frac{1}{C_{r0}}$, where F_{zr0} and F_{zf0} are the normal tyre loads at standstill and C_{r0} and C_{f0} are the cornering stiffnesses normalized by the standstill load.

Two distinctive velocities can be derived from (4.31). An oversteering car ($\zeta < 0$) has a *critical speed*[7] V_{cr} where the steer changes sign ($\delta_{ss} = 0$):

$$V_{cr} = \sqrt{-\frac{wg}{\zeta}}. \tag{4.33}$$

An understeering car ($\zeta > 0$) has a *characteristic speed* V_{ch}, where the steer angle is twice the kinematic or Ackermann steer angle ($\delta_{ss} = 2w/\mathcal{R}$):

$$V_{ch} = \sqrt{\frac{wg}{\zeta}}. \tag{4.34}$$

A third velocity is the *tangential speed* V_{tg}, which is the speed at which the longitudinal axis of the car is tangential to the trajectory ($v_{ss} = 0$):

$$V_{tg} = \sqrt{\frac{C_r bw}{am}}. \tag{4.35}$$

For $V < V_{tg}$ ($v_{ss} > 0$) the car travels on a turn 'nose out', while for $V > V_{tg}$ ($v_{ss} < 0$) the car travels 'nose in'.[8]

The above analysis holds only for small slip angles (a linear model). A more general definition of the understeer coefficient, which is valid for large slip angles, is

$$\zeta = \frac{\partial}{\partial(a_y/g)}\left(\delta_{ss} - \frac{w}{\mathcal{R}}\right), \tag{4.36}$$

which is identical to (4.30) in the linear case. Furthermore, the understeering/oversteering behaviour (i.e. the sign of ζ) can be deduced from the sign of the even simpler expression

$$\left.\frac{\partial\delta_{ss}}{\partial V}\right|_{\mathcal{R}=const}, \tag{4.37}$$

which deals with rates of change rather than absolute quantities. As a consequence, a car may start understeering at small slips, which implies (4.37)> 0 and $\alpha_f > \alpha_r$, then at large slips become oversteering, that is, (4.37)< 0, although $\alpha_f > \alpha_r$ hold true. This change of behaviour is related mainly to tyre nonlinearity.

These results have an interesting geometric interpretation that we now consider. The neutrally steering situation in which $\alpha_r = \alpha_f = \alpha$ is illustrated in Figure 4.2. Consider first that case in which the tyres are in pure non-slip rolling so that $\alpha_r = \alpha_f = 0$. In this case the lines connecting point C to the two tyre contact points are at right angles to the two wheel planes. If the front and rear tyres now operate with the same side slip angle, the instantaneous centre of rotation moves around the circumference of a circle as shown in Figure 4.2; this follows from the inscribed angle theorem. In this case the centre of rotation moves from C to C' as a result of tyre slip.

---

[7] This speed is 'critical' because above it the car is unstable.

[8] Under nose out conditions the projection of the velocity centre C falls behind the centre of mass G, while in nose-in conditions the projection of C falls ahead of G.

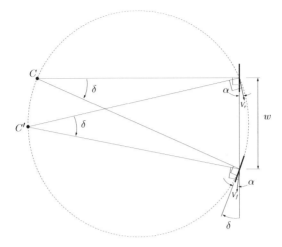

Figure 4.2: Neutral-steering single-track car model in which δ is the steering angle. The front- and rear-wheel tyre-contact velocity vectors are given by V_f and V_r respectively, with their corresponding side-slip angles given by α.

For any number of reasons the front- and rear-tyre side-slip angles may differ. Since the instantaneous centre of rotation is at the point of intersection of perpendiculars to the front- and rear-wheel velocity vectors, the centre of rotation moves from C' (as in Figure 4.2) to C'' (Figure 4.3) as the front-tyre slip angle reduces so that $\alpha_r > \alpha_f$, thereby producing an oversteering car. In order to maintain a constant-radius turn it follows from (4.31) that the steering angle has to be reduced as the vehicle's speed increases.

In an understeering car the tyre slip angles satisfy $\alpha_f > \alpha_r$ as shown in Figure 4.4. In this case the centre of rotation moves from C' (as in Figure 4.2) to C'' (Figure 4.4) as the front-tyre slip angle increases so that $\alpha_f > \alpha_r$. In order to hold a constant-radius turn in an understeering car, it follows from (4.31) that the steering angle has to be increased as the vehicle's speed increases.

Stability implications. The stability of this constant-velocity model has been studied extensively, for example in [8]. The poles of the system, which determine the car's stability, are the roots of the characteristic polynomial $\det(sI - A)$ with A given by (4.26). This gives

$$\lambda^2 + \frac{(a^2 + k^2)C_f + (b^2 + k^2)C_r}{mk^2 u_0}\lambda + \left(\frac{w^2 C_f C_r}{m^2 k^2 u_0^2} - \frac{C\nu}{mk^2}\right) = 0. \tag{4.38}$$

Following the Routh–Hurwitz criterion, the system will be stable if all the coefficients in (4.38) are positive. It is consequently clear that only an oversteering car can become unstable, since instability requires a positive value of $C\nu$ if the last term in (4.38) is to become negative. This occurs when the speed exceeds the critical value given by

$$u_0 > V_{crit} = \sqrt{\frac{w^2 C_f C_r}{mC\nu}}. \tag{4.39}$$

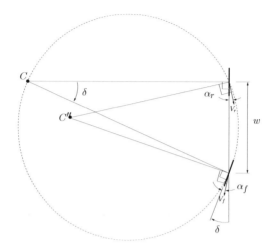

Figure 4.3: Oversteering single-track vehicle model. The front- and rear-wheel velocity vectors are given by V_f and V_r, with their corresponding side-slip angles satisfying $\alpha_r > \alpha_f$.

This is precisely the speed at which $\delta_{ss} = 0$, as derived above in (4.33).

To illustrate some of these ideas we consider a car with parameters given in Table 4.1. Two tyre sets have been chosen, in order to produce an understeering vehicle and an oversteering vehicle with a critical speed $V_{crit} \approx 70\,\text{m/s}$.

From the state-space model in (4.26) the transfer functions from the steering angle to each of the outputs are determined. While the system poles are determined by the characteristic equation, the zeros of each of the three transfer functions are given by

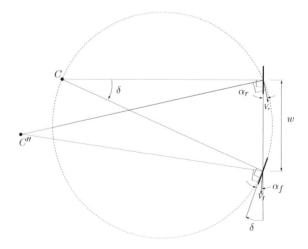

Figure 4.4: Understeering single-track vehicle, with the front- and rear-wheel side-slip angles satisfying $\alpha_F > \alpha_R$.

Table 4.1 Vehicle parameters.

Symbol	Description	Value
m	Car mass	$720\,\mathrm{kg}$
J	Moment of inertia about the z-axis	$450\,\mathrm{kg\,m^2}$
w	Wheelbase	$3.4\,\mathrm{m}$
a	Distance of mass centre from front axle	$1.8\,\mathrm{m}$
b	Distance of mass centre from rear axle	$1.6\,\mathrm{m}$
C_{f0}	Normalized front-tyre cornering stiffness (understeering)	$25\,\mathrm{rad^{-1}}$
C_{r0}	Normalized rear-tyre cornering stiffness (understeering)	$30\,\mathrm{rad^{-1}}$
C_{f0}	Normalized front-tyre cornering stiffness (oversteering)	$30\,\mathrm{rad^{-1}}$
C_{r0}	Normalized rear-tyre cornering stiffness (oversteering)	$25\,\mathrm{rad^{-1}}$

$$\{z_{a_y1}, z_{a_y2}\} = -\frac{bC_rw}{2Ju_0} \pm \sqrt{\left(\frac{bC_rw}{2Ju_0}\right)^2 - \frac{C_rw}{J}} \tag{4.40}$$

$$z_\beta = \frac{amu_0^2 - bC_rw}{Ju_0} \tag{4.41}$$

$$z_\omega = -\frac{C_rw}{mu_0a}. \tag{4.42}$$

The zeros z_{a_y} of the lateral acceleration transfer function will either be negative real, or will form a complex conjugate pair with a negative real part. The zero z_ω of the yaw velocity response transfer function will always be negative. At higher speeds, the zero z_β of the side-slip angle transfer function may become positive, resulting in a non-minimum phase response.

Figure 4.5 (a) shows the speed-dependent pole locations for the car with the parameters given in Table 4.1 and the understeering tyre combination. Both poles are non-oscillatory at low speed. As the speed increases, the poles move together along the

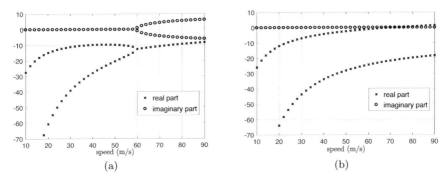

Figure 4.5: Real and imaginary parts of the simple car model eigenvalues for the parameters given in Table 4.1; (a) the understeering car, and (b) the oversteering car.

negative real axis, coalescing at $60\,\text{m/s}$ into an oscillatory complex-conjugate pair. The oscillatory frequency then increases with speed and the poles move towards the right-half plane. The case of the oversteering tyre combination is shown in Figure 4.5 (b); the poles are again non-oscillatory at low speed. As the speed increases, the poles migrate towards the right-half plane. At the critical speed (4.39) of $\approx 70\,\text{m/s}$ one pole enters the right-half plane, indicating instability. The tangential speeds (4.35) are $22\,\text{m/s}$ and $20\,\text{m/s}$ for the understeering and oversteering car respectively.

4.2.2   Accelerating and braking

It is shown in [160] that the vehicle model given in (4.26) can be considered slowly time-varying and therefore that the eigenvalues can be used to assess the vehicle's stability. As the vehicle accelerates (or decelerates), the frozen-time eigenvalues evolve with time and can be used to assess the time-localized stability of the car. These results are compared with the closed-form solution of the describing time-varying linear differential equation.

Closed-form solution.   The first-order differential equations (4.16) and (4.17) can be combined by eliminating the lateral velocity to produce a second-order differential equation in the yaw rate w. The analysis is further simplified by assuming that the steering angle is held at zero in order to obtain

$$u(t)^2\ddot{w} + Pu(t)\dot{w} + (Qu(t)^2 + R)w = 0, \tag{4.43}$$

where the parameters P, Q, and R are given by

$$P = \frac{-k^2 ma_x + (a^2 + k^2)C_f + (b^2 + k^2)C_r}{mk^2}, \quad Q = -\frac{C\nu}{mk^2}, \quad R = \frac{w^2 C_f C_r}{m^2 k^2}. \tag{4.44}$$

This time-varying differential equation has a structure reminiscent of a Bessel's differential equation, and a closed-form solution can be found in terms of Bessel functions. As demonstrated in [161], an equation with the form

$$u^2\frac{d^2 w}{du^2} + Pu\frac{dw}{du} + (Qu^2 + R)w = 0 \tag{4.45}$$

has solution

$$w = c_1 u^\rho(t) J_\nu(\alpha u(t)) + c_2 u^\rho(t) Y_\nu(\alpha u(t)), \tag{4.46}$$

in which J_ν and Y_ν are Bessel functions of the first and second kind, with c_1 and c_2 constants of integration. To convert (4.43) into (4.45), a change in the independent variable from t to u is required. The first- and second-order derivatives of w with respect to u are

$$\frac{dw}{dt} = \frac{dw}{du}\frac{du}{dt} = a_x\frac{dw}{du} \tag{4.47}$$

and

$$\frac{d^2 w}{dt^2} = \frac{d^2 w}{du^2}\left(\frac{du}{dt}\right)^2 + \frac{dw}{du}\frac{d^2 u}{dt^2} = a_x^2\frac{d^2 w}{du^2} \tag{4.48}$$

since the acceleration is assumed constant. The parameters ρ and α, and the order of the Bessel functions ν, can now be related to the coefficients of the differential equation and a_x by:

$$\rho = \frac{1 - \frac{P}{a_x}}{2}, \qquad \alpha = \sqrt{\frac{Q}{a_x^2}}, \qquad \nu = \sqrt{\frac{(1 - \frac{P}{a_x})^2}{4} - \frac{R}{a_x^2}}. \qquad (4.49)$$

If Q is negative, the argument of the Bessel functions is complex. In this case, in order to obtain real solutions, modified Bessel functions I_ν and K_ν of the first and second kind must be used to obtain

$$\omega = c_1 u^\rho(t) I_\nu(\alpha u(t)) + c_2 u^\rho(t) K_\nu(\alpha u(t)). \qquad (4.50)$$

The sign of Q is determined by the sign of $C\nu$ (4.20). However, unlike the constant-velocity case, this quantity is a function of the vehicle acceleration. A car might be oversteering at a constant velocity ($C\nu > 0$), but when the vehicle accelerates, the normal load transfer to the rear increases C_r and reduces simultaneously C_f, which may cause the car to understeer ($C\nu < 0$). Under braking the opposite occurs and $C\nu$ may remain positive (or it may become positive if the vehicle was initially understeering). As the sign of $C\nu$ changes, the sign-appropriate Bessel functions must be used.

If the velocity is assumed constant the yaw differential equation reduces to a linear second-order time-invariant equation with solution

$$\omega = c_1 e^{\lambda_1 t} + c_2 e^{\lambda_2 t}, \qquad (4.51)$$

where λ_1 and λ_2 are real eigenvalues. The constants c_1 and c_2, once again, come from initial conditions. When $C\nu$ is negative (understeering), and the speed sufficiently high, the eigenvalues may become complex. The time-invariant yaw response will differ from the time-varying solution over long time periods, and short-horizon time-invariant yaw responses must be pieced together. Suppose that the time horizon is divided into a sequence of subintervals, and that the time-invariant solution is computed on each subinterval with the terminal conditions from the previous interval used as the initial conditions for the current interval. As these subintervals decrease in duration the solution accuracy, as expected, improves.

The eigenvalues and system responses for constant $\pm 10\,\text{m/s}^2$ of acceleration/braking are shown in Figure 4.6 for the car parameters given in Table 4.1, with the oversteering tyre combination, and for an initial yaw velocity of 0.001 rad/s. When braking from 80 m/s to 20 m/s, one of the frozen-time eigenvalues changes sign at $V_{crit} \approx 40\,\text{m/s}$, which is much lower than the 70 m/s found for the constant-velocity case. The load transfer due to braking reduces the rear-tyre cornering stiffness, while increasing that in the front. This, in turn, leads to an accompanying reduction in the critical speed; see (4.39). The response of the time-varying system to a non-zero initial yaw velocity at 80 m/s is determined using the closed-form solution in (4.46). Initially, the yaw angle grows exponentially, typical of an unstable system. However, as the vehicle speed decreases, and the eigenvalues become negative, the yaw angle response slows and then

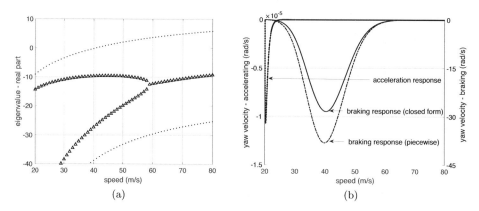

(a) (b)

Figure 4.6: (a) shows the real parts of the frozen-time eigenvalues of a vehicle accelerating in straight running at $10\,\mathrm{m/s^2}$ from $20\,\mathrm{m/s}$ to $80\,\mathrm{m/s}$ (shown as the curve formed from $\triangle$s), and a vehicle braking in straight running at $-10\,\mathrm{m/s^2}$ for vehicle speeds from $80\,\mathrm{m/s}$ to $20\,\mathrm{m/s}$ (shown dotted). (b) shows the yaw response to a non-zero initial yaw velocity at $20\,\mathrm{m/s}$ for the accelerating case, and at $80\,\mathrm{m/s}$ for braking.

decays. This behaviour suggests that a vehicle braking from $80\,\mathrm{m/s}$ may exhibit an undesirable 'burst' response in the $40\,\mathrm{m/s}$ speed range. Under acceleration the oversteering vehicle becomes understeering and the system response is stable. At higher speeds the eigenvalues are complex, which manifests in an underdamped yaw response to a perturbation.

4.2.3 Cornering

The straight-running analysis can be extended to the cornering case for both constant and variable speeds, but this must be done with care.

Constant-velocity case. The first complication relates to the fact that under cornering conditions the longitudinal and lateral dynamics given in (4.1) and (4.2) interact. The second complication relates to the fact that in cornering, the steady-state tyre side-slip angles are no longer zero and the linear tyre model given in (4.15) is not reasonably usable, unless the lateral accelerations are small. In these cases more complex tyre models of the form

$$F_y = f(\alpha, \kappa, F_z) \tag{4.52}$$

should be employed (see (3.84) and (3.109)–(3.110)); α, κ, and F_z are respectively the side-slip angle, the longitudinal slip, and the normal load. For present purposes we treat the tyre force laws as nonlinear functions of the three argument variables.

   If the EOMs in (4.1), (4.2), and (4.3) are linearized about a constant speed and a constant steady-state radius-of-turn, one obtains:

$$A = \begin{bmatrix} \frac{-C_f\delta_0(v_0+a\omega_0)}{mu_0^2} & \frac{C_f\delta_0}{mu_0}+\omega_0 & \frac{C_fa\delta_0}{mu_0}+v_0 \\ \frac{Cv_0+Cs\omega_0}{mu_0^2}-\omega_0 & -\frac{C}{mu_0} & \frac{-Cs}{mu_0}-u_0 \\ \frac{Csv_0+Cq^2\omega_0}{Ju_0^2} & \frac{-Cs}{Ju_0} & -\frac{Cq^2}{Ju_0} \end{bmatrix} \quad B = \begin{bmatrix} -\frac{C_f\delta_0+F_{fy0}}{m} \\ \frac{C_f-F_{fy0}\delta_0}{m} \\ \frac{C_fa-F_{fy0}a\delta_0}{J} \end{bmatrix}$$

$$\tag{4.53}$$

$$C = \begin{bmatrix} \frac{Cs\omega_0+v_0C}{mu_0^2} - \frac{C}{mu_0} & \frac{Cs}{mu_0} \\ 0 & 0 & 1 \\ -\frac{v_0}{u_0^2} & \frac{1}{u_0} & 0 \end{bmatrix} \quad D = \begin{bmatrix} \frac{C_f-F_{fy0}\delta_0}{m} \\ 0 \\ 0 \end{bmatrix}.$$

The outputs are again $[\hat{a}_y, \delta, \omega]$. For the front tyre one has

$$\begin{bmatrix} \frac{\partial F_{fy}}{\partial u} \\ \frac{\partial F_{fy}}{\partial v} \\ \frac{\partial F_{fy}}{\partial \omega} \\ \frac{\partial F_{fy}}{\partial \delta} \end{bmatrix} = \frac{\partial F_{fy}}{\partial \alpha_f}\Big|_{\alpha_{f0},\kappa_{f0},F_{fz0}} \begin{bmatrix} \frac{\partial \alpha_f}{\partial u} \\ \frac{\partial \alpha_f}{\partial v} \\ \frac{\partial \alpha_f}{\partial \omega} \\ \frac{\partial \alpha_f}{\partial \delta} \end{bmatrix}_{u_0,v_0,\omega_0,\delta_0} = C_f \begin{bmatrix} \frac{\omega_0 a+v}{u^2} \\ -\frac{1}{u_0} \\ -\frac{a}{u_0} \\ 1 \end{bmatrix}, \tag{4.54}$$

in which C_f is the gradient of the tyre force characteristic local to the steady-state operating point; C_f is operating point dependent. For the rear tyre one has

$$\begin{bmatrix} \frac{\partial F_{ry}}{\partial u} \\ \frac{\partial F_{ry}}{\partial v} \\ \frac{\partial F_{ry}}{\partial \omega} \\ \frac{\partial F_{ry}}{\partial \delta} \end{bmatrix} = \frac{\partial F_{ry}}{\partial \alpha_r}\Big|_{\alpha_{f0},\kappa_{f0},F_{fz0}} \begin{bmatrix} \frac{\partial \alpha_r}{\partial u} \\ \frac{\partial \alpha_r}{\partial v} \\ \frac{\partial \alpha_r}{\partial \omega} \\ \frac{\partial \alpha_r}{\partial \delta} \end{bmatrix}_{u_0,v_0,\omega_0,\delta_0} = C_r \begin{bmatrix} \frac{v-\omega_0 b}{u^2} \\ -\frac{1}{u_0} \\ -\frac{b}{u_0} \\ 0 \end{bmatrix}. \tag{4.55}$$

The operating conditions required to evaluate the linearized model are determined by the steady-state speed V and turn radius $\mathcal{R}$. Once V and $\mathcal{R}$ have been chosen, ω is determined by the kinematic relation:

$$\omega = \frac{V}{\mathcal{R}}. \tag{4.56}$$

The resultant vehicle speed is related to the lateral and longitudinal velocity magnitudes by

$$V^2 = v^2 + u^2. \tag{4.57}$$

Under steady-state conditions, the EOMs reduce to the following algebraic relations (for a RWD car)

$$0 = m\omega_0 v_0 - F_{fy0}\delta_0 + F_{rx0} \tag{4.58}$$
$$0 = -m\omega_0 u_0 + F_{fy0}\cos\delta_0 + F_{ry0} \tag{4.59}$$
$$0 = aF_{fy0}\cos\delta_0 - bF_{ry0}. \tag{4.60}$$

These five equations and the two nonlinear tyre models can be solved simultaneously to determine the steady-state values of u_0, v_0, ω_0, δ_0, as well as F_{rx0}, which is the (small) driving force required to maintain the vehicle at a constant velocity. This steady-state solution, together with the local tyre stiffnesses C_f and C_r, can now be substituted in (4.53) in order to determine the local steering dynamics and stability of the car.

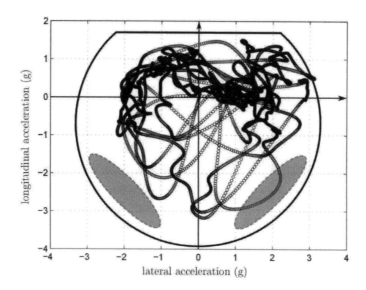

Figure 4.7: Sample G-G diagram of a Formula One car that shows the vehicle's lateral and longitudinal acceleration on a qualifying lap at Silverstone. The track contains eighteen corners, of which ten are right-handed; this accounts for the vehicle's positive-dominated lateral acceleration. In addition, the high-speed corners, out of which the driver accelerates, are also right-handed, which is consistent with vehicle's prolonged operation in the top-right quadrant of the diagram.

Acceleration and braking under cornering.   The operating characteristics of cars are often described in terms of a so-called G-G diagram (Figure 4.7). In this diagram the lateral acceleration of the vehicle is plotted against the longitudinal acceleration with a boundary representing the physical limits of the car. The engine power is the primary influence on the positive longitudinal acceleration limit, while the tyre grip is the key factor in determining the lateral and braking limits. A graphical overlay of recorded telemetry data can be used to analyse the driver's exploitation of the performance envelope. In racing, the more the driver is able to utilize the capability of the vehicle, the lower the lap time.

Handling and stability characteristics influence the driver's ability to operate the vehicle on the limit of performance. An unstable vehicle may undermine the driver's confidence, or in extreme cases make the vehicle undriveable. The highlighted regions in Figure 4.7 correspond to areas in which the vehicle is typically unstable and more difficult to control, thus requiring highly skilled drivers. Characterizing the vehicle stability across the G-G diagram is useful for predicting the extent to which the driver is able to exploit the vehicle's performance. The treatment of stability over the full operating range of the vehicle has produced a number of analysis techniques, but all have their limitations [162–164]

Straight-line accelerating and braking, and steady-state cornering, only represent the axes of the G-G diagram. To determine the stability over the rest of the operating regime it is necessary to develop a method of examining acceleration and braking under cornering. In this scenario, the vehicle is no longer in equilibrium, since the states are changing with time. It is possible to study the stability of an accelerating or braking vehicle by examining perturbations from some nominal motion [165]. To conduct this type of analysis the vehicle states along a nominal prescribed trajectory are required. These evolving states can be found by simulating the manoeuvre using the nonlinear vehicle model with feedback controllers that are used to generate the plant inputs that produce the required motion. This approach has a number of limitations: simulating the manoeuvre is a time-consuming process and determining vehicle stability is intended to be a way predicting the vehicle's response *without having to numerically integrate the equations of motion*! Furthermore, as the vehicle becomes unstable, and the braking and acceleration become more aggressive, the feedback controllers are less able to maintain the vehicle on the reference trajectory.

An alternative approach to determining the frozen-time vehicle states is to use a d'Alembert force system that acts as a surrogate for the real forces of inertia experienced under acceleration. In this approach, an accelerating or braking vehicle is approximated by one travelling at a constant velocity, but subjected to forces that capture the inertial effects of accelerating and braking. The advantage of this method is clear. The vehicle remains in equilibrium and the vehicle states can be determined from algebraic equations of motion, which now contain additional terms representing the apparent forces. The frozen-time eigenvalues of the system are computed at each point along a motion trajectory. The choice of manoeuvres used to 'fill out' the G-G diagram is not unique [166].

4.3 Timoshenko–Young bicycle

4.3.1 Introduction

The next introductory example is a simple bicycle model analysis due to Timoshenko and Young [167]; see Figure 4.8. The coordinates of the rear-wheel ground-contact point are (x, y) and are given in an inertial reference frame $Oxyz$. The vehicle's roll angle is ϕ and its yaw angle is ψ. The steer angle δ is measured between the front frame and the rear frame. The vehicle's mass m is concentrated at its mass centre that is a distance h from the line joining the two wheel ground-contact points, and a distance b in front of the rear-wheel ground-contact point. The acceleration due to gravity is denoted g, which points in the positive z-direction. The machine's wheelbase is w. Although wheels are drawn in Figure 4.8, they are both massless and inertia-less and so can be conceptually replaced by skates that can only slip in the direction in which they point (much like the Čhaplygin sleigh; see Section 2.8). To make the analysis easier, we assume that the radius of the front wheel is vanishingly small and that consequently the front-wheel ground-contact point is independent of the roll and steer angles.

4.3.2 Steering kinematics

The motion of the bicycle's wheels (skates) are constrained by a pure rolling assumption. Under this assumption the rear wheel must move in the direction of its line of

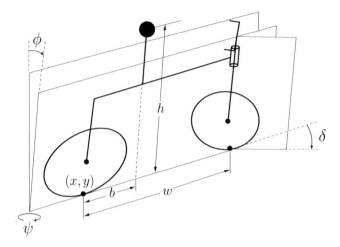

Figure 4.8: Timoshenko–Young bicycle. The yaw angle is denoted by ψ, the roll angle by ϕ, and the steer angle by δ. The mass centre is located at $(x + b, y - h)$ when the bicycle is upright and aligned with the positive x-axis. The wheelbase is w.

intersection with the ground plane. Thus

$$\dot{x} = uc_\psi \tag{4.61}$$

$$\dot{y} = us_\psi, \tag{4.62}$$

in which u is the vehicle's forward speed as measured at the rear ground-contact point, and s_ψ and c_ψ represent the sine and cosine of ψ. There are rolling constraints associated with the front wheel too.

Referring to Figure 4.9 we see that the bicycle's mass is located at

$$x_G = x + bc_\psi - hs_\psi s_\phi \tag{4.63}$$

$$y_G = y + bs_\psi + hc_\psi s_\phi \tag{4.64}$$

$$z_G = -hc_\phi, \tag{4.65}$$

while the front-wheel ground-contact point is located at

$$x^{fw} = x + wc_\psi \tag{4.66}$$

$$y^{fw} = y + ws_\psi. \tag{4.67}$$

The front-wheel ground-contact point velocity is thus

$$\dot{x}^{fw} = uc_\psi - w\dot{\psi}s_\psi \tag{4.68}$$

$$\dot{y}^{fw} = us_\psi + w\dot{\psi}c_\psi \tag{4.69}$$

in which (4.61) and (4.62) were used.

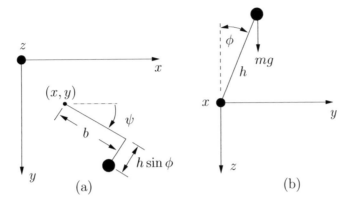

Figure 4.9: Timoshenko–Young bicycle. In Figure (a) the z-axis points into the page, while in Figure (b) the x-axis points into the page.

The bicycle's orientation in the inertial reference frame can be found using

$$\mathcal{R}_I = \mathcal{R}(e_z, \psi)\mathcal{R}(e_x, \phi) = \begin{bmatrix} c_\psi & -s_\psi c_\phi & s_\psi s_\phi \\ s_\psi & c_\psi c_\phi & -c_\psi s_\phi \\ 0 & s_\phi & c_\phi \end{bmatrix}. \tag{4.70}$$

The steering behaviour can be computed using

$$\mathcal{R}_S = \begin{bmatrix} c_\delta & -s_\delta & 0 \\ s_\delta & c_\delta & 0 \\ 0 & 0 & 1 \end{bmatrix}, \tag{4.71}$$

in which δ is the steering angle. If $\boldsymbol{\lambda}_B^{fw} = [0\,1\,0]^T$ is a unit vector aligned with the front wheel spindle in body-fixed coordinates, then

$$\boldsymbol{\lambda}_I^{fw} = \mathcal{R}_I \mathcal{R}_S \boldsymbol{\lambda}_B^{fw} \tag{4.72}$$

$$= \begin{bmatrix} -c_\psi s_\delta - s_\psi c_\phi c_\delta \\ -s_\psi s_\delta + c_\psi c_\phi c_\delta \\ s_\phi c_\delta \end{bmatrix} \tag{4.73}$$

is in the inertial reference frame. The projection of this vector onto the ground plane is orthogonal to the line of intersection between the front wheel and the ground plane. We call this vector $\boldsymbol{n}$:

$$\boldsymbol{n} = \begin{bmatrix} -c_\psi s_\delta - s_\psi c_\phi c_\delta \\ -s_\psi s_\delta + c_\psi c_\phi c_\delta \\ 0 \end{bmatrix}. \tag{4.74}$$

Under pure rolling the front wheel's ground-contact point velocity must follow the line of intersection between the wheel and the ground plane. Since the ground-contact point velocity and $\boldsymbol{n}$ must be orthogonal, there holds $[\dot{x}^{fw}, \dot{y}^{fw}, 0]^T \cdot \boldsymbol{n} = 0$, which gives

$$\dot{\psi} = \frac{vt_\delta}{wc_\phi}, \tag{4.75}$$

in which t_δ is $\tan\delta$. This equation shows how the yaw rate, and thus the radius of turn, is dictated (kinematically) by the steering angle.

We conclude this section by introducing the kinematic steering angle Δ for the Timoshenko–Young bicycle. Under the assumptions associated with nonholonomic rolling, the wheels must travel in the direction of the lines of intersection of the wheel planes and the ground plane. The kinematic steering angle is the angle between these lines of intersection, and therefore the angle between the front- and rear-wheel velocity vectors. Without loss of generality the rear wheel may be instantaneously aligned with the x-axis by setting $\psi = 0$. That is

$$\Delta = \tan^{-1}\left(\frac{\dot{y}^{fw}}{\dot{x}^{fw}}\right)\Bigg|_{\psi=0} = \tan^{-1}\left(\frac{t_\delta}{c_\phi}\right) \tag{4.76}$$

using (4.68), (4.69), and (4.75), where $t_\delta = \tan\delta$. The steering angle δ and the kinematic steering angle Δ are different when the bicycle is 'leaned over'. Since the instantaneous radius of curvature $\hat{\mathcal{R}}$ and the yaw rate are related by

$$u = \dot{\psi}\hat{\mathcal{R}}, \tag{4.77}$$

it follows from (4.75) that the instantaneous radius of curvature is given by $\hat{\mathcal{R}} = \frac{wc_\phi}{t_\delta}$, with the path curvature given by

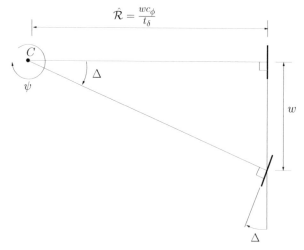

Figure 4.10: Ground-plane kinematics of the Timoshenko–Young model. The kinematic steering angle Δ is the angle between the line of intersection of the front wheel and the ground plane and the line of intersection of the rear wheel and the ground plane. The instantaneous centre of rotation is C.

$$C = \frac{1}{\hat{\mathcal{R}}} = \frac{t_\delta}{wc_\phi}. \tag{4.78}$$

The ground-plane kinematics of the Timoshenko–Young bicycle are shown in Figure 4.10. It follows from (4.78) that the radius of turn will 'tighten' if the bicycle is rolled over at fixed steering angle. A detailed study of motorcycle steering kinematics can be found in chapter 1 of [6].

4.3.3   Vehicle dynamics

We will determine the vehicle's equations of motion using Newton–Euler vector mechanics principles. To eliminate the reactions at the road–wheel contacts, we take moments around an axis that passes between the wheel ground-contact points. Without loss of generality we can consider an instant when the axis of moments is aligned with the x-axis. Taking moments gives

$$0 = gs_\phi - \ddot{z}_G s_\phi - \ddot{y}_G c_\phi. \tag{4.79}$$

It follows from (4.64) and (4.65) that

$$\ddot{y}_G|_{\psi=0} = u^2 C + b\ddot{\psi} - hs_\phi \dot{\phi}^2 + hc_\phi \ddot{\phi} - hs_\phi u^2 C^2 \tag{4.80}$$

$$\ddot{z}_G = hc_\phi \dot{\phi}^2 + hs_\phi \ddot{\phi}, \tag{4.81}$$

since

$$\dot{\psi} = uC$$

and

$$\ddot{y}|_{\psi=0} = u^2 C,$$

in which C is the instantaneous curvature of the rear-wheel ground-contact point trajectory. Substitution into (4.79) gives

$$h\ddot{\phi} = gs_\phi - c_\phi \left((1 - hCs_\phi)Cu^2 + b\ddot{\psi} \right). \tag{4.82}$$

An alternative version of (4.82), which is in terms of the steer angle rather than the yaw angle, can be obtained from (4.75) as

$$h\ddot{\phi} = gs_\phi - t_\delta \left(\frac{u^2}{w} + \frac{b\dot{u}}{w} + t_\phi \left(\frac{ub}{w}\dot{\phi} - \frac{hu^2}{w^2}t_\delta \right) \right) - \frac{bu\dot{\delta}}{wc_\delta^2}. \tag{4.83}$$

An inspection of either (4.82) or (4.83) shows that the Timoshenko–Young bicycle model is time-reversible and so is likely neutrally stable.

4.3.4   Lagrange's equation

The bicycle's kinetic energy is given by

$$T = \frac{m}{2}(\dot{x}_G^2(t) + \dot{y}_G^2(t) + \dot{z}_G^2(t)), \tag{4.84}$$

in which $\dot{x}_G$, $\dot{y}_G$, and $\dot{z}_G$ can be computed using (4.63), (4.64), and (4.65). The machine's potential energy is

$$V = -mgz_G, \tag{4.85}$$

and so the Lagrangian is given by

$$\mathcal{L} = T - V. \tag{4.86}$$

If we solve

$$0 = \frac{d}{dt}\left(\frac{\partial \mathcal{L}}{\partial \dot{\phi}}\right) - \left(\frac{\partial \mathcal{L}}{\partial \phi}\right) \tag{4.87}$$

for $\ddot{\phi}$, and substitute for the accelerations $\ddot{x}_G$, $\ddot{y}_G$, and $\ddot{z}_G$ that can be computed using (4.63), (4.64), and (4.65), then (4.82) again follows. Equation (4.83) can be found by substituting (4.75) as before.

As is explained in Section 2.2.8, any nonholonomic constraints are enforced by appending them to Lagrange's equations using Lagrange multipliers. In this case the time derivative of the roll coordinate does not appear in the constraint equations (4.68) and (4.69) and so the inclusion of Lagrange multipliers in (4.87) is unnecessary.

4.3.5   Linearized model

Equation (4.83) represents the nonholonomic Timoshenko–Young bicycle with control inputs δ and u. This equation can be linearized about a constant-speed straight-running condition in order to obtain a simple linear model that relates to small perturbation behaviour:

$$\ddot{\phi} = \frac{g}{h}\phi - \frac{u^2}{hw}\delta - \frac{bu}{wh}\dot{\delta}. \tag{4.88}$$

By taking Laplace transforms, we obtain from (4.88) the single-input single-output transfer function between δ and ϕ

$$H_{\phi\delta}(s) = -\frac{bu}{wh}\frac{s + u/b}{s^2 - g/h}, \tag{4.89}$$

which has a speed-dependent gain, a speed-dependent zero at $-u/b$, and fixed poles at $\pm\sqrt{g/h}$. The unstable pole corresponds to an inverted-pendulum-type capsize mode. The zero is in the left-half plane under forward-running conditions. It then moves through the origin into the right-half plane as the speed is reduced to zero, and then reverses in sign. Under backward-running conditions the right-half plane zero, which for some speeds comes into close proximity to the right-half plane pole, is associated with the control difficulties found in rear-steering bicycles [168].

Equation (4.89) tells us that in straight running the linearized Timoshenko-Young bicycle is always unstable with a fixed pole at $\sqrt{g/h}$; this is reminiscent of the unstable capsize mode associated with the simple inverted pendulum. This property does not carry over to even slightly more realistic models of the type considered in Section 4.4; in this case the poles of the linearized system become speed dependent and the vehicle model has a stable speed range. In more realistic models, certain of the vehicle's transfer functions have a right-half plane zero for positive speeds, resulting in a phenomenon known as 'counter steering'; more will be said about this in Chapter 5.

4.4 Lumped-mass bicycle

The ability of a bicycle to 'self balance' is one of its most important features. This self-stability property was not a feature of the Timoshenko–Young bicycle analysed in Section 4.3, and so the Timoshenko–Young model lacks at least one important dynamic feature of a practical bicycle. In order to begin the process of understanding this self-stability property, we will analyse briefly a point-mass bicycle introduced in [169] and illustrated in Figure 4.11. The key feature of this point-mass bicycle is that it is stable over a range of speed even though it has neither mechanical trail nor gyroscopic influences, which have for a long time been considered essential for a self-balancing two-wheeled vehicle. In broad terms, however, trail and gyroscopic moments are beneficial to stability.

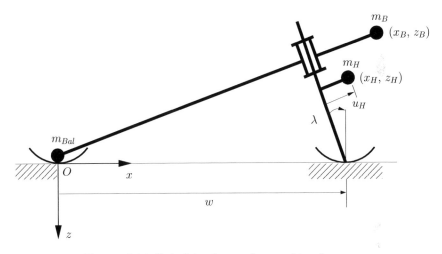

Figure 4.11: Primitive lumped-mass bicycle.

The primitive point-mass bicycle comprises three point masses, an inclined steering axis, and wheels without mass or inertia. From a dynamic modelling perspective the wheels may be treated as 'skates' that ensure that the vehicle proceeds along the lines of intersection between the ground plane and the two wheel planes. The rear frame has mass m_B and a ballast mass m_{Bal}, which is located at the rear-wheel ground-contact point. The ballast mass plays no role in the lateral dynamics, and serves only to prevent the rear wheel leaving the ground, thereby allowing the machine to pitch forward. The machine parameters we will use in this short study are given in Table 4.2.

4.4.1 Equations of motion

The first equation of motion comes from a momentum balance around an axis in the ground plane that is instantaneously aligned with the line of intersection between the rear frame and the ground plane:

$$T_B + g(u_H m_H \delta - (m_B z_B + m_H z_H)\varphi) = (m_B z_B^2 + m_H z_H^2)\ddot{\varphi}$$
$$- (m_B z_B + m_H z_H)\ddot{y}_R - (m_B x_B z_B + m_H x_H z_H)\ddot{\psi} - m_H u_H z_H \ddot{\delta}, \quad (4.90)$$

where $\ddot{y}_R$ is the lateral acceleration of the rear-wheel ground-contact point. The terms on the left-hand side include an external roll moment T_B, and gravitational terms relating to roll (φ) and steer influences (δ). The terms on the right of this equation relate respectively to the machine's roll acceleration, its lateral acceleration ($\ddot{y}_R$), its yaw acceleration, and the steering angle acceleration. The perpendicular distance betweeen the front frame mass m_H and the steer axis is given by

$$u_H = (x_H - w)\cos\lambda - z_H \sin\lambda,$$

as can be confirmed by a simple geometric calculation.

The second equation comes from an angular momentum balance around the steering axis.

$$T_{HB} + g(\varphi + \delta\sin\lambda)m_H u_H = m_H u_H \ddot{y}_R$$
$$- m_H u_H z_H \ddot{\varphi} + m_H u_H x_H \ddot{\psi} + m_H u_H^2 \ddot{\delta}. \quad (4.91)$$

The terms on the left-hand side of this equation contain forcing terms due to the steering moment T_{HB} and gravity-related roll and steer terms. The terms on the right-hand side of this equation describe the machine's lateral acceleration and inertial terms relating to roll, yaw, and steering respectively.

To obtain the equations that describe the bicycle's behaviour for small perturbations from straight running, we now need to recognize the nonholonomic constraints associated with the two road wheels (skates). The road-contact constraints allow the lateral position and its derivatives, and the yaw angle and its derivatives, to be expressed in terms of the steer angle. Figure 4.12 shows that the instantaneous radius of curvature is given by

$$\hat{\mathcal{R}} = \frac{w}{\delta\cos\lambda},$$

in which $\delta\cos\lambda$ is the (small) kinematic steering angle, which is the angle between the line of intersection of the rear wheel and the ground plane and the line of intersection

Table 4.2 Lumped-mass bicycle parameters.

Parameters	Symbol	Value
Wheelbase	w	1.0 m
Steer axis tilt angle	λ	$5\pi/180$ rad
Gravity	g	9.81 m/s^2
Rear frame		
Position of centre of mass	(x_B, y_B, z_B)	$(1.2, 0.0, -0.4)$ m
Mass	m_B	10 kg
Front frame		
Position of centre of mass	(x_H, y_H, z_H)	$(1.02, 0.0, -0.2)$ m
Mass	m_H	1 kg

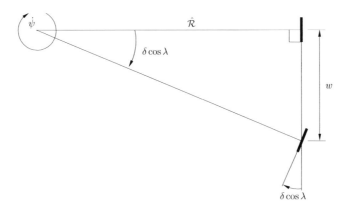

Figure 4.12: Bicycle ground-plane kinematics.

of the front wheel and the ground plane. For small angles, a handlebar steering angle of δ results in a kinematic steering angle of $\delta \cos \lambda$ and a (front-wheel) camber angle of $\delta \sin \lambda$. This means that the lateral acceleration $\ddot{y}_R$ may be expressed as

$$\ddot{y}_R = \frac{u^2}{\hat{\mathcal{R}}} = \frac{u^2 \delta \cos \lambda}{w},$$ (4.92)

where u is the velocity at the rear contact point. It is also clear that

$$\dot{\psi} = \frac{u}{\hat{\mathcal{R}}} = \frac{u \delta \cos \lambda}{w}$$

and so

$$\ddot{\psi} = \frac{u \dot{\delta} \cos \lambda}{w}.$$

The final roll equation comes from (4.90) and (4.91), once $\ddot{y}_R$ and $\ddot{\psi}$ have been eliminated. The resulting equations can be written in the form

$$M\ddot{q} + uC\dot{q} + g(K_0 + u^2 K_2)q = \begin{bmatrix} T_B \\ T_{HB} \end{bmatrix},$$ (4.93)

in which $q = [\phi, \delta]^T$, and the entries of the mass, damping, and stiffness matrices are given by

$$M_{11} = m_B z_B^2 + m_H z_H^2$$
$$M_{12} = -m_H u_H z_H$$
$$M_{21} = M_{12}$$
$$M_{22} = m_H u_H^2,$$

$$C_{11} = 0$$
$$C_{12} = -(m_B x_B z_B + m_H x_H z_H) \cos \lambda / w$$
$$C_{21} = 0$$
$$C_{22} = (m_H u_H x_H) \cos \lambda / w,$$

$$K_{011} = m_B z_B + m_H z_H$$
$$K_{012} = -m_H u_H$$
$$K_{021} = K_{012}$$
$$K_{022} = -m_H u_H \sin \lambda$$

and

$$K_{211} = 0$$
$$K_{212} = -(m_B z_B + m_H z_H) \cos \lambda / w$$
$$K_{221} = 0$$
$$K_{222} = m_H u_H \cos \lambda / w$$

respectively.

The cyclic coordinates, which determine the bicycle's (rear contact point) position and orientation, can be recovered a posteriori by integrating the nonholonomic constraint (4.93). The machine position can then be tracked by integrating

$$\dot{x}_P = u \cos \psi \qquad \dot{y}_P = u \sin \psi; \qquad (4.94)$$

small angles are not assumed.

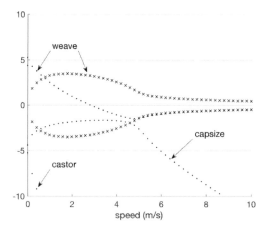

Figure 4.13: Stability properties of the primitive lumped-mass bicycle. The speed is varied from 0 to 10 m/s. The real parts of the eigenvalues are shown as dots, while the imaginary parts are shown as crosses.

Figure 4.13 shows the speed-dependent eigenvalues of the primitive bicycle that are computed using the characteristic equation associated with (4.93). The four zero-speed eigenvalues are located at ± 5.2484 and ± 4.3834 corresponding to inverted-pendulum-type capsize modes involving the whole machine and the steering assembly. The two unstable zero-speed modes immediately coalesce to form the so-called *weave mode*, which appears as a mixed roll-steer (fishtailing) oscillation. The two stable modes become the stable dynamic capsize mode and a (stable) castor mode. As the plot shows, this machine is stable for speeds above $2.8 \, \mathrm{m/s}$.

4.5 Wheel shimmy

Wheel shimmy can occur in everyday equipment such as shopping trolleys, wheeled stretchers, and wheelchairs. More important from our point of view is the occurrence of shimmy-related phenomena in road vehicles. The wheel shimmy modes in aircraft landing gear, automotive steering systems, and single-track vehicles can be violent and may lead to equipment damage and/or accidents. In March 1989 the left-hand main landing gear of a Fokker 100 aircraft failed in the landing at Geneva airport. From theoretical and experimental considerations it was concluded that any two-wheeled 'F.28-like' landing gear is essentially unstable [170]. Another comprehensive study of landing gear shimmy can be found in [87]. Original records of accelerations in high-speed hands-on bicycle shimmy are given in [86], which highlight how bicycle wobble (or shimmy) can be a frightening and dangerous phenomenon. The results show that shimmy is not attributable solely to the bicycle, but depends also on the rider. Another study of bicycle shimmy with experiments is given in [85]. The effect of frame compliance and rider mobility on scooter stability is investigated in [84]. In this work particular attention is paid to the wobble mode, because it may become unstable within the vehicle speed range. This article includes a summary of previous work and discusses the results of both numerical and experimental analyses. In practical situations these phenomena are complex and demand an analysis based on automated multibody modelling tools.

Castor systems comprise a spinning wheel, a kingpin bearing, and enough mechanical trail to provide a self-centring steering action. Our purpose here is to demonstrate that oscillatory instabilities can occur in relatively simple castor systems.

4.5.1 Simple case

A simple castor system commonly employed to demonstrate wheel shimmy, both experimentally and theoretically [8, 171], is shown in Figure 4.14. The wheel is axisymmetric and free to spin relative to its support forks; the wheel is deemed to have no spin inertia. The vertical moment of inertia of the wheel, fork, and kingpin assembly about its mass centre is J_z. The castor assembly is free to rotate around the kingpin through angle δ. The wheel has mechanical trail e and mass centre offset f with respect to the vertical kingpin bearing. The kingpin is free to translate laterally with displacement y from static equilibrium, while the whole assembly moves forward with constant speed v. The kingpin mounting has stiffness k, while the moving assembly has mass m.

Significant from the point of view of single-track vehicles, and aircraft nose-wheels, is the lateral compliance at the kingpin. If this compliance allows the assembly to rotate

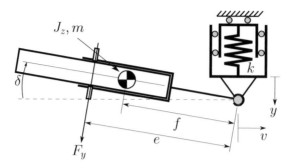

Figure 4.14: Plan view of a simple system capable of shimmy behaviours.

in roll about an axis well above the ground, as with a typical bicycle or motorcycle frame or aircraft fuselage, lateral motions of the wheel assembly are accompanied by camber changes. If, in addition, the wheel has spin inertia, gyroscopic effects come into play and may have an important influence on the shimmy behaviour. These effects are shown in [8] to create a second area of instability in the (v, e) space at higher speeds; v is the speed and e the mechanical trail. This phenomenon is known as 'gyroscopic shimmy'; gyroscopic shimmy has also been described as a by-product of wheel hop in rigid-axle road cars [172]. In a rigid-axle car, a steer angle velocity combines with the wheel angular momentum to produce a gyroscopic moment in the roll axis direction that couples the steer velocity with wheel hop: 'In this (shimmy) cycle, the tyre always meets the ground toed inwards, is swerved outward vigorously by the aligning torque, and is thrown into the air again toed out ready for the cycle to be repeated by the other wheel of the pair.' Olley [172] further asserts that the prevention of shimmy is a strong motivator for the introduction of independent front suspensions: 'Independent suspension simply inserts the entire sprung mass of the vehicle as a barrier between the two front wheels, so that the gyroscopic torque set up by the forced precession of one front wheel cannot be communicated to the other except through this mass.'

We will begin our study of shimmy with a system in which gyroscopic influences are absent. For small values of y and δ the equations of motion for the system in Figure 4.14 are found easily using Newton–Euler vector mechanics methods. To begin, we balance forces in the y-direction to obtain

$$m(\ddot{y} - f\ddot{\delta}) + ky - F_y = 0, \tag{4.95}$$

in which F_y represents the lateral tyre force. The tyre's (aligning) moment M_z has been neglected for simplicity. Next, we use a moment balance around the castor assembly mass centre to obtain

$$J_z\ddot{\delta} + (e - f)F_y + kyf = 0. \tag{4.96}$$

In order to model the tyre force we will use a simple slip-based model; see Chapter 3. The castor ground contact point velocity component orthogonal to the wheel plane is

$$V_{Sy} = \dot{y} - v\delta - e\dot{\delta}.$$

Table 4.3 Parameter values (in SI units) for the simple castor system illustrated in Figure 4.14.

J_z	m	e	f	k	σ	C_α
1.0	30.0	0.12	0.08	45,000	0.1	4,500

The lateral tyre slip angle is defined as

$$\alpha = -\frac{V_{Sy}}{v} = \delta + \frac{e\dot{\delta} - \dot{y}}{v}, \tag{4.97}$$

with the corresponding lateral tyre force given by the differential equation

$$\frac{\sigma}{v}\dot{F}_y + F_y = C_\alpha\alpha, \tag{4.98}$$

in which C_α is the tyre's cornering stiffness and σ its relaxation length (a space constant related to the delay in the force-generation mechanism); for present purposes C_α and σ can be treated as fixed tyre-related parameters. If we take Laplace transform of (4.95), (4.96), and (4.98) and assemble them into a single matrix equation, one obtains

$$\begin{bmatrix} ms^2 + k & -fms^2 & -1 \\ kf & s^2 J_z & e - f \\ C_\alpha s/v & -C_\alpha(1 + es/v) & 1 + \sigma s/v \end{bmatrix} \begin{bmatrix} y(s) \\ \delta(s) \\ F_y(s) \end{bmatrix} = 0, \tag{4.99}$$

with the system's characteristic equation given by the quintic polynomial

$$P(s) = \det\left(\begin{bmatrix} ms^2 + k & -fms^2 & -1 \\ kf & s^2 J_z & e - f \\ C_\alpha s/v & -C_\alpha(1 + es/v) & 1 + \sigma s/v \end{bmatrix}\right). \tag{4.100}$$

Suppose the parameters for the simple castor system are those given in Table 4.3. Since the characteristic polynomial (4.100) is quintic, one expects five speed-dependent root loci. This is indeed the case as is shown in Figure 4.15 (a)—the plot is symmetric with respect to the horizontal axis. There is one stable real locus, a high-frequency locus with a resonant frequency in the range 5.97–8.75 Hz, and an unstable low-frequency mode with resonant frequency in the range 2.86–3.85 Hz. This unstable low-frequency mode is the castor shimmy mode.

To better understand these modes we will now consider three special cases. In the first we consider the limit

$$\lim_{C_\alpha \to \infty} \frac{P(s)}{Ck} = a_3 s^3 + a_2 s^2 + a_2 s^2 + a_1 s + a_0$$

$$= \frac{(m(e - f)^2 + J_z)s^3}{kv} + \frac{m(e - f)s^2}{k} + \frac{e^2 s}{v} + e, \tag{4.101}$$

which corresponds to the case when the tyre's cornering stiffness goes to infinity thereby enforcing nonholonomic rolling. Assuming positive values for all the parameters and the speed, it follows from the Routh–Hurwitz criterion that the system will be stable provided $a_i > 0$ and

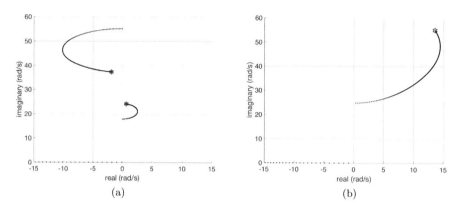

Figure 4.15: Speed-dependent root loci of the simple castor system shown in Figure 4.14. The speed is swept from 0 to 60 m/s; the hexagons corresponds to the high-speed end of the locus. The (a) figure represents the baseline configuration, while (b) represents the case with infinite cornering stiffness C_α.

$$\det\left(\begin{bmatrix} a_1 & a_0 \\ a_3 & a_2 \end{bmatrix}\right) = \frac{e}{vk}(mf(e-f) - J_z) > 0. \tag{4.102}$$

If condition (4.102) is satisfied, then $a_2 > 0$ is automatically satisfied. This illustrates the role played by the steering system geometry, and the mass and inertia properties of the moving assembly, in determining the stability, or otherwise, of the system. A root-locus plot for the infinite tyre cornering stiffness case with the parameter values given in Table 4.3 is shown in Figure 4.15 (b). Since condition (4.102) is not satisfied, an unstable system is predicted. The low-speed resonant frequency of the oscillatory mode in this system is of the order 25 rad/s $\approx$ 4.0 Hz, which matches that of the low-frequency mode in Figure 4.15 (a). The limit case in which $mf(e-f) = J_z$ is interesting. Using the fact that in this limit case the characteristic polynomial (4.101) must be of the form

$$a_3s^3 + a_2s^2 + a_2s^2 + a_1s + a_0 = a_3(s+p)(s^2 + \omega^2)$$
$$= a_3s^3 + a_3ps^2 + a_3\omega^2s + a_3p\omega^2,$$

where p is an unknown real root and ω an unknown resonant frequency. By equating coefficients in like powers of s, one can deduce that

$$\omega = \sqrt{ke/(m(e-f))}. \tag{4.103}$$

As expected, the kingpin stiffness, the mass of the castor assembly, and the castor geometry all influence the frequency of oscillation. The mass distribution $mf(e-f) = J_z$ ensures that the kingpin is at the centre of percussion for the tyre force,[9] and thus

---

[9]   This can be verified by solving (4.95) and (4.96) for $\ddot{\delta}$ and J_z in the case that $y = \ddot{y} = 0$.

the rolling constraint produces no force of reaction at the kingpin. Direct calculation using (4.95), (4.96), and $mf(e - f) = J_z$ gives

$$m\ddot{y} + \frac{key}{e - f} = 0, \tag{4.104}$$

reinforcing the fact that $w = \sqrt{ke/(m(e - f))}$.

In the second special case we consider

$$\lim_{k \to \infty} \frac{P(s)}{Ck} = b_3 s^3 + b_2 s^2 + b_2 s^2 + b_1 s + b_0$$

$$= \frac{\sigma(f^2 m + J_z)s^3}{Cv} + \frac{(f^2 m + J_z)s^2}{C} + \frac{e^2 s}{v} + e, \tag{4.105}$$

which corresponds to a stiff castor assembly. Assuming positive values for the speed and parameters, it follows from the Routh–Hurwitz criterion that the system will be stable provided $b_i > 0$ and

$$\det\left(\begin{bmatrix} b_1 & b_0 \\ b_3 & b_2 \end{bmatrix}\right) = e(e - \sigma)\frac{f^2 m + J_z}{C_\alpha v} > 0, \tag{4.106}$$

or that $e > \sigma$. In other words, shimmy cannot occur when the trail e is greater than the relaxation length σ. For smaller (and still positive) trail the system is oscillatory unstable (shimmy), while for negative trail the system is divergent unstable.[10] A root-locus plot for the stiff structure case with the parameter values given in Table 4.3 is shown in Figure 4.16 (a). Since condition (4.106) is satisfied, a stable system is predicted. After conversion to the Laplace transform domain, (4.97) and (4.98) give

$$F_y(s) = C_\alpha \left(\frac{v + se}{v + s\sigma}\right)\delta(s) \tag{4.107}$$

since $y(s) = 0$, which shows that stability in this case requires phase lead between $\delta(s)$ and the tyre force $F_y(s)$. In the limiting case that $e = \sigma$, $F_y = C_\alpha \delta$, which can be substituted into (4.95) and (4.96) to obtain

$$(J_z + mf^2)\ddot{\delta} + eC_\alpha \delta = 0. \tag{4.108}$$

This shows that the resonant frequency is given by

$$w = \sqrt{eC_\alpha/(J_z + mf^2)}. \tag{4.109}$$

In this case the tyre properties dictate both the conditions under which shimmy occurs and its frequency. The tyre relaxation length determines the onset, or otherwise, of shimmy, while the frequency of oscillation is dictated by the tyre's cornering stiffness.

---

[10] In the case the tyre aligning moment $M_z' = -tF_y$ is included, where t is the pneumatic trail, the divergent instability appears for negative trail lower than $-t$.

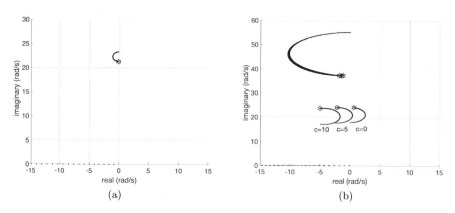

Figure 4.16: Root loci of the castor system shown in Figure 4.14 when (a) the castor has an infinite stiffness k, and (b) when a steer damper with different damping coefficients c is introduced. The speed is swept from 0 to 60 m/s with the hexagon at the high-speed end of the locus.

The Pirelli company reports on a tyre tester that relies on this result [173]. The test tyre is mounted in a fork trailing a rigidly mounted kingpin bearing and runs against a spinning drum to represent movement along a road. Following an initial steer displacement of the wheel assembly, the exponentially decaying steering vibrations are recorded, and the decrement yields the tyre relaxation length, while the frequency yields the cornering stiffness. Unlike other situations, the shimmy frequency is independent of speed.

In a third case, we consider a stiff castor assembly operating in concert with a stiff tyre. In the limit that both C_α and k go to infinity, both (4.101) and (4.105) give

$$P(s) = ekC_\alpha \left(\frac{es}{v} + 1 \right). \tag{4.110}$$

Importantly, shimmy cannot occur in this case when $e > 0$, since $P(s)$ has no right-half-plane roots. Under nonholonomic rolling conditions, following Laplace transformation, (4.97) gives

$$0 = \left(\frac{e}{v} s + 1 \right) \delta(s), \tag{4.111}$$

since the slip is zero.

As will soon be expanded on, a standard preventive measure for shimmy is the introduction of a castor assembly damper. This introduces a $c\dot{\delta}$ term in (4.96), where c is the castor damping coefficient. The stabilizing influence of this damper can be seen in Figure 4.16 (b). It is clear from this figure that the castor assembly with no damper has an unstable shimmy mode for all speeds. In the case that $c = 5$ Nm s/rad, the castor assembly is stable at low and high speeds, while becoming unstable at intermediate speeds. When $c = 10$ Nm s/rad, the shimmy mode is stable over the entire examined speed range. When damping is not included, the stability of the simple model is speed independent.

4.5.2   A more realistic setup

In this section we consider a steering system that resembles the front end of a bicycle, or motorcycle with a flexible frame. This system represents a self-centring castor assembly that is capable of shimmy—in this case gyroscopic influences will be present. The castor system in Figure 4.17 includes a twist axis SO with an associated twist angle ϕ; the twist axis is inclined at angle ϵ to the horizontal. The wheel assembly is free to rotate around a steering axis through angle δ. To make the modelling conceptually easier, we will assume the axis system $Oxyz$ is inertial with its origin located at O, which is at the junction of the twist and steer axes. The castor assembly is stationary with no translational freedoms, and the road moves backwards at speed v. The x-axis of $Oxyz$ is aligned with the twist axis, its y-axis points out of the page, and its z-axis is specified by $z = x \times y$, which is aligned with the steer axis. The unit vectors x and y are aligned with the x- and y-axes respectively (in $Oxyz$). The castor-fixed axis system $Ox'y'z'$ rotates through ϕ around the twist axis, and through angle δ around the steering axis. The wheel and wheel mount assembly has mass m_f, with its mass centre located at G, which has coordinates $(x_f, 0, z_f)$ in $Ox'y'z'$. The steer moment of inertia of the wheel frame assembly is I_{fz} (around G). The wheel has its centre at C, radius r_w, and spin inertia I_w. The wheel contacts the ground at P and the tyre is assumed to have no longitudinal slip and so $\dot{\xi} = -v/r_w$, where ξ is the castor spin angle.

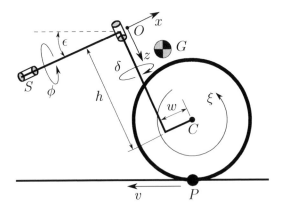

Figure 4.17: Castor with an inclined flexible mounting.

The equations of motion can be obtained using momentum-balancing arguments. Balancing the angular momentum around the twist axis (for small steer and twist angles) gives:

$$(I_{fx} + m_f z_f^2)\ddot{\phi} = m_f z_f x_f \ddot{\delta} + g m_f c_\epsilon (x_f \delta - z_f \phi) - (h + r_w c_\epsilon) F_y$$

$$-k_p \phi - \frac{v I_w \dot{\delta}}{r_w}. \tag{4.112}$$

The term on the left of (4.112) represents the moment due to the moment of inertia of the wheel mount assembly around the twist axis. The first term on the right-hand side of (4.112) is a product-of-inertia term. The second term on the right is a gravitational moment term. The third term is the moment produced by the tyre side force F_y. The fourth term is the twist-axis restoring moment, while the fifth and final term is a gyroscopic moment.

Balancing the angular momentum, for small steer and twist angles, around the steer axis gives

$$(I_{fz} + x_f^2 m_f)\ddot{\delta} = m_f x_f z_f \ddot{\phi} + g m_f s_\epsilon (x_f \delta - z_f \phi)$$

$$+(w - r_w s_\epsilon) F_y + \frac{v I_w \dot{\phi}}{r_w}. \tag{4.113}$$

The term on the left of (4.113) represents the moment of inertia of the wheel mount assembly around the twist axis (note the use of the parallel axis theorem), while the first term on the right-hand side of (4.113) is a product-of-inertia term. The second term on the right is a gravitational moment term. The third term is the moment produced by the tyre side force F_y, while the final term is a gyroscopic moment.

The tyre side force is generated from the side-slip and camber angles as follows,

$$F_y + \frac{\sigma}{v}\dot{F}_y = C_\alpha(\frac{\dot{\phi}}{v}(h + r_w c_\epsilon) + \frac{\dot{\delta}}{v}(r_w s_\epsilon - w) + c_\epsilon \delta - s_\epsilon \phi)$$

$$+C_\gamma(c_\epsilon \phi + s_\epsilon \delta). \tag{4.114}$$

The coefficient of the cornering stiffness C_α is the side-slip, while the coefficient of tyre's camber stiffness C_γ is the tyre's camber angle.

After taking Lapalce transforms, equations (4.112), (4.113), and (4.114) can be combined to gives the characteristic matrix

$$H(s) = \begin{bmatrix} s^2(I_{fx} + m_f z_f^2) + g m_f z_f c_\epsilon + k_p & \frac{s v I_w}{r_w} - s^2 m_f x_f z_f - g m_f x_f c_\epsilon & h + r_w c_\epsilon \\ \frac{s v I_w}{r_w} - g m_f z_f s_\epsilon + s^2 m_f x_f z_f & -s^2(I_{fz} + x_f^2 m_f) + g m_f x_f s_\epsilon & w - r_w s_\epsilon \\ C_\alpha\left(s_\epsilon - \frac{s(h + r_w c_\epsilon)}{v}\right) - C_\gamma c_\epsilon & C_\alpha\left(-c_\epsilon + \frac{s(w - r_w s_\epsilon)}{v}\right) - C_\gamma s_\epsilon & \frac{s\sigma}{v} + 1 \end{bmatrix} \tag{4.115}$$

in which s is the Laplace transform variable. The determinant of $H(s)$ is a quintic (in the Laplace variable s) with speed-dependent coefficients. A steering damper can be introduced by including a $-c_s\dot{\delta}$ term on the right-hand side of (4.113), and a corresponding $-sc_s$ term to the $(2,2)$ entry of (4.115), in which c_s is the steering damping coefficient. The five roots of $\det H(s)$ determine the system's stability. The parameters for the castor system considered here are given in Table 4.4.

Figure 4.18 shows the locus of the speed-dependent roots of $\det H(s)$ as a function of speed. In common with Figure 4.15, this figure has a single stable real root and

Table 4.4 Parameter set for the included castor.

$C_\alpha = 1500\,\text{N/rad}$	$C_\gamma = 150\,\text{N/rad}$	$h = 0.6\,\text{m}$	$I_w = 0.25\,\text{kgm}^2$
$I_{fz} = 0.1\,\text{kgm}^2$	$k_p = 1.0 \times 10^4\,\text{Nm/rad}$	$m_f = 15.0\,\text{kg}$	$r_w = 0.3\,\text{m}$
$\sigma = 0.1\,\text{m}$	$w = 0.1\,\text{m}$	$\epsilon = 0.45\,\text{rad}$	$x_f = 0.1\,\text{m}$
$z_f = 0.3\,\text{m}$	$g = 9.806\,\text{m}^2$	$I_{fx} = 1.0\,\text{kgm}^2$	

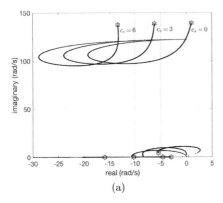 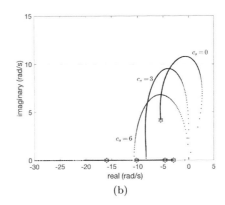

(a) (b)

Figure 4.18: (a) Root loci for the inclined castor system derived from (4.115); only the roots with nonnegative imaginary part are shown. The speed is swept from 0 to 80 m/s, with a hexagon marking the high-speed end of the complex loci. Steering damper values of $c_s = 0$ Nm/s, $c_s = 3$ Nm/s and $c_s = 6$ Nm/s are shown. (b) is a zoom view of the low-frequency shimmy mode.

two pairs of complex conjugate loci. The high-frequency locus, which begins on the imaginary axis at 122 rad/s (19.4 Hz), moves into the left-half plane as the wheel's angular velocity increases, and then becomes unstable at high speed with a frequency of 137 rad/s (21.8 Hz). The real locus also moves into the left-half plane as the speed increases. The low-frequency shimmy locus is more interesting from an engineering perspective and potentially more troublesome, because it is unstable over a large range of speeds. Figure 4.19 (b) is an expanded view of the shimmy loci in Figure 4.19 (a) for three different steering damper values. The (a) locus is the zero steering damper case, which is unstable for speeds between 0.1 m/s and 8.05 m/s. A damper of 3 Nm/s is introduced in the (b) locus with the system marginally stable at low speed. Finally, in the (c) locus a damper of 6 Nm/s is introduced, which stabilizes the shimmy mode over the entire speed range.

The angular displacement components (ϕ and δ) of the three modes are shown in Figure 4.19, which have been scaled so that the steer component δ is unity. In the case of the shimmy mode (a) the frame and steer perturbations are almost in phase with the magnitude of the frame twist angle roughly half that of the steer angle. In the case of the stable real mode (b) the frame and steer are in phase with the frame twist angle magnitude approximately 25 % of the steer angle magnitude. In the case of the high-frequency mode (c) the frame and steer perturbations are almost in anti-phase with the magnitude of the frame twist angle and roughly 15 % of the steer angle. The system stability is now speed dependent, whereas with the simple model it was not.

Reconciliation with the simple case. We will now reconcile, briefly, some of the features of the castor system in Figure 4.14 with those in the more complex system in Figure 4.17. To do this we begin by setting $\epsilon = 0$ in (4.115) to obtain

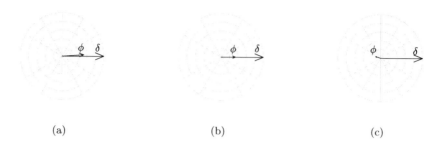

(a)   (b)   (c)

Figure 4.19: Mode shapes for the inclined castor system derived from (4.115) at $5\,\mathrm{m/s}$ and $c_s = 0\,\mathrm{Nm/s}$. The steering angle component (δ) of the mode shape is normalized to unity, with the castor assembly twist angle (ϕ) scaled the same way. Figure (a) is the unstable low-frequency shimmy mode in which δ and ϕ oscillate (almost) in phase at $1.64\,\mathrm{Hz}$. Figure (b) is the stable real mode in which the twist angles δ and ϕ converge in phase. Figure (c) is the stable high-frequency shimmy mode in which δ and ϕ oscillate in approximate anti-phase at $18.9\,\mathrm{Hz}$.

$$H(s) = \begin{bmatrix} s^2(I_{fx} + m_f z_f^2) + g m_f z_f + k_p \,\frac{s v I_w}{r_w} - s^2 m_f x_f z_f - g m_f x_f & h + r_w \\ \frac{s v I_w}{r_w} + s^2 m_f x_f z_f - s^2(I_{fz} + x_f^2 m_f) & w \\ -s C_\alpha \left(\frac{h+r_w}{v}\right) - C_\gamma \qquad C_\alpha\left(\frac{sw}{v} - 1\right) & \frac{s\sigma}{v} + 1 \end{bmatrix}.$$

(4.116)

In the case of 'stiff' mounting assemblies the difference between the translational structural freedom in Figure 4.14 and the rotational structural freedom in Figure 4.17 is immaterial. If the assembly in Figure 4.17 is deemed stiff, the system stability is determined by

$$\lim_{k_p \to \infty} \det(-H(s)/(k_p C_\alpha)) = \frac{\sigma(I_{fz} + m_f x_f^2)}{v C_\alpha} s^3 + \frac{I_{fz} + m_f x_f^2}{C_\alpha} s^2 + \frac{w^2}{v} s - w. \quad (4.117)$$

The first thing one observes is that w must be negative in order to achieve stable motion. This corresponds to the familiar 'flip' one observes when beginning to push a shopping trolley that has one of its castors pointing forwards. Once all the polynomial coefficients have the same sign, one requires

$$\det\left(\begin{bmatrix} a_1 & a_0 \\ a_3 & a_2 \end{bmatrix}\right) = \frac{w(I_{fz} + m_f x_f^2)}{v C_\alpha}(\sigma + w) > 0 \qquad (4.118)$$

as a result of the Routh–Hurwitz criterion, which is the same as saying $-w > \sigma$, with $w < 0$. This condition matches exactly the stability criterion given in (4.106). In the limit case that $\sigma = -w$, the frequency of oscillation is

$$\omega = \sqrt{\frac{\sigma C_\alpha}{I_{fz} + m_f x_f^2}}, \qquad (4.119)$$

which matches the expression given in (4.109). In the case where a 'stiff' castor assembly combines with a non-slipping tyre there obtains

$$\det(H(s)) = w(s\frac{w}{v} - 1), \tag{4.120}$$

which is stable for any $w < 0$. As with the simple configuration studied earlier, shimmy is not possible if the mount assembly is stiff and the tyre is non-slipping.

4.6 Unicycle

The unicycle is a human-powered, single-track, single-wheeled road vehicle that is more a plaything for the enthusiast than a practical means of transport. A number of wheel-based commuter devices have evolved from the rolling disc as it is described in Section 2.10. From a transport perspective the rolling disc's most direct descendant is the BC wheel, which is simply a wheel with pegs extending on either side of the wheel axle. In order to ride it you 'simply' balance on the pegs as the wheel is rolling. The name comes from the popular comic strip, *B.C.* by Johnny Hart, in which the characters commute by rolling on large stone wheels. The BC wheel is the most basic form of unicycle.

As was the case with the development of the bicycle, the next evolutionary step was the 'ultimate wheel', which includes pedals on either side of the wheel. This machine has no frame or seat but provides a means of introducing a propulsive moment. The vertical dynamics of this machine is obviously akin to the inverted pendulum with vertical stability control coming from the rider twisting and turning their body.

A conventional unicycle is illustrated in Figure 4.20. The machine has two components: a wheel and a rigid frame. The frame is free to yaw, roll, and pitch. Although the wheel shares common yaw and roll angles with the frame, it is free to spin relative to the frame. As is suggested by the sketch, the rider can be separated into a lower body that will be considered part of the frame and rigidly attached to it, while the rider's upper body that can 'swivel' relative to the lower body through additional yaw, roll, and pitch angular freedoms. This motion can be considered constrained by active control torques and passive structural spring-damper constraints. The road–wheel interaction can be modelled using a slipping tyre that generates longitudinal and lateral forces, and aligning and overturning moments. The road–wheel interaction can be modelled alternatively using nonholonomic constraints.

Figure 4.20: Unicycle and rider; picture taken from [174].

Scholarly articles on unicycle dynamics appear to be concentrated in the control systems literature. As a consequence of its inherent instability, the unicycle is considered a system worthy of study in a closed-loop control context [175–177]. The unicycle has been used as a case study in the development of autonomous robots, and in the emulation of human riding [174]. The unicycle has been used as an exemplar vehicle in tracking and motion-control studies [178], and in the study of the stabilization of non-holonomic systems [179]. Due to its presumed simplicity, the unicycle has been used in the study of nonlinear cyclic pursuit problems [180]. From a mechanics modelling perspective, this literature has a common shortcoming—the dynamic modelling of the vehicle is often simplistic.

A correct and interesting treatment of the unicycle can be found in [181]. Although the equations of motion were derived using multibody modelling software, they are not difficult to interpret in a classical Newton–Euler framework. Unlike any other treatment that the authors are aware of, realistic tyre forces and moments are given in the model presented. In the tyre description longitudinal-, lateral-, and turn-slips and, through them, shear forces and steering moment, and the rolling velocity are all included. The theory and results described provide a predictive capability for robotic unicycles, which can be used to guide the design of these machines.

5

The Whipple Bicycle

5.1 Background

The first substantial contribution to the theoretical bicycle literature was Whipple's seminal 1899 paper [2], which is arguably as contributory as anything written since (see Figure 5.1). This remarkable paper contains, for the first time, a set of nonlinear differential equations that describe the general motion of a bicycle and rider. Since the appropriate computing facilities were not available at the time, Whipple's general nonlinear equations could not be solved and consequently were not pursued beyond simply reporting them. Instead, Whipple studied a set of linear differential equations that correspond to small motions about a straight-running trim condition at a given constant speed. In the context of the nineteenth century, stability properties could be evaluated using the Routh criterion.

Whipple's model consists of two frames, namely the front frame and the rear frame, which are hinged together along an inclined steering-head assembly. The front and rear wheels are attached to the front and rear frames, respectively, and are free to rotate relative to them. The rider is described as an inert mass that is rigidly attached to the rear frame. The rear frame is free to roll and translate in the ground plane. Each wheel is assumed to be 'thin', and thus touches the ground at a single ground-contact point. The wheels, which are also assumed to be non-slipping, are modelled by holonomic constraints in the normal (vertical) direction and by nonholonomic constraints [41] in the longitudinal and lateral directions. From a mathematical perspective, the rolling disc (Section 2.10) and the bicycle wheels' contact dynamics are modelled in precisely the same way. There is no aerodynamic drag representation, no frame flexibility, and no suspension system—the rear frame is assumed to move at a constant speed. Since Whipple's linear straight-running model is fourth order, the corresponding characteristic polynomial is a quartic; see Figure 5.2.

Concurrent with Whipple's work, and apparently independently of it, Carvallo [54] derived the equations of motion for a free-steering bicycle linearized around a straight-running equilibrium condition. Klein and Sommerfeld [182] also derived equations of motion for a straight-running bicycle. Their slightly simplified model (as compared with that of Whipple) lumps the front wheel assembly mass into the front wheel. The main purpose of their study was to determine the effect of the gyroscopic moment due to the spinning front wheel on the machine's free-steering stability. They concluded that 'the gyroscopic effects, despite their smallness, are indispensable for the self-stability'. However, it was later pointed out that the analysis includes some algebra

Dynamics and Optimal Control of Road Vehicles. D. J. N. Limebeer and M. Massaro.
© D. J. N. Limebeer and M. Massaro 2018. Published in 2018 by Oxford University Press.
DOI: 10.1093/oso/9780198825715.001.0001

Figure 5.1: Francis John Welsh Whipple was born on March 17, 1876. He was educated at the Merchant Taylors' School and was subsequently admitted to Trinity College, Cambridge, in 1894. His university career was brilliant—he received his BA degree in mathematics in 1897 in the first class. Whipple received his MA in 1901 and an ScD in 1929. In 1899 he returned to the Merchant Taylors' School as a mathematics master; a post he held until 1914. Apart from his seminal work on bicycle dynamics, he made many other contributions to knowledge, including identities for generalized hypergeometric functions, several of which have subsequently become known as Whipple's identities and transformations. Reproduced with the permission of the National Portrait Galley, London.

mistakes, and the results reported are not valid for a general bicycle dataset [183].[1] While this moment does indeed stabilize the free-steering bicycle over a range of speeds, this effect is of only minor importance, because the rider can easily replace the stabilizing influence of the front wheel's gyroscopic precession with low-bandwidth rider control action [184].

An early attempt to introduce side-slipping and force-generating tyres into the bicycle literature appears in [185]. Other classical contributions to the theory of bicycle dynamics include [186] and [48]. The last of these references, in its original 1967 version, appears to contain the first analysis of the stability of the straight-running bicycle fitted with pneumatic tyres; several different tyre models are considered. Reviews of the bicycle literature from a dynamic modelling perspective can be found in [187]. The bicycle literature is comprehensively reviewed from a control theory perspective in [188], which also describes interesting bicycle-related experiments.

Some important and complementary applied work has been conducted in the context of bicycle dynamics. An attempt to build an unrideable bicycle (URB) is described in [184]. One of the URBs described had the gyroscopic moment of the front wheel

---

[1] Despite the calculation error, the conclusion is correct for the specific bicycle examined therein.

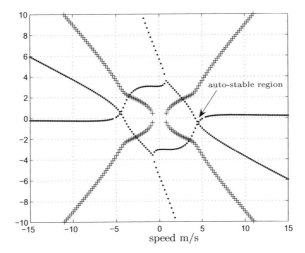

Figure 5.2: Stability properties of the 'Whipple bicycle'. Real (dots) and imaginary (crosses) parts of the eigenvalues of the straight-running Whipple bicycle model as functions of speed. Plot generated using equation (XXVIII) in [2], which was derived a century before the widespread availability of MATLAB!

cancelled by another which was counter-rotating. The cancellation of the front wheel's gyroscopic moment made little difference to the machine's apparent stability and handling qualities. It was also found that this riderless bicycle was unstable, an outcome that had been predicted theoretically in [182]. Three other URBs, described in [184], include various modifications to their steering geometry. These modifications include changes in the front wheel radius, and the magnitude and sign of the fork offset. Jones concluded that the castor trail is an important factor in the self-stability of single-track vehicles. However, it was later pointed out that the theoretical analysis therein reported is not valid for a general bicycle dataset [183].

After the results of [182] and [184], the gyroscopic effects and the trail have long been considered essential for the self-stability of bicycles (and single-track vehicles in general). A formal proof that this is not the case has been given recently in [169], where a self-stable bicycle with no gyroscopic effects and no castor trail has been analysed;[2] see Section 4.4.

Experimental investigations of bicycle dynamics have also been used in teaching [189].

---

[2] It is concluded that gyroscopic effects and castor trail, although not essential, are nevertheless beneficial to the stability of single-track vehicles of conventional design.

5.2 Bicycle model

As with Whipple's model, the bicycle model we consider here consists of two frames
and two wheels. Figure 5.3 shows the axis systems and geometric layout of the bicycle
model to be studied here The bicycle's rear frame assembly has a rigidly attached
rider and a rear wheel that is free to rotate relative to the rear frame. The front frame,
which comprises the front fork and handlebar assembly, has a front wheel that is free
to rotate relative to the front frame. The front and rear frames are attached using a
hinge that defines the steering axis. In the reference configuration, all four bodies are
symmetric relative to the bicycle midplane. As with Whipple's model, the non-slipping
road wheels are modelled by holonomic constraints in the normal (vertical) direction
and by nonholonomic constraints in the longitudinal and lateral directions. There is
no aerodynamic drag, no frame flexibility, no propulsion, and no rider control. The
dimensions and mechanical properties of the benchmark model are taken from [187]
and are presented in Table 5.1. All inertia parameters use the relevant body mass
centres as the origins for body-fixed axes. The axis directions are then chosen to
align with the inertial $Oxyz$ axes when the bicycle is in its nominal configuration,
as shown in Figure 5.3. Products of inertia I_{Bxz}, I_{Hxz}, and so on are defined as
$-\int\int xz\, m(x,z)dxdz$; see Section 2.7.

  The wheels are modelled as axisymmetric, with each wheel making a point contact
with the ground. The wheel-mass distributions need not be planar and so any positive
inertias are allowed provided $I_{xx} = I_{zz}$ and $I_{xx} + I_{zz} \geq I_{yy}$; $I_{xx} + I_{zz} = I_{yy}$ in the

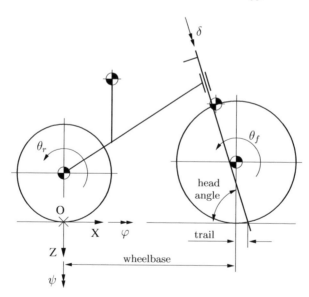

Figure 5.3: Whipple-type bicycle model with its degrees of freedom. The model comprises two frames pinned together along an inclined steering head. The rider is included
as a rigidly attached part of the rear frame. Each wheel is assumed to contact the road
at a single point.

Table 5.1 Bicycle parameters. The inertia matrices are referred to body-fixed axis systems that have their origins at the body's mass centre, and are aligned with the inertial reference frame $Oxyz$ when the machine is in its nominal configuration.

Parameters	Symbol	Value
Wheelbase	w	$1.02\,m$
Trail	t	$0.08\,m$
Head angle	α	$(2\pi)/5\,\mathrm{rad}$
Steer axis tilt angle	λ	$\pi/10\,\mathrm{rad}$
Gravity	g	$9.81\,m/s^2$
Rear wheel		
Radius	r_R	$0.3\,m$
Mass	m_R	$2\,kg$
Mass moments of inertia	$[I_{Rxx}, I_{Ryy}, I_{Rzz}]$	$[0.0603, 0.12, 0.0603]\,kg\,m^2$
Rear frame		
Position of centre of mass	$[x_B, y_B, z_B]$	$[0.3, 0.0, -0.9]\,m$
Mass	m_B	$85\,kg$
Mass moments of inertia	$\begin{bmatrix} I_{Bxx} & 0 & I_{Bxz} \\ & I_{Byy} & 0 \\ sym & & I_{Bzz} \end{bmatrix}$	$\begin{bmatrix} 9.2 & 0 & 2.4 \\ & 11 & 0 \\ sym & & 2.8 \end{bmatrix} kg\,m^2$
Front frame		
Position of centre of mass	$[x_H, y_H, z_H]$	$[0.9, 0.0, -0.7]\,m$
Mass	m_H	$4\,kg$
Mass moments of inertia	$\begin{bmatrix} I_{Hxx} & 0 & I_{Hxz} \\ & I_{Hyy} & 0 \\ sym & & I_{Hzz} \end{bmatrix}$	$\begin{bmatrix} 0.05892 & 0 & -0.00756 \\ & 0.06 & 0 \\ sym & & 0.00708 \end{bmatrix} kg\,m^2$
Front wheel		
Radius	r_F	$0.35\,m$
Mass	m_F	$3\,kg$
Mass moments of inertia	$[I_{Fxx}, I_{Fyy}, I_{Fzz}]$	$[0.1405, 0.28, 0.1405]\,kg\,m^2$

case of 'thin' planar wheels (equality follows from the perpendicular axis theorem; see Section 2.2.3).

This simple model is fully characterized by the twenty-five parameters listed in Table 5.1. Each parameter is defined in an upright reference configuration that has both wheels on a level ground plane and a zero steer angle. In the reference configuration the origin of the inertial coordinate system is at the rear-wheel ground-contact point O. We use the SAE vehicle dynamics sign convention that has the positive x-axis pointing forward, the positive z-axis pointing down and the positive y-axis pointing to the rider's right. Positive angles are defined in terms of a right-hand rule.

In the reference configuration the front-wheel ground-contact point is located at a distance w (the wheelbase) in front of the rear-wheel ground-contact point. The front-wheel ground-contact point trails a distance t behind the point where the steer axis intersects the ground. Although in standard bicycles and motorcycles the trail is also always positive, negative values of t can be contemplated. The steering head angle α is measured relative to the ground plane (in the machine's nominal configuration)—a steering head angle of $\pi/2$ would mean that the steering axis is vertical. The steer axis tilt angle λ is measured from the negative z-axis and $\pi/2 = \alpha + \lambda$. The steer axis location is defined implicitly by the wheelbase w, trail t, and steering head angle.

5.2.1   Model features

Before we get immersed in the details of finding the bicycle's (linearized) equations of motion, there are a number of properties of the model that can be established using informal arguments.

Degrees of freedom.   Let us begin by considering a bicycle floating in space. Under these conditions the main frame has six degrees of freedom—three translations and three rotations. The steering of the front frame introduces a seventh freedom, with two more freedoms associated with the rotations of the road wheels. At this point the accessible configuration space is nine-dimensional. Constraining the rear-wheel ground-contact point to the x–y plane removes the z-dimensional translational freedom of the main frame, thereby reducing the configuration space to eight dimensions. One more freedom is removed once the front-wheel ground-contact point is constrained to the ground plane. This constraint also introduces a closed kinematic loop that constrains the main body's pitch freedom. A kinematic analysis of the type to be given in Section 7.7.1 shows that the main-body pitch angle χ can be expressed in terms of the vehicle's roll and steer angles (φ and δ respectively) as follows

$$\chi = \chi_0 - \frac{t_n}{w}\delta \tan\varphi \tag{5.1}$$

in the case that the pitch and steer angle variations are 'small'; χ_0 is the main-body pitch angle in the nominal configuration. If the roll angle variations are also assumed 'small', the main-body pitch angle is constant to first order.

   Prior to considering the nonholonomic constraints associated with non-slipping wheels, the accessible configuration space is seven-dimensional. Since the four non-holonomic constraints associated with the non-slipping road wheels are velocity constraints, they do not reduce further the dimension of the accessible configuration space. By invoking manoeuvres akin to vehicle parking, one can still access the whole of the ground plane, arbitrary yaw angles, and arbitrary road–wheel position angles. The bicycle's accessible configuration space is seven-dimensional.

   Summarizing, the first two freedoms (x_R, y_R) specify the rear-wheel ground-contact point in an inertial coordinate system. The orientation of the main frame is defined relative to a global coordinate system in terms of a sequence of three rotations. The third and fourth freedoms are the yaw angle ψ and the roll angle φ—the pitch angle χ is constrained by a closed kinematic loop generated by the wheel ground-contact constraints. The front frame is free to rotate relative to the rear frame through the steer angle δ—this gives five degrees of freedom. Finally, there are two wheel rotation angles θ_R and θ_F. The bicycle configuration is therefore defined in terms of the seven (position) variables x_R, y_R, ψ, φ, δ, θ_R, and θ_F.

Velocity freedoms.   As was explained in some detail in the context of the rolling disc (Section 2.10), there are four nonholonomic constraints associated with the road wheels (two for each wheel). This reduces the number of velocity freedoms by four. The resulting three-dimensional, kinematically accessible velocity space can be characterized in terms of $\dot{\varphi}$, $\dot{\delta}$, and $\dot{\theta}_R$. The remaining four velocities $\dot{x}_R$, $\dot{y}_R$, $\dot{\psi}$, and $\dot{\theta}_F$ are made dependent by the rolling contact constraints.

State variables. Seven position variables and three velocity variables are necessary to describe the Whipple bicycle. In some models this number can be reduced further by removing the ignorable/cyclic coordinates (Section 2.3). Under the assumption of axisymmetric wheels, θ_R and θ_F can be treated as cyclic. The vehicle's dynamics are invariant under translation and rotation on the road plane and x_R, y_R, and ψ are cyclic too. The remaining position variables are φ and δ, together with three velocity variables $\dot{x}_R$, $\dot{\varphi}$, and $\dot{\delta}$. The final set of state variables is thus

$$[\dot{x}_R, \dot{\varphi}, \dot{\delta}, \varphi, \delta]^T. \tag{5.2}$$

It is often convenient to replace the longitudinal velocity $\dot{x}_R$ of the rear contact point with its absolute velocity $u = \dot{x}_R/\cos\psi$. Consequently the state vector is

$$[u, \dot{\varphi}, \dot{\delta}, \varphi, \delta]^T. \tag{5.3}$$

In constant-speed models the state-vector reduces even further to

$$[\dot{\varphi}, \dot{\delta}, \varphi, \delta]^T. \tag{5.4}$$

Energy conservation. In the analysis of the Čhaplygin sleigh given in Section 2.8, it was observed that in the nonlinear model any initial misalignment in the yaw angle would produce yaw oscillations that would die away until the sleigh was again aligned with the velocity vector. Since this is a conservative system, the (kinetic) energy associated with the yaw oscillations is transferred into the kinetic energy of forward motion; see Figure 2.14. In the linear model speed and yaw motions are decoupled and consequently the speed does not change, no energy transfer takes place, and the yaw energy is dissipated. Precisely the same phenomena occur in the Whipple bicycle. First, linearization ensures that the bicycle's speed remains unaltered, and any energy associated with oscillatory behaviour is simply dissipated (in the stable speed range). In contrast, if the nonlinear Whipple bicycle is running at some initial speed, with non-zero initial conditions in the roll and/or steer freedoms, then the associated oscillations will die away, transferring the associated energy into the forward velocity. This behaviour is seen in Figure 5.4, where there is an initial disturbance in the vehicle's roll angle velocity. This produces asymptotically stable roll and steer responses that transfer their energy into the forward speed, which increases by approximately 0.07 m/s in the given example.

Lateral dynamics decoupling. Consider the equations of motion describing the bicycle's small-perturbation dynamics relative to constant-speed straight running. In this case initial values of the vehicle's velocity, roll angle, and steer angle will evolve with time according to

$$\begin{bmatrix} u(t_f) \\ \varphi(t_f) \\ \delta(t_f) \end{bmatrix} = \begin{bmatrix} \Phi_{11}(t_f - t_0) & \Phi_{12}(t_f - t_0) & \Phi_{13}(t_f - t_0) \\ \Phi_{21}(t_f - t_0) & \Phi_{22}(t_f - t_0) & \Phi_{23}(t_f - t_0) \\ \Phi_{31}(t_f - t_0) & \Phi_{32}(t_f - t_0) & \Phi_{33}(t_f - t_0) \end{bmatrix} \begin{bmatrix} u(t_0) \\ \varphi(t_0) \\ \delta(t_0) \end{bmatrix},$$

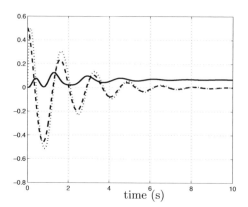

Figure 5.4: Nonlinear energy transfer properties of the straight-running Whipple bicycle. The free-rolling bicycle has an initial velocity of 4.5 m/s, which is in the stable speed range illustrated in Figure 5.5. There is an initial roll velocity is 0.5 rad/s. Plotted are the roll velocity (dashed) and steer angle velocity (dotted), and speed variation about its initial value of 4.5 m/s (solid).

in which $u(t_0)$, $\varphi(t_0)$, and $\delta(t_0)$ are initial values of the bicycle's speed, roll angle, and steering angle respectively. The matrix $\Phi(t_f - t_0)$ is the transition matrix associated with the straight-running linearized equations of motion. Since these equations are linear, negating the initial conditions will negate the final solution.

Consider the initial condition $\begin{bmatrix} 0 & \epsilon & 0 \end{bmatrix}$, which represents a perturbation to the machine's roll angle. This will cause the machine to either speed up or slow down. Either way, the same speed variation will occur in response to the condition $\begin{bmatrix} 0 & -\epsilon & 0 \end{bmatrix}$. It therefore follows that $\Phi_{12}(t_f - t_0) = 0$. It follows from a parallel argument relating to an initial steering angle perturbation $\begin{bmatrix} 0 & 0 & \epsilon \end{bmatrix}$ that $\Phi_{13}(t_f - t_0) = 0$. Now consider a small forward speed variation $\begin{bmatrix} \epsilon & 0 & 0 \end{bmatrix}$. The roll and steering angle will respond in some unspecified manner. Intuitively one would expect this same response from a negated version of this speed variation and so $\Phi_{21}(t_f - t_0) = 0$ and $\Phi_{31}(t_f - t_0) = 0$ must hold. If we now consider the unperturbed nonlinear equations of motion with initial speed $u(t_0)$, conservation of energy dictates that $u(t_f) = u(t_0)$ for all t_f. If one then applies a small initial speed variation, then this same speed variation will persist for all time and so $\Phi_{11}(t_f - t_0) = 1$, giving

$$\Phi(t_f - t_0) = \begin{bmatrix} 1 & 0 & 0 \\ 0 & \Phi_{22}(t_f - t_0) & \Phi_{23}(t_f - t_0) \\ 0 & \Phi_{32}(t_f - t_0) & \Phi_{33}(t_f - t_0) \end{bmatrix} \tag{5.5}$$

for all $t_f - t_0$.

These informal arguments demonstrate that to first order the in-plane and lateral dynamics are decoupled. Thus a bicycle with a constrained forward speed will have the same linearized equations of motion as one in which the forward speed is a free variable.

5.3   Linear in-plane dynamics

We begin by considering the in-plane dynamics of a bicycle in which the steering is locked and with the machine constrained in roll. Under these conditions a (rear-wheel) drive torque T_R is required to accelerate the machine's mass as well as spin up the angular momentum of both wheels. The assumed constraints preclude steer and roll angle contributions to the longitudinal dynamics. A force balance in the direction of motion and a moment balance around the wheel spin axes give the the following equations of motion:

$$m_T \dot{u} = F_{x_R} + F_{x_F} \tag{5.6}$$

$$I_{Ryy} \ddot{\theta}_R = F_{x_R} r_R + T_R \tag{5.7}$$

$$I_{Fyy} \ddot{\theta}_F = F_{x_F} r_F, \tag{5.8}$$

in which T_R is the drive/braking torque. Substitution of the expressions for F_{x_R} and F_{x_F}, derived in (5.7) and (5.8), into (5.6) gives

$$m_T \dot{v} = I_{Ryy} \frac{\ddot{\theta}_R}{r_R} - \frac{T_R}{r_R} + I_{Fyy} \frac{\ddot{\theta}_F}{r_F}. \tag{5.9}$$

Enforcing the no-slip conditions gives $\dot{u} = -\ddot{\theta}_R r_R$ and $\dot{u} = -\ddot{\theta}_F r_F$, which implies that $\ddot{\theta}_F = \ddot{\theta}_R r_R / r_F$ and

$$\left(r_R^2 m_T + I_{Ryy} + \left(\frac{r_R}{r_F} \right)^2 I_{Fyy} \right) \ddot{\theta}_R = T_R, \tag{5.10}$$

in which m_T is the total vehicle mass given by

$$m_T = m_R + m_F + m_B + m_H. \tag{5.11}$$

In the case that $T_R = 0$, the machine speed is constant and given by $u = -\dot{\theta}_R r_R$. It follows from (5.6) that the rear tyre longitudinal force F_{x_R} (assuming $F_{x_F} = 0$) must accelerate the total mass m_T of the vehicle, while (5.10) shows that the rear-wheel torque T_R must accelerate the total apparent mass of the vehicle that includes the inertial terms I_{Ryy}/r_R^2 and I_{Fyy}/r_F^2.

If we now consider small perturbations about the trim condition described in (5.10), and recollect from (5.5) that there is no cross-coupling between the longitudinal and lateral dynamics (to first order), we conclude that (5.10) also serves as a linear first-order small-perturbation equation for the vehicle's longitudinal dynamics. As with Chaplygin's sleigh, the linearization process destroys the system's conservation properties and removes the cross-coupling between the longitudinal and lateral dynamics; in the non-linear model the energy transfer between the longitudinal and lateral dynamics occurs via higher-order terms.

5.4   Linear out-of-plane dynamics

The treatment given here follows closely that given in [187]. Without loss of generality we can assume that the bicycle is rolling freely in a forward direction along the positive x-axis of a global inertial coordinate system. We will suppose also that the vehicle has been perturbed from a straight-running equilibrium—the perturbation is assumed 'small' so that linear behaviour can be assumed. The linearized equation of motion for the longitudinal dynamics has already been presented in (5.10). The external influences of importance include gravitational forces that act in the direction of the (positive) inertial z-axis, reaction forces at the two wheel ground-contact points, forces of constraint that act in the inertial $x - y$ plane and which correspond to non-slip rolling, a propulsion torque T_R that acts on the rear wheel, a steering torque T_{HB} that acts on the front frame and reacts on the rear frame, and a roll torque T_B that acts on the rear frame. The equations of motion will be derived using standard Newton–Euler-type angular momentum balance arguments; see Section 2.2 and (2.11).

5.4.1 Intermediate variables

Following [187] we will now introduce a number of intermediate variables that will be used in the sequel. Quantities associated with the four rigid bodies will be labelled as follows: a subscript 'R' will be used for parameters or variables relating to the rear wheel, a subscript 'B' will be used for the rear frame, a subscript 'F' will be used for the front wheel, and a subscript 'A' will be used for parameters or variables relating to the front frame assembly.

The combined mass of the bicycle is given in (5.11). When the bicycle is in its nominal configuration, the whole-machine mass centre is located at:

$$x_T = (x_B m_B + x_H m_H + w m_F)/m_T$$
$$z_T = (-r_R m_R + z_B m_B + z_H m_H - r_F m_F)/m_T.$$

Note that z_T is negative as are each of the coordinates of the individual mass centres.

The whole-machine moments and products of inertia are given next. These will be computed relative to the rear-wheel ground-contact point and along the inertial axes:

$$I_{Txx} = I_{Rxx} + I_{Bxx} + I_{Hxx} + I_{Fxx} + m_R r_R^2 + m_B z_B^2 + m_H z_H^2 + m_F r_F^2$$
$$I_{Txz} = I_{Bxz} + I_{Hxz} - m_B x_B z_B - m_H x_H z_H + m_F w r_F$$
$$I_{Tzz} = I_{Rzz} + I_{Bzz} + I_{Hxzz} + I_{Fzz} + m_B x_B^2 + m_H x_H^2 + m_F w^2.$$

A parallel set of calculations can be carried out for the front frame assembly. The mass of the front frame assembly is:

$$m_A = m_H + m_F, \tag{5.12}$$

while the front assembly mass centre is given by

$$x_A = (x_H m_H + w m_F)/m_A;$$
$$z_A = (z_H m_H - r_F m_F)/m_A.$$

The front assembly moments and products of inertia (with respect to the mass centre) are given next,

$$I_{Axx} = I_{Hxx} + I_{Fxx} + m_H(z_H - z_A)^2 + m_F(r_F + z_A)^2;$$
$$I_{Axz} = I_{Hxz} - m_H(x_H - x_A)(z_H - z_A) + m_F(w - x_A)(r_F + z_A);$$
$$I_{Azz} = I_{Hzz} + I_{Fzz} + m_H(x_H - x_A)^2 + m_F(w + x_A)^2.$$

A number of other quantities relating to the front frame geometry are also required. To begin, we will need the perpendicular distance between the front frame mass centre and the steering axis, which follows by simple planar geometry.[3] One of the equations of motion to follow comes from taking moments around the steering axis; we will therefore need the moment of inertia of the front assembly around this axis, which is computed by application of (2.14) with $\mathcal{R} = \mathcal{R}(e_y, \lambda)$ and (2.13) with $r = [u_A, 0, 0]^T$:

$$I_{A\lambda\lambda} = m_A u_A^2 + I_{Axx} \sin^2 \lambda + 2I_{Axz} \sin \lambda \cos \lambda + I_{Azz} \cos^2 \lambda. \qquad (5.14)$$

We will also require product-of-inertia terms associated with the moments generated around the inertial x- and z-axes by steering angular accelerations:

$$I_{A\lambda x} = -m_A u_A z_A + I_{Axx} \sin \lambda + I_{Axz} \cos \lambda$$
$$I_{A\lambda z} = m_A u_A x_A + I_{Axz} \sin \lambda + I_{Azz} \cos \lambda.$$

The expressions for these inertia terms can be obtained from (2.11) by taking moments around O considering (small) steer perturbations δ

$$\begin{bmatrix} \ddot{\delta} I_{A\lambda x} \\ 0 \\ \ddot{\delta} I_{A\lambda z} \end{bmatrix} = \frac{d\mathbf{H}_O}{dt} \qquad \mathbf{H}_O = \begin{bmatrix} I_{Axx} & 0 & I_{Axz} \\ 0 & I_{Ayy} & 0 \\ I_{Axz} & 0 & I_{Azz} \end{bmatrix} \begin{bmatrix} \dot{\delta} \sin \lambda \\ 0 \\ \dot{\delta} \cos \lambda \end{bmatrix} + r_{OA} \times P_A, \qquad (5.15)$$

where

$$r_{OA} = \begin{bmatrix} x_A \\ u_A \delta \\ z_A \end{bmatrix} \qquad P_A = \begin{bmatrix} 0 \\ m_A u_A \dot{\delta} \\ 0 \end{bmatrix}. \qquad (5.16)$$

---

3 Consider the similar triangles in the figure

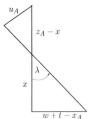

It follows that

$$\tan \lambda = \frac{w + t - x_A}{x} \qquad \sin \lambda = \frac{u_A}{z_A - x},$$

in which x is unknown. Equation (5.13) follows by eliminating x from the above equations (z_A is assumed negative),

$$u_A = (x_A - w - t) \cos \lambda - z_A \sin \lambda. \qquad (5.13)$$

The reaction force at the front-wheel ground-contact point will appear in the moment calculations given in the sequel. The sum of the front- and rear-wheel reaction forces is given by

$$N_F + N_R = gm_T.$$

When the bicycle is in its nominal configuration we also have

$$N_F = gm_T x_T / w \tag{5.17}$$

and so

$$N_R = gm_T(1 - \frac{x_T}{w}). \tag{5.18}$$

The rear- and front-wheel angular momenta along their y-axes, divided by the forward speed, form the gyrostatic coefficients:

$$S_R = \frac{I_{Ryy}}{r_R} \qquad S_F = \frac{I_{Fyy}}{r_F} \qquad S_T = S_R + S_F. \tag{5.19}$$

We conclude with two intermediate variables that appear in relation to the steering geometry:

$$\mu = \frac{t\cos\lambda}{w} \quad \text{and} \quad S_A = m_A u_A + \mu m_T x_T.$$

5.4.2   Equations of motion

Using (2.9), the moment balance (2.11) can be rewritten as

$$\sum_{i\in\text{bodies}} [\mathbf{r}_i \times \mathbf{a}_i m_i + I_i\dot{\boldsymbol{\omega}}_i + \boldsymbol{\omega}_i \times (I_i\boldsymbol{\omega}_i)] = \sum_{j\in\text{forces}} [\mathbf{r}_j \times \mathbf{F}_j] \tag{5.20}$$

in the case of an instantaneously stationary pivot point. The vectors $\mathbf{r}_i$ and $\mathbf{r}_j$ give the positions of the body mass centres and applied forces respectively, $\boldsymbol{\omega}_i$ are the angular velocities,[4] and $\mathbf{a}_i$ is the acceleration of the mass centres. Right-hand-side forces include gravitational forces and tyre forces, with normal loads given by (5.17) and (5.18) and lateral forces unknown. Three equations of motion will be computed: roll balance, yaw balance and steer balance. These equations will eventually be reduced to two, after elimination of the front-wheel lateral force and enforcement of the nonholonomic constraints.

The first equation of motion comes from a momentum balance around an axis $\boldsymbol{u}$ in the ground plane that is instantaneously aligned with the line of intersection between the rear frame and the ground plane. Note that the rear-wheel ground-contact point

---

[4]   The angular velocities related to the whole bike ω_T, front assembly ω_A, rear wheel ω_R, and front wheel ω_F, expressed in a frame yawed with the bicycle, are (see Section 2.5)

$$\omega_T = \begin{bmatrix} \dot\phi \\ 0 \\ \dot\psi \end{bmatrix} \quad \omega_A = \begin{bmatrix} \dot\phi + \dot\delta\sin\lambda \\ 0 \\ \dot\psi + \dot\delta\cos\lambda \end{bmatrix} \quad \omega_R = \begin{bmatrix} \dot\phi \\ -u/r_R \\ \dot\psi - \phi u/r_R \end{bmatrix} \quad \omega_F = \begin{bmatrix} \dot\phi + \dot\delta\sin\lambda + \delta\frac{u}{r_F}\cos\lambda \\ -u/r_F \\ \dot\psi + \dot\delta\cos\lambda - \phi u/r_F - \delta\frac{u}{r_F}\sin\lambda \end{bmatrix}.$$

falls on this line, but in general the front-wheel ground-contact point will not. In conformity with (5.20), and ignoring temporarily the rolling constraints, we obtain:

$$-m_T \ddot{y}_R z_T + I_{Txx}\ddot{\varphi} + I_{Txz}\ddot{\psi} + I_{A\lambda x}\ddot{\delta} + u\dot{\psi}S_T + u\dot{\delta}S_F \cos\lambda$$
$$= T_B - gm_T z_T \varphi + gu_A m_A \delta + g\mu x_T m_T \delta$$
$$= T_B - gm_T z_T \varphi + gS_A \delta. \tag{5.21}$$

The last two terms on the left-hand side are yaw- and steer-related gyroscopic effects. The terms on the right include an external roll moment T_B (e.g. a wind gust), a roll-related gravitational term, a steer-related gravitational term, and a term related to the reaction force at the front-wheel ground-contact point.[5]

The next equation comes from an angular momentum balance for the whole machine about an axis parallel to the inertial z-axis and that passes through the rear-wheel ground-contact point:

$$m_T x_T \ddot{y}_R + I_{Txz}\ddot{\varphi} + I_{Tzz}\ddot{\psi} + I_{A\lambda z}\ddot{\delta} - u\dot{\varphi}S_T - u\dot{\delta}S_F \sin\lambda = wF_{y_F}. \tag{5.22}$$

Again, the last two terms on the left-hand side are yaw- and steer-related gyroscopic terms. The only external force is the lateral front-wheel ground-contact reaction force F_{y_F}, which acts at the front-wheel ground-contact point. This force acts in the ground plane and is perpendicular to the line of intersection between the ground plane and the plane of symmetry of the front wheel.

The last equation comes from an angular momentum balance for the steering axis.

$$m_A u_A \ddot{y}_R + I_{A\lambda x}\ddot{\varphi} + I_{A\lambda z}\ddot{\psi} + I_{A\lambda\lambda}\ddot{\delta} + uS_F(\dot{\psi}\sin\lambda - \dot{\varphi}\cos\lambda)$$
$$= T_{HB} - tF_{y_F}\cos\lambda + g(\varphi + \delta\sin\lambda)S_A. \tag{5.23}$$

The right-hand side of this equation contains forcing terms from the steering moment T_{HB}, the front-wheel ground-reaction force, and steering-geometry-related roll

---

[5] The roll-related gravitational term is easy to visualize. The other terms are a little more subtle, as can be appreciated by studying the diagram of the upright bicycle with its steering assembly steered to the right.

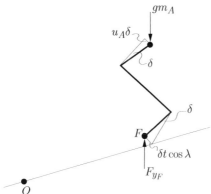

and steer terms. As we are only interested in first-order contributions, we consider separately the gravitational roll moments,[6] and then those due to steering.[7]

In order to obtain a final set of equations that describe the bicycle's dynamics for small perturbations from straight running we need to eliminate the front-wheel ground-contact force F_{y_F} from (5.22) and (5.23), and then recognize the nonholonomic rolling constraints associated with the two road wheels. In essence, the rolling constraints allow the lateral position and its derivatives, and the yaw angle and its derivatives to be expressed in terms of the steer angle. As explained earlier, the cyclic coordinates can then be recovered by integrating the nonholonomic rolling constraints. Assuming that the bicycle is travelling along the positive x-axis, the lateral velocity of the rear-wheel ground-contact point is

$$\dot{y}_R = u\psi \tag{5.24}$$

for small values of yaw angle. The lateral velocity of the front-wheel ground contact point can be recovered from the front wheel position. It follows from the figure in the

---

[6]   Consider the rolled machine in which the steering assembly is in its neutral (straight-ahead) position. In combination, the gravitational and front-wheel reaction forces produce a moment $gS_A\varphi$.

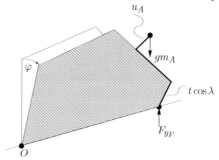

[7]   Consider in the figure the upright steered machine. The component of the gravitational force acting on the front frame assembly mass centre that is perpendicular to the steering axis produces moment $gm_A u_A \sin\lambda$, while the component of the front-wheel reaction force acting perpendicular to the steering axis produces a moment $tF_{y_F} \sin\lambda \cos\lambda$.

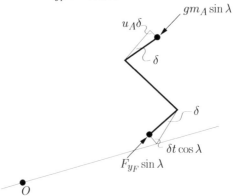

footnote on page 207 that the front wheel position is

$$y_F = y_R + w\psi - t \cos \lambda \delta. \tag{5.25}$$

For the front wheel, the body-fixed coordinate system is rotated with respect to the global z-axis by an amount $\psi + \delta \cos \lambda$. The lateral velocity of the front wheel is therefore given by

$$\frac{dy_F}{dt} = u(\psi + \delta \cos \lambda) \;\Rightarrow\; \dot{\psi} = \frac{u\delta + t\dot{\delta}}{w} \cos \lambda, \tag{5.26}$$

which gives

$$\ddot{\psi} = \frac{u\dot{\delta} + t\ddot{\delta}}{w} \cos \lambda, \tag{5.27}$$

$$\ddot{y}_R = \frac{u^2 \delta + v t \dot{\delta}}{w} \cos \lambda. \tag{5.28}$$

The final roll equation comes from (5.21), once $\ddot{y}_R$, $\ddot{\psi}$, and $\dot{\psi}$ have been eliminated using (5.26), (5.27), and (5.28) respectively. The steer equation comes from eliminating the front tyre side force F_{y_F} from (5.22) and (5.23) (which corresponds to using μ(5.22)+(5.23)), followed by the removal of $\ddot{y}_R$, $\ddot{\psi}$, and $\dot{\psi}$ as before. The resulting equations can be written in the form

$$M\ddot{q} + uC\dot{q} + (gK_0 + u^2 K_2)q = \begin{bmatrix} T_B \\ T_{HB} \end{bmatrix}, \tag{5.29}$$

where

$$q = \begin{bmatrix} \varphi \\ \delta \end{bmatrix}.$$

The entries of the mass, damping, and stiffness matrices are given by

$$M_{11} = I_{Txx}$$
$$M_{12} = I_{A\lambda x} + \mu I_{Txz}$$
$$M_{21} = M_{12}$$
$$M_{22} = I_{A\lambda\lambda} + 2\mu I_{A\lambda z} + \mu^2 I_{Tzz},$$

$$C_{11} = 0$$
$$C_{12} = \mu S_T + S_F \cos \lambda + \frac{I_{Txz}}{w} \cos \lambda - \mu m_T z_T$$
$$C_{21} = -(\mu S_T + S_F \cos \lambda)$$
$$C_{22} = \frac{I_{A\lambda z}}{w} \cos \lambda + \mu \left(S_A + \frac{I_{Tzz}}{w} \cos \lambda \right),$$

$$K_{011} = m_T z_T$$
$$K_{012} = -S_A$$
$$K_{021} = K_{012}$$
$$K_{022} = -S_A \sin \lambda,$$

and

$$K_{211} = 0$$

$$K_{212} = \frac{S_T - m_T z_T}{w} \cos \lambda$$

$$K_{221} = 0$$

$$K_{222} = \frac{S_A + S_F \sin \lambda}{w} \cos \lambda$$

respectively.

The cyclic coordinates, which determine the bicycle's position and orientation, can be recovered a posteriori by integrating the relevant nonholonomic constraint equations. Since these equations do not appear in (5.29), they need not conform to the small-perturbation assumption. The vehicle's yaw can be found by integrating (5.26). After that, the machine's global position can be found by integrating

$$\dot{x}_P = u \cos \psi \qquad \dot{y}_P = u \sin \psi, \tag{5.30}$$

which, unlike (5.24), do not assume small yaw angles.

5.4.3   Modal analysis

In order to study the modal behaviour of the Whipple bicycle as described in equation (5.29), we begin by introducing the matrix-valued polynomial

$$P(s, v) = s^2 M + suC + (u^2 K_2 + g K_0), \tag{5.31}$$

which is quadratic in both the forward speed u and in the Laplace variable s. The associated transfer function equation is

$$\begin{bmatrix} P_{11}(s) & P_{12}(s, u) \\ P_{21}(s, u) & P_{22}(s, u) \end{bmatrix} \begin{bmatrix} \varphi(s) \\ \delta(s) \end{bmatrix} = \begin{bmatrix} T_B(s) \\ T_{HB}(s) \end{bmatrix}, \tag{5.32}$$

where notably P_{11} is independent of u. When studying stability, the roots of the speed-dependent quartic equation

$$\det(P(s, u)) = 0 \tag{5.33}$$

must be analysed using numerical methods.[8]

Figure 5.5 shows the loci of the roots of (5.33) as functions of the forward speed. As with the simple inverted pendulum that has eigenvalues at $\pm \lambda_i$, a number of symmetries exist in the Whipple bicycle spectrum. At zero speed, if λ_i is an eigenvalue

---

[8]   A convenient approach is to cast (5.29) in the state-space form $\dot{x} = Ax + Bu$ and compute the eigenvalues of the A matrix given by

$$A = \begin{bmatrix} 0 & I \\ -M^{-1}(u^2 K_2 + g K_0) & -M^{-1} uC \end{bmatrix}. \tag{5.34}$$

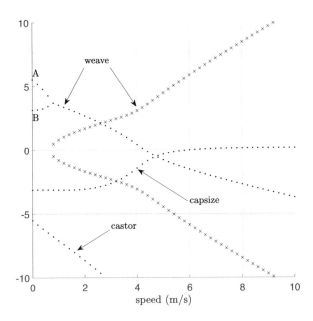

Figure 5.5: Whipple bicycle straight-running stability properties. Real and imaginary parts of the eigenvalues of the straight-running bicycle model as functions of speed. The dotted lines correspond to the real part of the eigenvalues and the crosses show the imaginary parts for the weave mode. The weave-mode eigenvalue stabilizes at $u_w = 4.3\,\text{m/s}$, while the capsize mode becomes unstable at $u_c = 6.0\,\text{m/s}$, giving the stable speed range $u_c \geq u \geq u_w$.

then so is $-\lambda_i$. It is also evident from (5.31) that if λ_i is an eigenvalue at speed u, then $-\lambda_i$ is an eigenvalue at speed $-u$. We will now examine each of the eigenvalues in a little more detail. The numerical values reported herein are obtained using the dataset in Table 5.1.

The Whipple bicycle has two important modes, namely, weave and capsize. The weave mode begins at zero speed with the two unstable eigenvalues marked A and B in Figure 5.5; these modes are the steer-capsize and body-capsize modes respectively. The eigenvector corresponding to the A-mode eigenvalue is steer dominated (hence the name steer-capsize) and has a steer-to-roll ratio of -36.9; the negative sign means that as the bicycle rolls (to the right, say), the steering rotates anti-clockwise as viewed from above. This shows that the motion associated with the A-mode is dominated by the front frame diverging towards full lock as the machine rolls over under gravity. Since real tyres make distributed contact with the ground, a real bicycle cannot be expected to behave in exact accordance with this prediction. The eigenvector components corresponding to the B-mode (body-capsize) eigenvalue has a steer-to-roll ratio of -0.55. The associated motion involves the rear frame toppling over, or capsizing, like an unconstrained inverted pendulum (to the right say), while the steering assembly rotates relative to the rear frame to the left. The term 'capsize' is used in three different contexts. The static and very low-speed capsizing of the bicycle front steered assembly and the main frame are associated with the points A and B, respectively, and the associated nearby loci; see Figure 5.5. The locus marked capsize in Figure 5.5 is associated with the higher-speed toppling and lateral translation of the vehicle. This mode crosses the stability boundary and becomes unstable when the matrix $u^2 K_2 + g K_0$ in (5.29) and (5.34) is singular.[9]

As the machine speed builds up from zero, the unstable (real) steer-capsize and body-capsize modes combine at approximately 0.6 m/s to produce the oscillatory fishtailing weave mode. The Whipple bicycle model predicts that the weave mode frequency is approximately proportional to speed for speeds above 0.6 m/s. In contrast, the capsize mode is a non-oscillatory motion, which when unstable, corresponds to the unsteered bicycle slowly toppling over at speeds above 6.0 m/s. From the perspective of bicycle riders and designers, this mode is only of minor consequence, since it is either an artefact of the model, or is easily stabilized by the rider. In practice, the capsize mode can also be stabilized using appropriately phased rider body motions, as is evident from hands-free riding. There is a third well damped mode called the castor mode, which describes the self aligning of the steering system—like the straightening of a supermarket trolley castor wheel.

In a measurement program an instrumented bicycle was used in [190] to validate the Whipple bicycle model described in [2] and [187]. The measurement data show close agreement with the model in the 3–6 m/s speed range; the weave mode frequency and damping agreement are noteworthy. The transition of the weave mode from the stable to the unstable speed ranges is also accurately predicted by the Whipple model. These measurements lend credibility to the idea that tyre and frame compliance effects can be neglected for benign manoeuvring in the 0–6 m/s range.

5.4.4   Gyroscopic influences

Gyroscopic precession is a favourite topic of conversation in bar-room discussions amongst motorcyclists. While it is not surprising that lay people have difficulty understanding these effects, inconsistencies also appear in the technical literature on single-track vehicle behaviour, as already alluded to in Section 5.1.

The experimental evidence is a good place to begin the process of understanding gyroscopic influences. Experimental bicycles whose gyroscopic influences are cancelled through the inclusion of counter-rotating wheels have been designed and built [184]. Other machines have had their gyroscopic influences exaggerated through the use of a high-moment-of-inertia front wheel [188]. In both cases the bicycles were found to be easily ridable. As with the stabilization of the capsize mode by the rider, the precession-cancelled bicycle appears to represent little more than a simple low-bandwidth challenge to the rider. As noted in [184], in connection with his precession-cancelled bicycle, '...Its "feel" was a bit strange, a fact I attributed to the increased moment of inertia about the front forks, but it did not tax my (average) riding skill even at low speed...'. When trying to ride this particular bicycle without hands, however, the

---

[9]   The matrix in (5.29) has a zero eigenvalue when $u^2 K_2 + g K_0$ is singular.

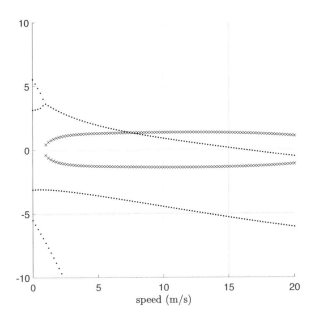

Figure 5.6: Gyroscopic stability properties. Real and imaginary parts of the eigenvalues of the straight-running Whipple bicycle with the gyroscopic moment associated with the front road wheel removed (by setting $I_{Fyy} = 0$). The dotted lines correspond to the real parts of the eigenvalues, while the crosses show the imaginary parts.

rider could only just keep it upright, because the vehicle seemed to lack balance and responsiveness.

In their theoretical work Klein and Sommerfeld [182] studied a Whipple-like quartic characteristic equation using the Routh criterion. While the Whipple bicycle model has a stable range of speeds, which Klein and Sommerfeld called the interval of auto-stability, this model with the spin inertia of the front wheel set to zero is unstable up to a speed of 16.2 m/s. This degraded stability can be seen in Figure 5.6, where the capsize mode remains stable with the damping increasing with speed (due to its stability, the capsize nomenclature may seem inappropriate in this case). In contrast, the weave mode is unstable for speeds below 16.4 m/s and the imaginary part is less than 1.4 rad/s. Klein and Sommerfeld attribute the stabilizing effect of front-wheel precession to a self-steering effect; as soon as a bicycle with spinning wheels begins to roll, the resulting gyroscopic moment due to the front wheel part of the $uC_{21}\dot{\varphi}$ term in (5.29) causes the bicycle to steer in the direction of fall. The front contact point consequently takes up a position below the steering assembly mass centre.

The Klein and Sommerfeld findings might leave the impression that gyroscopic effects are essential to auto-stabilization. However, as shown in [191] and discussed in Section 4.4, bicycles without trail or gyroscopic effects can auto-stabilize at modest speeds by adopting extreme mass distributions—but the design choices necessary do not make for a practical vehicle.

5.5   Control-theoretic implications

Locked steering behaviour. As was noted earlier, $P_{11}(s)$ in (5.32) in independent
of speed due to the fact that in (5.29) $C_{11} = 0$ and $K_{211} = 0$. This observation
is interesting from both a physical and a control theoretic perspective. Suppose the
steering degree of freedom is removed. In this case the steering angle $\delta(s)$ must be
set to zero in (5.32), and under the assumption of zero external moment T_B, the roll
freedom is described by

$$0 = P_{11}(s)\varphi(s)$$
$$= (s^2 I_{Txx} + gm_T z_T)\varphi(s). \tag{5.35}$$

The roots of $P_{11}(s)$ are given by

$$p_\pm = \pm\sqrt{\frac{-gm_T z_T}{I_{Txx}}}, \tag{5.36}$$

where m_T is the total mass of the bicycle and rider, z_T is the height of the combined
mass centre above the ground, and I_{Txx} is the roll moment of inertia of the entire
machine around the wheelbase ground line. For the data in Table 5.1 $p_\pm = \pm 3.1347$.

Since in this case the steering freedom is removed, the A-mode (see Figure 5.5) does
not appear. The vehicle's inability to steer also means that the weave mode disappears.
Instead, the machine's dynamics are fully determined by the speed-independent, whole-
vehicle capsize (inverted-pendulum) mode seen at point B in Figure 5.5 and given by
(5.36). Not surprisingly, motorcycles have a tendency to capsize at low speeds if the
once-common friction-pad steering is tightened down far enough to lock the steering
system; see [192].

5.5.1   A feedback-system perspective

Bicycle as a feedback system. We will now study a steering-related aspect of bicycle
behaviour and relate it to the 'locked steering' capsize mode evaluated in (5.36). As
Whipple [2] surmised, the rider's main control input is the steering torque. If we
suppose the external roll moment is zero, it follows from (5.32) that

$$\varphi(s) = -\frac{P_{12}(s, u)}{P_{11}(s)}\delta(s). \tag{5.37}$$

The second row of (5.32) gives

$$\delta(s) = \frac{-P_{21}(s, u)}{P_{22}(s, u)}\varphi(s) + \frac{1}{P_{22}(s, u)}T_{HB}(s). \tag{5.38}$$

Equations (5.37) and (5.38) are shown diagrammatically in the feedback configuration
given in Figure 5.7. Eliminating $\varphi(s)$ yields the closed-loop transfer function

$$H_{\delta T_{HB}}(s, u) = \frac{P_{11}}{\det(P)}(s, u). \tag{5.39}$$

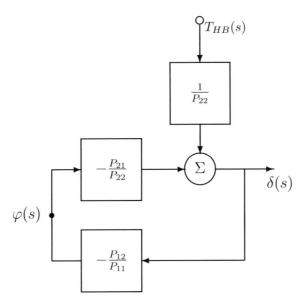

Figure 5.7: Block diagram of the Whipple bicycle model described in (5.32). The steer torque applied to the handlebars is $T_{HB}(s)$, $\varphi(s)$ is the roll angle, and $\delta(s)$ is the steer angle.

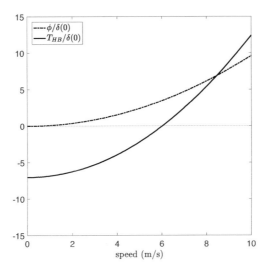

Figure 5.8: Roll angle to steer angle and steer torque to steer angle at equilibrium as a function of speed.

As the speed of the bicycle increases, the unstable poles associated with the static capsize modes coalesce to form the complex pole pair associated with the weave mode. As is clear from Figure 5.5, the weave mode is stable for speeds above 4.3 m/s. For typical datasets $\frac{\phi}{\delta}(0) > 0$; a positive steer angle corresponds to a positive roll angle under steady-state conditions. Also $\frac{\delta}{T_{HB}}(0) < 0$; a positive steer torque corresponds to a negative steer angle under steady-state conditions below the critical capsize speed (6 m/s with the current dataset). Above the critical capsize speed $\frac{\delta}{T_{HB}}(0) > 0$; see Figure 5.8. The zero response in the steer angle (from the steering torque) is due to the singularity of the stiffness matrix $u^2 K_2 + g K_0$ at the critical capsize speed.

The poles and zeros of $H_{\delta T_{HB}}(s, u)$, as a function of speed, are shown in Figure 5.9. Save for the pair of speed-independent zeros, this diagram contains the same information as that given in Figure 5.5. The zeros of $H_{\delta T_{HB}}(s, u)$, which derive from the roots of $P_{11}(s)$ as shown in (5.36), are associated with the speed-independent whole-vehicle capsize mode. The backward-running vehicle is seen to be unstable throughout the speed range, but this vehicle is designed for forward motion and, when running backwards, it has negative trail and a divergent castor action.

An explanation for the stabilization difficulties associated with backward-running bicycles centres on the positive zero fixed at $+\sqrt{g m_T z_T / I_{Txx}}$, which is in close proximity to a right-half-plane pole in certain speed ranges [168]. In control terms, this 'almost pole-zero cancellation' in the right-half plane constitutes an unstable 'almost uncontrollable' mode. In the case of an exact cancellation, this unstable uncontrollable mode becomes inaccessible to the rider.

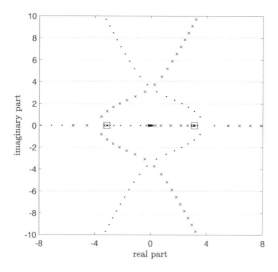

Figure 5.9: Poles and zeros of $H_{\delta T_{HB}}(s, u)$ as functions of speed. The speed u is varied between ± 10 m/s. The poles are shown as dots for forward speeds and crosses for reverse speeds. There are two speed-independent zeros shown as squares at ± 3.1347.

The right-half-plane (positive) zero has fundamental implications for the forward-running vehicle. It is well known that the step response of an asymptotically stable, strictly proper (i.e. more poles than zeros) transfer function exhibits an initial un-dershoot if the system has an odd number of positive zeros [193]. These systems also tend to be more difficult to control. The steer torque to steer angle transfer function (5.39) is non-minimum phase in the stable speed range. When applying a positive steer torque step, a positive steer angle response results that eventually settles to a negative steady-state value. This negative steady-state steer angle results in a negative steady-state yaw rate (5.26) and a negative roll angle. To summarize, in order to enter a counter-clockwise turn the rider needs to apply a clockwise steer torque. The non-minimum phase zero is the mathematical reason for this counter-steering behaviour.

Steering. The appreciation of the subtle nature of bicycle steering goes back over a hundred years. Archibald Sharp records (p. 222 of [3]) '...to avoid an object it is often necessary to steer for a small fraction of a second towards it, then steer away from it; this is probably the most difficult operation the beginner has to master ...'. While perceptive, such historical accounts make no distinction between steering torque control and steering angle control, they do not highlight the role played by the machine speed [194], and timing estimates are based on subjective impressions, rather than experimental measurement.

We will now examine the role of speed and steering torque in determining the vehicle's trajectory. The steer-torque-to-steer-angle response of the bicycle can be deduced from (5.39). Once the steer angle response is known, the small-perturbation yaw rate response for the model can be found using (5.26), with the lateral displacement then coming from (5.24). It then follows that the transfer function linking the lateral displacement to the steer angle is

$$H_{y_R\delta}(s,u) = \frac{u(u+st)\cos\lambda}{s^2 w},$$
(5.40)

and that the transfer function linking the lateral displacement to the steering torque is given by $H_{y_R T_{HB}} = H_{y_R\delta}(s,u)H_{\delta T_{HB}}(s,u)$, with $H_{\delta T_{HB}}(s,u)$ given in (5.39). This transfer function is used in the computation of responses to steering torque step inputs. The right-half-plane zero in $H_{\delta T_{HB}}(s,u)$ is likely to produce detectable non-minimum phase behaviours.

In order to study the bicycle's steering response at different speeds, including those outside the auto-stable speed range, it is necessary to introduce stabilizing rider control. In this study the rider will be emulated using the roll-angle-plus-roll-rate feedback law

$$T_{HB}(s) = r(s) + (k_\varphi + sk_{\dot\varphi})\varphi(s),$$
(5.41)

in which $r(s)$ is a reference torque input, and k_φ and $k_{\dot\varphi}$ are the roll and roll-rate feedback gains respectively. This feedback law can be combined with

$$H_{\varphi T_{HB}}(s,u) = -\frac{P_{12}}{\det P}(s,u)$$

to obtain the open-loop stabilizing steer-torque pre-filter

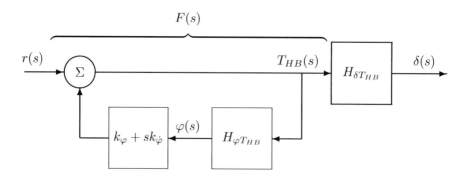

Figure 5.10: Steering torque pre-filter $F(s)$ described in (5.42). This filter is an open-loop realization of the roll-angle-plus-roll-rate feedback law described in (5.41). As readers familiar with control systems will realize, open- and closed-loop systems can be represented in equivalent ways if there are no disturbances and no modelling uncertainties.

$$F(s) = \frac{\det(P(s, u))}{\det(P(s, u)) + (k_\varphi + sk_{\dot\varphi})P_{12}(s, u)}, \qquad (5.42)$$

which maps the reference input $r(s)$ into the 'rider-produced' steering torque $T_{HB}(s)$ as shown in Figure 5.10. In the auto-stable speed range the stabilizing pre-filter is not needed and $F(s)$ can be set to unity in this case. The bicycle's steering behaviour can now be studied at speeds below, within, and above the auto-stable speed range. Prior to manoeuvring, the machine is in a constant-speed straight-running trim condition. In each of the three cases, the filtered steering torque, the roll angle, the steering angle, and the lateral displacement responses to a steering torque reference input $r(s)$ are shown in Figure 5.11. The magnitude of the torque step is $r = \varphi_0 \left(\frac{T_{HB}}{\varphi}(0) + k_\varphi \right)$, which is easily verified to give the same steady-state target roll angle φ_0.

The auto-stable case is considered first. In this case the clockwise (when viewed from above) steer torque reference is applied directly to the bicycle's steering system (see Figure 5.11). Following the steer torque input, the bicycle immediately rolls to the left (negative ϕ) in preparation for a left-hand turn. The machine initially steers to the right (positive δ) and the rear-wheel ground-contact point starts moving to the right (positive y_R). After approximately 0.6 s, the steer angle sign reverses, while the rear-wheel ground-contact point begins moving to the left after approximately 1.1 s. In order to turn to the left, one must steer to the right so as the make the machine roll to the left. This property of the machine to apparently roll in the wrong direction is sometimes referred to as 'counter-steering' [188, 195]. The non-minimum phase behaviour in the steer angle and lateral displacement responses is attributable to the right-half-plane zero in $H_{\delta T_{HB}}(s, u)$ given by the roots of $P_{11}(s) = 0$ and corresponding to the locked-steering whole-machine capsize mode as illustrated in (5.36). Towards the end of the simulation shown, the steer angle settles into an equilibrium condition, in which the

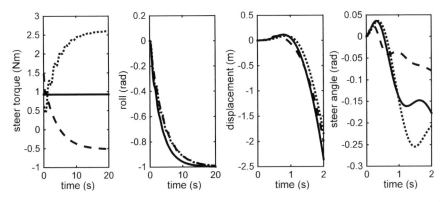

Figure 5.11: Response of the controlled Whipple bicycle to a steering torque reference $r = (T_{HB}/\varphi)(0) + k_\varphi$. The responses at the auto-stable speed of 5 m/s are shown solid; the pre-filter gains are $k_\varphi = k_{\dot\varphi} = 0$. The low-speed 4 m/s case, which is below the auto-stable speed range, is shown dotted; the stabilizing pre-filter gains are $k_\varphi = 2$ and $k_{\dot\varphi} = -1$. The high-speed 7 m/s case, which is above the auto-stable speed range, is shown dashed; the stabilizing pre-filter gains are $k_\varphi = -2$ and $k_{\dot\varphi} = -1$.

bicycle turns left in a circle with a fixed negative roll angle. In relation to the non-minimum phase response in the lateral displacement behaviour, the control difficulty arises if one rides near to a curb [195]: to escape, one has to go initially closer to the edge.

At speeds below the auto-stable range, active stabilizing steering must be utilized in order to prevent the machine from toppling over. In the low-speed (4 m/s) case the steer torque illustrated in Figure 5.11 is the step response of the pre-filter, which is the steer torque required to establish a steady turn. The output of the pre-filter, which crudely mimics the rider, is uni-directional apart from the superimposed weave-frequency oscillation required to stabilize the bicycle's unstable weave mode. In the case considered here, the steady-state steer torque is more than twice the auto-stable reference torque required to bring the machine to a steady-state roll angle of −1 rad. The steer angle and lateral displacement responses are similar to those obtained in the auto-stable case. If the trim speed is increased to the upper limit of the auto-stable range (in this case 6.0 m/s; see Figure 5.5), then the steady-state steering torque required to maintain an equilibrium steady-state turn falls to zero; this response is due to the singularity of the stiffness matrix $u^2 K_2 + gK_0$ at this speed.

At speeds above the auto-stable range, stabilizing rider intervention is again required. The interesting variation in this case is in the steering torque behaviour. This torque is initially positive and results in the vehicle rolling to the left. However, if this roll behaviour were left unchecked, the bicycle would topple over, and so to avoid the problem the steer torque immediately reduces and then changes sign after approximately 5 s. The steer torque then approaches a steady-state value of −0.5 Nm in order to stabilize the roll angle and maintain the counter-clockwise turn. This need to steer in one direction to initiate the turning roll angle response, and then to later apply an opposite steering torque that stabilizes the roll angle, is a high-speed phenomenon, providing an alternative interpretation of the 'counter-steering' phenomenon mentioned earlier. Counter-steering in the non-minimum phase response sense is always present, while in the second sense it is a high-speed[10] phenomenon only.

We will now argue that the pre-filter enforces this type of counter-steering for all stabilizing values of k_φ and $k_{\dot\varphi}$ and not only the particular numerical values chosen. First, the infinite frequency gain of $F(s)$ is unity; this follows from the fact that the numerator and denominator quartic coefficients (in s) are both given by $\det(M)$; see (5.31). Since k_φ and $k_{\dot\varphi}$ are stabilizing, all of the denominator polynomial coefficients of $F(s)$ are positive, as are all of the numerator polynomial coefficients in the auto-stable speed range. As the speed passes from the auto-stable range, $\det(u^2 K_2 + gK_0)$ changes sign as does the constant coefficient in the numerator of $F(s)$.[11] Therefore, at speeds above the auto-stable range, $F(s)$ has a negative steady-state gain thereby enforcing the sign reversal in the steering torque as observed in Figure 5.11.

We conclude this section by associating the Whipple bicycle's non-minimum phase response (in the steer angle) with its self-steering characteristics. To do this consider removing the bicycle's ability to self-steer by setting $\alpha = \pi/2$, $t = 0$, $I_{Axz} = 0$, $I_{Fyy} = 0$, and $x_H = w$. With these changes in place, it is easy to check that $P_{21}(s, u) = 0$, which means that (5.38) reduces to

$$H_{\delta T_{HB}} = \frac{1}{P_{22}(s, u)}$$
$$= \frac{1}{s I_{Azz}(s + u/w)}, \tag{5.43}$$

which is clearly minimum phase and represents the response one would expect when applying a torque to the steering assembly inertia in parallel with a speed-dependent 'steering damper'. The damping term is clearly speed-dependent and is related to the rear-wheel rolling constraints.

5.6   Model extensions

5.6.1   Acceleration

The simple model given in (5.29) can be extended to include the effects of acceleration and braking. This extended model takes the form

$$M\ddot{q} + uC\dot{q} + (gK_0 + a_x K_1 + uK_2)q = \begin{bmatrix} T_B \\ T_{HB} \end{bmatrix}, \tag{5.44}$$

in which the stiffness matrix K_1 introduces the effects of an acceleration a_x in the bicycle's longitudinal direction.

We begin from the in-plane dynamics, which are decoupled from the out-of-plane dynamics under the small-perturbation conditions; see Section 5.2.1. Force balance in

---

[10] In this instance we are referring to speeds above the capsize critical speed—if there is one.
[11] Suppose that $\det(P(s, u)) = \sum_{i=0}^{4} s^i a_i$, then $a_0 = \det(u^2 K_2 + gK_0)$, with $P(s, u)$ given by (5.31).

the longitudinal and vertical directions, and a moment balance with respect to the bicycle's mass centre, give

$$0 = F_{x_R} + F_{x_F} - m_T a_x \qquad (5.45)$$

$$0 = N_R + N_F - m_T g \qquad (5.46)$$

$$0 = -z_T (F_{x_R} + F_{x_F}) - x_T N_R + (w - x_T) N_F - I_{Ryy} \ddot{\theta}_R - I_{Fyy} \ddot{\theta}_F. \qquad (5.47)$$

Solving these equations for the normal loads N_R and N_F, the total longitudinal force $F_{x_R} + F_{x_F}$, and introducing the no-slip constraints $\ddot{\theta}_R = -a_x/r_R$ and $\ddot{\theta}_F = -a_x/r_F$ gives

$$N_R = m_T g \frac{w - x_T}{w} + \Delta N \qquad (5.48)$$

$$N_F = m_T g \frac{x_T}{w} - \Delta N \qquad (5.49)$$

$$F_{x_R} + F_{x_F} = m_T a_x, \qquad (5.50)$$

where

$$\Delta N = -\frac{a_x S_X}{w} \qquad S_X = m_T z_T - \frac{I_{Ryy}}{r_R} - \frac{I_{Fyy}}{r_F}. \qquad (5.51)$$

We are now ready to deal with the out-of-plane dynamics. As a result of the bicycle's acceleration, the yaw acceleration and lateral acceleration expressions given in (5.27) and (5.28) must be augmented by the additional acceleration-related terms

$$\Delta \ddot{\psi} = \frac{a_x \delta \cos \lambda}{w} \qquad \text{and} \qquad \Delta \ddot{y}_R = a_x \dot{\psi}. \qquad (5.52)$$

We must also recognize the front-wheel ground-contact longitudinal force F_{x_F}, which accelerates the front wheel. This is computed from the moment equilibrium with respect to the front-wheel centre in (5.8):

$$F_{x_F} = -\frac{S_F}{r_F} a_x \qquad (5.53)$$

in which S_F is given by (5.19).

We can now find the acceleration-related terms resulting from (5.51), (5.52), and (5.53) in the roll moment equation (5.21). The change in the front-wheel load, the change in the yaw angle acceleration, the change in the lateral acceleration, and the effect of the lateral (relative to the rear frame) component of F_{x_F} combine to introduce the following additional terms on the left-hand side of the roll balance equation (5.21):

$$\Delta T_R = a_x \delta \left[-\mu S_X + (I_{Txz}/w + S_F) \cos \lambda \right]. \qquad (5.54)$$

The yaw moment balance equation (5.22) also requires additional terms when the bicycle is accelerating. Changes in the yaw acceleration, the d'Alembert force on the front frame assembly, and the yaw-related effect of F_{x_F} must all be recognized. The combined effect of these influences is:

$$\Delta T_Y = \left[a_x \left(\frac{I_{Tzz}}{w} \cos \lambda - m_A u_A - S_F \sin \lambda \right) - F_{x_F} (w + t) \cos \lambda \right] \delta + a_x S_X \varphi. \quad (5.55)$$

Finally, the steer moment balance equation (5.23) must be modified to recognize acceleration-related forces. The combined influence of changes in the the yaw acceleration, changes in the front-wheel normal load, and the d'Alembert force acting on the front frame assembly result in the following change in the steer moment expression:

$$\Delta T_S = a_x \delta \left[\frac{I_{A\lambda z}}{w} \cos \lambda - m_A u_A \cos \lambda - \mu S_X \sin \lambda \right] - \mu a_x S_X \varphi. \quad (5.56)$$

Equations (5.54), (5.55), and (5.56) can be combined to produce the acceleration-related K_1 matrix given in (5.44) above. The first row of K_1 comes from (5.54), while the second row of K_1 is computed using $\mu(5.55) + (5.56)$.

The four entries of the K_1 matrix are

$$K_1(1, 1) = 0$$
$$K_1(1, 2) = -\mu S_X + (I_{Txz}/w + S_F) \cos(\lambda)$$
$$K_1(2, 1) = 0$$
$$K_1(2, 2) = -\mu S_X \sin(\lambda) - m_A u_A \mu + ((\mu I_{Tzz} + I_{A\lambda z})/w$$
$$+ S_F \mu(w + t - r_F \tan \lambda)/r_F - m_A u_A) \cos(\lambda).$$

In order to study the machine's behaviour for small perturbations from a straight-running accelerating trajectory, equation (5.44) can be solved numerically. Under accelerating conditions the machine's speed varies and so equation (5.44) is linear-time-varying, because uC and $u^2 K_2$ are time-varying. There is no background 'equilibrium' and the time-varying roots of the associated characteristic equation must be interpreted with care. As is well known from the linear system theory literature, the (frozen-time) eigenvalue–eigenvector pairs do not represent 'modes', and they do not necessarily determine the system's stability either. See [160, 196] and the references therein for further details.

5.6.2 Frame flexibility

A phenomenon known variously as 'speedman's wobble', 'speed wobble', or 'death wobble' is well known to cyclists [197, 198]. As the name suggests, wobble is a steering oscillation, not dissimilar to the shimmy oscillations that occur with supermarket trolley wheels, aircraft nose wheels, and in automobile steering systems. Documented records of this phenomenon in bicycles is sparse, but a survey suggested that shimmy at speeds between 4.5 and 9 m/s is unpleasant, while shimmy at speeds between 9 and 14 m/s is dangerous [198]. This study also suggested a wide spread of frequencies for these oscillations. with the most common being between 5 and 10 Hz. Experimental data for bicycle shimmy are reported in [85] and [86]. The first reports frequencies of approximately 7 Hz and speeds of approximately 5 m/s, while the latter reports frequencies of approximately 7 Hz and speeds in the range 10–17 m/s. The rotational frequency of the front wheel is sometimes close to the wobble frequency, so that forcing from wheel or tyre non-uniformity may be an added influence.

Rough surfaces may break the regularity of the wobble phenomenon thereby eliminating it, but an initial event is normally needed to trigger the problem. Shimmy usually appears during 'hands-off' riding, since the passive damping introduced by the rider is sufficient to damp the oscillations. These stabilizing effects have been reproduced theoretically [199, 200]. Hand-on shimmy is possible in extreme cases. Juden [198] advises: 'pressing one or both legs against the frame, while applying the rear brake' as a helpful practical procedure, if a wobble should commence. The possibility of accelerating out of a wobble is mentioned, suggesting the existence of a worst-case speed. The influences of loading are discussed, with special emphasis on the loading of steering-frame-mounted panniers. Sloppy wheel or steering head bearings and flexible wheels are thought to be contributory.

In an important paper from a practical viewpoint, Wilson-Jones [192] implies that wobble was a common motorcycling phenomenon in the 1950s. Machines of the period were usually fitted with a rider-adjustable friction-pad steering damper. The idea was that the rider should make the damper effective for high-speed running and ineffective for lower speeds; see also [201, 202]. Wilson-Jones offers the view that steering dampers should not be necessary for speeds under 45 m/s, indicating that, historically, wobble in motorcycles has been a high-speed problem. He also points to the dangers of returning from high speed to low speed, while forgetting to lower the preload on the steering damper.

The basic Whipple bike model cannot reproduce lightly damped and possibly unstable shimmy/wobble oscillations. In order to introduce these dynamics into the model, representing the flexible frame and/or the non-instantaneous tyre response is necessary; quantitatively accurate results require both. In this section we deal with the frame flexibility, while the tyre effect will be discussed in Section 5.6.3.

We will now introduce into the Whipple bicycle model a torsional frame flexibility that is perpendicular to the steering axis; see Figure 4.17. This lumped flexibility is used to represent the torsional compliance of the frame and the lateral compliance of the front fork. For the purposes of an initial study, the frame flexibility has been set at 2000 Nm/rad, at a distance 0.922 m from the front contact point, giving a lateral stiffness at the contact point of 2355 N/m. We have also introduced a structural damping of 20 Nm s/rad that acts in parallel with the string. The other data are again from Table 5.1. The eigenvalues of the straight-running linearized Whipple bicycle with frame flexibility are shown in Figure 5.12. As we would now expect, flexibility in the frame has introduced a new shimmy-related mode that is usually referred to as 'wobble'. When comparing Figures 5.12 and 5.5, we observe that the capsize and castor modes remain essentially unaltered. The weave-mode frequency range remains much as it did in the basic Whipple machine, although frame flexibility has a speed-dependent influence on the damping of the weave mode. At low speeds, frame flexibility has no effect, but at intermediate and high speeds the weave mode damping is compromised by the flexible frame, although this mode remains well damped. The wobble mode is new and attributable to frame flexibility. At low speeds this mode

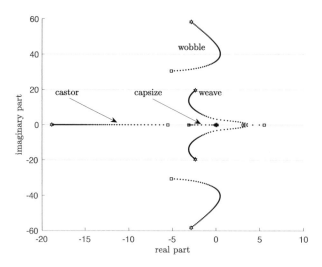

Figure 5.12: Stability properties of the Whipple bicycle with a flexible frame. The speed is varied from 0 to 20 m/s; the zero-speed end is represented by a square and the high-speed end by a hexagram. The torsional stiffness of the frame is 2000 Nm/rad and the frame's damping is 20 Nms/rad.

is well damped with a frequency just below 5 Hz. As the speed increases, the wobble mode frequency increases and its damping reduces until becoming unstable over an intermediate range of speeds, where the modal frequency is approximately 6.5 Hz. At higher speeds approaching 20 m/s the wobble mode frequency continues to increase as does its damping. This calculation has illustrated the possibility that bicycles can have a potentially dangerous wobble mode in the 6–8 Hz range at intermediate speeds.

5.6.3   Pneumatic tyres

There have been many studies into the influence of tyre characteristics on vehicular wheel shimmy; see in Section 3.1. Pacejka's PhD work [92] made an important contribution to the understanding of shimmy in general, concentrating particularly on the automotive context. Pacejka draws a distinction between 'gyroscopic shimmy', occurring in automobiles with beam axle suspensions, which was historically important in influencing the introduction of independent front suspensions for cars, and what he calls 'tyre shimmy' in which gyroscopic effects are not significant. In this latter case, the compliance of the tyre's carcass, together with the relatively high frequency of the oscillations, combine to introduce phase lags between wheel steering motions and tyre lateral forces and aligning moments, which affect the system stability markedly, as control engineers would appreciate. Pacejka drew attention to wheel-bearing clearance and kingpin friction as nonlinear influences on the small-amplitude behaviour, and tyre force saturation as determining the high-amplitude (limit-cycle) behaviour. Prior to Pacejka's research, tyre shimmy had been studied seriously only in an aeronautical context. Shimmy vibrations using an essential model were simulated and discussed in Section 4.5.

Up to this point in our study of the Whipple bicycle we have modelled the road–tyre rolling contact using velocity (non-holonomic) constraints. This approach is valid at low speeds, but becomes less reliable as the speed increases. Our focus in this section is to replace the road–tyre constraint model employed so far with the small-perturbation force-generating tyre representation of the kind introduced in Sections 3.7 and 3.8. A single-point-contact model is employed, where the lateral force response of the tyre due to steering, and therefore side slipping, is a dynamic response to the slip and camber angles of the tyre, which is modelled with a first-order relaxation equation as

$$\sigma \frac{dF_y}{ds} + F_y = F_z(C_\alpha \alpha + C_\varphi \varphi), \tag{5.57}$$

where σ is the relaxation length, F_z is the tyre normal load, α is the side-slip angle, and φ is the wheel's camber angle relative to the road. This model is used for both wheels, with the tyre parameters given in the following table:

Parameter	Value
$C_{\alpha r}$	14.0
$C_{\alpha f}$	14.0
$C_{\varphi r}$	1.0
$C_{\varphi f}$	1.0
σ_r	0.1 m
σ_f	0.1 m

The subscripts r and f are used to denote 'rear' and 'front' respectively. The products $C_{r/f1} = F_z C_{\alpha r/f}$ are the tyres' cornering stiffnesses, while the products $C_{r/f2} = F_z C_{\varphi r/f}$ are the tyres' camber stiffnesses.[12] It is now shown that the force-generating tyre model has a strong impact on the predicted properties of both wobble and weave.

---

[12] The following free body diagram represents an inverted-pendulum-type lumped-mass model of a bicycle under steady cornering:

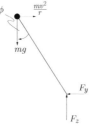

The forces acting on the bicycle's mass centre are the gravitational force mg and the centripetal force $\frac{mv^2}{r}$. Under equilibrium conditions the normal load is $F_z = mg$, while the tyre side force is $F_y = \frac{mv^2}{r}$. In the case of non-dissipative rolling the side-slip is zero and there is no corresponding side-slip-generated side force. Since

$$\frac{v^2}{rg} = \tan \phi \approx \phi$$

for small roll angles, and $F_y = F_z C_{\varphi r/f} \phi$, we conclude that under these conditions $C_{\varphi r/f} = 1.0$.

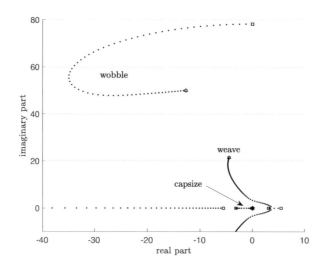

Figure 5.13: Stability properties of the Whipple bicycle with relaxed side-slipping tyres. The speed is varied from 0 to 20 m/s; the zero-speed end is represented by a square and the high-speed end by a hexagram.

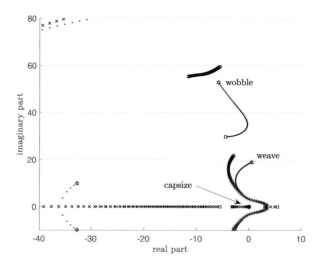

Figure 5.14: Stability properties of the Whipple bicycle with relaxed side-slipping tyres and a flexible frame. The speed is varied from 0 to 20 m/s; the zero-speed end is represented by a square and the high-speed end by a hexagram. The crossed loci represent a stiff-framed machine, while the dotted loci represent a more flexible vehicle. The properties illustrated here for the limited speed range of the bicycle are remarkably similar to those of the motorcycle, with its extended speed capabilities.

As can be seen in Figure 5.13, the introduction of side-slipping tyres also produces a wobble mode, which is not present in Figure 5.5. The resonant frequency of the very low-speed wobble mode is approximately 12.7 Hz, with low predicted damping. As the speed increases, the wobble mode frequency drops to approximately 8.0 Hz. The associated damping increases with speed and then reduces again as the machine speed approaches the top illustrated speed of 20 m/s. As with the flexible frame, side-slipping tyres have little impact on the weave mode at low speeds. However, as the speed increases, the relaxed side-slipping tyres cause a significant reduction in the intermediate and high-speed weave-mode damping.

Figure 5.14 illustrates the influence of a flexible frame in combination with relaxed side-slipping tyres; the characteristics of both stiff and more flexible (stiffness halved) vehicles are illustrated. This figure shows that the introduction of side-slipping tyres causes a reduction in the wobble-mode frequency range. As the frame stiffness increases, this effect becomes more and more pronounced. By extension from measured motorcycle behaviour, there is every reason to suspect that the accurate reproduction of bicycle wobble-mode behaviour requires a model that includes both relaxed side-slipping tyres and flexible frame representations.

6
Ride Dynamics

6.1 Background

As shown in Figure 1.8 in Chapter 1, suspension systems were part of road vehicle design from the early days of road transport. The obvious function of the suspension is to isolate the driver/rider from uneven road surfaces. Less obvious is its role in improving roadholding, improving driver/rider comfort, and reducing the fatigue damage to the vehicle that may result from hours of exposure to road-related vibrations. From a dynamics perspective, the suspension system acts as a compliant interface between the road wheels and the vehicle's main body. The main body, including the driver and passengers, can be thought of as a 'sprung mass',[1] because it is 'sprung', or isolated, from the road. The wheels, brakes, and suspension and frame components which are directly coupled to the wheels are 'unsprung masses', since through the tyres they are in direct contact with the road. In the Erasmus Darwin era the suspension system consisted of a sprung beam (axle) that extended across the width of the vehicle. Early

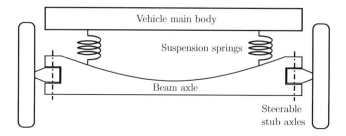

Figure 6.1: Simple beam axle suspension.

automobiles also used a beam axle design that was attached to the vehicle through springs that provide 'cushioning' (see Figure 6.1). Early leaf springs were replaced by coil springs and later parallel spring-damper combinations. Beam axle suspensions are 'dependent' suspension systems, since the wheels on each axle are rigidly linked. Later, 'independent' suspension systems were introduced, especially at the front axle, apparently to avoid gyroscopic shimmy problems [172].

---

[1] In more sophisticated models the driver and passengers are not deemed 'rigidly attached' to the chassis. Instead, the compliance of seats, and the driver and passengers' bodies, are included in the model.

Dynamics and Optimal Control of Road Vehicles. D. J. N. Limebeer and M. Massaro.
© D. J. N. Limebeer and M. Massaro 2018. Published in 2018 by Oxford University Press.
DOI: 10.1093/oso/9780198825715.001.0001

A wide variety of suspension systems have been made available by manufacturers for many years. In most cases developments are driven by perceived deficiencies in predecessor systems. In Chapter 7 the most common car independent suspensions, such as the MacPherson strut and double-wishbone suspension, are analysed. Motorcycle suspensions, such as the telescopic fork, and the duolever and telelever will also be discussed.

To restrict our treatment to a reasonable length, a number of simplifying assumptions will be made. Attention will be focused on low-frequency phenomena (20–30 Hz, the so-called *ride* frequency range), where vehicle and tyre belt structural dynamics are of little consequence. The models considered here will be of a general nature, and will not be dependent on any specific suspension layout. The stiffnesses and damping used will be at the road–tyre contact point—the computation of these quantities will be covered in Chapter 7. We will not consider suspension operation under nonlinear conditions such as those that might occur when a wheel leaves the road, or when suspension saturation occurs, or when the vehicle strikes a large object that causes large suspension displacements. Neither will we consider the operation of dampers and springs with asymmetric or nonlinear characteristics, although these are all important topics in their own right [203].

Introductory books on suspensions include [6, 7, 204], review articles on suspension design include [205–208], human perception of vehicle vibration is reviewed in [209], while suspension control is reviewed in [210, 211]. For the interaction between ride and handling see [212].

6.2   Road surface characteristics

The typical road vehicle is required to operate on a wide variety of road surfaces, where the road surface may have long and short spatial wavelengths as well as variable amplitudes. In other cases one may have to contend with more extreme road conditions that involve bumps, ruts, and potholes. While the suspension design task is in general directed towards the improvement of everyday performance properties, more extreme situations cannot be ignored. In many cases it is these more extreme situations that lead to safety-related incidents and customer dissatisfaction. In broad terms one can deal with typical suspension operation using linear analysis techniques, while more extreme events require a nonlinear investigation. As we will show, there are several analysis techniques that can be used in a linear setting, with the predominant nonlinear analysis tool being time domain simulation.

We assume that the road height is a function of time $z_r(t)$, with the time variability coming from translational movement over rough road surfaces at speed v. Since road roughness is a spatial phenomenon, the road height variable can also be expressed as a function of position $\hat{z}_r(x)$, in which x represents the vehicle's longitudinal displacement (in metres from some reference point); thus $\hat{z}_r(x) = \hat{z}_r(vt) = z_r(t)$. In the context of suspension analysis, road roughness has been modelled as a homogeneous and isotropic stochastic process since the 1950s [213,214]. For that reason road roughness is described as a measured spatial power spectral density (PSD) function; see ISO 8608:2016. If we suppose that the road's wave number is ν (rad/m) (or n (cycle/m)), then a widely used approximation to the measured road displacement PSD is

$$\Phi(\nu) = \Phi(\nu_0)\left(\frac{\nu}{\nu_0}\right)^{-w} \qquad \text{or} \qquad \Phi(n) = \Phi(n_0)\left(\frac{n}{n_0}\right)^{-w} \tag{6.1}$$

in which $\Phi(\nu)$ is in m^3/rad, $\Phi(n)$ is in $m^3/cycle$, ν_0 is usually $1\,rad/m$, n_0 is $0.1\,cycle/m$, and $w = 2$. Roads can be roughly classified according to $\Phi(\nu_0)$, which is typically between 4×10^{-6} and $4 \times 10^{-3}\,m^3$ (or $\Phi(n_0)$ between 64×10^{-6} and $65 \times 10^{-3}\,m^3$). The temporal angular frequency ω and the wave number ν are related by

$$\omega = v\nu \qquad \text{or} \qquad f = vn, \tag{6.2}$$

with $\nu = 2\pi n$ and $\omega = 2\pi f$.

We will often assume that the vehicle's rear axle will 'see' the same road as that seen by the front axle after a time delay l/v; l is the vehicle's length (wheelbase) and v its speed—this effect is called *wheelbase filtering*. For a vehicle travelling at a constant speed v, it may be shown that

$$\Phi(\omega) = \frac{1}{v}\Phi(\nu) \tag{6.3}$$

by equating mean square values through a change of variable.[2]

The generation of road profile descriptions can be obtained from the road PSD using standard inverse Fourier techniques [215, 216]. One of the earliest descriptions of a fully isotropic road profile is given in [217], where multi-dimensional Fourier transform is employed. Alternatively, rational approximation of the measured PSD can be used to generate the road profile [218, 219]; this is similar to some treatments of turbulence [220].

Example 6.1 Generate a road profile description for $L = 500\,m$ of type B road according to ISO 8608:2016. This standard gives $\Phi(n_0) = 64 \times 10^{-6}\,m^3$, $n_0 = 0.1\,cycle/m$, and $w = 2$; see the left-hand side of Figure 6.2, where the PSD is shown between $n_{min} = 0.01$ and $n_{max} = 3\,cycle/m$. The RMS value of the road profile height variations is $\sigma = \sqrt{\int_{n_{min}}^{n_{max}} \Phi(n)dn} = 0.008\,m$. A sample path profile can be obtained using the cosine series formula $z(x) = \sqrt{2}\sum_{k=1}^{N} A_k \cos(2\pi n_k x + \phi_k)$ [216], where $A_k = \sqrt{\Phi(n_k)\Delta n}$, Δn is the spatial frequency increment and ϕ_k is a random number in the range $[0, 2\pi]$; see the right-hand side of Figure 6.2. It is important to select a spatial sampling period Δx such that $1/\Delta x \geq 2n_{max}$. Alternatively the sample profile can be obtained as the real part of the inverse Fourier transform of $B_k = \sqrt{2}A_k e^{-i\phi_k}$ [216], where ϕ_k are again random numbers in the range $[0, 2\pi]$. In this case the generated length is related to the spatial frequency by $\Delta n = 1/L$. Yet another alternative is to obtain the sample profile as the output of the shaping filter $H = \frac{2\sqrt{\alpha}\sigma}{\alpha + i2\pi n}$ [218] in the case the input is a white-noise signal of unit intensity, since $HH^* = \Phi_a(n) = \frac{4\alpha\sigma^2}{\alpha^2 + (2\pi n)^2}$ is an approximation of $\Phi(n)$ given by (6.1), $\int_0^\infty \Phi_a(n)dn = \sigma^2$, and $\alpha \approx 0.1$ in order to make $\Phi_a(n)$ and $\Phi(n)$ coincident at n_0.

---

[2] If $\Phi(\cdot)$ is a single-sided PSD and thus $\int_0^\infty \Phi(n)\,dn = \int_0^\infty \frac{\Phi(n)}{2\pi}\,d\nu = \int_0^\infty \frac{\Phi(n)}{2\pi v}\,d\omega = \sigma^2$, where σ is the expected RMS.

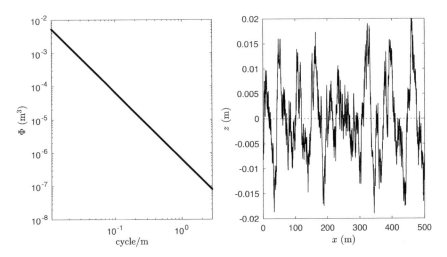

Figure 6.2: PSD of a type B road (left) and generated random profile (right).

We conclude the section with some remarks relating to road-surface models for two-track vehicles. While left- and right-track are usually statistically equivalent (that is, they have the same PSD that is of the form given in (6.1)), the profiles themselves are not the same. Information relating to this 'roll disturbance' can be described in terms of a *coherence function*. If one assumes that the road is isotropic, the coherence function can be derived from the road surface PSD. By definition, the inverse Fourier transform of the road PSD[3] gives the auto-correlation function $R(\Delta)$ between two points located a distance Δ apart. The cross-correlation function R_{LR}, between two points located on the two wheel tracks, which are separated by the track width t_w, and at distance Δ along the track, can be derived from $R_{LR} = R(\sqrt{\Delta^2 + t_w^2})$—this follows from the isotropy assumption. The Fourier transform of the cross-correlation function gives, by definition, the cross-spectral density. The ratio of the cross-spectral density and the PSD is the coherence function $\gamma(n)$. The coherence function resulting from an isotropy assumption can be derived numerically [221]. An alternative empirical model which shows a good fit with experimental data is [222]

$$\gamma(n) = e^{-\rho t_w n}, \tag{6.4}$$

in which $\rho > 0$ is a fitting coefficient, t_w is the axle width, and n is the spatial frequency as in (6.1).

---

3 The inverse Fourier of (6.1) is not calculable, because the related integral is unbounded. One fix is to assume that the PSD is constant up to some given frequency and zero thereafter [221]. The cut-off frequency can be selected 'high enough' so as not to affect the results sought and where (6.1) holds.

6.3   Design objectives

The suspension system is expected to deliver a number of dynamic improvements to the vehicle's behaviour, which will usually include enhanced driver and passenger comfort, enhanced roadholding and the proper functioning of the suspension within the available workspace. Ideally, this must be achieved on all the road surfaces one is likely to encounter, and at all speeds.

The first standard performance metric relates to 'passenger comfort', which is notoriously difficult to quantify objectively. Since driver comfort is related to fatigue, and the driver's ability to concentrate for extended periods, it is also safety related. Further complication derives from the fact that humans are all different and perceived levels of discomfort are both frequency and direction dependent. A combination of the vertical translational acceleration and the pitch acceleration of the sprung mass will be used as a measure of discomfort. In the context of motorcycles, minimizing the sprung-mass pitch acceleration is probably the more important of the two. Optimizing comfort will be taken as equivalent to minimizing the angular and vertical translational acceleration of the sprung body.

The second standard suspension design metric relates to the minimization of the tyre carcass deflections from their equilibrium values. Since near-constant tyre deflections lead to near-constant normal tyre loads, quasi-constant tyre operating conditions are ensured, and good roadholding properties achieved. If in an extreme case large tyre deflections were to occur, the tyre may lose contact with the road and tyre adhesion would thus be lost. Even when road contact is maintained, the average normal force is reduced [223]. Roadholding optimization relates to minimizing the unsprung body vibrations.

The third suspension-related design objective is conceptually straightforward and purely geometric in character. The suspension system must be designed so that it can function properly within the available workspace under all reasonable circumstances. 'Reasonable circumstances' would include such things as striking potholes at speed.

6.4   Single-wheel-station model

A complete investigation of suspension system behaviour requires complex models, which may include frame flexibilities, nonlinear suspension components, chain drive effects, nonlinear tyre models, and viscoelastic structural driver models. These sophisticated models provide simulation results that are in broad conformity with real machine behaviour, as well as making it possible to investigate the machine's dynamics at an early stage of development.

In order to appreciate how suspension systems work, and to evaluate their basic features, much simpler models can be used. A standard example of these simple linear models is the widespread single-wheel-station 'quarter car' model illustrated in Figure 6.3. These models ignore any cross-coupling effects with the other wheel-stations, and they only consider the vehicle's vertical (heave) dynamics. In this model the sprung mass is denoted m_s, the unsprung mass is denoted m_u, and the suspension system Y exerts a force F_s on the sprung and unsprung masses in conformity with the suspension velocity $\dot{z}_s - \dot{z}_u$. The tyre carcass is treated as a linear spring with stiffness k_t.

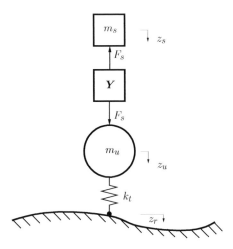

Figure 6.3: Single-wheel-station model.

The vertical component of the road roughness is denoted z_r, while the *absolute* vertical displacements of the sprung and unsprung masses are given by z_s and z_u respectively.

If the tyre remains in contact with the road, with z_s and z_u expressed in an inertial reference frame, it can be seen from Figure 6.3 that the equations of motion for the single-wheel-station system are

$$m_s \ddot{z}_s = -F_s \tag{6.5}$$
$$m_u \ddot{z}_u = F_s + k_t(z_r - z_u), \tag{6.6}$$

in which the suspension system is governed by

$$F_s = \mathbf{Y}(\dot{z}_s - \dot{z}_u), \tag{6.7}$$

in which $\mathbf{Y}$ is a general admittance, which is simply $\mathbf{Y} = c_s + \frac{k_s}{s}$ (in the Laplace domain) in the case of a typical parallel spring-damper suspension strut, where c_s and k_s are the suspension damping and stiffness respectively.

Example 6.2 Compute the modes of a single-wheel model with sprung mass $m_s = 120$ kg, unsprung mass $m_f = 15$ kg, suspension vertical stiffness $k_s = 15$ kN/m, suspension vertical damping $c_s = 1.0$ kNs/m, and tyre radial stiffness $k_t = 150$ kN/m. The modal frequencies and mode shapes are given by the eigenvalues and eigenvectors of the state-space A-matrix associated with (6.5)–(6.7).[4] The low-frequency mode is related to the sprung mass 'bouncing' on the suspension strut, while the high-frequency

---

[4] The unforced system can be written as $M\ddot{\mathbf{q}} + C\dot{\mathbf{q}} + K\mathbf{q} = 0$, where $M = \begin{bmatrix} m_s & 0 \\ 0 & m_u \end{bmatrix}$, $C = \begin{bmatrix} c_s & -c_s \\ -c_s & c_s \end{bmatrix}$,

$K = \begin{bmatrix} k_s & -k_s \\ -k_s & k_s + k_t \end{bmatrix}$, $\mathbf{q} = \begin{bmatrix} z_s \\ z_u \end{bmatrix}$, which can be converted in the state-space form $\dot{\mathbf{x}} = A\mathbf{x}$, where the

state-space matrix is given by $A = \begin{bmatrix} 0 & I \\ -M^{-1}K & -M^{-1}C \end{bmatrix}$.

(a)   (b)

Figure 6.4: Modes of the single-wheel vehicle model; the low-frequency mode is shown in (a), while the high-frequency mode is shown in (b).

mode is related to the unsprung mass 'bouncing' on the tyre radial stiffness. The nominal sprung mass mode is given by $\frac{1}{2\pi}\sqrt{k_s/m_s} = 1.78\,\text{Hz}$ (the damped quarter model frequency is 1.64 Hz), while the nominal hop mode is given by $\frac{1}{2\pi}\sqrt{k_t/m_u} = 15.92\,\text{Hz}$ (the damped quarter model frequency is 15.41 Hz), the difference between the nominal and quarter-car frequencies being attributable to damping influences and the participation of both degrees of freedom in each mode. The mode shapes are shown in Figure 6.4.

6.4.1 Invariant equation

Adding equations (6.5) and (6.6) gives

$$m_s\ddot{z}_s + m_u\ddot{z}_u = k_t(z_r - z_u), \tag{6.8}$$

which is independent of the suspension force F_s and is the well-known *invariant equation* associated with the quarter-car model. The Laplace transform of this equation gives

$$s^2 m_s z_s(s) + (k_t + s^2 m_u)z_u(s) = k_t z_r(s), \tag{6.9}$$

which we will now examine further.

In order to study the compromises associated with rider comfort, suspension movement, and tyre deflections, we introduce three single-input single-output transfer functions.

$$H_A = \frac{\ddot{z}_s(s)}{\dot{z}_r(s)} \tag{6.10}$$

$$H_{SD} = \frac{z_s(s) - z_u(s)}{\dot{z}_r(s)} \tag{6.11}$$

$$H_{TD} = \frac{z_u(s) - z_r(s)}{\dot{z}_r(s)}. \tag{6.12}$$

The first of these maps the vertical road velocity signal $\dot{z}_r$ into the sprung-mass heave acceleration $\ddot{z}_s$. The disturbance $\dot{z}_r$ is a white-noise process whose intensity is proportional to the vehicle's forward speed.[5] The transfer function H_A should be kept 'small' by the suspension in the interests of driver and passenger comfort. The transfer function H_{SD} is used to represent suspension deflections resulting from rough roads. This transfer function should also be kept small so that the suspension continues to operate within the available workspace. The last transfer function H_{TD} is used to represent variations in the tyre deflection (normal load). This transfer function should be kept small in order to reduce tyre load fluctuations and thus achieve good roadholding.

It can be shown that the invariant equation (6.9) enforces the following three relationships between H_A, H_{SD}, and H_{TD}

$$0 = m_s H_A + (k_t + s^2 m_u) H_{TD} + m_u s \tag{6.13}$$
$$= m_s s^2 H_{SD} + (k_t + s^2 (m_s + m_u)) H_{TD} + s(m_s + m_u) \tag{6.14}$$
$$= s^2 (k_t + m_u s^2) H_{SD} - (k_t + s^2 (m_u + m_s)) H_A + k_t s. \tag{6.15}$$

The first important conclusion to be drawn from these equations is that once any one of H_A, H_{SD}, and H_{TD} has been specified, the other two are determined automatically. It is thus impossible to make H_A, H_{SD}, and H_{TD} 'small' simultaneously across the operating bandwidth. These equations highlight important design trade-offs that must be made between ride quality, suspension displacements, and roadholding performance [224].

6.4.2 Interpolation constraints

We will show that equations (6.13)–(6.15) impose a number of interpolation constraints (fixed points) on H_A, H_{SD}, and H_{TD}, which must be satisfied for any suspension system that maps the workspace velocity $\dot{z}_s - \dot{z}_u$ into the suspension force F_s. Constraints of this type appear in Nevanlinna–Pick interpolation problems, which first appeared in engineering in the guise of broad-band matching [225], and then in a robust control context [226]. The connection between suspension dynamics and robust control is explored in [227].

Consider the case that $s = \pm j\omega_1$, where

$$\omega_1 = \sqrt{\frac{k_t}{m_u}} \tag{6.16}$$

is the tyre hop frequency. From (6.13) one sees that

$$H_A(\pm j\omega_1) = \mp j \frac{\sqrt{k_t m_u}}{m_s} \tag{6.17}$$

---

[5] The PSD $\Phi_v(\,\cdot\,)$ associated with the change in the vertical road displacement per unit distance travelled can be obtained from the displacement PSD as $\Phi_v(n) = \Phi(n)(2\pi n)^2$, or $\Phi_v(\nu) = \Phi(\nu)\nu^2$, where $\Phi(\,\cdot\,)$ is given by (6.1). It follows that $\Phi_v(\,\cdot\,)$ is a constant (white-noise) process and depends on the road type. The PSD associated with the road velocity fluctuation $\dot{z}_r$ is given by $\Phi_{\dot{z}_r}(\omega) = \Phi(\nu_0)\nu_0^2 v$, or $\Phi_{\dot{z}_r}(f) = \Phi(n_0)(2\pi n_0)^2 v$, and is thus a white-noise process with magnitude depending on the road type, and is proportional to the vehicle's travel speed v; see (6.2) and (6.3).

holds. In the case that $s = 0$, it follows from (6.15) that

$$\boldsymbol{H}_A(0) = 0. \tag{6.18}$$

Differentiating (6.15) with respect to s and setting $s = 0$ yields

$$\boldsymbol{H}'_A(0) = 1; \tag{6.19}$$

the prime denotes the derivative with respect to s.

If $s = \pm j\omega_2$, where

$$\omega_2 = \sqrt{\frac{k_t}{m_s + m_u}} \tag{6.20}$$

is the modal frequency corresponding to the sprung and unsprung masses oscillating on the tyre radial stiffness, it follows from (6.14) that

$$\boldsymbol{H}_{SD}(\pm j\omega_2) = \pm j\frac{m_s + m_u}{m_s}\sqrt{\frac{m_s + m_u}{k_t}}. \tag{6.21}$$

Setting $s = 0$ in (6.14) one obtains

$$\boldsymbol{H}_{TD}(0) = 0. \tag{6.22}$$

Differentiating (6.14) with respect to s and setting $s = 0$ yields

$$\boldsymbol{H}'_{TD}(0) = -\frac{m_s + m_u}{k_t}. \tag{6.23}$$

The conditions given in equations (6.17)–(6.23) restrict the achievable frequency responses associated with $\boldsymbol{H}_A$, $\boldsymbol{H}_{SD}$, and $\boldsymbol{H}_{TD}$ for any suspension system of the type described. This includes any passive, semi-active, or active system.

6.4.3   State-space analysis

We will demonstrate that there is advantage to be gained from studying suspension systems from the state-space perspective. This framework can be used, for example, to study the suspension model's high-frequency behaviour. It is also the preferred environment for optimization. As suggested in [228], we introduce the following state variables

$$\boldsymbol{x} = \begin{bmatrix} x_1 \\ x_2 \\ x_3 \\ x_4 \end{bmatrix} = \begin{bmatrix} z_s - z_u \\ \dot{z}_s \\ z_u - z_r \\ \dot{z}_u \end{bmatrix}, \tag{6.24}$$

which represent, respectively, the suspension deflection, the absolute velocity of the sprung mass, the tyre deflection, and the absolute velocity of the unsprung mass. The inputs are the road velocity disturbance $\dot{z}_r$ and the suspension strut force F_s. It follows from (6.5) and (6.6) that the state-space representation is

$$\dot{x} = Ax + \begin{bmatrix} B_1 & B_2 \end{bmatrix} \begin{bmatrix} \dot{z}_r \\ F_s \end{bmatrix} \tag{6.25}$$

with the state-space matrices given by

$$A = \begin{bmatrix} 0 & 1 & 0 & -1 \\ 0 & 0 & 0 & 0 \\ 0 & 0 & 0 & 1 \\ 0 & 0 & -k_t/m_u & 0 \end{bmatrix} \qquad \begin{bmatrix} B_1 & B_2 \end{bmatrix} = \begin{bmatrix} 0 & 0 \\ 0 & -1/m_s \\ -1 & 0 \\ 0 & 1/m_u \end{bmatrix}. \tag{6.26}$$

The system outputs are

$$\begin{bmatrix} y_1 \\ \dot{z}_s - \dot{z}_u \end{bmatrix} = \begin{bmatrix} C_1 \\ C_2 \end{bmatrix} x + \begin{bmatrix} \begin{bmatrix} D_{11} & D_{12} \\ D_{21} & D_{22} \end{bmatrix} \end{bmatrix} \begin{bmatrix} \dot{z}_r \\ F_s \end{bmatrix}, \tag{6.27}$$

where

$$C_2 = \begin{bmatrix} 0 & 1 & 0 & -1 \end{bmatrix} \qquad \text{and} \qquad \begin{bmatrix} D_{21} & D_{22} \end{bmatrix} = \begin{bmatrix} 0 & 0 \end{bmatrix}. \tag{6.28}$$

The matrices C_1 and $\begin{bmatrix} D_{11} & D_{12} \end{bmatrix}$ are selected in accordance with the problem under study. In the case that the output is $\ddot{z}_s$, we use

$$C_1 = \begin{bmatrix} 0 & 0 & 0 & 0 \end{bmatrix} \qquad \text{and} \qquad \begin{bmatrix} D_{11} & D_{12} \end{bmatrix} = \begin{bmatrix} 0 & \frac{-1}{m_s} \end{bmatrix} \tag{6.29}$$

to compute $H_A(s)$. In the case that the output is $z_s - z_u$ we use

$$C_1 = \begin{bmatrix} 1 & 0 & 0 & 0 \end{bmatrix} \qquad \text{and} \qquad \begin{bmatrix} D_{11} & D_{12} \end{bmatrix} = \begin{bmatrix} 0 & 0 \end{bmatrix} \tag{6.30}$$

to compute $H_{SD}(s)$. If the output is $z_u - z_r$, we invoke

$$C_1 = \begin{bmatrix} 0 & 0 & 1 & 0 \end{bmatrix} \qquad \text{and} \qquad \begin{bmatrix} D_{11} & D_{12} \end{bmatrix} = \begin{bmatrix} 0 & 0 \end{bmatrix} \tag{6.31}$$

to compute $H_{TD}(s)$.

Following the notation adopted in the robust control literature [229], the suspension can be thought of as a feedback arrangement in which P is the open-loop system, or the 'generalized plant', and the suspension admittance Y is the controller. This arrangement is illustrated in Figure 6.5. For present purposes it will be assumed that the suspension strut admittance is a rational transfer function of the form

$$Y = \frac{b_m s^m + b_{m-1} s^{m-1} + \cdots b_1 s + b_0}{s^n + a_{n-1} s^{n-1} + \cdots a_1 s + a_0}. \tag{6.32}$$

In the case that the suspension is passive, or made from passive mechanical components, Y will be a positive real function.[6] The generalized plant is described by

---

[6] A one-port mechanical element with applied force F and terminal velocity v is defined to be passive if $\int_{-\infty}^{T} F(t)v(t)\, dt \geq 0$ for all admissible F and v square integrable over $(-\infty, T]$. A rational transfer function relating the force and velocity is passive if it is positive-real, that is, if it is stable and with positive real part [71, 230, 231].

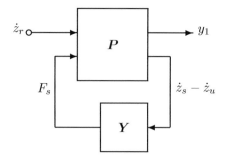

Figure 6.5: The single-wheel-station suspension as a feedback system. The suspension admittance function is given by Y.

$$\begin{bmatrix} y_1 \\ \dot{z}_s - \dot{z}_u \end{bmatrix} = \begin{bmatrix} P_{11} & P_{21} \\ P_{12} & P_{22} \end{bmatrix} \begin{bmatrix} \dot{z}_r \\ F_s \end{bmatrix}, \tag{6.33}$$

with F_s given by (6.7). The general open-loop model (6.33) can be used to find the closed-loop transfer functions in each of the various cases once Y has been specified.

In the case of driver/rider comfort, there holds

$$\boldsymbol{P}_A(s) = \begin{bmatrix} 0 & -\dfrac{1}{m_s} \\ -\dfrac{k_t}{s^2 m_u + k_t} & \dfrac{-(s^2(m_s+m_u)+k_t)}{s m_s(s^2 m_u + k_t)} \end{bmatrix}. \tag{6.34}$$

For the suspension-deflection problem one obtains

$$\boldsymbol{P}_{SD}(s) = \begin{bmatrix} -\dfrac{k_t}{s(s^2 m_u + k_t)} & \dfrac{-(s^2(m_s+m_u)+k_t)}{s^2 m_s(s^2 m_u + k_t)} \\ -\dfrac{k_t}{s^2 m_u + k_t} & \dfrac{-(s^2(m_s+m_u)+k_t)}{s m_s(s^2 m_u + k_t)} \end{bmatrix}, \tag{6.35}$$

and finally for the tyre-deflection problem there holds

$$\boldsymbol{P}_{TD}(s) = \begin{bmatrix} -\dfrac{s m_u}{s^2 m_u + k_t} & \dfrac{1}{(s^2 m_u + k_t)} \\ -\dfrac{k_t}{s^2 m_u + k_t} & \dfrac{-(s^2(m_s+m_u)+k_t)}{s m_s(s^2 m_u + k_t)} \end{bmatrix}. \tag{6.36}$$

The corresponding closed-loop transfer functions can be computed using the well-known (in the control systems community) linear fraction transformation formula [229]

$$H = P_{11} + P_{12}Y(I - P_{22}Y)^{-1}P_{21}. \tag{6.37}$$

We conclude this section by establishing the high-frequency behaviour (asymptotic roll-off rates) of each of the three suspension-related transfer functions in the case that $Y = c_s + \frac{k_s}{s}$. Substituting $s = \frac{1}{w}$ into $H_A(s)$, $H_{SD}(s)$, and $H_{TD}(s)$, and expanding as a truncated Taylor's series around $w = 0$, gives the following high-frequency asymptotic behaviours:

$$\boldsymbol{H}_A(\frac{1}{w}) \approx w^2 \frac{k_t c_s}{m_s m_u}; \tag{6.38}$$

$$\boldsymbol{H}_{SD}(\frac{1}{w}) \approx -w^3 \frac{k_t}{m_u}; \tag{6.39}$$

$$\boldsymbol{H}_{TD}(\frac{1}{w}) \approx -w. \tag{6.40}$$

This shows that $\boldsymbol{H}_{SD}(s)$ rolls off the fastest, while $\boldsymbol{H}_{TD}(s)$ rolls off the slowest.

6.4.4 Suspension optimization

The aim of this section is to use tools from control theory to optimize the parallel spring-damper suspension in the single-wheel-station setting. It should be recognized at the outset that this is a linear treatment and assumes that both the suspension spring and damper are linear.

By way of a brief summary of the background theory, the 2-norm of a system $\boldsymbol{H}$ is the expected root-mean-square (RMS) value of the output when the input is a realization of a unit-intensity white-noise process [229, 232]. That is, if the input is

$$w(t) = \begin{cases} \text{a unit-intensity white-noise process, } t \in [0, t_f] \\ 0 \quad \text{otherwise} \end{cases} \tag{6.41}$$

and the output is $\boldsymbol{z} = \boldsymbol{H}w$, the finite-horizon 2-norm of $\boldsymbol{H}$ is defined by

$$\|\boldsymbol{H}\|_{2,[0,t_f]}^2 = \mathcal{E}\left\{ \frac{1}{t_f} \int_0^{t_f} \boldsymbol{z}^T(t)\boldsymbol{z}(t)\, dt \right\}, \tag{6.42}$$

in which $\mathcal{E}(\,\cdot\,)$ is the expectation operator. If $\boldsymbol{H}$ is such that the integral remains bounded as $t_f \to \infty$ we obtain the infinite-horizon 2-norm of $\boldsymbol{H}$. If $\boldsymbol{H}$ is described by the constant state-space matrices A, B, and C, and $D = 0^7$, with A asymptotically stable, then $\|\boldsymbol{H}\|_2$ is finite and given by

$$\|\boldsymbol{H}\|_2^2 = \text{trace}(CLC^T), \tag{6.43}$$

where

$$AL + LA^T + BB^T = 0. \tag{6.44}$$

This theory is directly applicable to the single-wheel-station suspension problem, since $\dot{z}_r$ is a realization of a white-noise process; see footnote 5. In order to find an optimal compromise between the conflicting objective associated with the transfer functions given in (6.34)–(6.37), we minimize

$$J = \|\boldsymbol{H}_A\|_2^2 + q_1\|\boldsymbol{H}_{SD}\|_2^2 + q_2\|\boldsymbol{H}_{TD}\|_2^2, \tag{6.45}$$

where q_1 and q_2 are weighting factors on the suspension deflection and tyre deflection respectively. The weighting on $\|\boldsymbol{H}_A\|_2^2$ is set to unity without loss of generality.

---

[7] $D \neq 0$ results in an infinite 2-norm.

To begin the optimization process we set $F_s = \boldsymbol{Y}(\dot{z}_s - \dot{z}_u) = k_s x_1 + c_s(x_2 - x_4)$ (spring-damper strut) and form the closed-loop A and B matrices using (6.26) to obtain

$$
\begin{bmatrix} \dot{x}_1 \\ \dot{x}_2 \\ \dot{x}_3 \\ \dot{x}_4 \end{bmatrix} = \begin{bmatrix} 0 & 1 & 0 & -1 \\ -k_s/m_s & -c_s/m_s & 0 & c_s/m_s \\ 0 & 0 & 0 & 1 \\ k_s/m_u & c_s/m_u & -k_t/m_u & -c_s/m_u \end{bmatrix} \begin{bmatrix} x_1 \\ x_2 \\ x_3 \\ x_4 \end{bmatrix} + \begin{bmatrix} 0 \\ 0 \\ -1 \\ 0 \end{bmatrix} \dot{z}_r.
\tag{6.46}
$$

The output matrices corresponding to $\boldsymbol{H}_A$, $\boldsymbol{H}_{SD}$, and $\boldsymbol{H}_{TD}$ are:

$$
C_A = \begin{bmatrix} -k_s/m_s & -c_s/m_s & 0 & c_s/m_s \end{bmatrix}
\tag{6.47}
$$

$$
C_{SD} = \begin{bmatrix} 1 & 0 & 0 & 0 \end{bmatrix}
\tag{6.48}
$$

$$
C_{TD} = \begin{bmatrix} 0 & 0 & 1 & 0 \end{bmatrix}.
\tag{6.49}
$$

If L is the unique positive definite solution of the Lyapunov equation (6.44), in which the A and B matrices come from (6.46), it follows from (6.43) that the performance index (6.45) is given by

$$
J = \text{trace}\left(\text{diag}\begin{bmatrix} 1 & q_1 & q_2 \end{bmatrix} \begin{bmatrix} C_A \\ C_{SD} \\ C_{TD} \end{bmatrix} L \begin{bmatrix} C_A^T & C_{SD}^T & C_{TD}^T \end{bmatrix} \right).
\tag{6.50}
$$

A computer-assisted symbolic mathematics tool can be used to show that the symmetric solution L to the Lyapunov equation (6.44) has entries l_{ij}:

$l_{11} = (m_s + m_u)/(2c_s)$

$l_{12} = -(m_s + m_u)/(2m_s)$

$l_{13} = (k_s(m_s + m_u)^2 - m_s m_u k_t)/(2m_s k_t c_s)$

$l_{14} = -(m_s + m_u)/(2m_s)$

$l_{22} = (k_s^2(m_s + m_u)^2 + k_t((m_s + m_u)c_s^2 + m_s^2 k_s))/(2m_s^2 k_t c_s)$

$l_{23} = m_u/(2m_s)$

$l_{24} = (k_s^2(m_s + m_u)^2 + k_t((m_s + m_u)c_s^2 - m_s m_u k_s))/(2m_s^2 k_t c_s)$

$l_{33} = (k_s^2(m_s + m_u)^3 + k_t c_s^2(m_s + m_u)^2 + m_s m_u k_t(m_s k_t - 2(m_s + m_u)k_s))/(2m_s^2 k_t^2 c_s)$

$l_{34} = -1/2$

$l_{44} = (k_s^2(m_s + m_u)^2 + k_t(c_s^2(m_s + m_u) + m_s^2(k_t - k_s) - 2m_s m_u k_s))/(2m_s^2 k_t c_s).$ $\quad$ (6.51)

Another symbolic computation shows that

$$
\frac{\partial J}{\partial k_s} = 0 \qquad (\text{for any } c_s > 0)
$$

when

$$
k_s^{opt} = \frac{q_2 k_t m_s m_u}{k_t^2 + q_2(m_s + m_u)^2}.
\tag{6.52}
$$

Interestingly, this is a function of q_2 alone and so the suspension spring stiffness is unaffected by the cost associated with the suspension travel. Another symbolic computation shows that

$$\frac{\partial J}{\partial c_s} = 0, \tag{6.53}$$

when

$$\begin{aligned}
c_s^{opt} = &\{(k_s^2 k_t^2 (m_s + m_u) + q_1 k_t^2 m_s^2 (m_s + m_u) + q_2 k_s^2 (m_s + m_u)^3 \\
&+ q_2 k_t m_u m_s (k_t m_s - 2k_s (m_s + m_u))) \\
&/(k_t (k_t^2 + q_2 (m_s + m_u)^2))\}^{1/2}.
\end{aligned} \tag{6.54}$$

Example 6.3 Compute the optimal spring and damping coefficients for a single-wheel-station model with sprung mass $m_s = 120\,\text{kg}$, unsprung mass $m_f = 15\,\text{kg}$, and tyre radial stiffness $k_t = 150\,\text{kN/m}$ when the workspace weight is $q_1 = 70$ and the roadholding weight is $q_2 = 70 \times 10^5$. Then investigate the effect of varying q_1 and q_2 on the cost function J, and compare the transfer functions H_A, H_{SD}, and H_{TD} for the optimal configuration with those of a nominal configuration where $k_s = 15\,\text{kN/m}$ and $c_s = 0.5\,\text{kNs/m}$. Equations (6.52) and (6.54) give $k_s^{opt} = 12.6\,\text{kN/m}$ and $c_s^{opt} = 1.17\,\text{kNs/m}$. Figure 6.6 (a) shows the cost function (6.50) as a function of k_s and c_s in the case that $q_1 = 70$ and $q_2 = 70 \times 10^5$. It is instructive to show how the weighting factors in (6.45) influence the suspension parameter values. Equation (6.52) shows that the suspension stiffness is a function of q_2 (and not q_1). The way in which k_s^{opt} varies with q_2 is illustrated in Figure 6.6 (b) and demonstrates that improved roadholding requires an increased value of suspension spring stiffness. Figure 6.7 (a) shows the way in which the weighting factors in (6.45) influence the optimal suspension damping coefficient. In this example, varying q_1 has little influence on the suspension damping, while increasing the roadholding term q_2 leads to a hardening of the suspension by increasing the damper constant. Figure 6.7 (b) shows Bode magnitude plots of H_A for a parallel spring damper suspension with nominal component values given by $k_s = 15\,\text{kN/m}$ and $c_s = 0.5\,\text{kNs/m}$ together with the optimal component parameter settings $k_s^{opt} = 12.6\,\text{kN/m}$ and $c_s^{opt} = 1.17\,\text{kNs/m}$. Figure 6.7 (b) also shows

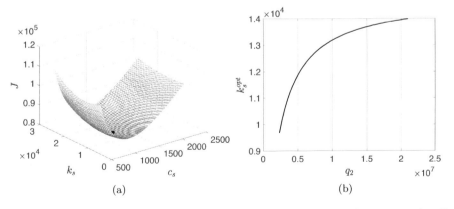

(a)   (b)

Figure 6.6: Cost function surface as a function of the suspension damping and stiffness (a), and optimal suspension spring stiffness as a function of q_2 (b).

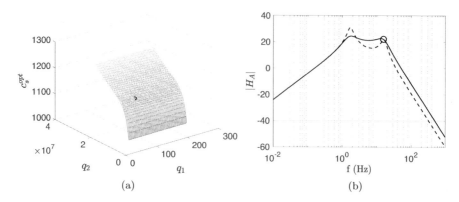

(a)   (b)

Figure 6.7: (a) Optimal suspension damping constant as a function of q_1 and q_2. (b) Bode plot of $\boldsymbol{H}_A$ for a nominal (dashed) and optimized (solid) suspension system. The interpolation point (6.17) is shown circled. Smaller magnitudes correspond to improved rider comfort.

the invariant point at the wheel hop frequency $\omega_1 = 15.9\,\text{Hz}$ as described in (6.17). It is clear that (6.18) is indicative of a band-pass system with a roll-up rate of 20 dB per decade as is predicted by (6.19). As expected from (6.38), the high-frequency roll-off rate is 40 dB per decade. Figure 6.8 (a) shows Bode magnitude plots for $\boldsymbol{H}_{SD}$ for the nominal and optimal suspensions described above. In this case there is an interpolation constraint at $\omega_2 = 5.3\,\text{Hz}$, as given by (6.21), with the 60 dB per decade roll-off rate predicted by (6.39). Figure 6.8 (b) shows Bode magnitude plots for $\boldsymbol{H}_{TD}$ for the

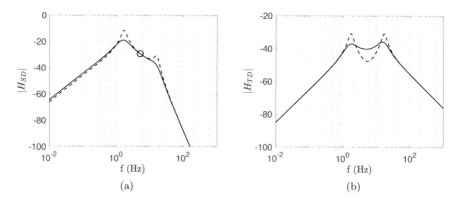

(a)   (b)

Figure 6.8: Bode plots for a nominal (dashed) and optimized (solid) suspension system of (a) $\boldsymbol{H}_{SD}$ (the interpolation point (6.21) is shown circled; smaller magnitudes correspond to decreased suspension deflection) and (b) $\boldsymbol{H}_{TD}$ (smaller magnitudes correspond to less tyre deflection and hence increased roadholding).

nominal and optimal suspensions described above. As predicted by (6.22), (6.23), and (6.40) the zero-frequency gain is zero, with 20 dB per decade roll-up and roll-off rates.

The interested reader is referred to [233] for the further analysis of analytically derived global optima of ride comfort and tyre grip (separately and in combination) for a single-wheel-station vehicle model and six different suspension layouts, comprising two springs, one damper, and possibly one inerter.

6.5   In-plane vehicle model

In this section we extend the single-wheel-station model to a suspension system comprising one sprung mass and two unsprung mass components. This type of model could be used for either motorcycles, or single-track car representations. This generalization allows the sprung mass to pitch as well as heave. This model also recognizes interactions that may exist between the front and rear wheel-stations. The single-track model that we will now analyse is shown in Figure 6.9. The sprung body has mass m_s and pitch moment of inertia I_y, the front unsprung body has mass m_f, and the rear unsprung body has mass m_r. The (sprung) mass centre is at distance b from the rear axle and distance a from the front axle, the front and rear tyres have radial stiffnesses k_{tf} and k_{tr} respectively. The system has four degrees of freedom: the heave z_s and pitch θ of the sprung mass, and the front and rear radial tyre deflections z_{uf} and z_{ur}. The displacements z_{sf} and z_{sr} represent the sprung-mass heave at the front and rear suspension points. Force balances on the sprung and unsprung masses, and a moment balance on the vehicle's sprung body, yield

$$m_s \ddot{z}_s = -F_r - F_f \tag{6.55}$$

$$I_y \ddot{\theta} = aF_f - bF_r \tag{6.56}$$

$$m_f \ddot{z}_{uf} = F_f + k_{tf}(z_{rf} - z_{uf}) \tag{6.57}$$

$$m_r \ddot{z}_{ur} = F_r + k_{tr}(z_{rr} - z_{ur}). \tag{6.58}$$

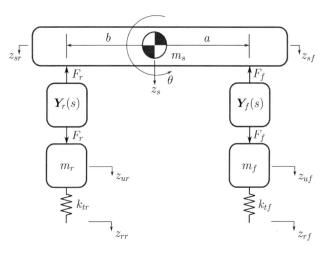

Figure 6.9: In-plane dynamics of a generic motorcycle or single-track car model.

In the Laplace domain the front and rear suspension forces are given by

$$F_f(s) = \mathbf{Y}_f(s)\left(s(z_s - a\theta - z_{uf})\right) \tag{6.59}$$

and

$$F_r(s) = \mathbf{Y}_r(s)\left(s(z_s + b\theta - z_{ur})\right) \tag{6.60}$$

respectively, where $\mathbf{Y}_f$ and $\mathbf{Y}_r$ are the front and rear suspension admittances. In the case of parallel spring-damper suspensions, these forces become

$$F_f(s) = (k_{sf} + sc_{sf})\left(z_s - a\theta - z_{uf}\right) \tag{6.61}$$

and

$$F_r(s) = (k_{sr} + sc_{sr})\left(z_s + b\theta - z_{ur}\right) \tag{6.62}$$

respectively. The front and rear suspension vertical stiffnesses and damping coefficients are k_{sf} and k_{sr}, and c_{sf} and c_{sr} respectively.

6.5.1 Invariant equations

By eliminating the suspension forces F_r and F_r from (6.55)–(6.58), one obtains invariant equations relating to the main body heave and pitch motions:

$$s^2 m_s z_s + s^2 m_f z_{uf} + s^2 m_r z_{ur} = k_{tf}(z_{rf} - z_{uf}) + k_{tr}(z_{rr} - z_{ur}) \tag{6.63}$$
$$s^2 I_y \theta - s^2 a m_f z_{uf} + s^2 b m_r z_{ur} = b k_{tr}(z_{rr} - z_{ur}) - a k_{tf}(z_{rf} - z_{uf}). \tag{6.64}$$

These equations give

$$m_s H_A^z + (s^2 m_f + k_{tf}) H_{TD}^F + (s^2 m_r + k_{tr}) H_{TD}^R = -s(m_f + e^{-s\tau} m_r) \tag{6.65}$$

$$I_y H_A^\theta + b(s^2 m_r + k_{tr}) H_{TD}^R - a(s^2 m_f + k_{tf}) H_{TD}^F = s(a m_f - b e^{-s\tau} m_r), \tag{6.66}$$

which have the same general form as (6.13). It is assumed that $z_{rf}(s) = z_r(s)$ and $z_{rr}(s) = e^{-s\tau} z_r(s)$, where $\tau = \frac{a+b}{v}$ is the wheelbase filtering delay. The transfer functions in (6.65) and (6.66) are

$$H_A^z = \frac{\ddot{z}_s(s)}{\dot{z}_r(s)} \quad H_A^\theta = \frac{\ddot{\theta}(s)}{\dot{z}_r(s)} \quad H_{TD}^F = \frac{z_{uf}(s) - z_r(s)}{\dot{z}_r(s)} \quad H_{TD}^R = \frac{z_{ur}(s) - e^{-s\tau} z_r(s)}{\dot{z}_r(s)}. \tag{6.67}$$

Counterparts to (6.14) can be derived by introducing the front- and rear-suspension transfer functions

$$H_{SD}^F = \frac{z_{sf}(s) - z_{uf}(s)}{\dot{z}_r(s)} \quad H_{SD}^R = \frac{z_{sr}(s) - z_{ur}(s)}{\dot{z}_r(s)}, \tag{6.68}$$

where

$$z_{sf} = z_s(s) - \theta(s)a \qquad \text{and} \qquad z_{sr} = z_s(s) + \theta(s)b. \qquad (6.69)$$

Direct computation using (6.63) and (6.64) gives

$$m_s b s^2 H_{SD}^F + m_s a s^2 H_{SD}^R + c_1 H_{TD}^F + c_2 H_{TD}^R = -c_3 \qquad (6.70)$$

$$I_y H_{SD}^R - I_y H_{SD}^F - c_4 H_{TD}^F + c_5 H_{TD}^R = -c_6 \qquad (6.71)$$

where

$$
\begin{aligned}
c_1 &= s^2(am_f + b(m_s + m_r)) + k_{tf}(a+b) \\
c_2 &= s^2(bm_r + a(m_s + m_r)) + k_{tr}(a+b) \\
c_3 &= s(bm_r + a(m_s + m_r))e^{-s\tau} + s(am_f + b(m_s + m_r)) \\
c_4 &= s^2(I_y + a(a+b)m_f) + a(a+b)k_{tf} \\
c_5 &= s^2(I_y + b(a+b)m_r) + b(a+b)k_{tr} \\
c_6 &= s(I_y + b(a+b)m_r)e^{-s\tau} + s(I_y + a(a+b)m_f).
\end{aligned}
$$

Counterparts to (6.15) can also be found

$$c_7 H_A^z + c_8 H_A^\theta - c_9 H_{SD}^F - c_{10} H_{SD}^R = c_{11} \qquad (6.72)$$

$$c_{12} H_A^z + c_{13} H_A^\theta + ac_9 H_{SD}^F - bc_{10} H_{SD}^R = c_{14} \qquad (6.73)$$

in which

$$
\begin{aligned}
c_7 &= s^2(m_s + m_r + m_f) + k_{tf} + k_{tr} \\
c_8 &= s^2(bm_r - am_f) + bk_{tr} - ak_{tf} \\
c_9 &= s^2(s^2 m_f + k_{tf}) \\
c_{10} &= s^2(s^2 m_r + k_{tr}) \\
c_{11} &= s(k_{tf} + e^{-\tau}k_{tr}) \\
c_{12} &= s^2(bm_r - am_f) + bk_{tr} - ak_{tf} \\
c_{13} &= s^2(a^2 m_f + b^2 m_r + I_y) + a^2 k_{tf} + b^2 k_{ktr} \\
c_{14} &= s(be^{-s\tau}k_{tr} - ak_{tf}).
\end{aligned}
$$

The system of six equations (6.65), (6.66), (6.70), (6.71), (6.72), (6.73) corresponds to (6.13)–(6.15) for the single-wheel model. Taken together, these equations have rank four and thus only two of the six transfer functions are independent. If, for example, H_{TD}^F and H_{TD}^R are pre-specified, (6.65) and (6.66) can be used to find H_A^z and H_A^θ. In the same way (6.70) and (6.71) can be used to find H_{SD}^F and H_{SD}^R. In this case (6.72) and (6.73) provide no additional information.

6.5.2 Mode shapes

A state-space model suitable for computing the system mode shapes can be found using (6.55)–(6.58), (6.61), and (6.62), with

$$q = \left[z_s, \, \theta, \, z_{uf}, \, z_{ur} \right]^T \qquad (6.74)$$

and no road forcing. This 're-packaging' exercise gives

$$M\ddot{q} + C\dot{q} + Kq = 0, \tag{6.75}$$

where the mass M, damper C, and stiffness K matrices are symmetric and given by

$$M = \begin{bmatrix} m_s & 0 & 0 & 0 \\ & I_y & 0 & 0 \\ & & m_{uf} & 0 \\ \text{sym} & & & m_{ur} \end{bmatrix} \quad C = \begin{bmatrix} (c_{sf} + c_{sr}) & (c_{sr}b - c_{sf}a) & -c_{sf} & -c_{sr} \\ & (c_{sf}a^2 + c_{sr}b^2) & c_{sf}a & -c_{sr}b \\ & & c_{sf} & 0 \\ \text{sym} & & & c_{sr} \end{bmatrix} \tag{6.76}$$

$$K = \begin{bmatrix} (k_{sf} + k_{sr}) & (k_{sr}b - k_{sf}a) & -k_{sf} & -k_{sr} \\ & (k_{sf}a^2 + k_{sr}b^2) & k_{sf}a & -k_{sr}b \\ & & k_{sf} + k_{tf} & 0 \\ \text{sym} & & & k_{sr} + k_{tr} \end{bmatrix}. \tag{6.77}$$

The model in (6.75) can be rewritten in state-space form as

$$\dot{x} = \begin{bmatrix} 0 & I \\ -M^{-1}K & -M^{-1}C \end{bmatrix} x. \tag{6.78}$$

Example 6.4 Find the mode shapes and modal frequencies of a motorcycle with sprung mass $m_s = 240$ kg and unsprung masses $m_f = 15$ kg and $m_r = 20$ kg (i.e. total mass 275 kg). The pitch moment of inertia is $I_y = 50$ kgm², the mass centre is located at $b = 0.7$ m from the rear axle and $a = 0.8$ m from the front axle. The front- and rear-suspension vertical stiffnesses are $k_{sf} = 15$ kN/m and $k_{sr} = 20$ kN/m respectively. The front- and rear-suspension vertical damping coefficients are $c_{sf} = 1.0$ kNs/m and $c_{sr} = 1.3$ kNs/m repectively. The front- and rear-tyre radial stiffnesses are $k_{tf} = k_{tr} = 150$ kN/m. The first two modes involve primarily the sprung body heave and pitch motions, while the remaining higher-frequency modes involve primarily the unsprung body motions (wheel hop modes). See Figure 6.10.

6.5.3   Reconciliation with single-wheel-station model

The usefulness of single-wheel-station models, such as that shown in Figure 6.3, stems from a combination of the simplicity of these models and their ability to capture some of the important suspension design issues. As we will now show, under some circumstances, single-wheel-station models represent perfectly the suspension dynamics of a complete motorcycle or single-track car model.

Consider the following sketch

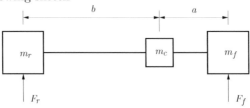

in which the whole-vehicle sprung mass m_s is decomposed into a rear point mass m_r, a front point mass m_f, and a third point mass m_c located at the vehicle's mass centre. These masses are used to capture simultaneously the vehicle's pitch inertia and its sprung mass properties. The front and rear suspension forces are F_f and F_r respectively. The rear mass is located a distance b behind the mass centre, while the front mass is located a distance a in front of it. In order for the three-mass system to be dynamically identical to an in-plane vehicle model with mass m_s and moment of inertia I_y, the three masses must sum to the total unsprung mass:

$$m_s = m_r + m_c + m_f. \tag{6.79}$$

The mass centre must be located at m_c and so

$$m_r b = m_f a. \tag{6.80}$$

must hold. The pitch inertia I_y (about the vehicle's mass centre) is given by

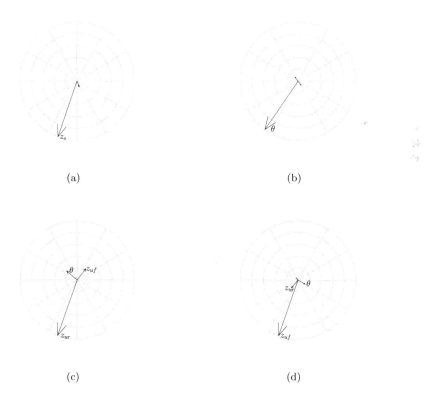

(a)

(b)

(c)

(d)

Figure 6.10: Vibration mode of the in-plane vehicle model: (a) bounce mode ($f = 1.75\,\text{Hz}$, $\zeta = 0.34$), (b) pitch mode ($f = 2.64\,\text{Hz}$, $\zeta = 0.59$), (c) rear-hop mode ($f = 12.49\,\text{Hz}$, $\zeta = 0.39$), and (d) front-hop mode ($f = 15.01\,\text{Hz}$, $\zeta = 0.34$).

$$I_y = m_r b^2 + m_f a^2. \tag{6.81}$$

Combining (6.81) and (6.80) gives

$$m_r = \frac{I_y}{b(a+b)} \quad \text{and} \quad m_f = \frac{I_y}{a(a+b)}, \tag{6.82}$$

with

$$m_c = m_s \left(1 - \frac{I_y}{abm_s} \right) \tag{6.83}$$

from (6.79) and (6.82). If the mass and the inertia properties of the real vehicle result in $I_y = abm_s$, then $m_c = 0$ and the vehicle's in-plane dynamics can be represented perfectly by two decoupled single-wheel-station models [213]. In the case that m_c is 'small', single-wheel-station models can be used to predict reasonably accurately the vehicle's low-frequency (long road wavelength and/or low speed) behaviour.

We conclude this section by proving formally that the condition $I_y = abm_s$ decouples the front and rear suspensions. We begin by rewriting (6.75) using the alternative position vector $\tilde{q} = [z_{sf}, z_{uf}, z_{sr}, z_{ur}]^T$, where $z_{sf} = z_s - \theta a$ and $z_{sr} = z_s + \theta b$ (see Figure 6.9), instead of the original vector $q = [z_s, \theta, z_{uf}, z_{ur}]^T$. The position vectors $\tilde{q}$ and q are related by

$$\tilde{q} = V^{-1} q \qquad V^{-1} = \begin{bmatrix} 1 & -a & 0 & 0 \\ 0 & 0 & 1 & 0 \\ 1 & b & 0 & 0 \\ 0 & 0 & 0 & 1 \end{bmatrix}. \tag{6.84}$$

The transformed mass, damping, and stiffness matrices related to $\tilde{q}$ are given by $\tilde{M} = V^T MV$, $\tilde{C} = V^T CV$ and $\tilde{K} = V^T KV$

$$\tilde{M} = \begin{bmatrix} \frac{m_s b^2 + I_y}{(a+b)^2} & 0 & \frac{m_s ab - I_y}{(a+b)^2} & 0 \\ 0 & m_f & 0 & 0 \\ \frac{m_s ab - I_y}{(a+b)^2} & 0 & \frac{m_s a^2 + I_y}{(a+b)^2} & 0 \\ 0 & 0 & 0 & m_r \end{bmatrix} \tag{6.85}$$

$$\tilde{C} = \begin{bmatrix} c_{sf} & -c_{sf} & 0 & 0 \\ -c_{sf} & c_{sf} & 0 & 0 \\ 0 & 0 & c_{sr} & -c_{sr} \\ 0 & 0 & -c_{sr} & c_{sr} \end{bmatrix} \qquad \tilde{K} = \begin{bmatrix} k_{sf} & -k_{sf} & 0 & 0 \\ -k_{sf} & k_{sf} + k_{tf} & 0 & 0 \\ 0 & 0 & k_{sr} & -k_{sr} \\ 0 & 0 & -k_{sr} & k_{sr} + k_{tr} \end{bmatrix}, \tag{6.86}$$

which show that $\tilde{M}_{13} = \tilde{M}_{31} = 0$ when $I_y = abm_s$. In this case the single-track vehicle suspension system can be treated as two independent single-wheel-station models.

6.5.4 Transfer functions

An alternative state-space model again based on (6.55)–(6.58), (6.61), and (6.62) can be used to find the transfer functions related to driver comfort, suspension deflection, and tyre deflection. In this case we use the state vector

$$\boldsymbol{x} = \left[\, z_{sf} - z_{uf},\, z_{sr} - z_{ur},\, z_{uf} - z_{rf},\, z_{ur} - z_{rr},\, \dot{z}_s,\, \dot{\theta},\, \dot{z}_{uf},\, \dot{z}_{ur} \,\right]^T \tag{6.87}$$

and the system input vector

$$\boldsymbol{u} = [\dot{z}_{rf},\, \dot{z}_{rr}]^T. \tag{6.88}$$

It can be shown by direct calculation that

$$\dot{\boldsymbol{x}} = (A_0 + F^T G)\boldsymbol{x} + B\boldsymbol{u}, \tag{6.89}$$

where

$$A_0 = \begin{bmatrix} 0 & 0 & 0 & 0 & 1 & -a & -1 & 0 \\ 0 & 0 & 0 & 0 & 1 & b & 0 & -1 \\ 0 & 0 & 0 & 0 & 0 & 0 & 1 & 0 \\ 0 & 0 & 0 & 0 & 0 & 0 & 0 & 1 \\ 0 & 0 & 0 & 0 & 0 & 0 & 0 & 0 \\ 0 & 0 & 0 & 0 & 0 & 0 & 0 & 0 \\ 0 & 0 & -\frac{k_{tf}}{m_f} & 0 & 0 & 0 & 0 & 0 \\ 0 & 0 & 0 & -\frac{k_{tr}}{m_r} & 0 & 0 & 0 & 0 \end{bmatrix}, \tag{6.90}$$

$$F = \begin{bmatrix} 0 & 0 & 0 & 0 & -\frac{1}{m_s} & \frac{a}{I_y} & \frac{1}{m_f} & 0 \\ 0 & 0 & 0 & 0 & -\frac{1}{m_s} & -\frac{b}{I_y} & 0 & \frac{1}{m_r} \end{bmatrix}, \tag{6.91}$$

$$G = \begin{bmatrix} k_{sf} & 0 & 0 & 0 & c_{sf} & -ac_{sf} & -c_{sf} & 0 \\ 0 & k_{sr} & 0 & 0 & c_{sr} & bc_{sr} & 0 & -c_{sr} \end{bmatrix}, \tag{6.92}$$

and

$$B = \begin{bmatrix} 0 & 0 & -1 & 0 & 0 & 0 & 0 & 0 \\ 0 & 0 & 0 & -1 & 0 & 0 & 0 & 0 \end{bmatrix}^T. \tag{6.93}$$

In the case of rider/driver comfort, the outputs of interest are the heave acceleration $\ddot{z}_s$ and the pich acceleration $\ddot{\theta}$. These quantities can be computed using

$$\begin{bmatrix} \ddot{z}_s \\ \ddot{\theta} \end{bmatrix} = \begin{bmatrix} -\frac{k_{sf}}{m_s} & -\frac{k_{sr}}{m_s} & 0 & 0 & -\frac{c_{sf}+c_{sr}}{m_s} & \frac{ac_{sf}-bc_{sr}}{m_s} & \frac{c_{sf}}{m_s} & \frac{c_{sr}}{m_s} \\ \frac{ak_{sf}}{I_y} & -\frac{bk_{sr}}{I_y} & 0 & 0 & \frac{ac_{sf}-bc_{sr}}{I_y} & -\frac{a^2c_{sf}+b^2c_{sr}}{I_y} & -\frac{ac_{sf}}{I_y} & \frac{bc_{sr}}{I_y} \end{bmatrix} \boldsymbol{x}. \tag{6.94}$$

The front and rear suspension deflections are given by

$$\begin{bmatrix} z_{sf} - z_{uf} \\ z_{sr} - z_{ur} \end{bmatrix} = \begin{bmatrix} 1 & 0 & 0 & 0 & 0 & 0 & 0 & 0 \\ 0 & 1 & 0 & 0 & 0 & 0 & 0 & 0 \end{bmatrix} \boldsymbol{x}. \tag{6.95}$$

Finally, we can compute the front and rear tyre deflections using

$$\begin{bmatrix} z_{uf} - z_{rf} \\ z_{ur} - z_{rr} \end{bmatrix} = \begin{bmatrix} 0 & 0 & 1 & 0 & 0 & 0 & 0 & 0 \\ 0 & 0 & 0 & 1 & 0 & 0 & 0 & 0 \end{bmatrix} \boldsymbol{x}. \tag{6.96}$$

Example 6.5 Using the motorcycle data given in Example 6.4, compute the frequency responses from the road vertical velocity to the heave and pitch accelerations, the suspension displacements, and the tyre deflections at 40 m/s using (6.89), (6.94), (6.95), and (6.96). Assume that the front and rear inputs are related by $z_{rf} = z_r$ and $z_{rr} = z_r e^{-s\tau}$, where $\tau = (a+b)/v$ is the wheelbase filtering time delay. Examine also the effect of halving and then doubling the nominal damper values. The required frequency response plots are shown in Figures 6.11 and 6.12. The lobes in the heave and pitch responses are associated with the wheelbase filtering. For pitch, minima occur at frequencies of $1/\tau, 2/\tau, 3/\tau, \ldots$ corresponding to wavelengths equal to the wheelbase or its submultiples—the road elevation excites the front and rear wheels in phase. In heave, minima occur at frequencies of $1/2\tau, 3/2\tau, 5/2\tau, \ldots$ corresponding to wavelengths equal to twice the wheelbase and its odd submultiples—the road excites the front and rear wheel in anti-phase. Now compute the expected RMS values of the vehicle accelerations, suspension deflections, and tyre deflections when the vehicle is travelling on a type B road (see example 6.1) at 40 m/s using the transfer functions given in Figures 6.11 (a), 6.11 (b), and 6.12 (a). The PSD of the response y is obtained from the well-known relationship $\Phi_y = |H_{y,\dot{z}_r}|^2 \Phi_{\dot{z}_r}$ [217], where $|H_{y,\dot{z}_r}|$ is the transfer function magnitude from the road velocity $\dot{z}_r$ to y; $\Phi_{\dot{z}_r} = \Phi(n_0)(2\pi n_0)^2 v = 1.01 \times 10^{-3}\,\mathrm{m^2/s}$ is the (constant) PSD of the road velocity fluctuation; see footnote 5. The expected RMS value of the output is obtained by integration of the PSD as $\sigma = \sqrt{\int_0^\infty \Phi_y df}$.[8] The expected RMS values of the heave acceleration $\sigma_{\ddot{z}}$ at 40 m/s, with the three different damping settings, are $\sigma_{\ddot{z}} = 1.30/1.30/1.67\,\mathrm{m/s^2}$; the expected RMS values of the pitch acceleration are $\sigma_{\ddot{\theta}} = 3.88/5.30/7.47\,\mathrm{rad/s^2}$; the expected RMS values of the front and rear suspension deflections are $\sigma_{z_{sf}-z_{uf}} = 0.008/0.005/0.004\,\mathrm{m}$ and $\sigma_{z_{sr}-z_{ur}} = 0.008/0.006/0.004\,\mathrm{m}$, respectively; and the expected RMS values of the front and rear tyre deflections are $\sigma_{z_{uf}-z_{rf}} = 0.003/0.002/0.003\,\mathrm{m}$ and $\sigma_{z_{ur}-z_{rr}} = 0.003/0.003/0.003\,\mathrm{m}$, respectively. The frequency responses from the road vertical velocity to the heave and pitch accelerations at 20 m/s are given in Figure 6.12 (b), where the effect of the increased wheelbase-filtering delay can be observed.

6.5.5   Interpolation constraints

The interpolation constraints examined in Section 6.4.2 are not restricted to single-input single-output feedback problems, or indeed the single-wheel-station suspension problem studied there. Our purpose here is to outline the source of interpolation constraints in the case of multi-input multi-output (MIMO) models, such as the two-wheel-station model, and to show how they might be computed. An understanding of these constraints is important, because they restrict the achievable performance of all suspension systems of the type shown in Figure 6.9. These constraints remain in force

---

[8]   The limits of integration used for the numerical results are $f_{min} = 0.40\,\mathrm{Hz}$ and $f_{max} = 120\,\mathrm{Hz}$, which correspond to $n_{min} = 0.01\,\mathrm{cycle/m}$ and $n_{max} = 3\,\mathrm{cycle/m}$ at $v = 40\,\mathrm{m/s}$. It can be verified that extending the integration range does not significantly affect the results.

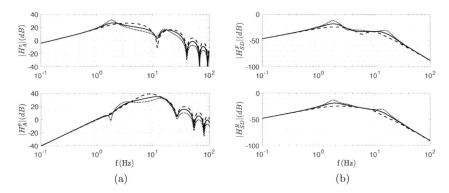

Figure 6.11: (a) Heave (above) and pitch (below) acceleration responses at 40 m/s; (b) Front (above) and rear (below) suspension deflections at 40 m/s. The solid lines represent the nominal damper values, the dotted lines represent half the nominal values, while the dashed lines represent double the nominal values.

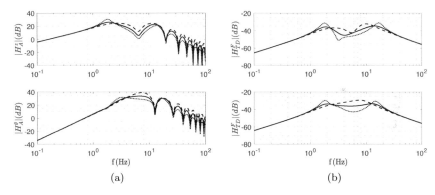

Figure 6.12: (a) Heave (above) and pitch (below) acceleration responses at 20 m/s; (b) Front (above) and rear (below) tyre deflections at 40 m/s. The solid lines represent the nominal damper values, the dotted lines represent half the nominal values, while the dashed lines represent double the nominal values.

even when contemplating active and semi-active systems, and the unwary designer risks spending significant effort trying to achieve impossible targets.

When analysing the interpolation constraints associated with the two-wheel-station problem, one approach is to use the methods developed in the robust control literature. Let us suppose again that the suspension is modelled as the feedback configuration illustrated in Figure 6.13, as described by equations (6.55)–(6.58), (6.61), and (6.62). An equivalent transfer-function representation is:

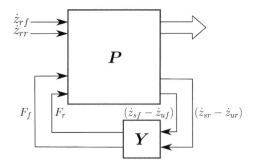

Figure 6.13: Single-track suspension as a feedback system. The generalized plant $\boldsymbol{P}$ represents the in-plane dynamics of the car/motorcycle and $\boldsymbol{Y}$ is the admittance function of the suspension system. The 'fat' arrow is an unspecified set of output variables.

$$\begin{bmatrix} \boldsymbol{z} \\ \boldsymbol{y} \end{bmatrix} = \begin{bmatrix} \boldsymbol{P}_{11} & \boldsymbol{P}_{12} \\ \boldsymbol{P}_{21} & \boldsymbol{P}_{22} \end{bmatrix} \begin{bmatrix} \boldsymbol{w} \\ \boldsymbol{u} \end{bmatrix} \tag{6.97}$$

with

$$\boldsymbol{u} = \boldsymbol{Y}\boldsymbol{y} \tag{6.98}$$

in which

$$\boldsymbol{w} = \begin{bmatrix} \dot{z}_{rf} \\ \dot{z}_{rr} \end{bmatrix}, \quad \boldsymbol{u} = \begin{bmatrix} F_f \\ F_r \end{bmatrix}, \text{ and } \boldsymbol{y} = \begin{bmatrix} \dot{z}_{sf} - \dot{z}_{uf} \\ \dot{z}_{sr} - \dot{z}_{ur} \end{bmatrix}.$$

The output vector $\boldsymbol{z}$ is determined by the particular problem under consideration (rider comfort, or tyre deflection and so on). Eliminating $\boldsymbol{u}$ and $\boldsymbol{y}$ from (6.97) and (6.98) yields the closed-loop transfer function matrix

$$\boldsymbol{H} = \boldsymbol{P}_{11} + \boldsymbol{P}_{12}\boldsymbol{Y}(\boldsymbol{I} - \boldsymbol{P}_{22}\boldsymbol{Y})^{-1}\boldsymbol{P}_{21}. \tag{6.99}$$

We will make use of standard theory relating to the parameterization of all stabilizing controllers in order to find the required two-wheel-station suspension interpolation constraints. An outline of the approach adopted is based on Theorem A.4.4 in [229].

Let $\boldsymbol{P}_{22} = \boldsymbol{N}_r \boldsymbol{D}_r^{-1} = \boldsymbol{D}_l^{-1} \boldsymbol{N}_l$ be stable rational right- and left-coprime factorizations of $\boldsymbol{P}_{22}$ [229], and let

$$\begin{bmatrix} \boldsymbol{V}_r & \boldsymbol{U}_r \\ -\boldsymbol{N}_l & \boldsymbol{D}_l \end{bmatrix} \begin{bmatrix} \boldsymbol{D}_r & -\boldsymbol{U}_l \\ \boldsymbol{N}_r & \boldsymbol{V}_l \end{bmatrix} = \begin{bmatrix} \boldsymbol{I} & 0 \\ 0 & \boldsymbol{I} \end{bmatrix}$$

be the generalized Bezout equation of the corresponding doubly coprime factorization [229]. Then $\boldsymbol{Y} = \boldsymbol{Y}_1 \boldsymbol{Y}_2^{-1}$, in which

$$\begin{bmatrix} \boldsymbol{Y}_1 \\ \boldsymbol{Y}_2 \end{bmatrix} = \begin{bmatrix} \boldsymbol{D}_r & -\boldsymbol{U}_l \\ \boldsymbol{N}_r & \boldsymbol{V}_l \end{bmatrix} \begin{bmatrix} \boldsymbol{Q} \\ \boldsymbol{I} \end{bmatrix}, \qquad \boldsymbol{Q} \in \mathcal{RH}_\infty \tag{6.100}$$

is the so-called Youla parametrization of all stable suspensions. In this description $\mathcal{RH}_\infty$ is the set of all stable rational transfer function matrices of appropriate dimensions. With $\boldsymbol{Y}$ given in (6.100), the suspension system is described by

$$H = P_{11} + P_{12}Y(I - P_{22}Y)^{-1}P_{21} = T_{11} + T_{12}QT_{21}, \qquad (6.101)$$

in which

$$\begin{bmatrix} T_{11} & T_{12} \\ T_{21} & 0 \end{bmatrix} = \begin{bmatrix} P_{11} - P_{12}U_l D_l P_{21} & P_{12}D_r \\ D_l P_{21} & 0 \end{bmatrix}. \qquad (6.102)$$

This is an affine parametrization of H in terms of Q. Note that each of the matrices in (6.102) is asymptotically stable and can be computed using standard state-space calculations [229]. The important point here is that *every* stabilizing, linear, time-invariant suspension can be represented by (6.101), and every linear, stable two-wheel-station problem can be described by

$$H = T_{11} + T_{12}QT_{21}, \qquad Q \in \mathcal{RH}_\infty, \qquad (6.103)$$

with the suspension transfer function coming from (6.100). These suspension transfer functions can be active or passive, and will typically involve interactions between the front- and rear-wheel systems. Since $Q \in \mathcal{RH}_\infty$, every closed right-half-plane zero of either T_{12} or T_{21} is *necessarily* also a zero of $T_{12}QT_{21}$ for every $Q \in \mathcal{RH}_\infty$. Now suppose there exist n zeros z_i with non-negative real parts such that

$$T_{21}(z_i)a_i = 0, \qquad i = 1, 2, \cdots, n_{21}$$

or

$$a_i^* T_{12}(z_i) = 0, \qquad i = n_{21} + 1, n_{21} + 2, \cdots, n.^9$$

It then follows that for these points

$$H(z_i)a_i = T_{11}(z_i)a_i = b_i, \qquad i = 1, 2, \cdots, n_{21} \qquad (6.104)$$

and

$$a_i^* H(z_i) = a_i^* T_{11}(z_i) = b_i^*, \qquad i = n_{21} + 1, n_{21} + 2, \cdots, n \qquad (6.105)$$

must hold. Conditions (6.104) and (6.105) are vector-valued interpolation conditions that must be satisfied by all active and passive suspension systems [234]. Equations (6.104) and (6.105) show that the invariant gains of the $H(z_i)$s in the a_i directions are $\|b_i\|$; in MIMO problems 'gains' have an associated input direction. Since T_{11} and T_{12} are output dependent, the interpolation conditions must be reviewed for each set of outputs under consideration. As was the case in the single-wheel-station problem [227], higher-order interpolation conditions, or interpolation constraints with multiplicities, may occur, see for example (6.18) and (6.19). The treatment of higher-order vector valued constraints can be found in [235]. All of the calculations involved in finding the interpolation constraints given in (6.104) and (6.105) can be carried out using routine linear algebraic techniques.

---

9 The transfer function matrices T_{12} and/or T_{21} lose rank at the z_is, which means that the stipulated vectors can be found; vectors a_i and a_i^* can be scaled to have unit magnitude.

6.6 Full-vehicle model

The two-degrees-of-freedom single-wheel-station model analysed in Section 6.4 represents one corner of a car, or half a motorcycle, and the sprung mass has a heave freedom only. In section 6.5 the single-track model was developed, which is representative of a single-track vehicle, or a single-track approximation of a two-track vehicle. In this model the sprung mass has heave and pitch freedoms. In this section we will derive (linear) equations of motion that can be used to model the ride response of a four-wheeled vehicle with independent suspensions and anti-roll bars. In this case the sprung mass is deemed 'flexible' and so it has heave, pitch, roll, and warp freedoms; warp occurs when a diagonally opposite wheel pair oscillates in phase. The warp freedom is removed when the sprung body is deemed 'rigid'.

The full-vehicle model consists of six bodies: a flexible sprung mass, which has total mass m_s, and comprises two rigid bodies interconnected by a torsional spring in the vehicle's body-fixed x-axis direction. The sprung-mass has torsional stiffness K_c between a front roll moment of inertia I_{xf} and rear roll moment of inertia I_{xr}. The pitch moment of inertia is I_y. There are four unsprung bodies, which have mass m_f on the front axle, and m_r on the rear axle. The sprung mass centre is at a distance b from the rear axle and at a distance a from the front axle. The front- and rear-wheel track widths are $2w_f$ and $2w_r$, respectively. The front and rear tyres have radial stiffnesses k_{tf} and k_{tr}, respectively. The front and rear suspensions have vertical stiffness and damping coefficients k_{sf} and k_{sr}, and c_{sf} and c_{sr}, respectively. The eight degrees of freedom are the sprung-mass heave z_s, the sprung-mass pitch angle θ, the sprung-mass roll angles ϕ_f and ϕ_r (the roll angles of the front and rear chassis), and the displacements of the four unsprung masses z_{fl}, z_{fr}, z_{rr}, and z_{rl}. The vertical road displacements under each of the four tyres are denoted z_{fl}^r, z_{fr}^r, z_{rr}^r, and z_{lr}^r.

Force and moment balances of each of the six bodies gives

$$m_s \ddot{z}_s = -F_{fl} - F_{fr} - F_{rl} - F_{rr} \tag{6.106}$$

$$I_{xf}\ddot{\phi}_f = w_f(F_{fl} - F_{fr}) - M_{af} - K_c(\phi_f - \phi_r) \tag{6.107}$$

$$I_{xr}\ddot{\phi}_r = w_r(F_{rl} - F_{rr}) - M_{ar} + K_c(\phi_f - \phi_r) \tag{6.108}$$

$$I_y\ddot{\theta} = a(F_{fr} + F_{fl}) - b(F_{rr} + F_{rl}) \tag{6.109}$$

$$m_f \ddot{z}_{fl} = F_{fl} + k_{tf}(z_{fl}^r - z_{fl}) - F_{af} \tag{6.110}$$

$$m_f \ddot{z}_{fr} = F_{fr} + k_{tf}(z_{fr}^r - z_{fr}) + F_{af} \tag{6.111}$$

$$m_r \ddot{z}_{rr} = F_{rr} + k_{tr}(z_{rr}^r - z_{rr}) + F_{ar} \tag{6.112}$$

$$m_r \ddot{z}_{rl} = F_{rl} + k_{tr}(z_{rl}^r - z_{rl}) - F_{ar}, \tag{6.113}$$

where the four suspension forces are given by

$$F_{fl} = k_{sf}(z_s - a\theta - w_f\phi_f - z_{fl}) + c_{sf}(\dot{z}_s - a\dot{\theta} - w_f\dot{\phi}_f - \dot{z}_{fl}) \tag{6.114}$$

$$F_{fr} = k_{sf}(z_s - a\theta + w_f\phi_f - z_{fr}) + c_{sf}(\dot{z}_s - a\dot{\theta} + w_f\dot{\phi}_f - \dot{z}_{fr}) \tag{6.115}$$

$$F_{rl} = k_{sr}(z_s + b\theta - w_r\phi_r - z_{rl}) + c_{sf}(\dot{z}_s + b\dot{\theta} - w_r\dot{\phi}_r - \dot{z}_{rl}) \tag{6.116}$$

$$F_{rr} = k_{sr}(z_s + b\theta + w_r\phi_r - z_{rr}) + c_{sf}(\dot{z}_s + b\dot{\theta} + w_r\dot{\phi}_r - \dot{z}_{rr}). \tag{6.117}$$

The subscripts fl, fr, rl, and rr denote front-left, front-right, rear-left, and rear-right, respectively. In equations (6.107) and (6.108), and equations (6.110)–(6.113), there are forces and moments produced by front- and rear-axle anti-roll bars. In the case of the front axle the moment on the sprung mass is

$$M_{af} = k_{af}\left(\phi_f - \frac{z_{fr} - z_{fl}}{2w_f}\right), \tag{6.118}$$

while

$$M_{ar} = k_{ar}\left(\phi_r - \frac{z_{rr} - z_{rl}}{2w_r}\right) \tag{6.119}$$

is the moment acting on the rear-axle sprung mass. The reaction forces on the front-axle unsprung masses are

$$F_{af} = \frac{M_{af}}{2w_f}, \tag{6.120}$$

while the reactions forces on the rear-axle unsprung masses are

$$F_{ar} = \frac{M_{ar}}{2w_r}. \tag{6.121}$$

6.6.1 Invariant equations

As with the invariant equation of the single-wheel-station model (Section 6.4.1), and the two invariant equations for the in-plane model (Section 6.5.1), four invariant equations can be obtained by eliminating the suspension forces in the heave, roll, and pitch equations of motion. These equations are independent of both the suspension struts and the anti-roll bar characteristics. Adding (6.106), (6.110), (6.111), (6.112), and (6.113) gives

$$m_s \ddot{z}_s + m_f(\ddot{z}_{fr} + \ddot{z}_{fl}) + m_r(\ddot{z}_{rr} + \ddot{z}_{rl}) = k_{tf}(z_{fl}^r + z_{fr}^r - z_{fl} - z_{fr}) + k_{tr}(z_{rr}^r + z_{rl}^r - z_{rr} - z_{rl}). \tag{6.122}$$

Computing $(6.109) + b((6.112) + (6.113)) - a((6.110) + (6.111))$ gives

$$I_y \ddot{\theta} + bm_r(\ddot{z}_{rl} + \ddot{z}_{rr}) - am_f(\ddot{z}_{fl} + \ddot{z}_{fr}) = bk_{tr}(z_{rr}^r + z_{rl}^r - z_{rl} - z_{rr}) - ak_{tf}(z_{fl}^r + z_{fr}^r - z_{fl} - z_{fr}). \tag{6.123}$$

Equation $(6.107) - w_f((6.110) - (6.111))$ gives

$$I_{xf}\ddot{\phi}_f + m_f w_f(\ddot{z}_{fr} - \ddot{z}_{fl}) = k_{tf}w_f(z_{fl} - z_{fr} + z_{fr}^r - z_{fl}^r) + K_c(\phi_r - \phi_f). \tag{6.124}$$

Finally, Equation $(6.108) + w_r((6.112) - (6.113))$ gives

$$I_{xr}\ddot{\phi}_r + m_r w_r(\ddot{z}_{rr} - \ddot{z}_{rl}) = k_{tr}w_r(z_{rl} - z_{rr} + z_{rr}^r - z_{rl}^r) + K_c(\phi_f - \phi_r). \tag{6.125}$$

These (four) invariant equations can be used to find twelve relationships between the twelve driver comfort, suspension deflection, and tyre deflection transfer functions

$$H_A^z, \; H_A^{\phi_f}, \; H_A^{\phi_r}, \; H_A^{\theta}, \; H_{SD}^{fl}, \; H_{SD}^{fr}, \; H_{SD}^{rl}, \; H_{SD}^{rr}, \; H_{TD}^{fl}, \; H_{TD}^{fr}, \; H_{TD}^{rl}, \; \text{and} \; H_{TD}^{rr}.$$

These details are left as an exercise for the reader. It can be verified that this system generates twelve equations of rank eight. Thus only four of the twelve transfer functions are independent; this is clearly be the case if the car is describable by four independent single-wheel-station models (see Section 6.6.3).

6.6.2 Mode shapes

A state-space model suitable for computing the system mode shapes can be found using the following generalized coordinates

$$\boldsymbol{q} = [z_s, \phi_f, \phi_r, \theta, z_{fl}, z_{fr}, z_{rr}, z_{rl}]^T. \tag{6.126}$$

A routine exercise, using (6.106) through to (6.118), gives

$$M\ddot{\boldsymbol{q}} + C\dot{\boldsymbol{q}} + K\boldsymbol{q} = 0, \tag{6.127}$$

where the mass matrix is M, the damping matrix is C, and the stiffness matrix is K. These matrices are given by

$$M = \text{diag}\left(\begin{bmatrix} m_s & I_{xf} & I_{xr} & I_y & m_{uf} & m_{uf} & m_{ur} & m_{ur} \end{bmatrix}\right), \tag{6.128}$$

$$C = \begin{bmatrix} 2(c_{sf}+c_{sr}) & 0 & 0 & 2(bc_{sr}-ac_{sf}) & -c_{sf} & -c_{sf} & -c_{sr} & -c_{sr} \\ & 2w_f^2 c_{sf} & 0 & 0 & w_f c_{sf} & -w_f c_{sf} & 0 & 0 \\ & & 2w_r^2 c_{sr} & 0 & 0 & 0 & -w_r c_{sr} & w_r c_{sr} \\ & & & 2(a^2 c_{sf}+b^2 c_{sr}) & ac_{sf} & ac_{sf} & -bc_{sr} & -bc_{sr} \\ & & & & c_{sf} & 0 & 0 & 0 \\ & \text{sym} & & & & c_{sf} & 0 & 0 \\ & & & & & & c_{sr} & 0 \\ & & & & & & & c_{sr} \end{bmatrix}, \tag{6.129}$$

and

$$K = \begin{bmatrix} 2(k_{sf}+k_{sr}) & 0 & 0 & 2(bk_{sr}-ak_{sf}) & -k_{sf} & -k_{sf} & -k_{sr} & -k_{sr} \\ & K_{22} & -K_c & 0 & K_{24} & K_{25} & 0 & 0 \\ & & K_{33} & 0 & 0 & 0 & K_{37} & K_{38} \\ & & & K_{44} & ak_{sf} & ak_{sf} & -bk_{sr} & -bk_{sr} \\ & & & & K_{55} & K_{56} & 0 & 0 \\ & \text{sym} & & & & K_{66} & 0 & 0 \\ & & & & & & K_{77} & K_{78} \\ & & & & & & & K_{88} \end{bmatrix}, \tag{6.130}$$

where

$$K_{22} = 2w_f^2 k_{sf} + K_c + k_{af} \tag{6.131}$$

$$K_{24} = k_{sf} w_f + \frac{k_{af}}{2w_f} \tag{6.132}$$

$$K_{25} = -K_{24} \tag{6.133}$$

$$K_{37} = -k_{sr} w_r + \frac{k_{ar}}{2w_r} \tag{6.134}$$

$$K_{38} = -K_{37} \tag{6.135}$$

$$K_{33} = 2w_r^2 k_{sr} + K_c + k_{ar} \tag{6.136}$$

$$K_{44} = 2(a^2 k_{sf} + b^2 k_{sr}) \tag{6.137}$$

$$K_{55} = k_{sf} + k_{tf} + \frac{k_{af}}{4w_f^2} \tag{6.138}$$

$$K_{56} = -\frac{k_{af}}{4w_f^2} \tag{6.139}$$

$$K_{66} = K_{55} \tag{6.140}$$

$$K_{77} = k_{sr} + k_{tr} + \frac{k_{ar}}{4w_r^2} \tag{6.141}$$

$$K_{78} = -\frac{k_{ar}}{4w_r^2} \tag{6.142}$$

$$K_{88} = K_{77}. \tag{6.143}$$

The un-driven model in (6.127) can be rewritten in state-space form for modal analysis purposes as

$$\dot{x} = Ax \tag{6.144}$$

where

$$x = \begin{bmatrix} q \\ \dot{q} \end{bmatrix} \qquad A = \begin{bmatrix} 0 & I \\ -M^{-1}K & -M^{-1}C \end{bmatrix}. \tag{6.145}$$

The generalized coordinates q are given in (6.126), M is given by (6.128), C is given by (6.129), and K is given by (6.130); I is an identity matrix of appropriate dimension.

Example 6.6 Find the modal frequencies and mode shapes for a car with sprung mass $m_s = 1,060$ kg, unsprung masses $m_f = m_r = 60$ kg (i.e. total mass 1,300 kg), roll moments of inertia $I_{xf} = I_{xr} = 300$ kgm^2, and pitch moment of inertia $I_y = 1,400$ kgm^2. The torsional stiffness between the front- and rear-half chassis is $K_c = 1.5 \times 10^6$ Nm/rad. The sprung mass centre is located $b = 1.3$ m from the rear axle and $a = 1.2$ m from the front axle. The front- and rear-wheel tracks are $2w_f = 2w_r = 1.6$ m. The front- and rear-suspension vertical stiffnesses (single corner) are $k_{sf} = 12.5$ kN/m and $k_{sr} = 17.5$ kN/m, respectively. The front-axle anti-roll-bar stiffness is $k_{af} = 17$ kNm/rad (there is no anti-roll bar at the rear, i.e. $k_{ar} = 0$). The front- and rear-tyre radial stiffnesses are $k_{tf} = k_{tr} = 200$ kN/m. The front- and rear-suspension damping coefficients (single corner) are $c_{sf} = 1.5$ kNs/m and $c_{sr} = 1.7$ kNs/m. Figure 6.14 can be obtained by computing the eigenvalues and eigenvectors of the matrix A in (6.145). The corresponding frequencies and damping ratios are: (a) $f = 1.01$ Hz, $\zeta = 0.38$, (b) $f = 1.26$ Hz, $\zeta = 0.39$, (c) $f = 1.41$ Hz, $\zeta = 0.31$, (d) $f = 9.04$ Hz, $\zeta = 0.25$, (e) $f = 9.06$ Hz, $\zeta = 0.22$, (f) $f = 9.17$ Hz, $\zeta = 0.24$, (g) $f = 9.60$ Hz, $\zeta = 0.21$, (h) $f = 15.76$ Hz, $\zeta = 0.03$. The three low-frequency modes (in the range 1–1.5 Hz) relate primarily to the sprung-mass heave, pitch, and roll motions; see Figure 6.14 (a), (b), and (c). The higher-frequency modes relate primarily to unsprung mass motions (wheel-hop modes) and chassis warping. In this case the wheel-hop modes are at frequencies around 9 Hz, while the chassis warp mode is at 16 Hz. The lowest-frequency wheel-hop mode involves primarily in-phase 'tramping' of the rear axle; see Figure 6.14 (d). Figure 6.14 (e) involves primarily in-phase tramping of the front axle. The modes in Figures 6.14 (f) and (g) involve primarily anti-phase tramping of the rear and front axles, respectively. The highest-frequency mode (at approximately 16 Hz) shown as Figures 6.14 (h) involves anti-phase warping of the sprung mass and anti-phase tramping of the front and rear axles. In the case that the roll bar is removed ($k_{af} = 0$), the roll modal frequency reduces to 1.16 Hz (from 1.41 Hz) and the front anti-phase wheel-hop mode reduces from 9.60 to 9.34 Hz; the other modal frequencies are largely unaffected.

6.6.3   Reconciliation with single-wheel-station model

As with the two-wheel-station model described in Section 6.5, under conditions to be described, the flexible-sprung-mass four-wheel-station model can be represented perfectly by four single-wheel-station models. The transformation

$$
V^{-1} = \begin{bmatrix}
1 & -w_f & 0 & -a & 0 & 0 & 0 & 0 \\
0 & 0 & 0 & 0 & 1 & 0 & 0 & 0 \\
1 & w_f & 0 & -a & 0 & 0 & 0 & 0 \\
0 & 0 & 0 & 0 & 0 & 1 & 0 & 0 \\
1 & 0 & w_r & b & 0 & 0 & 0 & 0 \\
0 & 0 & 0 & 0 & 0 & 0 & 1 & 0 \\
1 & 0 & -w_r & b & 0 & 0 & 0 & 0 \\
0 & 0 & 0 & 0 & 0 & 0 & 0 & 1
\end{bmatrix}
\qquad \tilde{q} = V^{-1} q
\tag{6.146}
$$

maps the original generalized coordinates

$$
q = \begin{bmatrix} z_s & \phi_f & \phi_r & \theta & z_{fl} & z_{fr} & z_{rr} & z_{rl} \end{bmatrix}
\tag{6.147}
$$

into

$$
\tilde{q} = \begin{bmatrix} z_{sfl} & z_{fl} & z_{sfr} & z_{fr} & z_{srr} & z_{rr} & z_{srl} & z_{rl} \end{bmatrix},
\tag{6.148}
$$

where z_{sfl}, z_{sfr}, z_{srr}, and z_{srl} are the strut suspension points on the sprung mass. It follows by direct computation that

$$
\tilde{M} = V^T M V = \text{Diag}\left(\begin{bmatrix} \frac{m_s b}{2(a+b)} & m_f & \frac{m_s b}{2(a+b)} & m_f & \frac{m_s a}{2(a+b)} & m_r & \frac{m_s a}{2(a+b)} & m_r \end{bmatrix}\right)
\tag{6.149}
$$

when

$$
I_y = abm_s \qquad I_f = m_s b w_f^2/(a+b) \qquad I_r = m_s a w_r^2/(a+b);
\tag{6.150}
$$

the mass matrix is given by (6.128).
   Another calculation shows that

$$
\tilde{C} = V^T C V = \text{Block Diag}\left(\begin{bmatrix} c_{sf} & -c_{sf} \\ -c_{sf} & c_{sf} \end{bmatrix} \begin{bmatrix} c_{sf} & -c_{sf} \\ -c_{sf} & c_{sf} \end{bmatrix} \begin{bmatrix} c_{sr} & -c_{sr} \\ -c_{sr} & c_{sr} \end{bmatrix} \begin{bmatrix} c_{sr} & -c_{sr} \\ -c_{sr} & c_{sr} \end{bmatrix}\right);
\tag{6.151}
$$

the damping matrix is given by (6.129).
   Finally, when

$$
k_{af} = 0 \qquad k_{ar} = 0 \qquad K_c = 0,
\tag{6.152}
$$

there holds

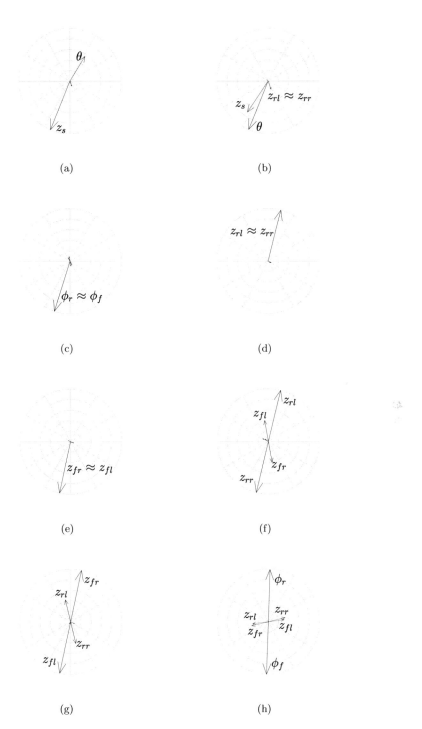

Figure 6.14: Vibration modes of the full-car model in Example 6.6.

$$\tilde{K} = V^T K V = \text{Block Diag} \left(\begin{bmatrix} k_{sf} & -k_{sf} \\ -k_{sf} & k_{sf} + k_{tf} \end{bmatrix} \begin{bmatrix} k_{sf} & -k_{sf} \\ -k_{sf} & k_{sf} + k_{tf} \end{bmatrix} \right.$$
$$\left. \begin{bmatrix} k_{sr} & -k_{sr} \\ -k_{sr} & k_{sr} + k_{tr} \end{bmatrix} \begin{bmatrix} k_{sr} & -k_{sr} \\ -k_{sr} & k_{sr} + k_{tr} \end{bmatrix} \right); \tag{6.153}$$

the stiffness matrix is given by (6.130). Thus, when the inertia conditions (6.150), and the stiffness conditions (6.152) are satisfied, the four-wheel-station problem behaves like four non-interacting single-wheel-station systems.

6.6.4   Transfer functions

The transfer functions for the full-vehicle model can be derived using $H = C(sI - A)^{-1}B + D$, where A is given by (6.145), and

$$B = \begin{bmatrix} 0 \\ M^{-1}B_0 \end{bmatrix} \qquad B_0 = \begin{bmatrix} 0 & 0 & 0 & 0 \\ 0 & 0 & 0 & 0 \\ 0 & 0 & 0 & 0 \\ 0 & 0 & 0 & 0 \\ k_{tf} & 0 & 0 & 0 \\ 0 & k_{tf} & 0 & 0 \\ 0 & 0 & k_{tr} & 0 \\ 0 & 0 & 0 & k_{tr} \end{bmatrix}. \tag{6.154}$$

The matrices C and D depend on the output(s) selected. Assuming that driver comfort is related to the sprung-mass accelerations

$$C_A = A(9:12,:) \quad \text{(Rows 9 to 12 of the A-matrix.)} \quad D_A = \begin{bmatrix} 0 & 0 & 0 & 0 \\ 0 & 0 & 0 & 0 \\ 0 & 0 & 0 & 0 \\ 0 & 0 & 0 & 0 \end{bmatrix}; \tag{6.155}$$

in the case of suspension travel

$$C_{ST} = \begin{bmatrix} 1 & -w_f & 0 & -a & -1 & 0 & 0 & 0 & 0 & 0 & 0 & 0 & 0 \\ 1 & w_f & 0 & -a & 0 & -1 & 0 & 0 & 0 & 0 & 0 & 0 & 0 \\ 1 & 0 & w_r & b & 0 & 0 & -1 & 0 & 0 & 0 & 0 & 0 & 0 \\ 1 & 0 & -w_r & b & 0 & 0 & 0 & -1 & 0 & 0 & 0 & 0 & 0 \end{bmatrix} \qquad D_{ST} = \begin{bmatrix} 0 & 0 & 0 & 0 \\ 0 & 0 & 0 & 0 \\ 0 & 0 & 0 & 0 \\ 0 & 0 & 0 & 0 \end{bmatrix}; \tag{6.156}$$

in the case of tyre deflections

$$C_{TD} = \begin{bmatrix} 0 & 0 & 0 & 0 & 1 & 0 & 0 & 0 & 0 & 0 & 0 & 0 & 0 \\ 0 & 0 & 0 & 0 & 0 & 1 & 0 & 0 & 0 & 0 & 0 & 0 & 0 \\ 0 & 0 & 0 & 0 & 0 & 0 & 1 & 0 & 0 & 0 & 0 & 0 & 0 \\ 0 & 0 & 0 & 0 & 0 & 0 & 0 & 1 & 0 & 0 & 0 & 0 & 0 \end{bmatrix} \qquad D_{TD} = \begin{bmatrix} -1 & 0 & 0 & 0 \\ 0 & -1 & 0 & 0 \\ 0 & 0 & -1 & 0 \\ 0 & 0 & 0 & -1 \end{bmatrix}. \tag{6.157}$$

As with the single-track model, the number of independent inputs can be halved using wheelbase filtering ideas. Under the assumption that the rear axle will see the same road elevation of the front axle after a time delay $(a+b)/v$, the system inputs are the two road elevations at the front axle

$$\boldsymbol{u} = [z^r_{fl}, u^r_{fr}]^T, \tag{6.158}$$

while the elevations at the rear axle are

$$z^r_{rl} = z_{fl}e^{-s\tau} \quad \text{and} \quad z^r_{rr} = z^r_{fr}e^{-s\tau}, \tag{6.159}$$

with $\tau = (a+b)/v$. Since the outputs $\boldsymbol{y}$ are related to the inputs $\boldsymbol{u}$ through the vehicle transfer function $\boldsymbol{H}$:

$$\begin{bmatrix} y_1 \\ \vdots \\ y_n \end{bmatrix} = \boldsymbol{H} \begin{bmatrix} u^r_{fl} \\ u^r_{fr} \end{bmatrix}. \tag{6.160}$$

The PSD response is given by

$$\begin{bmatrix} \Phi_{y_1 y_1} & \cdots & \Phi_{y_1 y_n} \\ \vdots & \ddots & \vdots \\ \Phi_{y_n y_1} & \cdots & \Phi_{y_n y_n} \end{bmatrix} = \boldsymbol{H} \begin{bmatrix} \Phi_{z^r_{fl} z^r_{fl}} & \Phi_{z^r_{fl} z^r_{fr}} \\ \Phi_{z^r_{fr} z^r_{fl}} & \Phi_{z^r_{fr} z^r_{fr}} \end{bmatrix} \boldsymbol{H}^* \tag{6.161}$$

where $*$ is the complex conjugate transpose. The various partitions of the road PSD are given by: $\Phi_{z^r_{fl} z^r_{fl}} = \Phi_{z^r_{fr} z^r_{fr}} = \Phi$, with Φ given by (6.1), and $\Phi_{z^r_{fl} z^r_{fr}} = \Phi_{z^r_{fr} z^r_{fl}} = \Phi\gamma$, with γ given by (6.4). The expected RMS response values can be obtained by integration to give $\sigma_i = \sqrt{\int \Phi_{y_i y_i} dn}$.

Example 6.7 Using the car data given in Example 6.6, compute the expected RMS response values associated with the driver comfort response (6.155), the workspace usage (6.156), and the roadholding response (6.157) when the vehicle is travelling on a type B road at 40 m/s (see example 6.1). Assume (6.159) holds between the front and rear axles, and that the coherence between the two road tracks can be expressed as (6.4) with $\rho = 3.5$. An application of (6.161) gives the response PSD, which is integrated to give the RMS values $\sigma = \sqrt{\int_0^\infty \Phi_y df}$ as follows (see footnote 8). For the bounce acceleration $\sigma_{\ddot{z}_s} = 0.59$ m/s^2; for the roll accelerations $\sigma_{\ddot{\phi}_f} = 0.83$ rad/s^2 and $\sigma_{\ddot{\phi}_r} = 0.75$ rad/s^2; for the pitch acceleration $\sigma_{\ddot{\theta}} = 0.73$ rad/s^2; for the front-left and front-right suspension travels $\sigma_{z_{sfl}-z_{fl}} = \sigma_{z_{sfr}-z_{fr}} = 0.007$ m; for the rear-left and rear-right suspension travels $\sigma_{z_{srl}-z_{rl}} = \sigma_{z_{srr}-z_{rr}} = 0.007$ m; for the front-left and front-right tyre deflections $\sigma_{z_{fl}-z^r_{fl}} = \sigma_{z_{fr}-z^r_{fr}} = 0.003$ m; and for the rear-left and rear-right tyre deflections $\sigma_{z_{rl}-z^r_{rl}} = \sigma_{z_{rr}-z^r_{rr}} = 0.003$ m. In the case that the inputs at the front axle are considered fully correlated, that is, $\gamma = 1$ in (6.4), the RMS value of the bounce acceleration increases to $\sigma_{\ddot{z}_s} = 0.67$ m/s^2, $\sigma_{\ddot{\theta}}$ increases to 0.89 rad/s^2, while $\sigma_{\ddot{\phi}_f}$ and $\sigma_{\ddot{\phi}_r}$ are both zero, since the roll dynamics are not excited in this case.

6.6.5   Rigid-chassis model

In the case that the sprung mass is treated as 'rigid', $\phi_f = \phi_r$, and the four sprung-mass suspension mounting points become co-planar. In this case the state-space dimension reduces to seven. This simplification can be achieved by introducing the transformation

$$T^T = \begin{bmatrix} 1 & 0 & 0 & 0 & 0 & 0 & 0 & 0 \\ 0 & 1 & 1 & 0 & 0 & 0 & 0 & 0 \\ 0 & 0 & 0 & 1 & 0 & 0 & 0 & 0 \\ 0 & 0 & 0 & 0 & 1 & 0 & 0 & 0 \\ 0 & 0 & 0 & 0 & 0 & 1 & 0 & 0 \\ 0 & 0 & 0 & 0 & 0 & 0 & 1 & 0 \\ 0 & 0 & 0 & 0 & 0 & 0 & 0 & 1 \end{bmatrix}. \tag{6.162}$$

The reduced-order mass matrix is given by

$$\tilde{M} = T^T M T = \mathrm{Diag}\left(\begin{bmatrix} m_s & I_{xf} + I_{xr} & I_y & m_{uf} & m_{uf} & m_{ur} & m_{ur} \end{bmatrix}\right). \tag{6.163}$$

The reduced-order damping matrix is

$$\tilde{C} = T^T C T = \begin{bmatrix} 2(c_{sf} + c_{sr}) & 0 & 2(c_{sr}b - c_{sf}a) & -c_{sf} & -c_{sf} & -c_{sr} & -c_{sr} \\ & 2(c_{sf}w_f^2 + c_{sr}w_r^2) & 0 & c_{sf}w_f & -c_{sf}w_f & -c_{sr}w_r & c_{sr}w_r \\ & & 2(c_{sf}a^2 + c_{sr}b^2) & c_{sf}a & c_{sf}a & -c_{sr}b & -c_{sr}b \\ & & & c_{sf} & 0 & 0 & 0 \\ & \text{sym} & & & c_{sf} & 0 & 0 \\ & & & & & c_{sr} & 0 \\ & & & & & & c_{sr} \end{bmatrix} \tag{6.164}$$

Finally, the reduced-order stiffness matrix is

$$\tilde{K} = T^T K T = \begin{bmatrix} 2(k_{sf} + k_{sr}) & 0 & 2(k_{sr}b - k_{sf}a) & -k_{sf} & -k_{sf} & -k_{sr} & -k_{sr} \\ & K_{22} & 0 & K_{24} & K_{25} & K_{26} & K_{27} \\ & & 2(k_{sf}a^2 + k_{sr}b^2) & k_{sf}a & k_{sf}a & -k_{sr}b & -k_{sr}b \\ & & & K_{44} & K_{45} & 0 & 0 \\ & \text{sym} & & & K_{55} & 0 & 0 \\ & & & & & K_{66} & K_{67} \\ & & & & & & K_{77} \end{bmatrix}, \tag{6.165}$$

where

$$K_{22} = 2(k_{sf}w_f^2 + k_{sr}w_r^2) + k_{af} + k_{ar}$$

$$K_{24} = k_{sf}w_f + \frac{k_{af}}{2w_f}$$

$$K_{25} = -K_{24}$$

$$K_{26} = -k_{sr}w_r - \frac{k_{ar}}{2w_r}$$

$$K_{27} = -K_{26}$$

$$K_{44} = k_{sf} + k_{tf} + \frac{k_{af}}{4w_f^2}$$

$$K_{45} = -\frac{k_{af}}{4w_f^2}$$

$$K_{55} = K_{44}$$

$$K_{66} = k_{sr} + k_{tr} + \frac{k_{ar}}{4w_r^2}$$

$$K_{67} = -\frac{k_{ar}}{4w_r^2}$$

$$K_{77} = K_{66}.$$

In the case of the reduced-order four-wheel model, the heave, and pitch-invariant equations (6.122) and (6.123), respectively, remain unaltered, while the roll-invariant equations (6.124) and (6.125) sum to give

$$(I_{xf} + I_{xr})\ddot{\phi} + w_f m_f(\ddot{z}_{fr} - \ddot{z}_{fl}) + w_r m_r(\ddot{z}_{rr} - \ddot{z}_{rl}) =$$
$$w_f k_{tf}(z_{fr}^r - z_{fl}^r - z_{fr} + z_{fl}) + w_r k_{tr}(z_{rr}^r - z_{rl}^r - z_{rr} + z_{rl}), \quad (6.166)$$

in which $\phi_f = \phi_r = \phi$.

6.7 Further analysis

The suspension models discussed in this chapter introduced many of the classical concepts relating to ride dynamics and suspension optimization; the examples given thus far focus on the ubiquitous parallel spring-damper suspension strut. Irrespective of the vehicle model used, the standard approach to suspension design is to define a suspension layout, usually comprising passive components, whose values are then optimized. This approach is restricted by the suspension topology and component types used. A more flexible approach is to optimize the suspension admittance, Y, without predefining a specific suspension layout. In this paradigm it would be usual to pre-assign the transfer function order and constrain it to be passive (positive real); see footnote 6. Once the optimized admittance function is obtained, network synthesis techniques can be used to derive a corresponding suspension layout, which will be a combination of springs that have admittance k/s (where k is the spring stiffness), dampers that have admittance c (where c is the damping coefficient), and possibly inerters that have admittance bs (where b is the inerters' inertance) [71].

Example 6.8 Assume that $Y = \frac{b_0 + b_1 s}{a_0 + a_1 s}$ is the positive-real suspension admittance resulting from an optimization process. Our objective here is to derive a corresponding passive network. One network can be obtained using a partial fraction decomposition of the admittance transfer function; $Y = \sum_{i=1}^n Y_i$ corresponds to a parallel connection of the Y_is. In the same way, if $1/Y = \sum_{i=1}^n 1/Y_i$, the total suspension impedance $Z = 1/Y$ can be obtained by connecting the impedances $1/Y_i$ in series. In the case $a_0 b_1 - a_1 b_0 < 0$ the admittance can be rewritten as follows

$$Y = \frac{b_1}{a_1} + \frac{a_1 b_0 - a_0 b_1}{a_1(a_1 s + a_0)} = \frac{b_1}{a_1} + \left(\frac{a_1^2 s}{a_1 b_0 - a_0 b_1} + \frac{a_0 a_1}{a_1 b_0 - a_0 b_1} \right)^{-1}$$

$$= \frac{b_1}{a_1} + \left(\frac{1}{\frac{a_1 b_0 - a_0 b_1}{a_1^2 s}} + \frac{1}{\frac{a_1 b_0 - a_0 b_1}{a_0 a_1}} \right)^{-1}$$

$$= c_1 + \left(\frac{1}{\frac{k}{s}} + \frac{1}{c_2} \right)^{-1}, \tag{6.167}$$

which represents a damper $c_1 = \frac{b_1}{a_1}$ in parallel with the series combination of a spring $k = \frac{a_1 b_0 - a_0 b_1}{a_1^2}$ and a damper $c_2 = \frac{a_1 b_0 - a_0 b_1}{a_0 a_1}$. In the case that $a_0 b_1 - a_1 b_0 > 0$, the admittance can be rewritten as

$$Y = \left(\frac{a_0 + a_1 s}{b_0 + b_1 s} \right)^{-1} = \left(\frac{a_1}{b_1} + \left(\frac{b_1(b_1 s + b_0)}{a_0 b_1 - a_1 b_0} \right)^{-1} \right)^{-1}$$

$$= \left(\frac{1}{b_1/a_1} + \left(\frac{b_1^2 s}{a_0 b_1 - a_1 b_0} + \frac{b_1 b_0}{a_0 b_1 - a_1 b_0} \right)^{-1} \right)^{-1}$$

$$= \left(\frac{1}{c_1} + (bs + c_2)^{-1} \right)^{-1}, \tag{6.168}$$

which represents a damper $c_1 = \frac{b_1}{a_1}$ connected in series with the parallel combination of an inerter $b = \frac{b_1^2}{a_0 b_1 - a_1 b_0}$ and a damper $c_2 = \frac{b_1 b_0}{a_0 b_1 - a_1 b_0}$. The two networks resulting from (6.168) and (6.167) replicate all first-order passive transfer functions.

Since much passive circuit theory relates to electrical networks, it is useful to be aware of the equivalence relationships between passive electrical and mechanical networks. This equivalence is as follows: inductor $\leftrightarrow$ spring, resistor $\leftrightarrow$ damper, capacitor $\leftrightarrow$ inerter [71]. In the 1930s it was shown that any (rational) positive-real function can be realized as the admittance of a network comprising resistors, capacitors, inductors, and transformers [236]. It was later realized that transformers were unnecessary in the synthesis of positive-real functions, and a procedure for deriving 'transformerless' network was proposed in [237]. At the end of the 1960s it was shown that any biquadratic admittance function of a one-port network, that can be realized with two reactive elements and an arbitrary number of resistors, can be realized with two reactive elements and three resistors [238]. Recent results have shown that the minimum number of elements in series-parallel networks realizing biquadratics is five (two reactive, three resistive) in the case of regular functions [239–242],[10] and between five and eight (six reactive and two resistive) in case of non-regular functions. The minimum number of elements required to realize any biquadratic admittance function is eight. In the case

---

[10] A positive-real function $Y(s)$ is defined to be regular if the smallest value of $\text{Re}(Y(s))$, or $\text{Re}(Y^{-1}(s))$, occurs at $s = 0$, or $s = \infty$ [239].

of bridge networks, the minimum number of elements reduces to seven (five reactive and two resistive). In the case that the analysis is restricted to first-order positive-real rational functions (which are regular), the minimum number of elements is three (one reactive and two resistive), and only two networks are necessary to replicate all cases; refer again to Example 6.8. Examples of optimization based on positive-real functions are reported in [243] for a single-wheel-station model, where $\mathcal{H}_2$ and $\mathcal{H}_\infty$ positive-real synthesis problems are formulated as bilinear matrix inequalities. These ideas were extended to the two-wheel-station problem in [244].

7
Advanced Vehicle Modelling

7.1 Background

Modern road vehicles are complex machines with a large number of subsystems, sensors, and distributed computing capabilities. Reduced development times, in combination with stringent cost constraints, make product developers increasingly reliant on high-fidelity models and simulation tools. These tools allow engineers to evaluate complete vehicles, vehicle components, or active controllers in complex driving environments.

Typical vehicle models describe the motion of multiple interconnected rigid bodies, and may include flexible components. Flexible finite-element-type modelling methods are often used to describe the chassis or other components of both cars and motorcycles. Common also is the use of model reduction techniques that retain only the modes of interest [245–247]. In many vehicle modelling studies rigid-body models with lumped stiffness suffice; we will confine our consideration to these cases.

The aim of this chapter is to discuss some of the modelling techniques used to replicate the important features of car, motorcycle, and driver dynamics. These more 'advanced' vehicle models will include a number of features that were neglected in models described in Chapters 4 and 5, including suspensions with anti-dive and/or anti-squat geometries, environmental influences such as aerodynamic effects, three-dimensional road geometries, and tyre models of the type described in Chapter 3.

We begin in Section 7.2 by discussing car and motorcycle trim conditions. Vehicular trim is of importance in its own right, and as part of the development of linearized models that consider small-perturbation conditions. Included under the 'trim' banner are lateral and longitudinal load-transfer, squat, dive, pitch, and roll behaviours. In Section 7.3 we consider three-dimensional road modelling, which is required for modelling all road vehicles. In order to represent roads with sufficient fidelity, and without needless complication, we will base our road modelling on the geometry of *ribbons*. One application of road modelling relates to the solution of optimal control problems such as minimum-time and minimum-fuel problems of the type discussed in Chapter 9. A closely related topic is vehicle positioning on the road, which is a kinematics topic treated in Section 7.4. A car model with a quasi-static suspension model, slipping tyres, and aerodynamics maps suitable for simulation on three-dimensional roads is discussed in Section 7.5. Driver and rider modelling is reviewed in Section 7.6, with a focus on cognitive, stretch-reflex, and passive responses. Section 7.7 addresses the problem of modelling motorcycles, as well as several of their subsystems such as tyres, structural flexibilities, suspensions, and powertrains.

Dynamics and Optimal Control of Road Vehicles. D. J. N. Limebeer and M. Massaro.
© D. J. N. Limebeer and M. Massaro 2018. Published in 2018 by Oxford University Press.
DOI: 10.1093/oso/9780198825715.001.0001

7.2   Vehicle trim

7.2.1   Longitudinal load transfer in cars and motorcycles

The longitudinal load transfer, and acceleration and deceleration limits, of a straight-running vehicle are considered in this section. Formulae that are based on longitudinal and vertical force balances under constant acceleration/braking, and a moment balance around the vehicle's centre of mass, are derived and used to shed light on some of the important issues; see Figure 7.1. Longitudinal and vertical force balances, and a pitch

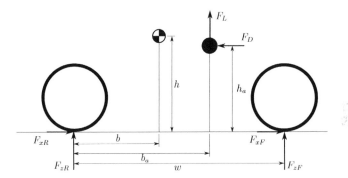

Figure 7.1: Forces acting on an accelerating/braking car.

moment balance, give

$$m\ddot{x}_G = F_{xF} + F_{xR} - F_D \tag{7.1}$$

$$0 = F_{zF} + F_{zR} - mg + F_L \tag{7.2}$$

$$I_F\dot{\omega}_F + I_R\dot{\omega}_R \pm I_e\dot{\omega}_e = F_{zR}b - F_{zF}\left(w - b\right) - \left(F_{xF} + F_{xR}\right)h$$
$$+ F_D\left(h - h_a\right) - F_L\left(b_a - b\right). \tag{7.3}$$

The mass of the vehicle is m, $\ddot{x}_G$ is the longitudinal acceleration of the vehicle's mass centre, I_F and I_R are the front- and rear-axle spin moments of inertia, $\dot{\omega}_F$ and $\dot{\omega}_R$ are the front- and rear-axle angular accelerations, I_e is the engine spin moment of inertia,[1] $\dot{\omega}_e$ is the engine's spin acceleration, F_D and F_L are the aerodynamic drag and lift forces,[2] h and h_a are the heights of the mass and aerodynamic centres respectively, w is the wheelbase, and b and b_a are the distances between the rear-axle ground-contact point and the centre of mass and the aerodynamic centre of pressure. The longitudinal

---

[1]   The engine spin axis is assumed parallel to the wheel axles. A positive sign is used when the engine spin direction is the same as that of the road wheels.

[2]   The drag and lift forces are computed using $F_D = \frac{1}{2}\rho C_D A\dot{x}_G^2$ and $F_L = \frac{1}{2}\rho C_L A\dot{x}_G^2$, respectively, where ρ is the density of air; C_D and and C_L are the drag and lift coefficients respectively. For cars $C_D A$ is in the range 0.5–1.0 m^2, A is in the range 1.8–2.2 m^2, while C_L is between -3 (in the case of high downforce related to wings) and 0.3 [248]; see also Section 7.5.6. In the case of motorcycles, $C_D A$ is in the range 0.3–0.6 m^2, A is in the range 0.3–0.9 m^2, and C_L is in the range 0.1–0.2 [6]; see also Section 7.7.5.

forces at the front and rear axles are F_{xF} and F_{xR} respectively. The normal forces on the front and rear axles are F_{zF} and F_{zR} respectively.

Equations (7.1)–(7.3) can be solved for the total longitudinal force $F_x = F_{xF} + F_{xR}$, and the front- and rear-axle vertical loads F_{zF} and F_{zR} respectively:

$$F_x = m\ddot{x}_G + F_D \tag{7.4}$$

$$F_{zF} = mg\frac{b}{w} - m\ddot{x}_G\frac{h}{w} - F_D\frac{h_a}{w} - F_L\frac{b_a}{w} - \frac{I_F}{w}\dot{\omega}_F - \frac{I_R}{w}\dot{\omega}_R \mp \frac{I_e}{w}\dot{\omega}_e \tag{7.5}$$

$$F_{zR} = mg\frac{w-b}{w} + m\ddot{x}_G\frac{h}{w} + F_D\frac{h_a}{w} - F_L\frac{w-b_a}{w} + \frac{I_F}{w}\dot{\omega}_F + \frac{I_R}{w}\dot{\omega}_R \pm \frac{I_e}{w}\dot{\omega}_e. \tag{7.6}$$

The first terms in (7.5) and (7.6) represent the static vertical loads on the axles, the second terms relate to load transfers due to the longitudinal acceleration of the vehicle (and attract a factor of h/w), the third terms are drag-related load-transfer terms (with h_a/w factors), the fourth terms are aerodynamic lift-force load-transfer terms,[3] the fifth and sixth terms represent load-transfer effects related to the wheel-spin inertias, and the seventh terms are the load transfer related to the engine-spin inertia. The wheel-spin inertias increase slightly the total load transfer to the rear axle. The engine-spin inertia may either increase (when revving in the wheels' spin direction) or reduce (when rotating in the direction opposite to that of the wheel-spin)[4] the total load transfer. The terms related to spin-inertia effects are usually small and will be neglected from now on.

In rear-wheel-drive (RWD) vehicles the longitudinal tyre force F_{xR} and dynamic normal load transfer $\hat{F}_{zR}$ at the rear axle (see equation (7.6)):

$$\hat{F}_{zR} = F_{zR} - mg\frac{w-b}{w}$$

$$= m\ddot{x}_G\frac{h}{w} + F_D\frac{h_a}{w} - F_L\frac{w-b_a}{w} + \overbrace{\frac{I_F}{w}\dot{\omega}_F + \frac{I_R}{w}\dot{\omega}_R \pm \frac{I_e}{w}\dot{\omega}_e}^{\text{Negligible}}$$

combine to produce a resultant force along the *load-transfer line*, the associated angle with respect to the horizontal is the *load-transfer angle*; see the left-hand side of Figure 7.2. When neglecting aerodynamic forces and spin-inertia influences, the longitudinal tyre force ($m\ddot{x}_G$) and the normal load transfer ($m\ddot{x}_G h/w$) generated under acceleration combine to give a resultant force at the load-transfer angle of

---

[3] Instead of applying F_D and F_L at the aerodynamic centre of pressure, one can apply F_D in the ground plane and two lift forces F_{Lf} and F_{Lr} at the front and rear axle respectively. Equations (7.5) and (7.5) show that

$$F_{Lf} = F_D\frac{h_a}{w} + F_L\frac{b_a}{w} \qquad F_{Lr} = -F_D\frac{h_a}{w} + F_L\frac{w-b_a}{w}. \tag{7.7}$$

Note that computing b_a and h_a from F_{Lf}, F_{Lr}, and F_D does not give a unique solution.

[4] A counter-rotating engine reduces slightly the risk of lifting the front wheel of a accelerating motorcycle [249].

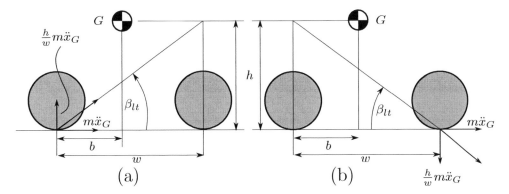

Figure 7.2: Load-transfer line (under acceleration) and the related load-transfer angle β_{tl} in the case that aerodynamic forces are neglected. (a) corresponds to a RWD, while figure (b) corresponds to a FWD vehicle.

$$\beta_{lt} = \arctan \frac{h}{w}. \tag{7.8}$$

The load-transfer line connects the rear-wheel ground-contact point and a point at the height of the centre of mass above the front axle; refer again to the left-hand side of Figure 7.2. In FWD vehicles, the load-transfer line joins the front-wheel ground-contact point and a point at the height of the centre of mass above the rear axle.

In the case of a vehicle under braking, load-transfer lines can be drawn at the front and rear axles, whose inclination angles depend on the force distribution from the front and rear axles; see Figure 7.3. Suppose that the ratio of the front- and rear-axle braking forces is defined as[5]

$$\gamma = \frac{F_{xF}}{F_{xR}}. \tag{7.9}$$

This ratio is called the *brake balance*, or *brake bias*. The load-transfer angles at the front and rear axles are given by

$$\beta_{lt}^{R} = \arctan\left(\frac{\gamma+1}{1}\frac{h}{w}\right) \qquad \beta_{lt}^{F} = \arctan\left(\frac{\gamma+1}{\gamma}\frac{h}{w}\right). \tag{7.10}$$

These load-transfer lines intersect at the height of the centre of mass, and at a distance $\frac{w}{\gamma+1}$ from the rear contact point.[6] This intersection point is sometimes called the *braking centre* or *acceleration centre* [250], and is located halfway between the axles when $\gamma = 1$, and coincides with the centre of mass when $\gamma = \frac{w}{b} - 1$.

---

[5] An alternative definition employed by some authors is $\hat{\gamma} = \frac{F_{xF}}{F_{xF}+F_{xR}} = \frac{\gamma}{\gamma+1}$.

[6] In a reference frame centred on the rear-wheel ground-contact point, the rear-axle load-transfer line is described by $z = -\tan\beta_{lt}^{R}x$, while the front-axle load-transfer line is given by $z = \tan\beta_{lt}^{F}(x-w)$, with β_{lt}^{R} and β_{lt}^{F} given by (7.10). Eliminating z from these equations gives $x = \frac{w}{\gamma+1}$, which when introduced in either the rear or front load lines gives $z = -h$.

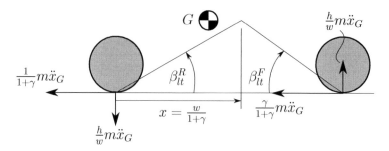

Figure 7.3: Load-transfer lines and related angles β_{tl} under braking (aerodynamic forces neglected).

The peak longitudinal acceleration of a vehicle is limited by the engine power and tyre friction. If we assume for simplicity that all the engine power P_e is used to generate the longitudinal tyre force (i.e. we neglect losses in the powertrain and tyres), we have from (7.4) that

$$\frac{\ddot{x}_G}{g} = \frac{P_e}{mg}\frac{1}{\dot{x}_G} - \frac{F_D}{mg} \tag{7.11}$$

and so the aerodynamic drag F_D reduces the maximum achievable acceleration. Also, given the maximum engine power P_e, the higher the speed $\dot{x}_G$ the lower the achievable maximum acceleration $\ddot{x}_G/g$ (in g s).

The maximum acceleration can also be restricted by the tyre force μF_z; μ is the friction coefficient and F_z is the tyre normal load. In the case of an AWD vehicle, (7.4) can be rewritten as

$$\frac{\ddot{x}_G}{g} = \mu\frac{F_{zF}}{mg} + \mu\frac{F_{zR}}{mg} - \frac{F_D}{mg}. \tag{7.12}$$

Following the introduction of the normal loads (7.5)–(7.6) into (7.12), one obtains

$$\frac{\ddot{x}_G}{g} = \mu\left(1 - \frac{F_L}{mg}\right) - \frac{F_D}{mg}. \tag{7.13}$$

If the aerodynamic forces are neglected, and the AWD vehicle has both axles engaged at their limits, there holds

$$\frac{\ddot{x}_G}{g} = \mu. \tag{7.14}$$

In the case of a FWD vehicle, the second term on the right-hand side of (7.12) does not apply, and the acceleration limit is

$$\frac{\ddot{x}_G}{g} = \frac{\mu}{w + \mu h}\left(b - \left(h_a + \frac{w}{\mu}\right)\frac{F_D}{mg} - b_a\frac{F_L}{mg}\right). \tag{7.15}$$

In the case of a RWD vehicle, the first term on the right-hand side of (7.12) does not apply, and the acceleration limit becomes

$$\frac{\ddot{x}_G}{g} = \frac{\mu}{w - \mu h}\left(w - b + \left(h_a - \frac{w}{\mu}\right)\frac{F_D}{mg} - (w - b_a)\frac{F_L}{mg}\right). \tag{7.16}$$

A comparison between (7.15) and (7.16) shows that the drag force is detrimental to FWD vehicles, while it is beneficial to the friction-related acceleration limit of RWD vehicles. Aerodynamic lift forces are always detrimental to the acceleration limit. Raising the height of the centre of mass is beneficial to RWD vehicles, while is detrimental to FWD vehicles. The maximum acceleration performance of a RWD vehicle is obtained when the centre of mass is over the rear axle. In general, RWD vehicles have higher friction-related acceleration limits as compared with FWD vehicles.[7] Not unexpectedly, AWD vehicles are superior to both RWD and FWD vehicles, as shown by (7.13).

Since braking torques are applied to each wheel of most road vehicles, vehicular braking analysis is identical to the acceleration analysis of an AWD vehicle. The friction-related maximum deceleration, under the assumption of optimal engagement of both axles, follows from (7.13) and is given by

$$-\frac{\ddot{x}_G}{g} = \mu\left(1 - \frac{F_L}{mg}\right) + \frac{F_D}{mg}. \tag{7.17}$$

Optimal braking occurs when the longitudinal braking force normalized by the normal load at the front axle, is the same as the longitudinal braking force normalized by the normal load at the rear axle. That is,

$$\frac{F_{xF}}{F_{zF}} = \frac{F_{xR}}{F_{zR}}. \tag{7.18}$$

The optimal brake ratio is obtained by substituting (7.5) and (7.6) into (7.18) to obtain

$$\gamma_{opt} = \frac{b - \frac{\ddot{x}_G}{g}h - \frac{F_D}{mg}h_a - \frac{F_L}{mg}b_a}{w - b + \frac{\ddot{x}_G}{g}h + \frac{F_D}{mg}h_a - \frac{F_L}{mg}(w - b_a)}. \tag{7.19}$$

Since $\ddot{x}_G < 0$ under braking, (7.19) increases as the deceleration increases. That is, the ratio of the front-axle brake force to the rear-axle brake force increases as the severity of braking increases.

Motorcycles typically have independent front and rear brakes and the brake balance is determined by the rider. In most passenger cars the brake balance is fixed and therefore only optimal for some fixed nominal deceleration. For deceleration levels above optimal, the rear axle is over-engaged and oversteering can occur due to high levels of rear-axle tyre slippage. For deceleration levels below optimal, the front axle is over-engaged and understeering can occur. Hard braking in wet conditions may result in front-axle sliding, while hard braking on dry road may result in rear-axle slippage (the tyre friction limit is reached for a deceleration level above optimal).

---

[7] If aerodynamic forces are neglected, and we assume that $b = w/2$ for a fair comparison, the ratio of the RWD acceleration limit (7.16) to the FWD acceleration limit (7.15) is $(w + \mu h)/(w - \mu h) > 1$.

In the case of motorcycles, high h/w values lead to large load transfers and it is possible to lift the front wheel under acceleration (the *wheelie*), or lift the rear tyre while braking (the *stoppie*).

The wheelie acceleration limit is computed using (7.5) by setting $F_{zF} = 0$, which gives

$$\frac{\ddot{x}_G}{g} = \frac{b}{h} - \frac{F_D}{mg}\frac{h_a}{h} - \frac{F_L}{mg}\frac{b_a}{h}. \tag{7.20}$$

The aerodynamic drag and lift force reduce the wheelie-related acceleration limit. When the aerodynamic terms are neglected, the acceleration limit is related to the position of the centre of mass, by the b/h factor. High acceleration limits are obtained by moving the centre of mass forwards and downwards, that is, large b and small h. Motorcycles equipped with a low-power engine, or when they are running on a low-friction surface, may not be able to reach the acceleration limit given in (7.20).

The stoppie limit is computed by setting $F_{zR} = 0$ in (7.5), to give

$$-\frac{\ddot{x}_G}{g} = \frac{w - b}{h} + \frac{F_D}{mg}\frac{h_a}{h} - \frac{F_L}{mg}\frac{w - b_a}{h}. \tag{7.21}$$

It is clear that aerodynamic drag increases the stoppie-related deceleration limit, while aerodynamic lift has an opposite effect. When the aerodynamic terms are neglected, the maximum deceleration is related to the position of the centre of mass through the factor $(w - b)/h$. High deceleration limits are obtained by moving the centre of mass rearwards and downwards, that is, large $w - b$ and small h. Motorcycles running on a low-friction surface may not be able to reach the deceleration limit given in (7.21).

In sum, the vehicle's acceleration limit is determined by the engine power (7.11), the tyre friction limit (7.13)–(7.16), or the wheelie limit (7.20), whichever is smallest. The braking limit is set by the braking torque, which is usually large by design, the tyre friction limit (7.17) and the stoppie limit (7.21), whichever is smallest. Better estimates of the acceleration limit can be obtained with more complex models that include the suspension (the suspension may move the mass centre when accelerating or braking). Other influences could include the engine map, the gearbox, and so on.

A number of typical design parameters can be selected on the basis of the above relationships, while neglecting aerodynamic and spin-inertia influences. In this way speed-independent design parameters such as the geometric and inertial quantities may be selected.

Example 7.1 Compute the optimal brake ratio and acceleration limits at $1\,\mathrm{m/s}$ and $50\,\mathrm{m/s}$ for a motorcycle with $m = 275\,\mathrm{kg}$ (including rider), $w = 1.50\,\mathrm{m}$, $b = 0.70\,\mathrm{m}$, $h = h_a = 0.65\,\mathrm{m}$, $C_D A = 0.35\,\mathrm{m}^2$, $\rho = 1.2\,\mathrm{kg/m}^3$ ($F_D = 0.5\rho C_D A\dot{x}_G^2$), $C_L A = 0$ ($F_L = 0.5\rho C_L A\dot{x}_G^2$), $P_e = 150\,kW$, and $\mu = 1.23$. The inverse of the optimal brake ratio (7.19) is shown in Figure 7.4. At high speeds the brake balance at the rear wheel increases significantly as compared with lower-speed scenarios. The engine-related acceleration limits (7.11) are $55g$ and $0.91g$ at $1\,\mathrm{m/s}$ and $50\,\mathrm{m/s}$ respectively, the friction-related acceleration limits are $1.40g$ and $1.21g$ (7.16), the wheelie-related acceleration limits are $1.08g$ and $0.88g$ (7.20), the friction-related braking limits (7.17) are $-1.23g$ and

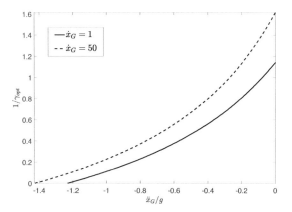

Figure 7.4: Reciprocal of the optimal braking ratio γ_{opt} as a function of the normalized deceleration $\ddot{x}_G/g$.

$-1.43g$, and the stoppie-related braking limits (7.21) are $-1.23g$ and $-1.43g$ (the friction and stoppie limits coincide in this example, because $\mu = (w - b)/h$ and $h = h_a$).

7.2.2   Squat, dive, and pitch

A vehicle's acceleration determines the load on the front and rear axles (see (7.5) and (7.6)), and these forces will usually be accompanied by suspension travel. The suspension geometry may be designed to reduce this motion, in order to prevent rear-end 'squat' under acceleration (anti-squat behaviour), or front-end 'dive' under braking (anti-dive behaviour). The related fundamentals will be discussed in this section for both cars and motorcycles.

   Key quantities used in the analysis of suspension behaviour are *load-transfer lines*, which are given by (7.10) in the case that aerodynamic forces are neglected, and *anti-squat* or *anti-dive* lines (collectively known as *support lines* [250]), which are the lines along which load-transfer forces produce no suspension travel. Sometimes the anti-squat and anti-dive lines/angles are simply called squat and dive lines/angles—hereafter these names should be considered synonymous.

Rear end in motorcycles (chain transmission).   We begin our analysis with the rear suspension of an accelerating motorcycle equipped with a chain-drive transmission by studying the equilibrium of the swingarm. Figure 7.5 shows that the moment balance around the swingarm pivot P (the moment due to the weight of the swingarm is neglected, because it is small relative to the other moments) is

$$M_s = F_{zR}l_s \cos\theta - F_{xR}(r_R + l_s \sin\theta) + F_c(r_W - l_s \sin(\theta - \mu_u)),   (7.22)$$

in which $M_s = F_s\tau_s$ is the restoring moment produced by the suspension system. The suspension force is F_s, and τ_s is the velocity ratio between the suspension travel and

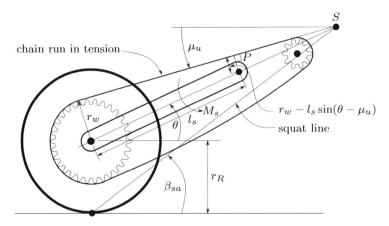

Figure 7.5: Squat angle β_{sa}, squat line, and squat point S in the case of a chain transmission.

the swingarm angular rotation.[8] The swingarm length is l_s, θ is the swingarm angle, r_R the rear-tyre radius, and F_c is the chain force acting on the upper chain run that has an angle μ_u.[9] If the rear-wheel spin inertia is neglected, a moment balance around the wheel pivot gives $F_c = F_{xR} r_R / r_w$, which can be substituted into (7.22) to give

$$M_s = F_{zR} l_s \cos\theta - F_{xR} l_s \sin\theta - F_{xR}\frac{r_R}{r_w} l_s \sin(\theta - \mu_u). \tag{7.23}$$

At standstill, when $F_{xR} = 0$, (7.23) simplifies to

$$M_s^{st} = mg\frac{w - b}{w} l_s \cos\theta. \tag{7.24}$$

Subtracting (7.24) from (7.23) gives

$$\hat{M}_s = \hat{F}_{zR} l_s \cos\theta - F_{xR} l_s \left(\sin\theta + \frac{r_R}{r_w}\sin(\theta - \mu_u)\right), \tag{7.25}$$

where $\hat{M}_s = M_s - M_s^{st}$ and $\hat{F}_{zR} = F_{zR} - mg(w - b)/w$ (see (7.6)) are the moment and load variations with respect to static conditions. In the case that $\hat{M}_s = 0$, there is no change in the swingarm trim relative to static conditions. When $\hat{M}_s > 0$ the suspension compresses relative to static conditions, while $\hat{M}_s < 0$ produces an extension of the suspension. In the case $\hat{M}_s = 0$ (7.25) can be rewritten as

$$\frac{\hat{F}_{zR}}{F_{xR}} = \frac{\sin\theta + \frac{r_R}{r_w}\sin(\theta - \mu_u)}{\cos\theta}, \tag{7.26}$$

which defines the anti-squat condition. The tangent of the load transfer angle β_{lt} is given by $\frac{\hat{F}_{zR}}{F_{xR}}$; see Section 7.2.1. When $\hat{M}_s = 0$, the tangent of the anti-squat angle β_{sa} is

---

[8]   Several motorcycle suspensions will be analysed in Section 7.7.6.
[9]   The kinematic analysis leading to the computation of chain angles will be given in Section 7.7.7.

given by the right-hand side of (7.26). Under the assumption that aerodynamic forces are negligible, the load-transfer angle can be expressed in purely kinematic terms, $\tan\beta_{lt} = h/w$ (7.8). In the case of a chain transmission, the anti-squat line joins the rear contact point and a point S, which is at the intersection of the swingarm centre line and the upper chain line;[10] see Figure 7.6 (a).

In order to monitor the rear suspension behaviour under acceleration, we define the *anti-squat ratio*:[11]

$$\mathcal{A} = \frac{\tan\beta_{sa}}{\tan\beta_{lt}}. \tag{7.27}$$

The suspension compresses (extends) when $\mathcal{A} < 1$ ($\mathcal{A} > 1$), while retaining its static value when $\mathcal{A} = 1$.

The case of the braking motorcycle is now analysed, with a view to determining the support angle. No suspension motion is obtained when the load-transfer line at the rear wheel during braking is aligned with the support line. Under braking the chain is either slack (the engine is disengaged and the braking effort comes from the rear brake), or the lower chain run is in tension (the engine is 'braking' the rear wheel). It should be remembered that with motorcycles, moderate to firm braking must be conducted predominantly on the front wheel, as suggested by (7.19).

When braking with a slack chain (7.22) holds with $F_c = 0$. Subtracting (7.24) from (7.22) gives

$$\hat{M}_s = \hat{F}_{zR} l_s \cos\theta - F_{xR}(r_R + l_s \sin\theta), \tag{7.28}$$

where $\hat{F}_{zR} < 0$ (rear tyre load reduces) and $F_{xR} < 0$ (rear tyre brake force). Solving (7.28) with $\hat{M}_s = 0$ for $\frac{\hat{F}_{zR}}{F_{xR}}$ gives the slack-chain support angle

$$\tan\beta_{sa} = \frac{(r_R + l_s \sin\theta)}{l_s \cos\theta}. \tag{7.29}$$

In this case the support line connects the rear-wheel ground-contact point and the swingarm pivot; see Figure 7.6 (b).

When the lower chain run is in tension, the suspension moment balance (7.22) is

$$M_s = F_{zR} l_s \cos\theta - F_{xR}(r_R + l_s \sin\theta) - F_c(r_w - l_s \sin(\mu_l - \theta)). \tag{7.30}$$

A moment balance around the rear-wheel centre, while neglecting the rear-wheel spin inertia, gives $M_b + F_c r_w + F_{xR} r_R = 0$, where M_b is the rear brake moment (generated by the callipers) and thus (7.30) can be rewritten as

---

[10] For a reference frame centred on the rear-wheel ground-contact point, the swingarm line is described by $z = -x\tan\theta - r_R$, with the chain line described by $z = -x\tan\mu_u - r_R - r_w/\cos\mu_u$. By eliminating z, one obtains the x-coordinate of the squat point as $x_S = r_w \cos\theta/\sin(\theta - \mu_u)$, which can be substituted into the swingarm line to obtain $z_S = -(r_w \sin\theta + r_R \sin(\theta - \mu_u))/\sin(\theta - \mu_u)$. The tangent of the anti-squat angle is thus $\tan\beta_{sa} = -z_S/x_S$.

[11] Some authors use the term *squat-line* instead of anti-squat line, and define the *squat ratio* as $\mathcal{R} = \tan\beta_{lt}/\tan\beta_{sa}$ [6].

$$M_s = F_{zR}l_s \cos\theta - F_{xR}l_s \left(\sin\theta + \frac{r_R}{r_w}\sin(\mu_l - \theta)\right) + M_b\left(1 - \frac{l_s}{r_w}\sin(\mu_l - \theta)\right). \quad (7.31)$$

Setting $M_b = 0$ (pure engine braking) gives

$$\hat{M}_s = \hat{F}_{zR}l_s \cos\theta - F_{xR}l_s\left(\sin\theta + \frac{r_R}{r_w}\sin(\mu_l - \theta)\right). \quad (7.32)$$

The support line is computed by solving (7.32) with $\hat{M}_s = 0$ for $\frac{\hat{F}_{zR}}{F_{xR}}$ to give the following support angle in the engine-braking case

$$\tan\beta_{sa} = \frac{\sin\theta + \frac{r_R}{r_w}\sin(\mu_l - \theta)}{\cos\theta}. \quad (7.33)$$

In this case the squat point S is at the intersection of the swingarm centre line and the lower chain run line, as can be shown using the arguments given in footnote 10; see Figure 7.6 (c).

In sum, different brake strategies result in different squat point locations and thus different rear-suspension equilibria. Whenever the load-transfer line is below the support line (i.e. when $\beta_{lt}^R < \beta_{sa}$) the rear suspension compresses under braking. Reducing the brake bias (7.9) increases the suspension compression, because β_{lt}^R is reduced.

Rear end in motorcycles (shaft transmissions). The swingarm equilibrium in the case of shaft transmission is similar to the slack-chain case. Subtracting (7.24) from (7.22) with $F_c = 0$ (no chain force) gives

$$\hat{M}_s = \hat{F}_{zR}l_s \cos\theta - F_{xR}\left(r_R + l_s \sin\theta\right). \quad (7.34)$$

The related support angle is obtained by solving (7.34) for $\hat{F}_{zR}/F_{xR}$ when $\hat{M}_s = 0$ to give β_{sa} in (7.29). In this case the squat point S is again coincident with the swingarm pivot; see Figure 7.6 (b).

With a typical motorcycle dataset, the anti-squat ratio, when accelerating the vehicle in the shaft transmission case, tends to be significantly greater than one ($\beta_{sa} \gg \beta_{lt}$). Thus large suspension extensions would be experienced with a shaft transmission in combination with a standard swingarm. For this reason, most motorcycles with shaft drive transmissions are equipped with a four-bar linkage in place of a swingarm. In this way squat ratios of approximately unity can be obtained; the squat point is at the intersection of the two rocker extensions; see Fig 7.6 (d).

Rear end in cars. There is a wider variety of suspension layouts in cars as compared with motorcycles. In each case, for small displacements around an equilibrium, the side projection of the suspension motion can be represented as a rotation around an instantaneous velocity centre of a 'virtual swingarm' with length l_s. Before going into a detailed analysis, we must distinguish between solid drive (or 'live') axles and independent axle suspensions.

The suspension motion of a RWD car with a solid-drive-axle configuration is dynamically equivalent to a motorcycle with a shaft-drive transmission; see Figure 7.7 (a).

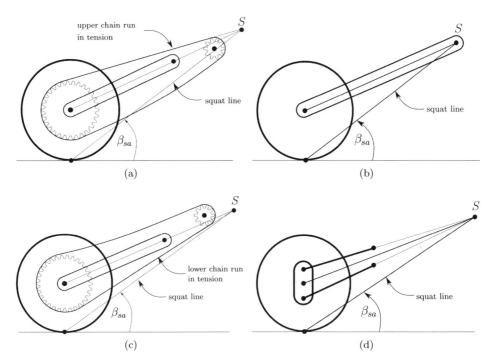

Figure 7.6: Squat line, squat point S, and squat angle β_{sa} in the case of (a) swingarm with upper chain in tension; (b) swingarm with shaft drive; (c) swingarm with lower chain in tension; (d) four-bar linkage with shaft drive.

For that reason (7.34) still applies, with l_s the distance between the wheel centre and the instantaneous velocity centre of the wheel carrier, and the support line joins the rear-wheel ground-contact point and the velocity centre [113, 251]. When the load-transfer line has the same inclination angle as the support line, zero suspension travel is obtained under acceleration.

In the case of a RWD car with an independent suspension, where the differential is attached to the chassis, (7.34) no longer holds and must be replaced by

$$\hat{M}_s = \hat{F}_{zR} l_s \cos\theta - F_{xR}\left(r_R + l_s \sin\theta\right) + M_p, \qquad (7.35)$$

where M_p is the externally applied propulsive moment. A moment balance around the wheel centre, under the assumption of negligible spin inertia, gives $M_p = F_{zR} r_R$, and (7.35) becomes

$$\hat{M}_s = \hat{F}_{zR} l_s \cos\theta - F_{xR} l_s \sin\theta. \qquad (7.36)$$

The inclination of the anti-squat line is obtained from (7.36) when $\hat{M}_s = 0$:

$$\tan\beta_{sa} = \frac{\hat{F}_{zR}}{F_{xR}} = \tan\theta. \qquad (7.37)$$

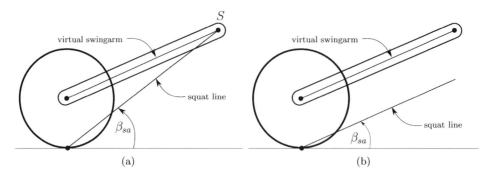

Figure 7.7: Squat line and squat angle β_{sa} in the case of cars with (a) solid axle with a differential mounted on it; (b) independent suspension with the differential mounted on the chassis.

This shows that the anti-squat line is now parallel to the 'virtual swingarm'; see Figure 7.5. Full anti-squat behaviour, that is, zero suspension travel under acceleration, is obtained when the load-transfer line has the same inclination as the 'virtual swingarm'. When braking on the front wheel, (7.34) again holds, since the braking moment acts on the wheel and reacts on the unsprung mass (not the chassis) and the support angle is given by (7.29).

Front end in motorcycles.   Motorcycles are almost invariably RWD. Under braking, the front suspension of a motorcycle tends to 'dive', and the reasoning used when discussing motorcycle rear-end equilibria may again be applied. The support line will lie between the front-wheel ground-contact point and the velocity centre of the front assembly with respect to the chassis.

   With a motorcycle equipped with a telescopic front fork, the velocity centre is at infinite distance and the dive line is normal to the sliding axis of the suspension; see Figure 7.8 (a). In the case of a motorcycle equipped with a four-bar linkage, such as Duolever (Sections 7.7.6), the velocity centre is at the intersection of the two rocker extensions; Figure 7.8 (b).

   When the load transfer is along the support line, which in this case is the *dive line*, there is no suspension compression under braking—full anti-dive behaviour is thereby achieved. The anti-dive ratio is defined as the anti-squat ratio in (7.27); a value greater than one is associated with an extension of the suspension. Suspensions such as the Duolever in Figure 7.8 (b) can be designed for good anti-dive behaviour, since the location of the rocker extension line intersection can be moved by changing the linkage geometry.

Front end in cars.   Again, for small displacements around an equilibrium, the motion of the front wheels projected on the vehicle's plane of symmetry can be thought of as a rotation around a velocity centre. When accelerating, the solid-axle case and the independent-suspension case must be treated separately. In the first case the support line (often called anti-lift line) connects the velocity centre of the suspended wheel assembly with the ground-contact point, while in the second case the support line

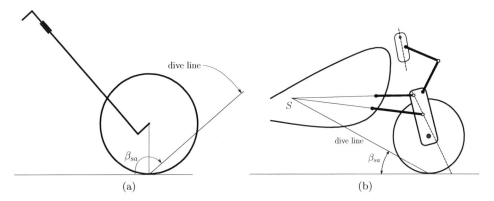

Figure 7.8: Dive line and dive angle β_{sa} in the case of motorcycles with (a) telescopic front fork; (b) four-bar linkage.

is parallel to a virtual swingarm, and through the ground-contact point—the same argument used for the rear wheels. The anti-lift ratio is defined as the anti-squat ratio in (7.27) and a value greater than one is now associated with suspension compression during acceleration. In front-wheel braking the anti-dive line joins the front-wheel ground-contact point and the velocity centre, since the braking torque is internal to the unsprung mass.

Pitch. The equilibrium positions of the front and rear suspensions determine the pitch of the vehicle under acceleration and braking. If $\hat{z}_F$ and $\hat{z}_R$ represent the front- and rear-wheel displacements (with respect to static trim), the pitch is approximately

$$\mu = \frac{\hat{z}_F - \hat{z}_R}{w}; \tag{7.38}$$

$\hat{z} > 0$ is related to suspension extension, and when $\mu > 0$ the vehicle pitches rearwards. For small displacements around a static trim, the suspension elastic moment can be written as $\hat{M}_s \approx -k_\theta \hat{\theta}$, where k_θ is a torsional spring stiffness and $\hat{\theta}$ is the rotation of the (front or rear) virtual swingarm with respect to the static trim. The equivalent vertical stiffness at the rear wheel can be obtained from the equivalence of potential energies

$$\frac{1}{2}k_z\hat{z}^2 = \frac{1}{2}k_\theta\hat{\theta}^2 \rightarrow k_z = k_\theta \left(\frac{1}{l_s\cos\theta}\right)^2, \tag{7.39}$$

where $\hat{z}/\hat{\theta} = l_s\cos\theta$ is the velocity ratio between the vertical travel of the wheel carrier and the rotation of the suspension's virtual swingarm.

We will now consider the pitching behaviour of an accelerating RWD vehicle. The rear-end suspension characteristics depend on the suspension layout, as determined by (7.25), (7.34), or (7.36), but they can all be written in the following common form:

$$-k_{zR}\hat{z}_R = F_{xR}\left(\tan\beta_{lt} - \tan\beta_{sa}^R\right), \tag{7.40}$$

where $\tan \beta_{lt}$ is given by (7.8), $\tan \beta_{sa}$ depends on the suspension layout, and (7.39) is used to express $\hat{M}_s$ in terms of $\hat{z}$. The equilibrium at the front end can be written as

$$k_{zF}\hat{z}_F = F_{xR}\tan \beta_{lt}, \tag{7.41}$$

since $F_{xF} = 0$. This means that the pitch (7.38) of the accelerating RWD vehicle is given by

$$\mu = \frac{F_{xR}}{w}\left(\tan \beta_{lt}\left(\frac{1}{k_{zF}} + \frac{1}{k_{zR}}\right) - \frac{\tan \beta_{sa}^R}{k_{zR}}\right). \tag{7.42}$$

No pitching will occur when

$$\tan \beta_{sa}^R = \left(1 + \frac{k_{zR}}{k_{zF}}\right)\tan \beta_{lt}. \tag{7.43}$$

Observe that full anti-squat systems (i.e. $\beta_{sa}^R = \beta_{lt}$) do not result in pitch-free suspensions. In order to keep the vehicle level, it follows from (7.43) that it is necessary to compensate for front-end lift. If we assume that $k_{zR} \approx k_{zF}$,[12] full anti-pitch behaviour is obtained when $\tan \beta_{sa}^R = 2\tan \beta_{lt}$.

In the case of an accelerating FWD vehicle, rear-end equilibria are determined by

$$-k_{zR}\hat{z}_R = F_{xF}\tan \beta_{lt}, \tag{7.44}$$

while the front-end equilibria are given by

$$k_{zF}\hat{z}_F = F_{xF}\left(\tan \beta_{lt} - \tan \beta_{sa}^F\right). \tag{7.45}$$

Substituting (7.44) and (7.45) into (7.38) gives the pitch of an accelerating FWD vehicle:

$$\mu = \frac{F_{xF}}{w}\left(\tan \beta_{lt}\left(\frac{1}{k_{zF}} + \frac{1}{k_{zR}}\right) - \frac{\tan \beta_{sa}^F}{k_{zF}}\right). \tag{7.46}$$

The zero-pitch condition is

$$\tan \beta_{sa}^F = \left(1 + \frac{k_{zF}}{k_{zR}}\right)\tan \beta_{lt}, \tag{7.47}$$

which again simplifies to $\tan \beta_{sa}^F = 2\tan \beta_{lt}$ in the case that $k_{zR} \approx k_{zF}$; a zero-lift configuration does not result in zero-pitch behaviour.

---

[12] Typical values for motorcycles are of the order $20\,\text{kN/m}$, while in cars values range from this figure to an order of magnitude greater in the case of high-downforce race cars.

The pitching behaviour of a braking vehicle is determined by its dive behaviour at the front axle and the lift characteristics of the rear axle. The equilibria at the rear and front axles are determined by

$$k_{zR}\hat{z}_R = -F_{xR}\left(\tan\beta_{lt}^R - \tan\beta_{sa}^R\right) \tag{7.48}$$

$$-k_{zF}\hat{z}_F = -F_{xF}\left(\tan\beta_{lt}^F - \tan\beta_{sa}^F\right), \tag{7.49}$$

where the load-transfer angles $\beta_{lt}^{R,F}$ are given by (7.10), while the support angles $\beta_{sa}^{R,F}$ depend on the suspension layout. The pitch is again determined using (7.38) to give

$$\mu = \frac{F_{xF}}{w}\left(\frac{1}{k_{zF}}\left(\tan\beta_{lt}^F - \tan\beta_{sa}^F\right) + \frac{1}{\gamma k_{zR}}\left(\tan\beta_{lt}^R - \tan\beta_{sa}^R\right)\right); \tag{7.50}$$

full anti-pitch behaviour is obtained when $\mu = 0$.

Example 7.2 Suppose a car has wheelbase $w = 2.5\,\text{m}$, mass $m = 1,300\,\text{kg}$, centre-of-mass height $h = 0.50\,\text{m}$, and front- and rear-wheel radii $r_F = r_R = 0.30\,\text{m}$. The equivalent vertical spring stiffnesses at the rear and front axles are $k_{zR} = 35\,\text{kN/m}$ and $k_{zF} = 25\,\text{kN/m}$. The rear and front suspension virtual pivots are located at height $h_{SR} = h_{SF} = 0.37\,\text{m}$ from ground, while the distances from the rear and front axles are $b_{SR} = a_{SF} = 0.70\,\text{m}$ respectively, and the braking ratio is $\gamma = 60/40$. Compute the anti-squat, anti-lift, and anti-dive ratios and pitch angle when accelerating/braking at $\ddot{x}_G = \pm0.5\,\text{g}$, in the case of RWD and FWD vehicles, in the case of solid 'live' axles and independent suspensions. Neglect aerodynamic influences. For both RWD and FWD vehicles the load-transfer angle under acceleration is $\tan\beta_{lt} = 0.2$ using (7.8). The longitudinal force under acceleration is $m\ddot{x}_G = 6377\,\text{N}$. The support angle of the accelerating RWD car is $\tan\beta_{sa} = 0.53$ from (7.29) in the case of 'live' axles and $\tan\beta_{sa} = 0.1$ using (7.37) in the case of independent suspensions, which gives anti-squat ratios of 264% and 50% respectively using (7.27). The corresponding pitch angles are $-0.2°$ and $+1.6°$ using (7.42). Since $b_{SR} = a_{SF}$ and $h_{SR} = h_{SF}$, the support angles of the accelerating FWD car are identical to those of the accelerating RWD vehicle. Therefore the anti-lift ratios are 264% and 50% respectively with the corresponding pitch angles given by $-1.1°$ and $+1.4°$; see (7.46). When braking with both RWD and FWD, the load-transfer angles (7.10) at the rear and front axles are $\tan\beta_{lt}^R = 0.50$ and $\tan\beta_{lt}^F = 0.33$ respectively, while the support angles are $\tan\beta_{sa}^R = \tan\beta_{sa}^F = 0.53$. The configuration is 159% anti-dive at the front and 106% anti-lift at the rear—this is an extreme scenario. The braking force is $-6377\,\text{N}$. The pitch angle (7.50) is $0.7°$.

7.2.3   Lateral load transfer in cars

The lateral load transfer and acceleration limit of a vehicle under steady turning (i.e. constant speed V and constant lateral acceleration $a_y = V^2/R$, where R is the cornering radius) will be considered in this section. We will begin by considering a single axle before progressing to the whole vehicle. We will develop the idea of a *roll centre* (for a single axle), and then the notion of a *roll axis* for the whole car. We will consider force and kinematic interpretations of the roll centre. We will assume that the sprung part of the car (i.e. everything 'above' the suspension; the main chassis) rolls through a

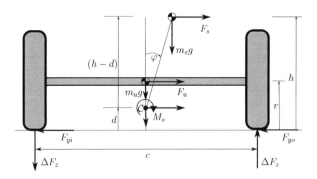

Figure 7.9: Force interpretation of the roll centre $\mathcal{C}$.

small angle with respect to the unsprung mass (i.e. everything 'below' the suspension; the front and rear axle and wheel assemblies). From a modelling perspective, we will assume that the sprung mass is joined to the unsprung mass through revolute joints at the front- and rear-axle roll centres. The front-wheel steering angle will be assumed negligible throughout the analysis.

Roll centre—force interpretation. In the force interpretation, the roll centre is defined as a point in the vertical plane above the wheel centres on a given axle, to which the application of a lateral force produces no suspension roll. In order to develop this idea we will make use of Figure 7.9, which shows the roll centre position in a left-hand turn when viewed from the rear; the wheel axle and tyres are assumed rigid.

The car's lateral acceleration a_y produces lateral forces on the sprung and unsprung masses given by

$$F_s = a_y m_s$$
$$F_u = a_y m_u$$

respectively; m_s and m_u are the car's sprung and unsprung masses respectively. The force on the sprung mass can be moved to the roll centre $\mathcal{C}$ by introducing the suspension moment

$$M_s = m_s(h - d)(a_y + g\varphi) \tag{7.51}$$

in which d and h are the heights of the roll centre and sprung mass respectively.

A moment balance around the projection of $\mathcal{C}$ on the ground shows that the load transfer has three terms that relate to the unsprung-mass lateral force, the sprung-mass lateral force, and the correcting moment M_s as given by

$$\Delta F_z = \frac{F_u r}{c} + \frac{F_s d}{c} + \frac{M_s}{c} \tag{7.52}$$

$$= \frac{F_u r}{c} + \frac{F_s h}{c} + \frac{m_s g(h - d)\varphi}{c}. \tag{7.53}$$

As the wheel radius r increases, the load transfer due to the unsprung mass increases; in some analyses the influence of the unsprung mass is neglected. As the height of the

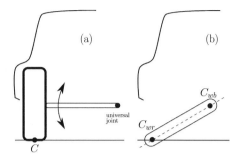

Figure 7.10: Roll centre in the case of a primitive rear suspension.

sprung mass centre increases, the load transfer due the sprung mass increases. The car rolls towards the outside of the turn when the mass centre of the sprung mass is above the roll centre, there is no roll when the mass centre is at the roll centre, and the car rolls towards the inside of the turn when the mass centre of the sprung mass is below the roll centre.

Roll centre—kinematic interpretation.   In the kinematic interpretation the roll centre is the instantaneous centre of rotation of the sprung mass with respect to the ground, or alternatively the centre of rotation of the unsprung mass with respect to the sprung mass when a pure moment is applied on the unsprung mass. Consider the simple (swing-axle) suspension system illustrated in Figure 7.10. In the (a) part of the figure we see the left-hand suspension rotating about a universal joint that is fixed to the differential, which in turn is fixed to the car body. The universal joint (replaceable by a revolute joint in this two-dimensional analysis) moves so as to keep the wheel in contact with the ground. In the (b) part of the figure the suspension, wheel, and wheel carrier are abstracted as a link that has instantaneous centres of rotation C_{wr} and C_{wb}. Centre C_{wr} is a point fixed in the road about which the wheel–axle unit rotates, which is usually assumed to be in the centre of the tyre contact patch. The link is also pinned to the differential and the car body and rotates around the point C_{wb}. We want to know the instantaneous centre C_{br} of the car body with respect to the road. The Aronhold–Kennedy theorem [252] says that this third instantaneous centre lies on the straight line joining C_{wr} and C_{wb}. By symmetry, the roll centre is at the intersection of this line and the car's vertical longitudinal plane of symmetry.

By making use of these ideas, the roll centres of a MacPherson, and a double-wishbone suspension can be found, as shown in Figure 7.11. The instantaneous centre of rotation of the wheel and wheel carrier in the road is the ground-contact point C, while the instantaneous centre of rotation of the wheel and wheel carrier in the car body is S. The car's roll centre is the intersection point of the line through C and S and the car's longitudinal plane of symmetry.

Car roll centre.   In this section we extend the axle-based notion of a roll centre to the entire car. Let us now assume that the vehicle's sprung mass is joined to the front and rear unsprung masses through two revolute joints, whose common axis of rotation is called the *roll axis* of the sprung mass. For present purposes the car's roll

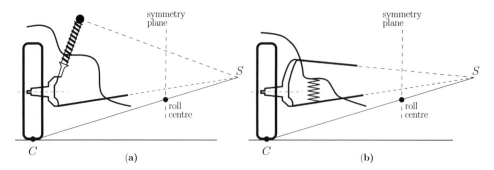

(a)   (b)

Figure 7.11: Roll centre in the case of (a) MacPherson strut and (b) double-wishbone suspension.

axis is assumed to pass through the roll centres of the front and rear axles. Referring to Figure 7.12, the roll axis has height d_F and d_R above the front and rear axles respectively, and height

$$d = \frac{d_F b + d_R (w - b)}{w} \tag{7.54}$$

above the projection of the mass centre of the sprung mass onto the ground plane.

Under the assumption that the roll angle φ and the roll axis inclination angle α are both small, a moment balance around $\mathcal{C}$ gives

$$m a_y (h - d) = (K - mg (h - d)) \varphi, \tag{7.55}$$

where m is the sprung mass (usually approximately the mass of the car; herein assumed equal), h is the height of the centre of mass, and K is the roll stiffness of the whole suspension system, which arises from the parallel combination of the front suspension stiffness K_F and rear suspension stiffness K_R:

$$K = K_F + K_R. \tag{7.56}$$

Equation (7.55) can be solved for the car roll angle

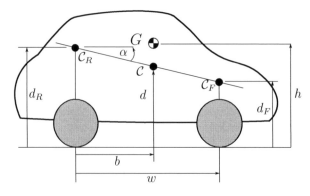

Figure 7.12: Car model for lateral load transfer.

$$\varphi = ma_y \frac{h-d}{K - mg\,(h-d)} = ma_y \frac{h-d}{\tilde{K}}, \tag{7.57}$$

where $\tilde{K}$ is the effective roll stiffness; we will again assume that $\tilde{K} \approx K$. Another related index is the roll rate, which is defined as $\partial\varphi/\partial a_y = m(h-d)/\tilde{K}$.

Moment balances for the front and rear unsprung masses with respect to the points C_F and C_R on the roll axis are

$$K_F\varphi = \Delta F_{zF}c_F - F_{yF}d_F \tag{7.58}$$
$$K_R\varphi = \Delta F_{zR}c_R - F_{yR}d_R, \tag{7.59}$$

in which c_F and c_R are the front- and rear-wheel track widths respectively, ΔF_{zF} and ΔF_{zR} are the front and rear lateral load transfers respectively, and F_{yF} and F_{yR} are the front and rear lateral forces, that is, $F_{yF} = F_{yFo} + F_{yFi}$ and $F_{yR} = F_{yRo} + F_{yRi}$, where 'i' and 'o' mean inner and outer respectively.[13] A lateral force balance together with a moment balance around the centre of mass give

$$F_{yF} = ma_y \frac{b}{w} \tag{7.60}$$

$$F_{yR} = ma_y \frac{w-b}{w}. \tag{7.61}$$

When substituting (7.57), (7.60), and (7.61) into (7.58) and (7.59) one obtains

$$\Delta F_{zF} = ma_y \left(\frac{K_F}{K}\frac{h-d}{c_F} + \frac{d_F}{c_F}\frac{b}{w} \right) \tag{7.62}$$

$$\Delta F_{zR} = ma_y \left(\frac{K_R}{K}\frac{h-d}{c_R} + \frac{d_R}{c_R}\frac{w-b}{w} \right). \tag{7.63}$$

The first terms on the right-hand sides relate to the roll stiffness distribution between the front and rear axles; the second terms are geometric.

The total lateral load transfer is computed by combining the front and rear load transfers in (7.62) and (7.63) to give

$$\Delta F_{zR}c_R + \Delta F_{zF}c_F = ma_y h, \tag{7.64}$$

or

$$\Delta F_z = ma_y \frac{h}{\tilde{c}} \tag{7.65}$$

with

$$\tilde{c} = \frac{\Delta F_{zR}c_R + \Delta F_{zF}c_F}{\Delta F_z}. \tag{7.66}$$

The constant $\tilde{c}$ is the equivalent wheel track width, with $\Delta F_z = \Delta F_{zR} + \Delta F_{zF}$ (in the case that $c_F = c_R = c$, $\tilde{c} = c$).

---

[13] An uneven lateral force distribution between the inner and outer tyre on an axle does not affect the derivation, since only the total lateral force on the axle matters.

To summarize, the total lateral load transfer is independent of the roll stiffness and depends only on the $h/\tilde{c}$ ratio; a reduced load transfer is obtained by lowering the car's mass centre and/or widening the car's track width. The load-transfer distribution between the front and rear axles is determined by the front and rear roll stiffness and some geometric parameters; harder suspensions lead to higher roll stiffness.

Jacking.   The lateral forces on the inner and outer tyres of an axle need not be equal. Suppose that the outer lateral tyre force is given by $F_{yo} = \bar{F}_y + \Delta F_y$, while the inner lateral tyre force is given by $F_{yi} = \bar{F}_y - \Delta F_y$, where $\bar{F}_y = (F_{yo} + F_{yi})/2$. This lateral force difference is associated with the possible 'lifting' of cars fitted with independent suspensions as a consequence of lateral load transfer (e.g. when turning); this effect is called *suspension jacking*. Taking moments around the velocity centre of the outer wheel carrier (see Figure 7.11) gives

$$\hat{M}_{so} = \Delta F_z b_{RC} - (\bar{F}_y + \Delta F_y)h_{RC}$$
$$= (\Delta F_z b_{RC} - \bar{F}_y h_{RC}) - \Delta F_y h_{RC}, \tag{7.67}$$

where $\hat{M}_{so}$ is the outer-suspension elastic moment relative to static conditions (positive in compression). The quantities b_{RC} and h_{RC} are the lateral distance and height of the roll centre with respect to the tyre contact point. Similarly, for the inner wheel,

$$\hat{M}_{si} = (\Delta F_z b_{RC} - \bar{F}_y h_{RC}) + \Delta F_y h_{RC}, \tag{7.68}$$

where $\hat{M}_{si}$ is the inner-suspension elastic moment (negative in compression). The first components in (7.67) and (7.68) are associated with the roll motion (identical for both suspensions), while the second component is associated with *jacking*, which results in both an extension of suspension and a reduction in the vehicle's track width.

Lateral acceleration limit.   We will now show that lateral load transfer is detrimental to a car's steady-cornering performance. A lateral force balance gives

$$ma_y = (F_{yFo} + F_{yRo}) + (F_{yFi} + F_{yRi})$$
$$= F_{yo} + F_{yi}, \tag{7.69}$$

where the F_{yo}s represent the lateral forces due to the outside tyres, while the F_{yi}s represent the lateral forces due to the inside tyres. The lateral force depends on the road–tyre friction coefficient μ and normal load; the friction coefficient usually reduces with the load, see equation (3.96). Under static conditions, the normal load F_z is split equally between the outer and inner pairs of tyres; longitudinal symmetry will be assumed here and so $F_{zo} = F_{zi} = F_z/2$. Equation (7.69) can be rewritten as

$$ma_y = \mu F_{zo} + \mu F_{zi} \tag{7.70}$$
$$= \mu_0 (1 - \alpha \Delta F_z) \left(\frac{F_z}{2} + \Delta F_z \right) + \mu_0 (1 + \alpha \Delta F_z) \left(\frac{F_z}{2} - \Delta F_z \right)$$
$$= \mu_0 \left(F_z - 2\alpha(\Delta F_z)^2 \right), \tag{7.71}$$

where μ_0 is the friction coefficient at nominal load and ΔF_z is the total lateral load transfer in (7.65); the lateral load transfer ΔF_z always reduces the maximum lateral acceleration:[14]

$$a_y = \frac{\mu_0(F_z - 2\alpha(\Delta F_z)^2)}{m}. \tag{7.72}$$

In the case that the friction coefficient is not load dependent, the lateral load transfer has no effect on the maximum lateral acceleration. In the case that aerodynamic loading is negligible (i.e. $F_z = mg$), the maximum acceleration expression becomes

$$\frac{a_y}{g} = \mu_0. \tag{7.73}$$

In other words, the tyre–road friction coefficient determines the maximum theoretical achievable lateral acceleration of the car.[15] Note that the same results have been obtained for the friction-related deceleration limit (7.14)

Cars with a large h/c ratio may have their lateral acceleration limited by *rollover* rather than the tyre friction limit (7.73). This limit may be estimated quasi-statically by taking limits around the heavily loaded tyre ground-contact point, while assuming no normal load on the other tyre. Setting $\Delta F_z = mg/2$ and $\tilde{c} = c$ in (7.65) gives

$$\frac{a_y}{g} = \frac{c}{2h}, \tag{7.74}$$

which shows that a large track width c and low mass-centre height h reduce the rollover propensity. More precise estimates of rollover limits can be obtained by modelling the suspension, steering, tyres, and road banking under dynamic conditions. Special attention must be given to rollover induced by a tyre hitting an object such as a curb—the so-called *tripped* conditions [113, 253, 254].

Example 7.3 Suppose a car with has wheelbase $w = 2.5\,$m, mass $m = 1,300\,$kg, centre-of-mass height $h = 0.5\,$m, distance of mass centre from the rear axle $b = 1.3\,$m, track width $c = c_F = c_R = 1.6\,$m, roll stiffness balance $K_F/K = 0.55$, front-axle roll stiffness $K_F = 33\,$kNm/rad, roll axis height $d = d_F = d_R = 0.1\,$m, tyre friction coefficient $\mu_0 = 1$, and $\alpha = 10^{-4}\,$N^{-1}. Compute the lateral load transfer under steady turning with a lateral acceleration of $a_y = 0.5\,g$. The normal load on each of the front tyres is 3,314 N, while that on the rear tyres is 3,060 N; use half of the first components on the right-hand sides of (7.5) and (7.6). The total lateral load transfer is 1,992 N (7.65); the lateral load transfer on the front axle is 1,087 N (7.62) while that on the rear axle is 905 N (7.63). The roll angle of the car is $\varphi = 2.7^\circ$ (7.57) and the roll rate is $\partial\varphi/\partial a_y = 5.3^\circ/g$. The lateral acceleration limit is $a_y = 0.83\,g$ and is given by (7.72) and (7.65)—in this condition $\Delta F_z = 3,303\,$N. This figure becomes $a_y = 1\,g$ (7.73) in the case that the tyre friction coefficient is not load dependent.

---

[14] The coefficient α is related to p_{y4} in (3.96) by $\alpha = -p_{y4}/p_{y3}F_{z0}$, where $p_{y3} = \mu_0$ and $p_{y4} < 0$ are the coefficients related to the outer or inner pairs of tyres, while $F_{z0} = F_{zo} = F_{zi}$.

[15] We call this lateral acceleration 'theoretical', because it assumes that all the tyres reach the friction limit simultaneously. In practice the first tyre that reaches the friction limit sets the maximum acceleration limit.

7.2.4 Roll trim in motorcycles

The equilibrium roll angle of a motorcycle under steady turning conditions, that is, a constant speed V and constant lateral acceleration a_y, is considered next. To keep the analysis simple, aerodynamic and gyroscopic influences are neglected, and the front and rear tyres are assumed aligned with the central plane of the chassis.

Referring to the left-hand part of Figure 7.13, in the case of thin-disk tyres, a moment balance around the ground-contact point is

$$0 = -ma_y h \cos\varphi + mgh \sin\varphi, \tag{7.75}$$

where $a_y = V^2/R_G$ is the lateral acceleration; and V the velocity of the vehicle's mass centre and R_G the distance between the centre of mass and the centre of the turn. The height of the centre of mass is h, φ is the roll angle, m is the mass of the vehicle and rider, and g is the gravity. Equation (7.75) can be solved for the equilibrium roll angle to give

$$\varphi_0 = \arctan\frac{a_y}{g}. \tag{7.76}$$

In the case of finite-width tyres with crown radius $\rho_F = \rho_R = \rho$, the roll balance becomes

$$0 = -ma_y\left(\rho + (h - \rho)\cos\varphi\right) + mg\left(h - \rho\right)\sin\varphi, \tag{7.77}$$

which can be written in the standard form

$$A\sin x + B\cos x + C = 0. \tag{7.78}$$

The various constants are given by

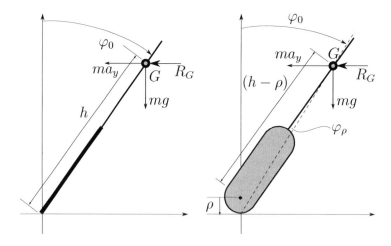

Figure 7.13: Motorcycle models for roll trim: thin-disk tyre (left) and finite-width toroidal tyre (right).

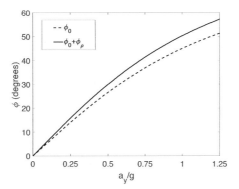

Figure 7.14: Roll angle as a function of the lateral acceleration in the case of thin-disk tyres (dashed line) and finite-width tyres (solid line). The parameters used are $\rho = 0.075\,m$ and $h = 0.650\,m$.

$$A = mg\,(h - \rho) \qquad B = -ma_y\,(h - \rho) \qquad C = -ma_y\rho \qquad x = \varphi. \qquad (7.79)$$

Equation (7.78) has solution[16]

$$\varphi = \arctan \frac{a_y}{g} + \arcsin \frac{a_y\rho}{(h - \rho)\sqrt{g^2 + a_y^2}} \qquad (7.81)$$

$$= \varphi_0 + \varphi_\rho.$$

A comparison between (7.76) and (7.81) shows that for a given lateral acceleration a_y, the roll angle increases as the tyre crown radius increases; see Figure 7.14. The equilibrium roll angle is independent of the height of the loaded vehicle mass centre (motorcycle plus rider) in the case of thin-disk tyres. In the case of finite-width tyres, the roll angle increases as the mass-centre height reduces.

---

[16] The solution is obtained normalizing by $\sqrt{A^2 + B^2}$ and introducing the variable β as follows:

$$\frac{A}{\sqrt{A^2 + B^2}}\sin x + \frac{B}{\sqrt{A^2 + B^2}}\cos x = \frac{-C}{\sqrt{A^2 + B^2}}$$

$$\cos \beta \sin x + \sin \beta \cos x = \frac{-C}{\sqrt{A^2 + B^2}}$$

$$\sin (x + \beta) = \frac{-C}{\sqrt{A^2 + B^2}}$$

which has solution

$$x = -\beta + \arcsin \frac{-C}{\sqrt{A^2 + B^2}}$$

$$= \arctan (-B, A) + \arcsin \frac{-C}{\sqrt{A^2 + B^2}}. \qquad (7.80)$$

We conclude this section by noting that in steady-state cornering the vehicle weight and the centrifugal force combine to produce a suspension load of $mg/\cos\varphi > mg$, and thus an additional suspension compression is expected under cornering.

7.3   Road modelling

Road models are required in a number of applications areas including virtual reality environments, transportation management and planning systems, and automated vehicle navigation systems. These models are also required in vehicular optimal control problems so that road-related path constraints can be incorporated into the problem [120, 255]. Road-related path constraints are enforced by ensuring that the vehicle remains in contact with the road and that its mass centre, or its road wheels, remain within the track boundaries. In the treatment given here *ribbons* or *strips* are central to our road models and provide a differential-geometric description of general navigable road surfaces [256]. We will assemble a number of geometric ideas that will be used to describe three-dimensional roads. All of these concepts can be found in the classical differential geometry literature; see for example [257–259]. We will also make use of the properties of rotating frames as given in Section 2.5.

7.3.1   Three-dimensional curves

Consider the parametric representation of the *arc of a curve*

$$\mathcal{C} = \left\{ \mathbf{x}(s) = [x(s)\ y(s)\ z(s)]^T \in \mathbb{R}^3 : s \in [s_0, s_f] \right\}, \tag{7.82}$$

in which the parameter s is the *arc length* of $\mathcal{C}$. We assume that each element of $d^3\mathbf{x}/ds^3$ is continuous and that $d^2\mathbf{x}/ds^2$ is not identically 0 for any s. The curve is deemed *simple* if any point on the curve corresponds to a single value of s.

A *moving trihedron* associated with $\mathcal{C}$ is defined by a triple of unit vectors $\mathbf{t}(s)$, $\mathbf{p}(s)$, and $\mathbf{b}(s)$ that form a right-handed coordinate system with (moving) origin $\mathbf{x}(s)$. The unit vector $\mathbf{t}$ is tangent to $\mathcal{C}$ and is defined by $\mathbf{t} = \mathbf{x}'$; the prime represents a derivative with respect to the arc length s. Since $\mathbf{t} \cdot \mathbf{t} = 1$, $\mathbf{t} \cdot \mathbf{t}' = 0$ and so $\mathbf{t}'$ is orthogonal to $\mathbf{t}$. The unit vector $\mathbf{p}$ is the *principal normal* to $\mathcal{C}$ and is defined as $\mathbf{p} = \mathbf{t}'/|\mathbf{t}'|$. The unit vector $\mathbf{b}$ is called the *binormal* to the curve and is defined by the vector product $\mathbf{b} = \mathbf{t} \times \mathbf{p}$. A pictorial representation of $\mathcal{C}$ and its moving trihedron is shown in Figure 7.15. The *curvature vector* of $\mathcal{C}$ is defined as the rate of change of the tangent vector, $\boldsymbol{\kappa} = \mathbf{t}'$. The *curvature* is the scalar product $\kappa = \boldsymbol{\kappa} \cdot \mathbf{p}$, which is a measure of the deviation of $\mathcal{C}$ from a straight line; it follows that $\boldsymbol{\kappa} = \kappa\mathbf{p}$, where $\kappa = |\mathbf{t}'|$. The *radius of curvature* is defined as $\rho = 1/\kappa$. The *centre of curvature* is given by $\mathbf{c} = \mathbf{x} + \rho\mathbf{p}$. The *torsion* of $\mathcal{C}$ measures the direction and extent to which the curve deviates from its *osculating plane* $\mathbf{t}\,\mathbf{p}$. The torsion is defined as $\tau = -\mathbf{p} \cdot \mathbf{b}'$.

The orientation of the moving trihedron is defined by the three-dimensional rotation matrix

$$\mathcal{R} = \begin{bmatrix} \mathbf{t}\ \mathbf{p}\ \mathbf{b} \end{bmatrix}. \tag{7.83}$$

The curvature and torsion define the *angular rate* of the moving trihedron through the kinematic equations known as the *Frenet–Serret formulae*

$$\mathcal{R}' = \mathcal{R} \begin{bmatrix} 0 & -\kappa & 0 \\ \kappa & 0 & -\tau \\ 0 & \tau & 0 \end{bmatrix} = \mathcal{R}S(\mathbf{\Omega}^{\mathcal{C}}); \qquad (7.84)$$

which is obtained using (2.169), with the skew-symmetric matrix S defined in (2.154) and $\mathbf{\Omega}^{\mathcal{C}}$ the moving trihedron's angular rate (in radians per metre and *not* radians per second!). Using (2.146), each column of (7.84) can be interpreted kinematically as $\mathbf{v} = \mathbf{\Omega}^{\mathcal{C}} \times \mathbf{x}$, where $\mathbf{x}$ is a position of the origin of the moving trihedron. By direct calculation using (7.83) and (7.84), $\mathbf{t}' = \kappa\mathbf{p}$, $\mathbf{p}' = \tau\mathbf{b} - \kappa\mathbf{t}$, and $\mathbf{b}' = -\tau\mathbf{p}$.

The angular rate vector of the moving trihedron is known as the *Darboux* vector and is given by

$$\mathbf{\Omega}^{\mathcal{C}} = \tau\mathbf{t} + \kappa\mathbf{b}. \qquad (7.85)$$

Once the curvature vector $\mathbf{\Omega}^{\mathcal{C}}$ is given as a function of the curvilinear abscissa s, by imposing initial conditions on the position $\mathbf{x}(s_0)$ and the orientation $\mathcal{R}(s_0)$, the curve $\mathcal{C}$ is generated by

$$\begin{cases} \mathcal{R} = \displaystyle\int_{s_0}^{s} \mathcal{R}' \, ds, \\ \mathbf{x} = \displaystyle\int_{s_0}^{s} \mathbf{t} \, ds, \end{cases} \qquad (7.86)$$

in which $\mathbf{t}$ is the first column of $\mathcal{R}$ in (7.83).

7.3.2 Ribbons

We will represent the road by a geometric object called a *ribbon*, which can be created by augmenting the curve $\mathcal{C}$ with notions of 'width' and 'twist'. Both of these quantities will be defined in terms of a unit *camber* vector $\mathbf{n}$, which lies in the plane of the ribbon

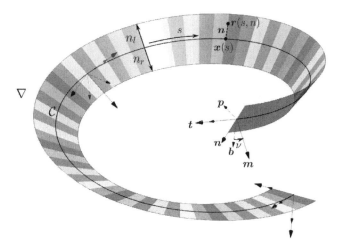

Figure 7.15: A road represented by a ribbon ∇ that is generated by a spine curve $\mathcal{C}$.

and is orthogonal to the tangent vector $\mathbf{t}$; see Figure 7.15. We will represent the ribbon parametrically as:

$$\nabla = \left\{ \mathbf{r}(s,n) = \mathbf{x}(s) + \mathbf{n}(s)n \in \mathbb{R}^3 : s \in [s_0, s_f], n \in [n_l(s), n_r(s)] \right\}, \tag{7.87}$$

in which $\mathbf{x}$ is the ribbon's *spine curve*, s is the arc length, and n is the lateral offset in the ribbon plane; the origin of the moving trihedon travels along $\mathbf{x}$. The left- and right-hand road boundaries are described by n_l and n_r respectively. Lateral displacements on the ribbon surface are measured from the spine in the direction of unit vector $\mathbf{n}$. The width of the ribbon at each s is given by $|n_r - n_l|$. This parameterization will accommodate variable-width ribbons that need not be symmetric around the spine curve $\mathcal{C}$. If $\mathbf{r}(s,n)$ is a point in ∇, then we will assume that there is a one-to-one mapping between this point and the parameter domain. Curves on ∇ corresponding to constant s, or constant n, are called *coordinate curves*; the coordinate curve corresponding to $n = 0$ is the spine of the ribbon. The spine is modelled as a simple curve $\mathcal{C}$ and can be described in terms of the geometric entities introduced in Section 7.3.1. The camber vector lies in the plane normal to the spine's tangent vector $\mathbf{t}$ and can therefore be described in terms of $\mathbf{p}$, $\mathbf{b}$, and a *twist angle* ν (around $\mathbf{t}$) as follows

$$\mathbf{n} = \mathbf{p}\cos\nu + \mathbf{b}\sin\nu. \tag{7.88}$$

The *twist rate* is ν'. A unit vector $\mathbf{m}$ that is normal to the ribbon surface is defined to complete a right-handed axis system with $\mathbf{t}$ and $\mathbf{n}$, whence $\mathbf{m} = \mathbf{t} \times \mathbf{n}$. See Figure 7.15 for an illustration of a ribbon and these quantities.

The Frenet–Serret formulae (7.84) can be used to describe the spine, but they do not have the flexibility to describe the ribbon itself. Developing again a kinematic interpretation of the problem, a moving triad $\mathbf{t}\,\mathbf{n}\,\mathbf{m}$, which we call the *ribbon trihedron*, is used to describe the evolution of the spine and the ribbon itself—note that the spine trihedron was previous defined using the moving triad $\mathbf{t}\,\mathbf{p}\,\mathbf{b}$. The ribbon trihedron is sometimes called the *Darboux frame* of ∇ on $\mathcal{C}$.

The angular rate of the reference frame fixed to the ribbon trihedron is given by the sum of the Darboux vector (7.85) and the twist rate ν'

$$\Omega^{\mathcal{C}} = [\tau + \nu']\,\mathbf{t} + \kappa\mathbf{b}. \tag{7.89}$$

The ribbon angular rate vector Ω can be expressed in the ribbon trihedron through a coordinate transformation from the curve triad $\mathbf{t}\,\mathbf{p}\,\mathbf{b}$ to the ribbon triad $\mathbf{t}\,\mathbf{n}\,\mathbf{m}$ using $\mathcal{R}(e_x, \nu)^T$ given by (2.157)

$$\Omega^{\nabla} = \begin{bmatrix} 1 & 0 & 0 \\ 0 & \cos\nu & \sin\nu \\ 0 & -\sin\nu & \cos\nu \end{bmatrix} \begin{bmatrix} \tau + \nu' \\ 0 \\ \kappa \end{bmatrix} = \begin{bmatrix} \tau + \nu' \\ \kappa\sin\nu \\ \kappa\cos\nu \end{bmatrix}, \tag{7.90}$$

where Ω^{∇} is expressed in the ribbon trihedron or Darboux frame. The angular rate vector Ω^{∇} is thus a vector in $\mathbb{R}^3$, which has a skew-symmetric matrix representation $S(\Omega^{\nabla})$:

$$\Omega^{\nabla} = \begin{bmatrix} \Omega_x \\ \Omega_y \\ \Omega_z \end{bmatrix}, \quad S(\Omega^{\nabla}) = \begin{bmatrix} 0 & -\Omega_z & \Omega_y \\ \Omega_z & 0 & -\Omega_x \\ -\Omega_y & \Omega_x & 0 \end{bmatrix}, \tag{7.91}$$

where Ω_x, Ω_y, and Ω_z are known as the *relative torsion*, the *normal curvature*, and the *geodesic curvature*, respectively. The matrix $S(\mathbf{\Omega}^\nabla)$, also known as the *Cartan matrix* [260], is a generalization of the Frenet–Serret formulae to the case of ribbons and

$$\mathcal{R}' = \mathcal{R}S(\mathbf{\Omega}^\nabla). \tag{7.92}$$

As before, once the generalized curvature vector $\mathbf{\Omega}^\nabla$ and the widths n_r, n_l are given as a function of the curvilinear abscissa s, the ribbon can be computed by integration, given the initial conditions $\mathbf{x}(s_0)$ and $\mathcal{R}(s_0)$:

$$\begin{cases} \mathcal{R} = \displaystyle\int_{s_0}^{s} \mathcal{R}' \, ds, \\ \mathbf{r} = \displaystyle\int_{s_0}^{s} \mathbf{t} \, ds + n\mathbf{n}, \end{cases} \tag{7.93}$$

in which $\mathcal{R}'$ is given by (7.92). The generalized Frenet–Serret equations are used as a differential-geometric description of ribbons in terms of their generalized curvature vector $\mathbf{\Omega}^\nabla$ and the width parameters.

7.3.3 Euler angles

The ribbon-based model of the road is given in (7.93), with the evolution of the ribbon trihedron as a function of arc length coming from integrating a 3×3 rotation matrix $\mathcal{R}$. As explained in Section 2.5, $\mathcal{R}$ can be expressed using three rotations—the related angles are the Euler angles, which we herein define as θ, μ, and ϕ. As in most mechanics problems, the idea is to replace the 3×3 matrix $\mathcal{R}$ in (7.92) with three scalar differential equations in the Euler angles.

The relationships between the θ, μ, and ϕ and κ, τ, and ν are obtained when noting that the ribbon angular rate given by $\Omega^{\mathcal{C}}$ (7.89) or Ω^∇ (7.90) can also be expressed in terms of the Euler angles using (2.166) or (2.169) respectively, with $\mathcal{R} = \mathcal{R}(\theta, \mu, \phi)$. In the z–y–x (yaw, pitch, and roll) case,

$$\mathcal{R} = \mathcal{R}(e_z, \theta)\mathcal{R}(e_y, \mu)\mathcal{R}(e_x\phi) = \begin{bmatrix} c_\theta c_\mu & c_\theta s_\mu s_\phi - s_\theta c_\phi & c_\theta s_\mu c_\phi + s_\theta s_\phi \\ s_\theta c_\mu & s_\theta s_\mu s_\phi + c_\theta c_\phi & s_\theta s_\mu c_\phi - c_\theta s_\phi \\ -s_\mu & c_\mu s_\phi & c_\mu c_\phi \end{bmatrix}, \tag{7.94}$$

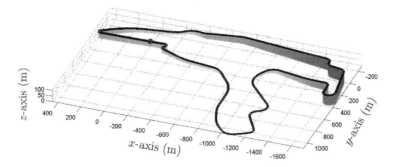

Figure 7.16: Ribbon-based model of the 7 km Circuit de Spa-Francorchamps: the start/finish line is marked with a black circle.

with $\mathcal{R}(e_i, \cdot)$ given by (2.156)–(2.158). The following relationships are obtained from $S(\mathbf{\Omega}^\nabla) = \mathcal{R}^T\mathcal{R}'$:

$$\theta' = \kappa\frac{\cos(\nu - \phi)}{\cos\mu} \tag{7.95}$$

$$\mu' = \kappa\sin(\nu - \phi) \tag{7.96}$$

$$\phi' = \tau + \nu' + \kappa\tan\mu\cos(\nu - \phi). \tag{7.97}$$

Using (7.95)–(7.97), the kinematic parameterization of the ribbon (7.93) becomes

$$\begin{cases} \theta(s) &= \displaystyle\int_{s_0}^s \theta'\,ds, \\[2mm] \mu(s) &= \displaystyle\int_{s_0}^s \mu'\,ds, \\[2mm] \phi(s) &= \displaystyle\int_{s_0}^s \phi'\,ds, \\[2mm] \mathbf{r}(s, n) &= \displaystyle\int_{s_0}^s \mathbf{t}\,ds + n\mathbf{n}, \end{cases} \tag{7.98}$$

where $\mathbf{t}$ and $\mathbf{n}$ are the first and second columns, respectively, of the rotation matrix $\mathcal{R}$ as given in (7.94).

The initial conditions $\theta(s_0)$, $\mu(s_0)$, and $\phi(s_0)$ are found by solving the inverse problem for a given $\nabla(s_0)$, for example, $\tan\theta = \mathcal{R}_{21}/\mathcal{R}_{11}$, $\tan\phi = \mathcal{R}_{32}/\mathcal{R}_{33}$, and $\sin\mu = -\mathcal{R}_{31}$, where $\mathcal{R}_{ij}$ means row i and column j of $\mathcal{R}$.

Example 7.4 The differential geometric properties of the three-dimensional ribbon-based model for the Circuit de Spa-Francorchamps (Spa) is shown in Figure 7.16. The track features were estimated using measured GPS data for the road boundaries [256]. The model is then computed by solving an optimal control problem of the kind described in Section 8.6.7. Figure 7.17 (a) shows the track's inclination angle (μ) and

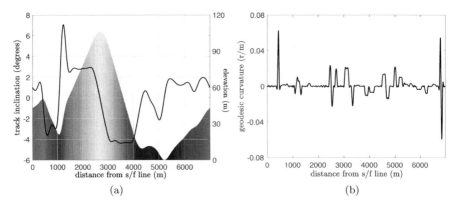

Figure 7.17: Track inclination and elevation (a) and geodesic curvature (b) of the Circuit de Spa-Francorchamps as functions of the distance from the start/finish line.

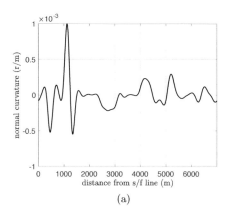
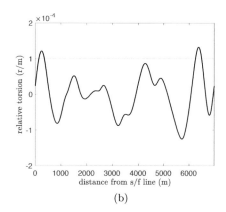

Figure 7.18: Normal curvature (a) and relative torsion (b) of the Circuit de Spa-Francorchamps as functions of the distance from the start/finish line.

spine elevation as a function of elapsed distance s. The start/finish line is approximately 43 m above the track's lowest point, with the full-track elevation change being approximately 106 m. Figure 7.17 (b) shows the track's geodesic curvature as a function of elapsed distance s. The first positive spike in the geodesic curvature Ω_z corresponds to the sharp right-hand hairpin (La source) following the start/finish line (the track is raced in a clockwise direction). Notable also is the positive–negative doublet in the geodesic curvature at approximately 6,825 m, which corresponds to the right–left chicane immediately before the start/finish line; the other track corners can be identified in much the same way. The normal curvature Ω_y describes the way in which the track elevation changes with elapsed distance, while the relative torsion Ω_x describes the way in which the track camber varies with elapsed distance. These curves are shown in Figure 7.18, with the track's camber (ϕ) shown in Figure 7.19.

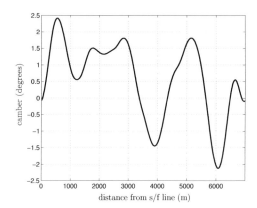

Figure 7.19: Camber angle of Circuit de Spa-Francorchamps as a function of the distance from the start/finish line.

7.4   Vehicle Positioning

The track and vehicle (bicycle, car, or motorcycle) kinematics can be modelled using the differential-geometric framework developed in Section 7.3. We assume that the road surface is a plane that is orthogonal to the road-surface-normal vector m as defined in Section 7.3.2 and Figure 7.15. This ribbon-derived plane will move with the vehicle and generates a road surface that heaves, pitches, and rolls under the vehicle.

A track segment is illustrated in Figure 7.20 and shows an inertial axis system, a point s (the curvilinear abscissa) that travels with the car along the spine of the track, and a vector n, which is perpendicular to the tangent vector t and that points towards the projection on the ground plane of a fixed point G_p on the vehicle—this point could be the vehicle's mass centre, for example.

The road normal m is perpendicular to t and n and defines the moving road tangent plane—the ribbon Darboux frame. The magnitude of n is the perpendicular distance between the track's spine and the vehicle's fixed point G_p. The angle ξ is the yaw angle of the vehicle relative to the spine of the track t. The track's spatial curvature vector or angular rate (in radians per metre) is given by (7.89) or (7.90). The track's angular velocity vector is given by $\boldsymbol{\omega} = \boldsymbol{\Omega}\frac{ds}{dt}$, where $\boldsymbol{\omega} = [\omega_x\,\omega_y\,\omega_z]^T$; in the unit-speed case $\boldsymbol{\omega} = \boldsymbol{\Omega}$, since $\frac{ds}{dt} = 1$.

The next kinematic relationships we will require relate to the way in which the vehicle progresses down the road. Suppose that the absolute velocity of the vehicle point G_p in its body-fixed coordinate system is $\boldsymbol{v} = [u\,v\,w]^T$—these variables will derive from the integration of the equation of motion; see Section 4.2. The distance between the vehicle's fixed point G_p and the origin of the Darboux frame is $\boldsymbol{n} = [0\,n\,0]^T$ in Darboux coordinates assuming that G_p lies in the plane of the road.

We can now express the absolute velocity of G_p in the Darboux frame using (2.144)

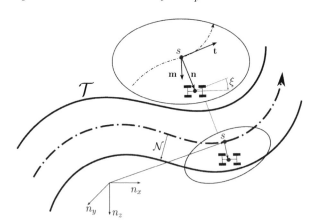

Figure 7.20: Differential-geometric description of a track segment $\mathcal{T}$. The track half-width is $\mathcal{N}$, with ξ the vehicle's yaw angle relative to the spine's tangent direction.

$$\begin{bmatrix} \dot{s} \\ 0 \\ 0 \end{bmatrix} + \boldsymbol{\omega} \times \boldsymbol{n} + \begin{bmatrix} 0 \\ \dot{n} \\ 0 \end{bmatrix} = \mathcal{R}(\boldsymbol{e}_z, \xi)\boldsymbol{v}, \tag{7.99}$$

where the first term on the left-hand side is the velocity of the Darboux frame origin, the second term is the transferred velocity and the third term is the relative velocity of G_p in the Darboux frame. The right-hand-side term is the absolute velocity of G_p, where $\mathcal{R}(\boldsymbol{e}_z, \xi)$ is given by (2.156) and represents the yawing of the vehicle relative to the spine of the track. Equation (7.99) can be rewritten as follows:

$$\begin{bmatrix} \dot{s} \\ \dot{n} \\ 0 \end{bmatrix} = \begin{bmatrix} n\omega_z + u\cos\xi - v\sin\xi \\ u\sin\xi + v\cos\xi \\ w - n\omega_x \end{bmatrix}. \tag{7.100}$$

The first row of (7.100) gives the speed of the origin of the Darboux frame in its tangent direction and can be written as

$$\dot{s} = \frac{u\cos\xi - v\sin\xi}{1 - n\Omega_z}, \tag{7.101}$$

since $\omega_z = \dot{s}\Omega_z$ in the variable-speed case. This equation is used to generate $S_f(s) = \frac{dt}{ds}$, which transforms 'time' as the independent variable into the 'elapsed distance' as the independent variable. In order to fulfil this function, $S_f(s)$ and its inverse must be non-zero everywhere on the spine curve. The second row of (7.100) describes the way in which the vehicle moves normal to the spine:

$$\dot{n} = u\sin\xi + v\cos\xi. \tag{7.102}$$

The third row of (7.100) describes the absolute vertical velocity of G_P as a result of the road camber changes, and follows from the fact that the vehicle cannot have a vertical velocity component in the $\boldsymbol{m}$ direction. This gives

$$w = n\,\omega_x. \tag{7.103}$$

If the 'distance travelled' is selected as the independent variable, then a complete set of kinematics equations includes (7.101), which generates $S_f(s)$, (7.95)–(7.97), and

$$\frac{dn}{ds} = S_f(s)\,(u\sin\xi + v\cos\xi) \tag{7.104}$$

which comes from (7.102).

The angular velocity of the vehicle in its body-fixed frame $\bar{\boldsymbol{\omega}}$ is equal to the angular velocity of the road expressed in the vehicle-fixed frame, that is, $\mathcal{R}(\boldsymbol{e}_z, \xi)^T\boldsymbol{\omega}$, plus the angular velocity of the vehicle relative to the road, again expressed in the vehicle-fixed frame, that is, $[0, 0, \dot{\xi}]^T$—the summation rule holds for angular velocities; Section 2.5. The following expression is thus obtained

$$\bar{\boldsymbol{\omega}} = \mathcal{R}(\boldsymbol{e}_z, \xi)^T\boldsymbol{\omega} + \begin{bmatrix} 0 \\ 0 \\ \dot{\xi} \end{bmatrix} \tag{7.105}$$

which gives

$$
\begin{bmatrix} \bar{\omega}_x \\ \bar{\omega}_y \\ \bar{\omega}_z \end{bmatrix} = \begin{bmatrix} \omega_x \cos \xi + \omega_y \sin \xi \\ \omega_y \cos \xi - \omega_x \sin \xi \\ \omega_z + \dot{\xi} \end{bmatrix}.
\tag{7.106}
$$

We will use the vehicle dynamics equations to find $\bar{\omega}_z$, with the vehicle yaw angle ξ deduced from the third row of (7.106) by integrating

$$
\frac{d\xi}{ds} = S_f(s)\bar{\omega}_z - \Omega_z.
\tag{7.107}
$$

Summarizing, the road is described in terms of the generalized curvature vector $\Omega(s)$, that is, the curvature $\kappa(s)$, the torsion $\tau(s)$, and the twist $\nu(s)$. The position and orientation of the vehicle on the road are given in terms of $n(s)$ and $\xi(s)$ by integrating (7.104) and (7.107)—u, v, w, and $\bar{\omega}_z$ will come from the integration of the vehicle's equations of motion. Since the car has no suspension system, the vehicle's angular velocities $\bar{\omega}_x$ and $\bar{\omega}_y$ depend on the road angular velocities ω_x, ω_y, and ξ; see (7.106). Examples of simulation of the combined road-vehicle model will be shown in Chapter 9 in the context of minimum time manoeuvring.

7.5   Car modelling

One of the early references relating to car dynamic models is [261]. An early model that predicts the steering response of a road car is developed in [262], where steering oscillations are also discussed. This paper shows a high level of agreement between steady-state measurements and theoretical calculations. An analysis of the influence of tyre characteristics on the behaviour of a car undergoing manoeuvres is given in [263]. Simulation results are used to assess the influence of variations in the lateral and longitudinal tyre stiffnesses, and the coefficient of friction at the tyre–road interface on the vehicle's steering and braking responses. Overviews of motor-vehicle dynamics modelling are available in [206, 264].

In this section we derive the equations of motion of a four-wheeled vehicle model that is suitable for optimal control calculations on a three-dimensional road of the type described in Section 7.3. The model includes slipping tyres and aerodynamic maps that depend on a quasi-static suspension model. The equations describing the dynamics of the car can be derived using either the Newton–Euler (Section 2.2.2) or the Lagrange equations (Section 2.2.8).

It follows from (2.144) and (7.103) that the absolute velocity of the car's mass centre (expressed on the vehicle's body-fixed coordinate system) is given by

$$
\boldsymbol{v}_B = \boldsymbol{v} + \bar{\boldsymbol{\omega}} \times \boldsymbol{h}
\tag{7.108}
$$

$$
= \begin{bmatrix} u \\ v \\ n\omega_x \end{bmatrix} + \begin{bmatrix} \bar{\omega}_x \\ \bar{\omega}_y \\ \bar{\omega}_z \end{bmatrix} \times \begin{bmatrix} 0 \\ 0 \\ -h \end{bmatrix} = \begin{bmatrix} u - h\bar{\omega}_y \\ v + h\bar{\omega}_x \\ n\omega_x \end{bmatrix},
\tag{7.109}
$$

where $\boldsymbol{v}$ is the velocity of the origin G_p of the vehicle-fixed frame of Section 7.4, $\bar{\boldsymbol{\omega}}$ the vehicle angular velocity, and h is the height of the vehicle mass centre above the ground. The vertical velocity w of the vehicle-fixed frame has been written in terms of the road angular velocity component ω_x and the lateral position with respect to the centre line n.

Newton–Euler equations require the computation of the accelerations of the car. These are derived from the velocities $\boldsymbol{v}_B$ and $\bar{\boldsymbol{\omega}}$, which are both defined in the body-fixed (non-inertial) frame. Application of (2.6) and (2.11) using (2.10) gives[17]

$$M\left(\dot{\boldsymbol{v}}_B + \bar{\boldsymbol{\omega}} \times \boldsymbol{v}_B\right) = F_B + Mg\mathcal{R}^T\boldsymbol{e}_z \tag{7.110}$$

$$I_B\dot{\bar{\boldsymbol{\omega}}} + \bar{\boldsymbol{\omega}} \times (I_B\bar{\boldsymbol{\omega}}) = M_B, \tag{7.111}$$

in which M is the car's mass and I_B its inertia with respect to the mass centre in the body-fixed coordinate system (assumed to be diagonal $I_B = \mathrm{diag}(I_x\ I_y\ I_z)$), $F_B = [F_x\ F_y\ F_z]^T$ and $M_B = [M_x\ M_y\ M_z]^T$ are the external force and moment, and $\mathcal{R} = \mathcal{R}(\boldsymbol{e}_z, \theta)\mathcal{R}(\boldsymbol{e}_y, \mu)\mathcal{R}(\boldsymbol{e}_x, \phi)\mathcal{R}(\boldsymbol{e}_z, \xi)$ represents the orientation of the vehicle-fixed frame with respect to the ground, with θ, μ, ϕ the Euler angles related to the road Darboux frame (Section 7.3.3) and ξ the relative angle between the car and the road (Section 7.4). The gravitational acceleration acting on the car's mass centre is

$$g\mathcal{R}^T\boldsymbol{e}_z = g\begin{bmatrix} \sin\xi\sin\phi\cos\mu - \cos\xi\sin\mu \\ \sin\xi\sin\mu + \cos\xi\sin\phi\cos\mu \\ \cos\phi\cos\mu \end{bmatrix}. \tag{7.112}$$

The car's equations of motion can now be assembled from (7.110), (7.111), and (7.112) as follows:

$$\dot{u} = (v + h\bar{\omega}_x)\bar{\omega}_z - n\omega_x\bar{\omega}_y + h\dot{\bar{\omega}}_y$$
$$+ g\left(\sin\xi\sin\phi\cos\mu - \cos\xi\sin\mu\right) + F_x/M \tag{7.113}$$

$$\dot{v} = n\omega_x\bar{\omega}_x - (u - h\bar{\omega}_y)\bar{\omega}_z - h\dot{\bar{\omega}}_x$$
$$+ g\left(\sin\xi\sin\mu + \cos\xi\sin\phi\cos\mu\right)) + F_y/M \tag{7.114}$$

$$\dot{\bar{\omega}}_z = \left((I_x - I_y)\bar{\omega}_x\bar{\omega}_y + M_z\right)/I_z, \tag{7.115}$$

in which F_x, F_y, and M_z are, respectively, the longitudinal and lateral tyre and aero-dynamic forces, and the z-axis tyre moment acting on the car. These quantities are illustrated in Figure 7.21 and are given by

$$F_x = \cos\delta(F_{frx} + F_{flx}) - \sin\delta(F_{fry} + F_{fly}) + F_{rrx} + F_{rlx} + F_{ax} \tag{7.116}$$

$$F_y = \cos\delta(F_{fry} + F_{fly}) + \sin\delta(F_{frx} + F_{flx}) + F_{rry} + F_{rly} \tag{7.117}$$

$$M_z = a\left(\cos\delta(F_{fry} + F_{fly}) + \sin\delta(F_{frx} + F_{flx})\right) - b(F_{rry} + F_{rly})$$
$$+ w_f\left(\sin\delta(F_{fry} - F_{fly}) + \cos\delta(F_{flx} - F_{frx})\right)$$
$$+ w_r\left(F_{rlx} - F_{rrx}\right), \tag{7.118}$$

where F_{ax} is the aerodynamic drag force (more detail will be given in Section 7.5.6).

---

[17] The transformation of the time derivative between inertial and body-fixed reference frames is (chapter IV in [42])

$$\left.\frac{d\,\cdot}{dt}\right|_I = \left.\frac{d\,\cdot}{dt}\right|_B + \bar{\boldsymbol{\omega}} \times \cdot$$

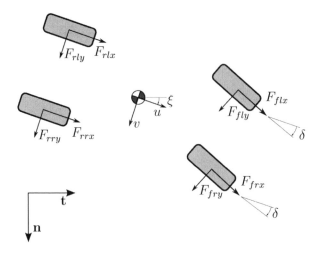

Figure 7.21: Tyre force system. The car's yaw angle is ξ with respect to the Darboux frame, which is defined in terms of the vectors t and n.

If spinning road wheels are included in the model, they can be described by equations of the form

$$I_{wij}\dot{\omega}_{ij} = T_{ij} - R_i F_{ijx} \qquad i = f, r \qquad j = l, r. \tag{7.119}$$

The moment of inertia of each wheel about its axis of rotation is I_{wij}, the radius of each wheel is R_i; the wheel torque is T_{ij} (positive when propulsive and negative when braking) and the wheel angular velocities are given by ω_{ij} ($i = f, r$ $j = l, r$).

The time domain equations of motion (7.113), (7.114), and (7.115), can be expressed in terms of the elapsed distance as follows:

$$\frac{du}{ds} = S_f \dot{u} \tag{7.120}$$

$$\frac{dv}{ds} = S_f \dot{v} \tag{7.121}$$

$$\frac{d\bar{\omega}_z}{ds} = S_f \dot{\bar{\omega}}_z. \tag{7.122}$$

The angular accelerations $\dot{\bar{\omega}}_y$ in (7.113) and $\dot{\bar{\omega}}_x$ in (7.114) can be derived from (7.106). In many cases the Darboux frame $\dot{\omega}_x$ and $\dot{\omega}_y$ can be neglected, and so

$$\dot{\bar{\omega}}_x \approx \dot{\xi}(\omega_y \cos \xi - \omega_x \sin \xi) \tag{7.123}$$

$$\dot{\bar{\omega}}_y \approx -\dot{\xi}(\omega_y \sin \xi + \omega_x \cos \xi). \tag{7.124}$$

7.5.1 Tyre friction

The tyre frictional forces can be modelled using empirical formulae of the type described in Chapter 3. The tyre's longitudinal slip is described by a longitudinal slip coefficient κ, while the lateral slip is described by a slip angle α, where

$$\kappa = -\left(1 - \frac{R_i\omega_{ij}}{u_w}\right),\tag{7.125}$$

$$\tan\alpha = -\frac{v_w}{u_w},\tag{7.126}$$

where R_i is the wheels' radius and ω_{ij} the wheels' spin angular velocity (non-negative in normal running conditions). The terms u_w and v_w are the absolute speed components of the wheel centre in a wheel-fixed coordinate system. The four wheel slip angles are given by

$$\begin{aligned}
\alpha_{rr} &= -\arctan\left(\frac{v-\bar\omega_z b}{u-\bar\omega_z w_r}\right),\\
\alpha_{rl} &= -\arctan\left(\frac{v-\bar\omega_z b}{u+\bar\omega_z w_r}\right),\\
\alpha_{fr} &= -\arctan\left(\frac{v+\bar\omega_z a}{u-\bar\omega_z w_f}\right)+\delta,\\
\alpha_{fl} &= -\arctan\left(\frac{v+\bar\omega_z a}{u+\bar\omega_z w_f}\right)+\delta,
\end{aligned}\tag{7.127}$$

with the longitudinal slip ratios given by

$$\begin{aligned}
\kappa_{rr} &= -\left(1-\frac{R\omega_{rr}}{u-\bar\omega_z w_r}\right),\\
\kappa_{rl} &= -\left(1-\frac{R\omega_{rl}}{u+\bar\omega_z w_r}\right),\\
\kappa_{fr} &= -\left(1-\frac{R\omega_{fr}}{\cos\delta(u-\bar\omega_z w_f)+\sin\delta(v+\bar\omega_z a)}\right),\\
\kappa_{fl} &= -\left(1-\frac{R\omega_{fl}}{\cos\delta(u+\bar\omega_z w_f)+\sin\delta(v+\bar\omega_z a)}\right).
\end{aligned}\tag{7.128}$$

Equations (7.127) should be compared with (4.5)–(4.6) corresponding to the single-track car model. Sometimes alternative expressions for the front slips are reported,[18] which are derived using the trigonometric identity $\arctan A+\arctan B = \arctan\frac{A+B}{1-AB}$.

7.5.2 Load transfer

The ability of the tyres to generate side forces is influenced by the normal loads acting on them. In order to compute the tyre loads as the car moves down the road, we balance the forces acting on the car normal to the road, and then balance moments around the body-fixed roll and pitch axes, x_b and y_b, respectively; see Figure 7.22. These calculations must recognize the gravitational, inertial, centripetal, and aerodynamic forces acting on the car, as well as the gyroscopic moments.

The vertical force balance on the car comes from the z-axis component of (7.110)

$$\begin{aligned}
n\dot\omega_x &= \left(F_{rrz}+F_{rlz}+F_{frz}+F_{flz}+F_{az}^f+F_{az}^r\right)/M\\
&+g\cos\phi\cos\mu+u\bar\omega_y-v\bar\omega_x-h\left(\bar\omega_x^2+\bar\omega_y^2\right)-\omega_x\dot n,
\end{aligned}\tag{7.129}$$

in which the $F_{\cdot\cdot z}$s are the vertical tyre forces acting on each of the four wheels (all the $F_{\cdot\cdot z}$s are negative under parked equilibrium conditions), F_{az}^f and F_{az}^r are the

---

[18] $\alpha_{fr} = \arctan\left(\frac{\sin\delta(\bar\omega_z w_f-u)+\cos\delta(\bar\omega_z a+v)}{\cos\delta(u-\bar\omega_z w_f)+\sin\delta(\bar\omega_z a+v)}\right)$ and $\alpha_{fl} = \arctan\left(\frac{\cos\delta(\bar\omega_z a+v)-\sin\delta(\bar\omega_z w_f+u)}{\cos\delta(\bar\omega_z w_f+u)+\sin\delta(\bar\omega_z a+v)}\right)$.

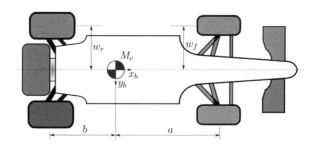

Figure 7.22: Plan view showing some of the vehicle's basic geometric parameters.

aerodynamic downforces at the front and rear axles, the seventh term is the force due to gravity, and the eighth and ninth terms are centripetal forces. In many cases the $\dot{\omega}_x$ term in (7.129) can be neglected.

Balancing moments (7.111) around the car's mass centre in the x_b direction gives

$$I_x \dot{\bar{\omega}}_x = w_r(F_{rrz} - F_{rlz}) + w_f(F_{frz} - F_{flz}) - hF_y$$
$$+(I_y - I_z)\bar{\omega}_z\bar{\omega}_y. \tag{7.130}$$

The first two terms are the roll moments produced by the vertical tyre forces, the third term is the roll moment produced by the lateral tyre forces F_y in (7.117), and the fourth term is a gyroscopic moment acting in the x_b direction; see Figure 7.22. In many cases the $\dot{\bar{\omega}}_x$ term in (7.130) can be approximated by (7.123).

Balancing moments (7.111) around the car's mass centre in the y_b direction gives

$$I_y \dot{\bar{\omega}}_y = b(F_{rrz} + F_{rlz} + F_{az}^r) - a(F_{frz} + F_{flz} + F_{az}^f) + hF_x$$
$$+(I_z - I_x)\bar{\omega}_z\bar{\omega}_x. \tag{7.131}$$

The first two terms represent the pitching moments produced by the vertical tyre forces and downforces, the third term is the pitching moment produced by the longitudinal force F_x in (7.116), and the fourth term is a gyroscopic moment acting in the y_b direction. In many cases the $\dot{\bar{\omega}}_y$ term in (7.131) can be approximated by (7.124).

Equations (7.129), (7.130), and (7.131) constitute a set of three linear equations in four unknowns. This structure is statically indeterminate (or hyperstatic) and so a fourth equation is required in order to make a unique solution possible. A unique solution for the tyre loads can be obtained by adding a suspension-related roll balance relationship, in which the lateral load difference across the front axle is some fraction of the whole

$$F_{frz} - F_{flz} = D(F_{frz} + F_{rrz} - F_{flz} - F_{rlz}), \tag{7.132}$$

where $D \in [0, 1]$. If this fourth equation is recognized, and equations (7.129), (7.130), (7.131), and (7.132) are linearly independent, a unique solution results. A comparison between (7.132), and (7.62) and (7.65), with $c_F = c_R = c$, gives

$$D = \frac{K_F}{K}\frac{h-d}{h} + \frac{d_F}{w}\frac{b}{h}. \tag{7.133}$$

7.5.3   Non-negative tyre loads

The forces satisfying equations (7.129), (7.130), (7.131), and (7.132) are potentially both positive and negative. Negative forces are indicative of vertical reaction forces, while positive forces are fictitious 'forces of attraction'. Since the model being considered here has no pitch, roll, or heave freedoms, none of the wheels is free to leave the road, while simultaneously keeping faith with (7.129) to (7.132); an approximation is thus required.

To cater for the possible 'positive force' ('light wheel') situation within an optimal control environment we introduce the tyre normal load vector $\bar{\mathbf{F}}_{\mathbf{z}} = [\bar{F}_{flz}, \bar{F}_{frz}, \bar{F}_{rlz}, \bar{F}_{rrz}]^T$, which may contain positive components. These are set to zero by defining $\mathbf{F}_{\mathbf{z}} = \min(\bar{\mathbf{F}}_{\mathbf{z}}, \mathbf{0})$, where the minimum function $\min(\cdot, \cdot)$ is interpreted element-wise. It is clear that $\bar{\mathbf{F}}_{\mathbf{z}}$ and $\mathbf{F}_{\mathbf{z}}$ will be equal unless at least one entry of $\bar{\mathbf{F}}_{\mathbf{z}}$ is positive (i.e. non-physical). Since the model must respect the laws of mechanics, equations (7.129), (7.130), and (7.131) must be enforced unconditionally. In contrast, we assume that the solution to (7.132), which is only an approximate representation of the suspension system, can be 'relaxed' in the event of a tyre normal load sign reversal.

Equations (7.129), (7.130), and (7.131) are arranged in matrix form as

$$A_1 \mathbf{F}_{\mathbf{z}} = \mathbf{c}, \tag{7.134}$$

while (7.132) can be written as

$$A_2 \mathbf{F}_{\mathbf{z}} = \mathbf{0}. \tag{7.135}$$

The entries in the matrices A_1 and A_2 and the vector $\mathbf{c}$ can be assembled from (7.129), (7.130), (7.131), and (7.132). In order to deal with the 'light wheel' situation, we combine (7.134) and (7.135) as

$$\begin{bmatrix} A_1 & 0 \\ 0 & A_2 \end{bmatrix} \begin{bmatrix} \mathbf{F}_{\mathbf{z}} \\ \bar{\mathbf{F}}_{\mathbf{z}} \end{bmatrix} = \begin{bmatrix} \mathbf{c} \\ \mathbf{0} \end{bmatrix} \tag{7.136}$$

in which $\mathbf{F}_{\mathbf{z}}$ in (7.135) has been replaced by $\bar{\mathbf{F}}_{\mathbf{z}}$. If there is a 'light wheel', the mechanics equations (7.129), (7.130), and (7.131) will be satisfied by the non-positive forces $\mathbf{F}_{\mathbf{z}}$, while the roll balance equation (7.132) is satisfied by the now fictitious forces $\bar{\mathbf{F}}_{\mathbf{z}}$ that contain a 'force of attraction'. It is clear that the tyre normal loads have to satisfy (7.136), which will be treated as constraints within an optimal control environment. The 'light wheel' modelling is covered in more detail in [255].

7.5.4   Wheel torque distribution

In order to optimize the vehicle's performance, one needs to control the torques applied to the individual road wheels. In some applications, including Formula One racing, the braking system is designed so that equal pressure is applied to the brake callipers of each axle, with the braking pressures between the front and rear axles satisfying a pre-specified design ratio (7.9). The drive torques applied to the driven wheels are controlled by a differential mechanism.

Brakes. We equate equal brake calliper pressures with equal braking torques when neither wheel on a particular axle is locked. If one wheel is locked, the braking torque applied to the locked wheel may be lower than that applied to the rolling wheel. In the case of the front wheels, assuming negligible spin inertia, this constraint might be modelled as

$$0 = \max(\omega_{fr}, 0)\max(\omega_{fl}, 0)(F_{frx} - F_{flx}),\tag{7.137}$$

in which ω_{fr} and ω_{fl} are the angular velocities of the front right and front left wheel, respectively. If either road wheel 'locks up', the corresponding angular velocity will be non-positive and the braking torque constraint (7.137) becomes inactive. The rear wheels are treated similarly, with $(\cdot)_{rr,rl}$ replacing $(\cdot)_{fr,fl}$ in (7.137).

Differential. One might want to assume that the drive torque is delivered to the driven heels through a limited-slip differential. In the case of a RWD vehicle one may wish to stipulate

$$R_r(F_{lrx} - F_{rrx}) = -k_d(\omega_{lr} - \omega_{rr}),\tag{7.138}$$

in which ω_{lr} and ω_{rr} are the rear-wheel angular velocities, R_r is the rear-wheels' radius, and k_d is a torsional damping coefficient. The special cases of an open and a locked differential correspond to $k_d = 0$ and k_d arbitrarily large respectively. The speed-sensing differential in (7.138), where the torque bias on the driven axle is related to wheels' speed difference, could be replaced by a torque-sensing differential, in which the torque bias is related to the torque input, with possibly different behaviours in driving and coasting conditions [166, 265].

7.5.5   Suspensions

A variety of different car suspensions have been devised. An introductory discussion of vehicular suspension systems can be found in [266]. A number of independent suspensions are surveyed in [267], where the synthesis of suspension linkages is studied, including their kinematic structure, their dimensions, and their compliance properties. The well-known MacPherson strut suspension is modelled in [268] as a three-dimensional kinematic mechanism; both linear and nonlinear analyses are provided. A MacPherson suspension mechanism model is also developed in [269].

In Section 7.2.3 we presented a simple graphical method for computing the roll centre of MacPherson and double-wishbone suspensions. In this section we focus on the analytic modelling of a typical Formula One suspension. Among the objectives of the analysis are the determination of the position and orientation of the wheel as the suspension compresses/extends, and the computation of the vertical force F_z on a wheel generated by the spring-damper force F in the suspension strut. The relationship between F_z and F can be obtained using the virtual power principle. This gives

$$F_z\dot{z} = F\dot{l},\tag{7.139}$$

where z is the wheel vertical travel and l is the suspension strut travel. Thus

$$F_z = F\left(\frac{\dot{l}}{\dot{z}}\right) = F\tau,\tag{7.140}$$

where τ is the *velocity ratio*. The suspension is called *progressive* when τ increases as the suspension compresses.

Formula One suspension. Figure 7.23 shows the twenty-four elements of a Formula One car suspension. The wishbones are shown solid, the pushrods are shown dashed, and the trackrods are shown dotted. The ends of the wishbones, pushrods and track-rods are all marked with hexagrams. A detailed kinematic and quasi-static analysis of a typical multi-link suspension will now be given.

The kinematics of a suspension system are determined by its motion constraints. We will now focus on Figure 7.24 (a), which shows the suspension of the front-left wheel. Rods one to four make up the double-wishbone suspension elements. Rod five is the steering trackrod, which connects the wheel carrier to the steering rack. Rod six is the suspension push bar, which connects the wheel carrier to a suspension strut rocker mechanism mounted within the car's body; see Figure 7.25. In the following analysis all the joints in Figure 7.24 are treated as spherical.

In order to describe the suspension motion constraints in detail, we introduce car-fixed and wheel-carrier-fixed reference frames with origins $\boldsymbol{O}_C$ and $\boldsymbol{O}_W$ respectively. As shown in Figure 7.24 (a), points $\boldsymbol{P}_1$ to $\boldsymbol{P}_4$ are fixed in the car chassis and are connected to $\boldsymbol{Q}_1$ to $\boldsymbol{Q}_4$ that are fixed in the wheel carrier using fixed-length wishbones of length l_i $i = 1, \cdots, 4$. Points $\boldsymbol{P}_5$ and $\boldsymbol{Q}_5$ are at opposite ends of the steering trackrod with $\boldsymbol{Q}_5$ fixed in the wheel carrier. Point $\boldsymbol{P}_5$ is connected to the steering rack and moves in the car body (y-direction) as the steering-wheel position is varied. Points $\boldsymbol{P}_6$ and $\boldsymbol{Q}_6$ are at opposite ends of the suspension pushrod with $\boldsymbol{Q}_6$ fixed in the wheel carrier. Point $\boldsymbol{P}_6$ is connected to a rocker mechanism and moves in the car body in sympathy with movements of the pushrod; see Figure 7.25. The steering trackrod and suspension pushrod have lengths l_5 and l_6 respectively. This system has two degrees of freedom corresponding to changes in the steering angle and to changes in the angular position of the suspension rocker mechanism. The suspension rockers are also connected to a

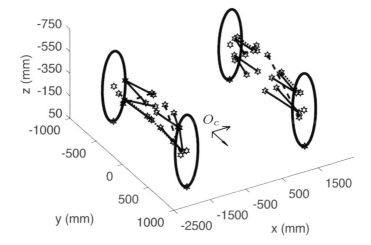

Figure 7.23: The suspension system of a typical Formula One car.

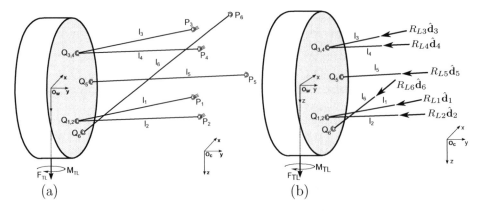

Figure 7.24: Front-left suspension configuration of a typical Formula One car with a pushrod suspension system (a) and free body diagram of the front-left wheel carrier (b).

parallel heave spring and damper combination, a torsion bar, and an anti-roll bar; the damper is not considered in the quasi-static analysis presented here. A tyre moment $\boldsymbol{M}_{TL}$ is applied to the wheel carrier body, with a tyre force $\boldsymbol{F}_{TL}$ applied at the wheel's ground-contact point.

In the following analysis all vectors will be expressed in the car reference frame. In the case a vector is resolved in the wheel carrier frame the following notation will be used $(\cdot)^W$. Six motion constraints derive from the fixed lengths of the six suspension elements. If we suppose that $\boldsymbol{P}_i$ and $\boldsymbol{Q}_i$ are vector representations of the corresponding points in Figure 7.24 (a), then the magnitudes of each of the six differences $\boldsymbol{P}_i - \boldsymbol{Q}_i$ (lengths) must be fixed too. That is,

$$\|\boldsymbol{P}_i - \boldsymbol{Q}_i\|^2 = l_i^2 \qquad i = 1, \cdots, 6, \tag{7.141}$$

in which the lengths l_i are constant. A steering angle δ results in a displacement of the steering rod mount point $\boldsymbol{P}_5$ in the y-axis direction, which is proportional to the radius r of the steering pinion

$$\boldsymbol{P}_5 = \begin{bmatrix} 0 & -r\delta & 0 \end{bmatrix}^T + \boldsymbol{P}_5^0; \tag{7.142}$$

$\boldsymbol{P}_5^0$ is the straight-running position of $\boldsymbol{P}_5$. These equations determine a hypersurface over which the twelve points in Figure 7.24 (a) must move.

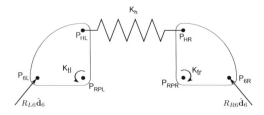

Figure 7.25: Pushrod and rocker setup; the heave spring has stiffness K_h.

There are six degrees of relative motion movement that are constrained by (7.141). The translational freedoms (relative to the car) of the wheel carrier reference frame origin $\boldsymbol{O}_W$ are

$$\boldsymbol{D} = \begin{bmatrix} \Delta x & \Delta y & \Delta z \end{bmatrix}^T. \tag{7.143}$$

The rotational freedoms of the wheel carrier relative to the vehicle's chassis are described in terms of the elementary rotations $\mathcal{R}(\boldsymbol{e}_x,\phi)$, $\mathcal{R}(\boldsymbol{e}_y,\theta)$, and $\mathcal{R}(\boldsymbol{e}_z\psi)$, which are described by (2.156)–(2.157). When employing the yaw–roll–pitch (ψ–ϕ–θ) convention

$$\boldsymbol{X} = \mathcal{R}(\boldsymbol{e}_z,\psi)\mathcal{R}(\boldsymbol{e}_x,\phi)\mathcal{R}(\boldsymbol{e}_y,\theta)\boldsymbol{X}^W + \boldsymbol{D} \tag{7.144}$$

describes a point $\boldsymbol{X}^W$, which is fixed in the wheel carrier, in the chassis frame. Equations (7.141) and (7.144) provide six motion constraints that determine the kinematic behaviour of the suspension system.

As shown in Figure 7.25, the left- and right-hand sides of the suspension system are coupled through a heave spring (and a parallel damper which is not shown) and a rocker assembly. The right- and left-hand rockers are mounted on torsion bars with stiffnesses K_{tr} and K_{tl} respectively. The two rockers are also coupled by an anti-roll bar that resists chassis roll. This is represented by a moment $(\alpha_R + \alpha_L)K_{arb}$ applied to each rocker, in which α_R and α_L are the right- and left-hand rocker angles. The anti-roll bar stiffness is denoted by K_{arb}. In Figure 7.25 R_{L6} and R_{R6} are the magnitudes of the right- and left-hand pushrod forces, $\boldsymbol{P}_{6L}$ and $\boldsymbol{P}_{6R}$ are the pushrod attachment points, $\boldsymbol{P}_{HL}$ and $\boldsymbol{P}_{HR}$ are the end points of the heave spring, $\boldsymbol{R}_{RPL}$ and $\boldsymbol{R}_{RPR}$ are the rocker pivot points, and K_h is the heave spring stiffness.

A moment balance around the pin joint $\boldsymbol{P}_{RPL}$ gives

$$0 = ((\boldsymbol{P}_{6L} - \boldsymbol{P}_{RPL}) \times F_{L6}\hat{\boldsymbol{d}}_6 + (\boldsymbol{P}_{HL} - \boldsymbol{P}_{RPL}) \times F_{heave}\hat{\boldsymbol{d}}_{hs}) \cdot \boldsymbol{n}_{RL} \\ - K_{tl}\alpha_L - K_{arb}(\alpha_L + \alpha_R) \tag{7.145}$$

in which $\boldsymbol{n}_{RL}$ is the left rocker pivot rotation axis, $\hat{\boldsymbol{d}}_6$ is a unit vector in the direction of l_6, $\hat{\boldsymbol{d}}_{hs}$ is a unit vector in the heave spring direction pointing from $\boldsymbol{P}_{HL}$ to $\boldsymbol{P}_{HR}$ and F_{heave} is the heave spring force magnitude.

In the same way, one can take moments around $\boldsymbol{P}_{RPR}$ to obtain

$$0 = ((\boldsymbol{P}_{6R} - \boldsymbol{P}_{RPR}) \times F_{R6}\hat{\boldsymbol{d}}_6 - (\boldsymbol{P}_{HR} - \boldsymbol{P}_{RPR}) \times F_{heave}\hat{\boldsymbol{d}}_{hs}) \cdot \boldsymbol{n}_{RR} \\ - K_{tr}\alpha_R - K_{arb}(\alpha_L + \alpha_R) \tag{7.146}$$

in which $\hat{\boldsymbol{d}}_6$ in (7.146) must be interpreted in the context of the right-hand suspension assembly. The heave spring force F_{heave} is a function of the change in its length, and is given by

$$F_{heave} = k_h \Delta L \tag{7.147}$$

in which $\Delta L = \|\boldsymbol{P}_{HR} - \boldsymbol{P}_{HL}\| - L_0$, with L_0 the unloaded length. In general, the heave spring stiffness tends to 'harden' as it is compressed, thus giving a nonlinear relationship between the spring force and spring travel.

The quasi-static equations of motion describing the suspension system are determined using force and moment balances with the aid of the free body diagram shown in Figure 7.24 (b). We will suppose that the reaction force acting at point $\boldsymbol{Q}_i$ has magnitude R_{Li} and acts in direction $\hat{\boldsymbol{d}}_i$. It should be noted that although the reaction forces are vector-valued quantities, it is only the magnitudes of these vectors that are unknown, since their directions are determined by the kinematic constraints. The external forces acting on the wheel carrier are the tyre force $\boldsymbol{F}_{TL}$, and the gravitational force due to the wheel carrier's mass m_L, which is given by

$$\boldsymbol{F}_{Lg} = \begin{bmatrix} 0 & 0 & m_L g \end{bmatrix}^T. \tag{7.148}$$

Summing forces on the left-hand wheel carrier gives

$$0 = \boldsymbol{F}_{TL} + \boldsymbol{F}_{Lg} + R_{L1}\hat{\boldsymbol{d}}_1 + R_{L2}\hat{\boldsymbol{d}}_2 +$$
$$R_{L3}\hat{\boldsymbol{d}}_3 + R_{L4}\hat{\boldsymbol{d}}_4 + R_{L5}\hat{\boldsymbol{d}}_5 + R_{L6}\hat{\boldsymbol{d}}_6, \tag{7.149}$$

in which R_{L6} is the force acting on the left-hand rocker. This equation provides three scalar equations in the six left-hand unknown reaction force magnitudes. A similar force balance on the right-hand front wheel carrier gives

$$0 = \boldsymbol{F}_{TR} + \boldsymbol{F}_{Rg} + R_{R1}\hat{\boldsymbol{d}}_1 + R_{R2}\hat{\boldsymbol{d}}_2 + R_{R3}\hat{\boldsymbol{d}}_3 +$$
$$R_{R4}\hat{\boldsymbol{d}}_4 + R_{R5}\hat{\boldsymbol{d}}_5 + F_{R6}\hat{\boldsymbol{d}}_6, \tag{7.150}$$

where R_{R6} is the force acting on the right-hand suspension rocker. In this case

$$\boldsymbol{F}_{Rg} = \begin{bmatrix} 0 & 0 & m_R g \end{bmatrix}^T \tag{7.151}$$

with the $\hat{\boldsymbol{d}}_i$s interpreted in a like manner in the context of the right-hand front wheel. This equation provides another three equations in the six right-hand reaction force magnitudes.

As we will now show the remaining equations come from moment balance calculations. To begin, we define a set of vectors that point from point $\boldsymbol{Q}_1$, which is fixed in the wheel carrier, to points $\boldsymbol{Q}_i$ $i = 2, \cdots, 6$, $\boldsymbol{O}_W$ and the tyre ground-contact point $\boldsymbol{Q}_{GC}$. These vectors are given by $\boldsymbol{d}_{12}$, $\boldsymbol{d}_{13}$, $\boldsymbol{d}_{14}$, $\boldsymbol{d}_{15}$, $\boldsymbol{d}_{16}$, $\boldsymbol{d}_{1W}$, and $\boldsymbol{d}_{1GC}$ respectively. With the exception of $\boldsymbol{d}_{1GC}$, these vectors are known and fixed in the wheel carrier. In order to find the vector $\boldsymbol{d}_{1GC}$, which represents a moving point in the wheel frame, we need to study the geometry of the ground-contact point. To that end we introduce the vector $\hat{\boldsymbol{e}}_w^W = [0\ 1\ 0]^T$, which is fixed in the wheel carrier, and points in the direction of the wheel spindle. The wheel spindle unit vector, expressed in the car frame, is thus given by

$$\hat{\boldsymbol{e}}_w = \mathcal{R}(\boldsymbol{e}_z, \psi)\mathcal{R}(\boldsymbol{e}_x, \phi)\mathcal{R}(\boldsymbol{e}_y, \theta)\hat{\boldsymbol{e}}_w^W. \tag{7.152}$$

If the pitch and roll of the car relative to the road are 'small', the unit road-normal expressed in the chassis frame can be approximated by $\hat{\boldsymbol{e}}_{rn} = [0\ 0\ 1]^T$. This means that the vector pointing from the wheel centre to the ground-contact point is given by

$$\boldsymbol{r}_w = \frac{(\hat{\boldsymbol{e}}_{cw} \times \hat{\boldsymbol{e}}_{rn}) \times \hat{\boldsymbol{e}}_{cw}}{\|(\hat{\boldsymbol{e}}_{cw} \times \hat{\boldsymbol{e}}_{rn}) \times \hat{\boldsymbol{e}}_{cw}\|} R_{fw} \tag{7.153}$$

and so $\boldsymbol{d}_{1GC} = \boldsymbol{r}_w + \boldsymbol{d}_{1W}$; R_{fw} is the loaded tyre radius, which depends on the tyre force $\boldsymbol{F}_{TL}$ and is usually determined by measurement.

Taking moments around $\boldsymbol{Q}_1$ in Figure 7.24 (B) gives

$$\begin{aligned}
0 = {}& \boldsymbol{M}_{TL} + \boldsymbol{d}_{1GC} \times \boldsymbol{F}_{TL} + \boldsymbol{d}_{1W} \times \boldsymbol{F}_{Lg} + R_{L2}(\hat{\boldsymbol{d}}_2 \times \boldsymbol{d}_{12}) \\
& + R_{L3}(\hat{\boldsymbol{d}}_3 \times \boldsymbol{d}_{13}) + R_{L4}(\hat{\boldsymbol{d}}_4 \times \boldsymbol{d}_{14}) + R_{L5}(\hat{\boldsymbol{d}}_5 \times \boldsymbol{d}_{15}) \\
& + R_{L6}(\hat{\boldsymbol{d}}_6 \times \boldsymbol{d}_{16}).
\end{aligned} \tag{7.154}$$

A similar calculation for the front-right wheel gives

$$\begin{aligned}
0 = {}& \boldsymbol{M}_{TR} + \boldsymbol{d}_{1GC} \times \boldsymbol{F}_{TR} + \boldsymbol{d}_{1W} \times \boldsymbol{F}_{Rg} + R_{R2}(\hat{\boldsymbol{d}}_2 \times \boldsymbol{d}_{12}) \\
& + R_{R3}(\hat{\boldsymbol{d}}_3 \times \boldsymbol{d}_{13}) + R_{R4}(\hat{\boldsymbol{d}}_4 \times \boldsymbol{d}_{14}) + R_{R5}(\hat{\boldsymbol{d}}_5 \times \boldsymbol{d}_{15}) \\
& + R_{R6}(\hat{\boldsymbol{d}}_6 \times \boldsymbol{d}_{16})
\end{aligned} \tag{7.155}$$

in which the vectors $\hat{\boldsymbol{d}}_2, \cdots, \hat{\boldsymbol{d}}_6$, $\boldsymbol{d}_{1GC}$, and $\boldsymbol{d}_{12}, \cdots, \boldsymbol{d}_{16}$ must be interpreted in an analogous manner in the context of the right-hand suspension.

In sum, equations (7.141) are used to determine the right- and left-hand Euler angles and the right- and left-hand wheel carrier displacements in (7.143); twelve variables in total. Equations (7.145) and (7.146) determine the two rocker angles. Equations (7.149), (7.150), (7.154), and (7.155) determine the twelve right- and left-hand pushrod force magnitudes. Equations (7.141), (7.145), (7.146), (7.149), (7.150), (7.154), and (7.155) can be solved for the twenty-six unknowns associated with each axle as a function of the tyre force $\boldsymbol{F}_T$ and tyre moment $\boldsymbol{M}_T$ using a nonlinear algebraic equation solver.

Figure 7.26 illustrates the typical force-displacement behaviour of the front-left suspension system as a function of the normal tyre load; the front-right tyre is assumed unloaded [270], that is, the inputs are $\boldsymbol{F}_{TL} = [0, 0, F_{flz}]^T$, $\boldsymbol{M}_{TL} = \boldsymbol{M}_{TR} = \boldsymbol{F}_{TR} = [0, 0, 0]^T$, and $\delta = 0$. It is evident from Figure 7.26 that the wheel's camber and toe angles vary very little as a result of normal load variations. This figure also shows that the suspension movement resulting from normal load variations is almost entirely in the vertical direction.

7.5.6 Aerodynamic maps

The suspension analysis in Section 7.5.5 describes the kinematics of the car as a function of suspension's motion. Computable quantities include the front- and rear-axle ride heights and the car's roll angle. These quantities are illustrated in Figure 7.27. The axle roll angles are computed as the wheel height difference divided by the axle track, while the heave is computed as the average height. If the suspension analysis is included in the car model, (7.132) can be replaced by the rigid-body assumption

$$\phi_f = \phi_r, \tag{7.156}$$

where ϕ_f and ϕ_r are the roll angles at the front and rear axles.

In race cars, where aerodynamic wings are allowed, an important suspension influence relates to the vehicle's aerodynamic performance. In the case of hard suspensions, such as those used in race cars, suspension movements are 'small' and 'nearly linear' models can be used. Table 7.1 shows how the front- and rear-axle right heights and roll

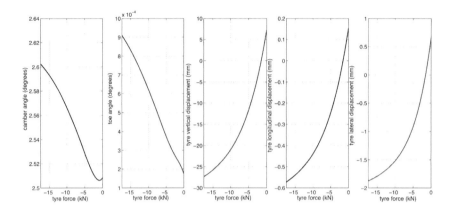

Figure 7.26: Variations in the front-left wheel carrier camber angle ϕ in (7.144), toe angle ψ in (7.144), and the ground-contact point as a function of tyre vertical load F_{flz}.

angles vary with left- and right-wheel normal loads F_l and F_r; in the case given, only linear and quadratic terms in the normal loads are required to describe the suspension motion [270]. More complex meta models may be derived that give suspension motions as a function of the longitudinal and normal tyre forces.

The external forces acting on the car come from the tyres and from aerodynamic influences. A dynamic car model can be used to compute the normal tyre loads $F_{rrz} \cdots F_{flz}$, and the steering and side-slip angles δ and $\beta = \arctan(v/u)$, respectively. As shown in Figure 7.28, these data are supplied to the suspension and aero models to produce the aerodynamic coefficients, which are then dynamically updated. The front- and rear-axle downforce coefficients are given by C_L^f and C_L^r respectively, while the drag coefficient is C_D.

The aerodynamic downforces in (7.129) and (7.131) are computed using

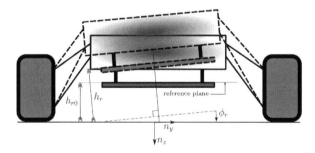

Figure 7.27: Suspension-induced roll angle ϕ_r and ride height h_r at the rear axle; the car is driving into the page and into a right-hand bend. The rear-axle ride height in the nominal configuration is h_{r0} and is measured relative to the reference plane (fixed in the car).

Table 7.1 Multivariate polynomial coefficients (meta model) of the front- and rear-axle ride height (in mm) and roll angle (in rad). The left- and right-wheel normal loads F_l and F_r, respectively, are given in kN.

	1	F_l	F_r	F_l^2	$F_l F_r$	F_r^2
h_f	-21.52	-2.513	-2.513	-0.08324	0.1126	-0.08324
ϕ_f	0	0.00851	-0.00851	0.000256	0	-0.0002562
h_r	-89.04	-4.1305	-4.1305	-0.09450	0.1119	-0.09450
ϕ_r	0	0.0112	-0.0112	6.1×10^{-05}	0	-6.1×10^{-05}

$$F_{az}^f = \frac{1}{2}\rho C_L^f A u^2,\tag{7.157}$$

and

$$F_{az}^r = \frac{1}{2}\rho C_L^r A u^2,\tag{7.158}$$

which are applied at the centres of the front and rear axles, while the drag force in (7.116) is given by

$$F_{ax} = -\frac{1}{2}\rho C_D A u^2,\tag{7.159}$$

which is applied in the rod plane. In (7.157), (7.158), and (7.159) A is the car's frontal area, ρ is the air density, and u is the air velocity over the car. The drag has a negative sign, since it acts in the negative x-axis direction. The drag and downforce coefficients are typically derived from track and wind-tunnel measurements, with representative values given in Table 7.2.

Table 7.2 Downforce and drag coefficients and sensitivities.

C_L^f	C_L^r	C_D	Input
1.2000	0.9494	0.9455	Nominal Value
10.0000	1.0000	0.1000	h_f (m)
-5.0000	-22.2222	-0.1000	h_r (m)
0	-222.2222	0	h_r^2 (m^2)
-5.1294	-10.2588	0	β^2 (rad^2)
-0.1641	-0.1641	0	δ^2 (rad^2)
-8.2070	-32.8281	0	ϕ^2 (rad^2)

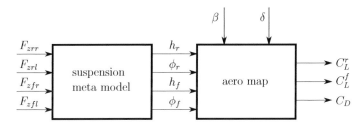

Figure 7.28: Quasi-static aero–suspension model. The dynamic model generates the normal tyre loads, and the steering and side-slip angles, while the quasi-static suspension model generates the aerodynamic drag and downforce coefficients.

Example 7.5 Consider a race car with mass $m = 660\,\text{kg}$, wheelbase $w = 3.4\,\text{m}$, and with the centre of mass $b = 1.6\,\text{m}$ from the rear axle. Compute the ride heights and the drag and lift coefficients at standstill using the data given in Tables 7.1 and 7.2. Repeat the computation when the loads on the front and rear tyres increase by 3.330 kN and 3.142 kN respectively; the total downforce is twice the weight of the car. Finally, compute the effect of a side-slip angle of $\beta = 6°$ and a roll angle of $\phi_f = \phi_r = 2°$ on the drag and lift coefficients. At standstill, the normal loads on the front are rear tyres are $F_{flz} = F_{frz} = -1.523\,\text{kN}$ and $F_{rlz} = F_{rrz} = -1.713\,\text{kN}$, respectively, which gives ride heights of $h_f = -14\,\text{mm}$ and $h_r = -75\,\text{mm}$ at the front and rear axles respectively. The corresponding drag coefficient is $C_D = 0.95$, while the front- and rear-axle downforce coefficients are $C_L^f = 1.44$ and $C_L^r = 1.35$ respectively. When downforce effects are included, the loads increase to $F_{flz} = F_{frz} = -4.853\,\text{kN}$ and $F_{rlz} = F_{rrz} = -4.855\,\text{kN}$, the ride heights reduce to $h_f = +1.6\,\text{mm}^{19}$ and $h_r = -51\,\text{mm}$, and the drag remains $C_D = 0.95$, while the lift coefficients become $C_L^f = 1.47$ and $C_L^r = 1.51$. A side-slip angle of $6°$ reduces the down-force coefficients to $C_L^f = 1.41$ and $C_L^r = 1.39$, while an additional $2°$ roll angle reduces the lift to $C_L^f = 1.40$ and $C_L^r = 1.35$.

7.6 Driver modelling

In certain scenarios, particularly in the case of motorcycles, it is necessary to augment the vehicle model with some form of driver/rider representation. Examples include driving 'on the limit' in racing, the stabilization of motorcycles in low-speed manoeuvring, and obstacle avoidance. There are three main physiological functions that characterize a driver's steering response that we will now discuss: (i) the cognitive response of the brain, (ii) the muscle-related stretch reflex response, and (iii) the purely passive (intrinsic) response [199].

The cognitive response is related primarily to the driver's path-following role and comprises two parts: the feedforward steering action based on the previewed road path and the learned vehicle behaviour, and a feedback steering action based on the vehicle's sensed motion. From the control point of view, the use of a receding horizon strategy with both linear quadratic (LQ) preview control (e.g. [272] and [273, 274] for

---

[19] The ride height is computed with respect to a reference plane that is attached to the vehicle chassis; this zero-height datum is shown as a dot-dash line in the sketch below:

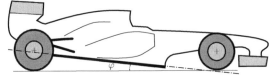

Below the reference plane is a 'plank' (shown brown), which must be 10 mm thick (drawing 7 in [271]). The lower surface of the plank is at a height of $+10\,\text{mm}$, which is still above the road surface. As soon as the car pitches forward, through angle φ, say, the reference plane can lie 'under' the road at the front axle, but the car's bodywork is designed to avoid physical contact with the road under these circumstances. A ride height of zero does not mean that the front of the car is under the road.

cars and motorcycles respectively), and model predictive control (MPC) (e.g. [275] and [276, 277] for cars and motorcycles respectively) have been shown to be effective for the simulation of the path-following task. Indeed, the driver/rider action of these controllers consists of a feedforward component and a feedback component. While the feedforward component is open loop, whose frequency content can reach significant values, the feedback component has limited bandwidth (1–2 Hz) due to the brain processing time [278]. It is likely that the vehicle driver uses the noisy sensed signals to estimate the current vehicle state in order generate the feedback control. An attempt at modelling the effect of human sensory noise on path-following steering control has been reported in [279]. Another option is to deal with the cognitive response using a nonlinear optimal control framework; such approach is discussed in Chapters 8 and 9. Proportional integral derivative (PID) controllers have been used as well [280], but are generally not suitable for demanding manoeuvres, where path following is combined with significant speed variations (e.g. hard braking into a curve), especially in the case of motorcycles.

The driver stretch reflex response and the passive response—the so-called neuromuscular-system dynamics (NMSD)—are important when it comes to both vehicle stability (especially in the case of two-wheeled vehicles) and to the study of sophisticated steering manoeuvres (especially in the case of four-wheeled vehicles). In the case of motorcycles, NMSD describe the effects of holding the handlebars. Driver–vehicle stability properties 'hands-on' are different from driver–vehicle stability properties 'hands-off', especially in the case of motorcycles, as will be discussed in Section 7.7.9. In [281] and [200] the NMSD of car drivers and motorcycle riders have been identified in terms of transfer functions $H(s)$ between the steering torque $T(s)$ and the steering angle $\delta(s)$; $H(s) = \delta(s)/T(s)$. In the frequency range of interest (up to 10 Hz), the driver–car system $H_c(s)$ exhibits one resonance, while the rider–motorcycle system $H_m(s)$ presents two resonances. For cars, a single-degree-of-freedom system model is deemed sufficient (the steering rotation being the only significant degree of freedom), while in the case of motorcycles a model with at least two degrees of freedom is required (steering and the rotation of the torso about the spine; see Figure 7.29). The following expressions can be used

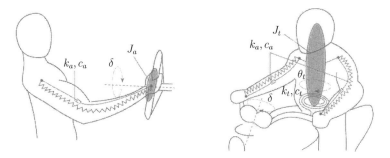

Figure 7.29: Car driver (left) and motorcycle rider (right).

$$H_c(s) = \frac{1}{J_a s^2 + c_a s + k_a} \qquad H_m(s) = \frac{J_t s^2 + (c_t + c_a)\, s + (k_t + k_a)}{(J_t s^2 + c_t s + k_t)(c_a s + k_a)}. \quad (7.160)$$

The steering torque is computed at the interface between the driver's hands and the steering wheel/handlebar and so the inertia of the steering system is not included in (7.160). Moreover, in H_c all the driver inertia involved in the vibration dynamics (mainly that of the arms and hands) is accounted for in J_a, which experiences the steering rotation δ, while in H_m all the rider's inertia (mainly the torso) is accounted in J_t, which experiences rotation θ_t. Average driver properties are available for both cases, both in relaxed and tensed condition for the car driver, and only in the relaxed condition for the motorcycle rider [199, 200, 281]; Table 7.3. When comparing the transfer functions, it is evident that both systems have spring-like behaviour at low frequency, although the static gains ($1/k_a$ and $(k_t + k_a)/k_t k_a$) are significantly different; the gain of the car driver in a tensed condition is close to that of the motorcycle rider in a relaxed state. Different muscles are involved in each system; the car driver is mainly excited in the roll axis, while the motorcycle rider is mainly excited in the yaw axis (Figure 7.29). Also, the radius of the steering wheel is smaller than the distance between the hand grips on the handlebar and the steering axis. The main $H(s)$ resonance, for the given dataset, is at 0.81 Hz for the relaxed car driver and 3.89 Hz for the tensed driver, while the motorcycle rider has a damped resonance at 1.62 Hz.

It is possible to separate the stretch-reflex response from the passive response. It is evident that when the driver is completely relaxed, there is no control input from the limb dynamics; think of an unconscious person whose arms can be moved without significant resistance. In contrast, a conscious person is able to control the position of their limbs, by activating the cognitive and stretch-reflex responses. The reflex response is related to the difference between the desired and current limb configurations, with a limit in the reconfiguation bandwidth of approximately 3 Hz [278]. It is assumed that the reflex response is activated by the difference between the desired and current steering angles. The investigation in [278] highlights the fact that the passive (intrinsic) muscle response is damper-like. Putting aside the cognitive control, it is assumed that all of the spring-like limb responses arise from stretch-reflex action, while the rest is passive. A general structure for a vehicle-driver model can be found in [199].

In the case of motorcycles, an additional stretch-reflex/passive element of the rider's response is related to the lateral and roll dynamics. Following the above arguments, these dynamics can again be modelled restraining the rider's lateral and roll motions with (properly tuned) spring-damper elements. Further details appear in Section 7.7.4.

Table 7.3 Average driver and rider parameters.

car driver (relaxed)	car driver (tense)	parameter	motorcycle rider (relaxed)
0.094	0.094	J_a (kgm^2)	–
0.59	1.12	c_a (Nms/m)	19.28
3.4	59.6	k_a (Nm/rad)	1053.4
–	–	J_t (kgm^2)	0.644
–	–	c_t (Nms/m)	4.79
–	–	k_t (Nm/m)	75.8

In sum, when simulating a manoeuvre with no unexpected driver responses, the stretch-reflex dynamics are not involved, while the passive dynamics and brain are contributory. When unexpected motion and vibration are present, the stretch-reflex response is involved. In practice, when aiming at the estimation of the stability properties of a vehicle in a hands-on configuration, the (properly tuned) spring-damper elements depicted in Figure 7.29 should be included in the mathematical model; see Section 7.7.9.

An extensive review on car driver modelling can be found in [282], while bicycle and motorcycle driver modelling are reviewed in [283]. A recent review on NMSD modelling is reported in [284].

7.7 Motorcycle modelling

From a mathematical modelling perspective single-track vehicles are multibody systems and include bicycles, motorcycles, and motor scooters, all of which have broadly similar dynamic properties. Some of the early contributions to the subject are discussed in Chapter 5. While this historically important work represented a promising start, these early models are only usable in straight running at low speed, and they fail to reproduce a number of important dynamic phenomena. We will now review a number of two-wheeled vehicle modelling enhancements.

7.7.1 Tyre contact geometry

Tyre contact geometry in the context of bicycles and motorcycles is complex. This complexity stems largely from the fact that motorcycle and bicycle tyres are profiled and may operate through a large range of roll angles.

In Section 5.2.1 we showed that the pitch angle of a rigid-wheel single-track vehicle model is determined by the solution of a closed kinematic loop that derives from the need to have both wheels in contact with the ground. We will now study this kinematic loop in detail and show that the pitch angle is a function of both the roll and steer angles. This combination of rolling and steering means that the tyre contact point can move laterally and well as circumferentially over the tyre's surface. The need to solve this closed kinematic loop for the pitch angle does, in a sense, 'disappear' when suspension systems and/or tyre carcass flexibility is considered. Various tyre carcass models are discussed at the end of this section.

Rigid tyres: pitch angle kinematic loop. If the vehicle is deemed to have a stiff suspension and stiff tyres, one my wish to neglect these compliances. We will now consider the pitch-determining closed kinematic loop in detail in the case of a toroidal tyre. The 'thin disc' wheel model is captured as a special case.

Suppose that the front and rear tyres are toroidal with crown radii ρ_F and ρ_R respectively. Other parameters that determine the closed kinematic loop conditions are the front and rear tyre radii r_F and r_R, the distance u_R between the rear-wheel centre and the steer axis given by (7.186), the distance u_F between the front-wheel centre and the steer axis given by (7.185), and the distance d between the wheel centres projected on to the steer axis (7.187); Figure 7.30. The yaw–roll–pitch convention (Section 2.5) will be used for the chassis orientation.

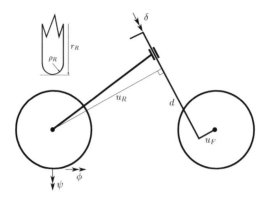

Figure 7.30: Bicycle geometry with toroidally profiled tyres.

The position of the front-wheel centre relative to the rear-wheel ground-contact point can be expressed in terms of the following eight 4×4 transformation matrices; see (Section 2.5)

$$\mathcal{T}_R = \mathcal{T}_1 \mathcal{T}_2 \mathcal{T}_3 \mathcal{T}_4 \mathcal{T}_5 \mathcal{T}_6 \mathcal{T}_7 \mathcal{T}_8 \qquad (7.161)$$

where

$$\mathcal{T}_1 = \begin{bmatrix} 1 & 0 & 0 & x_R \\ 0 & 1 & 0 & y_R \\ 0 & 0 & 1 & -\rho_R \\ 0 & 0 & 0 & 1 \end{bmatrix} \qquad \mathcal{T}_2 = \begin{bmatrix} \mathcal{R}(e_z, \psi) & 0 \\ 0 & 1 \end{bmatrix} \qquad \mathcal{T}_3 = \begin{bmatrix} \mathcal{R}(e_x, \varphi) & 0 \\ 0 & 1 \end{bmatrix}$$

$$\mathcal{T}_4 = \begin{bmatrix} 1 & 0 & 0 & 0 \\ 0 & 1 & 0 & 0 \\ 0 & 0 & 1 & -(r_R - \rho_R) \\ 0 & 0 & 0 & 1 \end{bmatrix} \qquad \mathcal{T}_5 = \begin{bmatrix} \mathcal{R}(e_y, \chi) & 0 \\ 0 & 1 \end{bmatrix} \qquad \mathcal{T}_6 = \begin{bmatrix} 1 & 0 & 0 & u_R \\ 0 & 1 & 0 & 0 \\ 0 & 0 & 1 & 0 \\ 0 & 0 & 0 & 1 \end{bmatrix}$$

$$\mathcal{T}_7 = \begin{bmatrix} \mathcal{R}(e_z, \delta) & 0 \\ 0 & 1 \end{bmatrix} \qquad \mathcal{T}_8 = \begin{bmatrix} 1 & 0 & 0 & u_F \\ 0 & 1 & 0 & 0 \\ 0 & 0 & 1 & d \\ 0 & 0 & 0 & 1 \end{bmatrix}.$$

The transformation $\mathcal{T}_1$ is a pure translation and locates the rear-wheel crown centre vertically above the rear-wheel contact point x_R, y_R. The rotations $\mathcal{T}_2$ and $\mathcal{T}_3$ describe the vehicle's yaw ψ and roll φ, respectively.[20] The translation $\mathcal{T}_4$ positions the rear-wheel centre in an arbitrary vehicle configuration; the rear-wheel radius is r_R. The rotation in $\mathcal{T}_5$ recognizes the pitch rotation χ of the vehicle. The translation $\mathcal{T}_6$ locates the steer axis, $\mathcal{T}_7$ introduces a steering rotation δ, and $\mathcal{T}_8$ is a translation to the front-wheel centre. The rotation matrices $\mathcal{R}(\,\cdot\,, \,\cdot\,)$ have been defined in (2.156)–(2.158).

---

[20] There is no loss of generality in assuming that the rear frame is aligned with the inertial x-axis resulting in $\psi = 0$; the $\mathcal{T}_2$ rotation can thus be ignored.

Alternatively, starting from the front-wheel contact point, the following sequence of four 4×4 transformation matrices

$$\mathcal{T}_F = \mathcal{T}_9 \mathcal{T}_{10} \mathcal{T}_{11} \mathcal{T}_{12} \tag{7.162}$$

where

$$\mathcal{T}_9 = \begin{bmatrix} 1 & 0 & 0 & x_F \\ 0 & 1 & 0 & y_F \\ 0 & 0 & 1 & -\rho_F \\ 0 & 0 & 0 & 1 \end{bmatrix} \qquad \mathcal{T}_{10} = \begin{bmatrix} \mathcal{R}(e_z, \psi_F) & 0 \\ 0 & 1 \end{bmatrix}$$

$$\mathcal{T}_{11} = \begin{bmatrix} \mathcal{R}(e_x, \varphi_F) & 0 \\ 0 & 1 \end{bmatrix} \qquad \mathcal{T}_{12} = \begin{bmatrix} 1 & 0 & 0 & 0 \\ 0 & 1 & 0 & 0 \\ 0 & 0 & 1 & -(r_F - \rho_F) \\ 0 & 0 & 0 & 1 \end{bmatrix}$$

locates the front-wheel crown centre in an arbitrary configuration. As before, the translation $\mathcal{T}_9$ locates the centre of the front-tyre crown, $\mathcal{T}_{10}$ describes the yaw rotation ψ_F of the front wheel, $\mathcal{T}_{11}$ represents the roll rotation φ_F of the front wheel, and $\mathcal{T}_{12}$ represents the translation to the front-wheel centre.

The orientation of the front-wheel spin axis can be determined by the vector $[0, 1, 0]^T$ in frame $\mathcal{T}_R$, or in frame $\mathcal{T}_F$. These expressions must be equivalent and so

$$\mathcal{T}_R \begin{bmatrix} 0 \\ 1 \\ 0 \\ 0 \end{bmatrix} = \mathcal{T}_F \begin{bmatrix} 0 \\ 1 \\ 0 \\ 0 \end{bmatrix}. \tag{7.163}$$

The z-axis component (third row) of the two alternative vector expressions gives

$$f_1 = \sin \varphi_F = c_1 \sin \chi + c_2 \cos \chi + c_3, \tag{7.164}$$

in which

$$c_1 = \cos \varphi \sin \delta \tag{7.165}$$
$$c_2 = 0 \tag{7.166}$$
$$c_3 = \sin \varphi \cos \delta, \tag{7.167}$$

as a first constraint equation.

The height of the front-wheel centre can be computed from the (3,4) entries of the matrices $\mathcal{T}_F$ in (7.162) and $\mathcal{T}_R$ in (7.161). This provides a second loop constraint equation, which can be written in the form

$$f_2 = \cos \varphi_F = c_4 \sin \chi + c_5 \cos \chi + c_6 \tag{7.168}$$

where

$$c_4 = \frac{\cos \varphi \, (u_R + u_F \cos \delta)}{r_F - \rho_F} \qquad (7.169)$$

$$c_5 = \frac{-d \cos \varphi}{r_F - \rho_F} \qquad (7.170)$$

$$c_6 = \frac{\rho_R - \rho_F + \cos \varphi \, (r_R - \rho_R) - u_F \sin \varphi \sin \delta}{r_F - \rho_F}. \qquad (7.171)$$

Following [285], the two loop equations can be combined as $(7.164)^2 + (7.168)^2$ to eliminate the left-hand-side terms in φ_F; a quadratic expression in $\sin \chi$ and $\cos \chi$ is thereby obtained. We then use the trigonometric identities $\cos \chi = (1 - \tan^2 \frac{\chi}{2})/(1 + \tan^2 \frac{\chi}{2})$ and $\sin \chi = 2 \tan \frac{\chi}{2}/(1 + \tan^2 \frac{\chi}{2})$ to obtain

$$a_4 \left(\tan \frac{\chi}{2} \right)^4 + a_3 \left(\tan \frac{\chi}{2} \right)^3 + a_2 \left(\tan \frac{\chi}{2} \right)^2 + a_1 \left(\tan \frac{\chi}{2} \right) + a_0 = 0 \qquad (7.172)$$

where

$$a_4 = (c_3 - c_2)^2 + (c_6 - c_5)^2 - 1 \qquad (7.173)$$

$$a_3 = 4 \, (c_1 \, (c_3 - c_2) + c_4 \, (c_6 - c_5)) \qquad (7.174)$$

$$a_2 = 4 \, (c_1^2 + c_4^2) + 2 \, (c_3^2 - c_2^2 + c_6^2 - c_5^2 - 1) \qquad (7.175)$$

$$a_1 = 4 \, (c_1 \, (c_3 + c_2) + c_4 \, (c_6 + c_5)) \qquad (7.176)$$

$$a_0 = (c_3 + c_2)^2 + (c_6 + c_5)^2 - 1. \qquad (7.177)$$

The 'thin' wheel case is easily recovered from the above by setting $\rho_F = 0$ and $\rho_R = 0$ to obtain

$$c_4 = \frac{\cos \varphi \, (u_R + u_F \cos \delta)}{r_F} \qquad (7.178)$$

$$c_5 = \frac{-d \cos \varphi}{r_F} \qquad (7.179)$$

$$c_6 = \frac{r_R \cos \varphi - u_F \sin \varphi \sin \delta}{r_F}. \qquad (7.180)$$

One of the earliest expressions for the pitch angle is reported in [286] that considered a single-track vehicle with 'thin' (knife-edge) wheels—the resulting expression was again quartic. A number of other simplified expressions have been proposed over the years. In [287] both the pitch[21] and steer angles are assumed 'small', and [288, 289] assumed that the pitch angle variation is 'small'.

Following [285], an approximated expression that is valid for small pitch variation can be obtained by expanding f_1 and f_2 as power series around the pitch angle in

---

[21] More precisely, the pitch variation with respect to the nominal value.

the nominal configuration χ_0. It follows from the loop closure expressions (7.164) and (7.168) that

$$f_1 \approx f_1|_{\chi=\chi_0} + \left.\frac{\partial f_1}{\partial \chi}\right|_{\chi=\chi_0} (\chi - \chi_0) = \sin \varphi_F \tag{7.181}$$

$$f_2 \approx f_2|_{\chi=\chi_0} + \left.\frac{\partial f_2}{\partial \chi}\right|_{\chi=\chi_0} (\chi - \chi_0) = \cos \varphi_F. \tag{7.182}$$

This gives

$$\chi = \chi_0 - \left.\frac{f_1^2 + f_2^2 - 1}{2\left(f_1 \frac{\partial f_1}{\partial \chi} + f_2 \frac{\partial f_2}{\partial \chi}\right)}\right|_{\chi=\chi_0}. \tag{7.183}$$

Under the additional assumption that the steering angle is 'small', (7.183) reduces to (5.1).

Example 7.6 A numerical example is given in Figure 7.31, where the exact expression (obtained from the solution of the quartic (7.172)) is compared with the linearized expression (7.183). An inspection of Figure 7.31 shows that there is a reduction in the pitch angle χ (i.e. a lowering of the front frame), when steering in the upright position ($\varphi = 0$). When steering at a positive roll angle, the pitch angle has a minimum $\chi_{\min}$ at a positive steer angle (δ). The results have been obtained using the dataset in Table 5.1 with $\rho_R = 0.020\,m$, $\rho_F = 0.015\,m$, and $\chi_0 = 0.1\pi$.

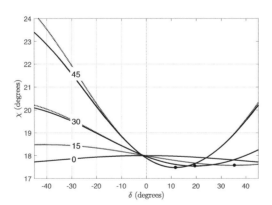

Figure 7.31: Pitch angle as a function of the steer angle δ for different roll angles φ. The exact expression given in (7.172) is shown as a continuous line, while the approximate solution (7.183) is shown dotted. The black points represent minima.

Simple relationships. In the case of the upright bike, analytical expressions for some important dependent parameters can be computed in terms of the parameters given in Table 5.1 and the sketch given in Figure 7.32.

The *normal trail* t_n is the perpendicular distance from the front-wheel ground-contact point to the steering axis and is given by

$$t_n = t \cos \lambda \tag{7.184}$$

in the vehicle's nominal configuration. The perpendicular distance between the front-wheel centre and the steering axis, sometimes called the *wheel offset*, is given by

$$u_F = r_F \sin \lambda - t_n. \tag{7.185}$$

The perpendicular distance between the rear-wheel centre and the steering axis is given by

$$u_R = (w + t) \cos \lambda - r_R \sin \lambda. \tag{7.186}$$

The distance between the wheel centres projected onto the steering axis is given by

$$d = (r_R - r_F) \cos \lambda + w \sin \lambda, \tag{7.187}$$

with a closely related parameter given by

$$p = d + r_F \cos \lambda + t \sin \lambda. \tag{7.188}$$

In the following analysis we will assume that u_r and u_F remain fixed under telescopic front fork compression/extension movements, while d and p both vary. It is clear from Figure 7.32 that

$$u_R \sin \lambda - p \cos \lambda + r_R = 0, \tag{7.189}$$

which can be solved for the angle λ, assuming a variable p, using (7.80):

$$\lambda = \arcsin \left(\frac{-r_R}{\sqrt{u_R^2 + p^2}} \right) + \arctan(p, u_R). \tag{7.190}$$

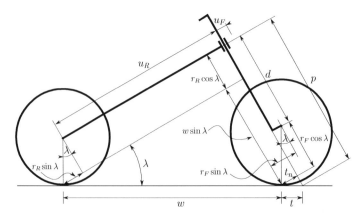

Figure 7.32: Telescopic suspension.

The resulting changes in the trail and wheelbase are given by

$$t = (r_F \sin \lambda - u_F)/(\cos \lambda) \tag{7.191}$$

and

$$w = u_r \cos \lambda + p \sin \lambda - t \tag{7.192}$$

respectively.

It is also possible to compute analytically an expression of the front-wheel yaw angle in terms of the handlebar steer angle δ and the vehicle roll angle φ. As the machine rolls and steers, the front wheel yaw angle varies according to

$$\psi_F = \arctan \left(\frac{\cos \lambda \sin \delta}{\cos \varphi \cos \delta - \sin \varphi \sin \lambda \sin \delta} \right), \tag{7.193}$$

which can be derived from the three-dimensional analysis given in Section 7.7.1 using the ratio of the $(1,1)$ and $(2,1)$ entries of (7.163).[22]

Example 7.7 Using the parameters in Table 5.1, we will study the effect of telescopic fork length variations. In the vehicle's nominal configuration it is easily checked using (7.185)–(7.187) and (7.190) that $p = 0.6252$ m, $u_r = 0.9535$ m, $u_F = 0.0321$ m, $\lambda = 18°$, and $d = 0.2676$ m. Figure 7.33 illustrates the effect of p variations on the steering tilt angle (a), the trail (b), and the wheelbase (c). As one would expect, each of the three variables increase, as the telescopic forks increase in length. Figure 7.33(d) shows the influence of the suspension on the kinematic steering angle, which was computed using (7.193).

Flexible tyres. The introduction of force-generating tyres marked a major advance in the modelling of road vehicles. As explained in Chapter 3, tyre force-generating properties are heavily influenced by the tyre's normal load and so in high-fidelity vehicle models these loads and their points of application (see Section 3.5.1 and Figure 3.11 (b)) have to be computed step by step as the equations of motion are solved.

There are two main modelling options when it comes to representing tyre carcass geometry and flexibility. The first introduces a spring k_n (and possibly a damper in parallel) to account for the tyre compliance in a direction normal to the road plane [290]; see Figure 7.34 (a). The second method makes use of two springs: one in the tyre radial direction k_r and one in the tyre lateral direction k_l [291]; see Figure 7.34 (b). In this case an equivalent spring stiffness k_n in a direction normal to the road surface, when the tyre has a camber angle φ, can be computed as a combination of the radial and lateral stiffness[23]

$$k_n = k_r \left(\cos \varphi \right)^2 + k_l \left(\sin \varphi \right)^2. \tag{7.196}$$

---

[22] In the case that $\lambda = 0$, (7.193) reduces to the expression of the kinematic steering angle Δ obtained with the Timoshenko–Young model in (4.76). Another commonly used expression is obtained by neglecting the second term in the denominator of (7.193) to give $\psi_F \approx \arctan \left(\frac{\cos \lambda \tan \delta}{\cos \varphi} \right)$—the steering 'gain' is attenuated ($\psi_F < \delta$) when $\varphi < \lambda$, and amplified when $\varphi > \lambda$.

[23] This equivalence is readily obtained from a potential energy balance

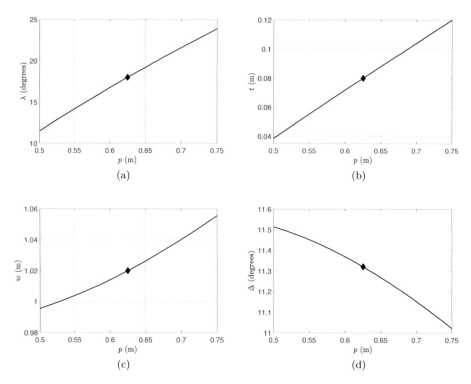

Figure 7.33: Influence of variations in p on the steering tilt angle λ (a), the trail t (b), the wheelbase w (c), and kinematic steering angle Δ for $\delta = 10°$ and $\phi = 30°$ (d); the nominal values are marked with a diamond. The p parameter is varied by $\pm 20\%$ of its nominal value.

In Figure 7.34 the tyre's crown centre is labelled r_{cc}, while r_{tc} represents an extensible string that represents the tyre carcass compliance; the compressed length of this spring is indicative of the normal load. It is evident that the ground-contact point r_{tc} will move to the outer extremity of the tyre as it is rolled over into a corner. In advanced models the normal load (the spring extension) is computed step by step with the aid of force and moment balances on the wheel.

To help appreciate the tyre-contact geometry in three dimensions, we make use of Figure 7.35, which shows a motorcycle tyre with lean angle φ. Point O_n is the original

$$\frac{1}{2} k_n s_n^2 = \frac{1}{2} k_r s_r^2 + \frac{1}{2} k_l s_l^2, \tag{7.194}$$

where s_r, s_l, s_n are the spring deflections in the radial, lateral, and normal directions. These deflections are related by

$$s_r = s_n \cos \varphi \qquad s_l = s_n \sin \varphi, \tag{7.195}$$

which can be be introduced into (7.194) to give (7.196).

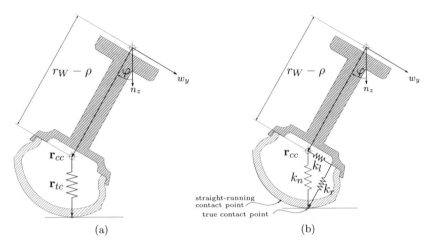

(a)   (b)

Figure 7.34: Two-dimensional tyre-to-ground contact geometry showing the lateral movement of the ground-contact point around the tyre carcass together with (a) the tyre structure represented by a vertical flexibility and (b) with radial and lateral flexibilities.

of an inertial coordinate system, with the associated orthogonal basis vectors given by $\boldsymbol{n}_x$, $\boldsymbol{n}_y$, and $\boldsymbol{n}_z$. Vector $\boldsymbol{w}_y$ is a unit vector in the wheel spindle direction, while $\boldsymbol{r}_{wl}$ is a vector that represents the line of intersection between the ground plane and the wheel's plane of symmetry. The first thing to observe is that

$$\boldsymbol{r}_{wl} = \frac{\boldsymbol{w}_y \times \boldsymbol{n}_z}{\|\boldsymbol{w}_y \times \boldsymbol{n}_z\|} \tag{7.197}$$

is a unit vector along the ground plane line of intersection; although $\boldsymbol{w}_y$ and $\boldsymbol{n}_z$ are unit vectors $\boldsymbol{w}_y \times \boldsymbol{n}_z$ is not. The tyre's crown centre is located by

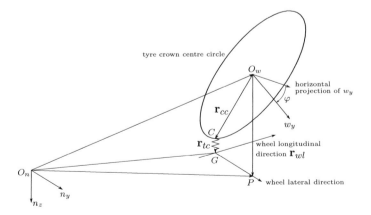

Figure 7.35: Flexible tyre ground-contact geometry.

$$\boldsymbol{r}_{cc} = (r_W - \rho)(\boldsymbol{r}_{wl} \times \boldsymbol{w}_y)$$

$$= (r_W - \rho)\frac{(\boldsymbol{w}_y \times \boldsymbol{n}_z) \times \boldsymbol{w}_y}{\|\boldsymbol{w}_y \times \boldsymbol{n}_z\|}, \tag{7.198}$$

with the ground-contact point directly below it;[24] this expression will be familiar from the car suspension study in (7.153).

Once the normal load and the longitudinal and lateral tyre slips are known (at each integration step), tyre models of the type given in Chapter 3 can be used to compute the tyre's force and moment system. It is important to note that with a model of the type shown in Figure 7.34 (b), there is no need to include a relaxation equation in order to account for the lagging behaviour of lateral forces. This behaviour arises automatically from the lateral spring as explained in Section 3.8.

7.7.2 Side-slipping relaxed tyres

An early attempt to introduce side-slipping and force-generating tyre models into the bicycle literature appears in [186]. Side-slipping tyres facilitate the reproduction of an important oscillatory phenomenon known as 'wobble', which is the name given to shimmy (see Section 4.5 and the discussion therein) in single-track vehicles. Relaxed tyres (Section 5.6.3), as opposed to instantaneous force-generating tyres, make an additional contribution to the accurate modelling of shimmy, allowing these oscillations to become unstable in the normal speed range [292]. Relaxed side-slipping tyres predict correctly the experimentally measured frequency of hands-off wobble, but at low and intermediate speeds (<100 kph) wobble instability is not correctly captured due to poor damping predictions. The accurate predictions of experimentally measured hands-off shimmy damping is better obtained by including frame compliance, as will be discussed in Section 7.7.3. The motorcycle model in [292] provides the minimum level of complexity required for predicting the capsize, weave, and wobble modes.[25] Sharp's model is reminiscent of Whipple's analysis in terms of rider modelling and the frame degrees of freedom.

7.7.3 Structural flexibility

Motivated by the known deterioration in the steering behaviour resulting from torsional compliance between the wheels, [293] extends the model of [292] by allowing the rear wheel to camber relative to the rear frame. In this model the frame flexibility freedom is constrained by a parallel spring-damper arrangement. It was found that swingarm flexibility had very little influence on the capsize and wobble modes, but reduces the weave mode damping at medium and high speeds. The removal of the damping associated with the swingarm flexibility made no material difference to these findings.

Low- and intermediate-speed-wobble mode damping has been associated with front fork compliance and shows improved behaviour with stiffer forks [294]. It is also shown

---

[24] Vectors $\boldsymbol{r}_{wl}$ and $\boldsymbol{w}_y$ are always orthogonal to each other and so no new normalization is required.
[25] The general characteristics of weave and capsize were discussed in Section 5.4.3, and those of wobble in Section 5.6.3.

in [294] that stiffening the rear frame increased the damping of the weave mode. The discrepancy between theory and experiment, with respect to the damping of the wobble mode as a function of speed, is substantially removed by the results in [295] and [296], where mathematical models are extended to include front frame compliances. The compliance of the front frame, which is a combination of the lateral bending of the fork and the torsional compliance of main frame close to the steering head, plays a key role in the stability characteristics of wobble. Stiff configurations, such as those found in sports machines, are prone to high-speed (straight-running) wobble instability, while compliant configurations are prone to low-speed wobble instability. The gyroscopic effects introduced by the frame flexibility appear to be the main wobble-stabilizing influence [84]. The analysis can be extended to include rear-frame flexibility. Bending compliance of the rear of the machine may be beneficial to high-speed weave stability, while torsional compliance is in general detrimental.

Additional contributions relating to the effect of chassis structural flexibilities on motorcyce dynamics can be found in [291, 297–299]. To model the structural properties of a motorcycle, a single lumped torsional flexibility between the front frame and the main chassis is usually employed [290, 300].

The engine-to-rear-wheel slip dynamics (i.e. the powertrain dynamics) may be influenced by structural flexibilities. These are especially important when seeking to model chatter phenomena in racing motorcycles [301–303]. These dynamics may also affect the performance of traction control systems [304, 305].

7.7.4 Rider

Unlike in cars, the rider is a substantial component of a two-wheeled vehicular system, since the ratio of the rider weight to the motorcycle weight is typically in the range 0.25–0.35. Rider movement can therefore be expected to influence the vehicle dynamics. The way the rider moves is related to the dynamics discussed in Section 7.6. In early bicycle and motorcycle models, rider dynamics were neglected and the rider is represented as an inert mass rigidly attached to the rear frame [2, 292].

One of the earliest attempts to include the rider's stretch-reflex/passive response is reported in [306, 307], where the rider's lower body is represented as an inert mass attached to the rear frame, while the upper body is represented as an inverted pendulum that has a single roll freedom constrained by a parallel spring damper arrangement.

This model has been extended to include a lateral degree of freedom in [308], which includes experimentally identified parameters of the spring-damper elements. Values vary from rider to rider, but the average values of [308] have become a reference for successive stability analyses. The results are given in terms of natural frequencies f_l and f_r, and the damping ratios ζ_l and ζ_r, from which the values of the spring k_l and k_r, and the damping constants c_l and c_r, are obtained given the rider mass and inertia:

$$k_l = (2\pi f_l)^2 m \qquad\qquad c_l = 4\pi f_l \zeta_l m \qquad\qquad (7.199)$$
$$k_r = (2\pi f_r)^2 \left(J_u + m_u h_u^2\right) + m_u g h_u \qquad c_r = 4\pi f_r \zeta_r \left(J_u + m_u h_u^2\right). \quad (7.200)$$

The subscripts l and r are related to the lateral and roll motions respectively. The mass of the rider is m, m_u is the mass of the rider's upper body (from head to hip), J_u is the roll moment of the inertia of the rider's upper body with respect to its centre of mass, h_u is the distance between the roll axis of the upper body and the upper-body mass centre, and g is gravity. The average frequency and damping ratio for the lateral motion are 4.0 Hz and 0.321, while for roll they are 1.27 Hz and 0.489. It is suggested that $m_u/m \approx 0.33$ and $h_u \approx 0.4\,m$, although different values have been suggested by various authors.

The straight-running stability of a combined motorcycle–rider model, which focuses on the frame flexibilities and the rider's dynamic characteristics, is studied in [308]. The frequency and damping ratios of the wobble and weave modes are calculated at various speeds and compared with results obtained from experiments conducted with four motorcycles of various sizes. It was found that the rider's oscillatory characteristics influence both wobble and weave. The parameters relating to the rider's upper body motion are most influential on weave, while the parameters associated with the rider's lower body influence primarily the wobble mode. In design, given the variabilities in rider parameters that can be expected, it is important that the machine be designed to be insensitive to changes in these parameters.

Additional experimental data are reported in [309], where thirty-five different riders were tested to find the locations of their mass centres (both in normal and forward-leaned configurations), their related moments of inertia, and typical resonant frequencies and damping ratios. Natural frequencies and damping ratios for the lateral motion are in the range 3.5–4.0 Hz and 0.1–0.3 respectively, while for the roll motion the ranges are 0.8–1.5 Hz and 0.3–0.6. It was also found that the vertical location of the whole rider mass centre is (on average) 0.26 m–0.20 m from the seat surface in normal/leaned position, and from 0.09 m–0.23 m from the rider's trochanter in a normal/leaned position. Finally, the roll moment of inertia of the whole rider is roughly proportional to the rider's mass, the proportionality constant being in the range 0.114–0.077 for the normal/leaned configuration. In the case of the yaw moment of inertia, the proportionality constant is between 0.040 and 0.042 for the normal/leaned configuration.

An alternative to the one- or two-DOF passive rider models discussed above is the use of more sophisticated multibody models. An example is the complex rider model introduced in [310], which comprises twelve rigid bodies representing the upper and lower body, the upper and lower arms, and the upper and lower legs, with appropriate mass and inertia properties. The various rider model masses are restrained by linear springs and dampers so that rider motions such as steering, rolling, pitching, weight shifting, and knee gripping are possible. Rider control actions associated with these degrees of freedom are also modelled, using proportional control elements.

When it comes to the role of the rider as an active controller, the main input in bicycles/motorcycles is the steering torque (as opposed to the steering angle), while the body motion provides slower responses (it is used as a secondary input to assist with steering) [283, 311, 312]. At very low speeds, a stabilizing knee motion in bicycles has been observed [313].

7.7.5   Aerodynamic forces

The importance of aerodynamic forces on the performance and stability of motorcycles at high speeds was demonstrated in [314]. Wind tunnel data were obtained for the steady-state aerodynamic forces acting on a wide range of motorcycle–rider configurations. It appears that the effects of aerodynamic side forces, yawing moments, and rolling moments on the lateral stability of production motorcycles are minor. However, the drag, lift, and pitching moments contribute significantly to changes in the posture of the machine on its suspension and to the tyre loads.

In [315] the effect of a handlebar fairing was examined experimentally; positive aerodynamic damping is obtained and a stabilizing effect on wobble stability was thus postulated. The effect of top boxes was investigated experimentally in [316]: the tests show that top boxes shed eddies of a fairly well-defined frequency. Resonance with the wobble frequency is possible for speeds of approximately 35 m/s, while the weave frequency is generally too low to be affected. In [317] the aerodynamic characteristics of different motorcycles were compared, while in [318] the effect of the rider was discussed.

In addition to the contributions mentioned above, there are several other more recent studies on motorcycle aerodynamics, for example [319–322]. The rider's posture usually has a significant effect on drag. In advanced applications it is common to use different drag coefficients when the vehicle accelerates (rider leaning on the tank), and when the vehicle is under braking (rider in upright position). When simulating motorcycle dynamics and stability, typically only steady-state forces are modelled. These forces significantly affect the tyre loads, which in turn influence the tyre responses and thus the vehicle stability.

7.7.6 Suspensions

The addition of suspension systems introduces two degree of freedom and two related modes, which are usually called *bounce* and *pitch*. The former is mainly a vertical translation of the chassis, while the latter is mainly a chassis pitch mode.[26] Adding tyre radial flexibilities into the model introduces a further two modes: *front hop* and *rear hop*. The former is mainly a front-wheel vibration along on its radial stiffness, while the latter is mainly an oscillation of the rear wheel on its radial stiffness; see Chapter 6. These modes are relatively insensitive to speed variations, and are decoupled from the out-of-plane modes (weave, wobble, and capsize) described in Chapter 5 under straight running. Variabilities, such as they are, in these in-plane modes derive from speed-dependent aerodynamic loading.

The kinematics of common motorcycle suspensions will be analysed in this section. Rear suspensions are two-dimensional single-degree-of-freedom mechanisms, while front suspensions are three-dimensional two-degree-of-freedom mechanisms. Suspension systems can be modelled directly link by link and joint by joint, or with the aid of a separate kinematic 'pre-analysis'. This pre-analysis gives accurate analytic (or numerical) relationships between the suspension position and velocity, and changes in the length (and rate of change of length), of the suspension strut. Once this kinematic relationship is known, changes in the suspension position, and changes in the suspension velocity, can be used to find the restoring force/moment being applied to the wheel carrier. The relationship found in the pre-analysis can usually be replaced by

---

[26] It can be shown that the bounce is a purely vertical motion, and that the pitch is a pure rotation around the centre of mass when $k_f(w - b) - k_r b = 0$, where k_f and k_r are the front and rear suspension stiffnesses at the contact points; see Chapter 6.

an approximate functional relationship tailored to the suspension's expected range of movement; a low-order polynomial will often suffice. Direct literal modelling is the simpler approach, but it will provide equations of motion that will integrate slowly, since the simulation will have to solve the closed kinematic loop equations at each integration step.

In the case of motorcycle rear suspensions, the key issue is to determine the moment M_s applied to the swingarm pivot generated by the spring-damper force F of the suspension strut. The relationship between M_s and F can be obtained using the virtual power principle. This gives

$$M_s \dot{\theta} = F \dot{l}, \tag{7.201}$$

where θ is the swingarm rotation and l is the suspension-strut travel. Thus

$$M_s = F \left(\frac{\dot{l}}{\dot{\theta}} \right) = F\tau, \tag{7.202}$$

where τ is the *velocity ratio*. The steering of the front wheel has to be factored into the analysis when analysing the front suspension.

Cantilever rear suspension. In the simple cantilever rear suspension the suspension struct is attached between the swingarm and the chassis; see Figure 7.36. The kinematic analysis is carried out using the closed kinematic loop $p_2\,p_3\,p_4$. It follows from the cosine rule that

$$
\begin{aligned}
l_{34} &= \sqrt{l_{23}^2 + l_{24}^2 - 2l_{23}l_{24}\cos\left(\theta + \gamma_0 + \pi - \theta_{24}\right)} \\
&= \sqrt{l_{23}^2 + l_{24}^2 + 2l_{23}l_{24}\cos\left(\theta + \gamma_0 - \theta_{24}\right)}.
\end{aligned} \tag{7.203}
$$

The velocity ratio is obtained by differentiating (7.203) to obtain

$$\tau = \frac{\dot{l}_{34}}{\dot{\theta}} = -\frac{l_{23}l_{24}}{l_{34}}\sin\left(\theta + \gamma_0 - \theta_{24}\right). \tag{7.204}$$

Unitrack-like rear suspension. The Unitrack-like rear suspension is illustrated in Figure 7.37. The key feature is that the spring-damper unit (monoshock) is attached to the chassis and to the rocker $p_2\,p_3\,p_4$ of the four-bar linkage $p_4\,p_3\,p_5\,p_6$.

Referring to Figure 7.37, one observes that points p_4, p_6, and p_7 are fixed in the main frame. The constant lengths l_2 and l_5, and the constant angle ϕ_0 describe the geometry of a swinging link (rocker), l_3 is the length of a connecting rod that connects the swingarm to the swinging link, and l_4 is a fixed length determined by the swingarm design. The monoshock spring-damper unit has (variable) length l_1. The angle between the swingarm axis $p_1\,p_6$ and the horizontal is θ, γ_0 is the (constant) angle between l_4 and the swingarm axis, while δ describes the angular position of the swinging link.

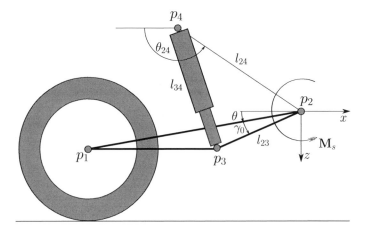

Figure 7.36: Cantilever rear suspension with the spring-damper unit connected between the chassis and swingarm.

For the closed loop p_6–p_4–p_3–p_5–p_6, both the x and z displacements must sum to zero. This means that

$$0 = x_4 - x_6 - l_2 \cos \delta - l_3 \cos \xi + l_4 \cos (\theta + \gamma_0) \tag{7.205}$$

$$0 = z_4 - z_6 + l_2 \sin \delta - l_3 \sin \xi - l_4 \sin (\theta + \gamma_0). \tag{7.206}$$

Also, the position of p_2 relative to the fixed point p_4 is determined by

$$x_2 = x_4 - l_2 \cos \delta + l_5 \cos(\phi_0 + \delta) \tag{7.207}$$

$$z_2 = z_4 + l_2 \sin \delta - l_5 \sin(\phi_0 + \delta). \tag{7.208}$$

Setting

$$c_1 = x_4 - x_6 + l_4 \cos (\theta + \gamma_0) \tag{7.209}$$

$$c_2 = z_4 - z_6 - l_4 \sin (\theta + \gamma_0) \tag{7.210}$$

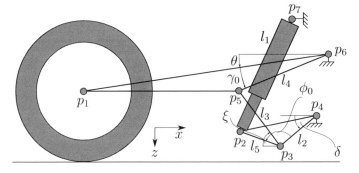

Figure 7.37: Unitrack-like rear suspension with the spring-damper unit between the chassis and the rocker of a four-bar linkage.

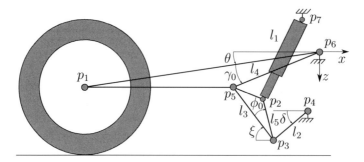

Figure 7.38: Pro-Link-like rear suspension with the spring-damper unit between the chassis and the connecting rod of a four-bar linkage.

in (7.205) and (7.206) respectively, and then squaring and adding in order to eliminate the ξ terms, gives

$$0 = c_1^2 + c_2^2 + l_2^2 - l_3^2 + 2c_2 l_2 \sin \delta - 2c_1 l_2 \cos \delta, \tag{7.211}$$

which can be solved using (7.80) to give the rocker angle δ as a function of the swingarm angle θ

$$\delta = \arcsin \left(\frac{-c_1^2 - c_2^2 - l_2^2 + l_3^2}{2 l_2 \sqrt{c_1^2 + c_2^2}} \right) + \arctan(c_1, c_2). \tag{7.212}$$

The corresponding suspension length is given by

$$l_1 = \sqrt{(x_2 - x_7)^2 + (z_2 - z_7)^2}, \tag{7.213}$$

with x_2 and z_2 given by (7.207) and (7.208) respectively; l_1 is thus a function of θ—$l_1 = l_1(\theta)$, say. The velocity ratio $\tau = \dot{l}_1(\theta)/\dot{\theta}$ can be computed by differentiating (7.213).

Example 7.8 Suppose a Unitrack suspension has the following link lengths: $l_2 = 0.040\,\text{m}$, $l_3 = 0.025\,\text{m}$, $l_4 = 0.051\,\text{m}$, $l_5 = 0.016\,\text{m}$, with point locations $x_4 = 0$, $z_4 = 0.030\,\text{m}$, $x_6 = z_6 = 0$, $x_7 = 0$, $z_7 = -0.030\,\text{m}$, and fixed angles $\phi_0 = 0.675\,\text{rad}$, $\gamma_0 = 0.197\,\text{rad}$. When the swingarm rotation is $\theta = 0$, $c_1 = 0.050, c_2 = 0.020$ and the swingarm link rotation (7.212) is $\delta = 0.072\,\text{rad}$. The related suspension-strut length (7.213) is $l_1 = 0.059\,\text{m}$, with $x_2 = -0.028\,\text{m}$, $z_2 = 0.022\,\text{m}$ given by (7.207) and (7.208). The suspension travel can be fitted (e.g. for θ in the range 0–0.3 rad) with the second-order polynomial $l_1 \approx 10^{-2}\,(5.913 + 3.771\,\theta - 0.980\,\theta^2)$ to give a norm of residuals $0.002 \times 10^{-2}\,\text{m}$. The velocity ratio is $\tau \approx 10^{-2}(3.771 - 1.960\,\theta)$. The linkage is thus slightly progressive.

Pro-Link-like rear suspension.   This suspension is illustrated in Figure 7.38. Its key feature is a spring-damper unit attached to the chassis and to the connecting rod $p_2\,p_3\,p_5$ of the four-bar linkage $p_4\,p_3\,p_5\,p_6$. The Pro-Link-like suspension system represents a variation on unitrack-like designs. The closed kinematic loop $p_6\,p_4\,p_3\,p_5\,p_6$ is identical to that of the previous suspension and (7.205) and (7.206) remain valid. In this case, the position of attachment point p_2 relative to the fixed point p_4 is determined by

$$x_2 = x_6 - l_4 \cos(\theta + \gamma_0) + l_3 \cos \xi - l_5 \cos(\phi_0 + \xi) \tag{7.214}$$

$$z_2 = z_6 + l_4 \sin(\theta + \gamma_0) + l_3 \sin \xi - l_5 \sin(\phi_0 + \xi). \tag{7.215}$$

Squaring and adding (7.205) and (7.206) in order to eliminate the δ terms gives

$$0 = c_1^2 + c_2^2 + l_3^2 - l_2^2 - 2c_2 l_3 \sin \xi - 2c_1 l_3 \cos \xi \tag{7.216}$$

in which c_1 and c_2 are again given by (7.209) and (7.210). Equation (7.216) can be solved using (7.80) to obtain the connecting-rod angle ξ as a function of the swingarm angle θ

$$\xi = \arcsin\left(\frac{-c_1^2 - c_2^2 - l_3^2 + l_2^2}{2 l_3 \sqrt{c_1^2 + c_2^2}}\right) + \arctan(c_1, -c_2). \tag{7.217}$$

The length of the spring-damper unit is again given by (7.213), with x_2 and z_2 given by (7.214) and (7.215). The suspension unit extension is thus a known function of the swingarm rotation θ, and the velocity ratio can be computed by differentiation as before.

Example 7.9 Consider a Pro-Link suspension with the same link lengths (l_2, l_3, l_4, l_5), point locations (p_4, p_6, p_7), and fixed angles ϕ_0, γ_0 as used in Example 7.8. When the swingarm rotation is $\theta = 0$, $c_1 = 0.050$ and $c_2 = 0.020$, to give a connecting-rod rotation, using (7.217), of $\xi = 1.155\,\text{rad}$. The corresponding suspension length (7.213) is $l_1 = 0.059\,\text{m}$, with $x_2 = -0.036\,\text{m}$, $z_2 = 0.017\,\text{m}$ given by (7.214) and (7.215). A second-order polynomial fitting the curve with norm of residuals $0.002 \times 10^{-2}\,\text{m}$ is $l_1 \approx 10^{-2}\,(5.940 + 3.813\,\theta - 1.251\,\theta^2)$. The related velocity ratio is $\tau \approx 10^{-2}\,(3.813 - 2.502\,\theta)$.

Unit-Pro-Link/Full-Floater-like rear suspension. The key feature of this suspension type is a spring-damper unit that has no chassis attachment points and usually appears in the two variants illustrated in Figure 7.39.

In the case of Unit-Pro-Link-type suspensions the monoshock unit is attached between the connecting rod and the swingarm, and the kinematic analysis is almost identical to that of the Pro-Link-like suspensions. The suspension-strut length l_1 is again given by (7.213), with p_2 given by (7.214) and (7.215) and ξ given by (7.217). The only difference with respect to the Pro-Link layout is related to the point p_7, which is no longer fixed to the chassis, but attached to the swingarm point p_7, and this can be computed in terms of the swingarm angle as follows:

$$x_7 = x_6 - l_6 \cos(\eta_0 - \theta) \tag{7.218}$$

$$z_7 = z_6 - l_6 \sin(\eta_0 - \theta), \tag{7.219}$$

where l_6 is a constant length and η_0 is a constant angle. The velocity ratio is computed by differentiating $l_1(\theta)$ with respect to θ.

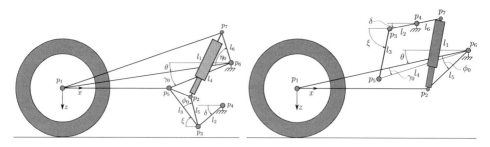

Figure 7.39: Left: rear suspension with the spring-damper unit between the swingarm and the connecting rod of a four-bar linkage (Unit-Pro-Link). Right: spring-damper between the swingarm and the rocker of a four-bar linkage (Full Floater).

In the case of fully floating rear suspension systems the suspension strut is attached between a rocker and the swingarm, and the rocker angle δ is given by (7.212). The suspension travel l_1 is defined using (7.213) with p_2 (on the swingarm) and p_7 (on the rocker) given by

$$x_2 = x_6 - l_5 \cos(\theta + \phi_0) \tag{7.220}$$
$$z_2 = z_6 + l_5 \sin(\theta + \phi_0) \tag{7.221}$$
$$x_7 = x_4 + l_6 \cos \delta \tag{7.222}$$
$$z_7 = z_4 - l_6 \sin \delta. \tag{7.223}$$
$$\tag{7.224}$$

Equations (7.220)–(7.224) depend on the swingarm angle θ and rocker angle $\delta(\theta)$ only. Thus the spring-damper unit length l_1 is a function of θ. The velocity ratio is computed by derivation as in the previous cases.

Telescopic front suspension.   Unlike the rear suspensions of two-wheeled vehicles, the front suspensions of bicycles and motorcycles are three-dimensional mechanisms; their three-dimensionality derives from the steering freedom.

An important property of motorcycle front suspensions is their *anti-dive* characteristic; see Section 7.2.2. Figure 7.40 shows the dive characteristics of the telescopic fork under study. For illustrative purposes the steering tilt angle has been increased to $\lambda = 25°$ with the chassis held fixed; P is fixed in the chassis and is thus stationary. Point P_c represents the front-wheel spin axis, which moves as the suspension compresses. The variation in the front-wheel ground-contact point position is also shown. When the motorcycle undergoes braking, a load-transfer force $\boldsymbol{F}_l$ and a braking force $\boldsymbol{F}_x$ are produced with $\boldsymbol{F}_R$ the resultant; the load-transfer angle is (7.10) (in the case that aerodynamic forces are neglected). The resultant force can be decomposed into a dive-producing force $\boldsymbol{F}_d$, which acts along the suspension's sliding axis (inclined by the support angle β_{sa}), and an orthogonal force component $\boldsymbol{F}_\perp$. This compression effect is more pronounced as the sliding axis becomes more closely aligned with the resultant force $\boldsymbol{F}_R$.

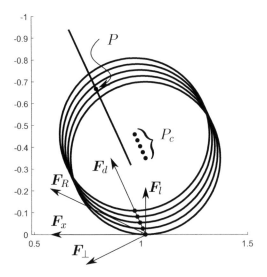

Figure 7.40: Telescopic suspension compression for $\lambda = 25°$. Point P is fixed in the frame, while P_c locates the wheel centre as the suspension compresses. The load-transfer force is $\boldsymbol{F}_l$, the braking force is $\boldsymbol{F}_x$, and their resultant is $\boldsymbol{F}_R$. The dive-producing force is $\boldsymbol{F}_d$, and $\boldsymbol{F}_R = \boldsymbol{F}_d + \boldsymbol{F}_\perp$.

Duolever front suspension. The BMW Duolever front-wheel suspension is essentially a four-bar linkage, in which two trailing links are attached via revolute joints to the chassis, and by spherical joints to the wheel carrier. The suspension scheme is illustrated in Figure 7.41; R and S denote revolute and spherical joints respectively.

The trailing links steer the wheel carrier. A central suspension strut is linked to the lower trailing link at T_2 and the main chassis at T_1. A second linkage mechanism is used to attach the four-bar linkage that supports the wheel carrier to the handlebars [323], a feature originally patented by the British inventor Norman Hossack in the 1980s. The

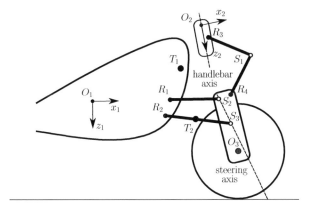

Figure 7.41: Duolever suspension scheme.

Duolever is a three-dimensional mechanism that consists of $n_B = 7$ bodies (chassis, wheel carrier, handlebar, lower rocker between R_2 and S_3, upper rocker between R_1 and S_2, and upper links between R_4 and S_1 and between R_3 and S_1), $n_R = 5$ revolute joints; the steering freedom represents a fifth revolute joint. There are $n_S = 3$ spherical joints as marked in Figure 7.41. The Chebychev–Grübler–Kutzbach formula [324] shows that the system has two degrees of freedom (suspension travel and steering)

$$n_{DOF} = 6(n_B - 1) - 5n_R - 3n_S = 2. \tag{7.225}$$

The idea is to 'separate' the steering and suspension actions, improve the torsional rigidity of the front wheel carrier, and, with a proper design layout, improve the anti-dive properties of the suspension system.

A kinematics analysis can be carried out using a chassis-fixed reference frame with origin O_1, in which the x- and z-coordinates of R_1 and R_2 are fixed. We will also make use of a handlebar-fixed frame with origin O_2, in which the x- and z-coordinates of R_3 are fixed, and a wheel-carrier-fixed reference frame with origin O_3, in which the x- and z- coordinates of S_2, S_3, and R_4 are fixed. The point S_1 can be defined using a rotation α_3 of the constant length vector R_4S_1 in the wheel carrier frame, or with a rotation α_2 of the constant length vector R_3S_1 in the handlebar frame. The frames are defined using the yaw–roll–pitch (ψ–ϕ–μ) convention; (2.159). The position and orientation of the chassis frame are given, the position and rotation of the handlebar frame are given, but the handlebar steer angle δ and the position and orientation of the wheel carrier frame are unknown. At this stage there are nine unknowns, namely $\delta, x_3, y_3, z_3, \psi_3, \phi_3, \mu_3, \alpha_2, \alpha_3$, and as many constraints are necessary to solve the kinematic problem:

(1) the length of the lower rocker of the four-bar linkage is constant, that is ($S_3 - R_2$)·($S_3 - R_2$) = $L_{R_2S_3}^2$; S_3 and R_2 are position vectors;
(2) the lower rocker of the four-bar linkage operates in the chassis x–z plane, that is, ($S_3 - R_2$)·$y_1 = 0$;
(3) a constant length for the upper rocker of the four-bar linkage, that is ($S_2 - R_1$)·($S_2 - R_1$) = $L_{R_2S_3}^2$;
(4) the upper rocker operates in the chassis x–z plane, that is ($S_2 - R_1$)·$y_1 = 0$;
(5) the absolute x-component of S_1 defined in the wheel carrier using α_3 must coincide with the absolute x-component of S_1 defined in the handlebar using α_2;
(6) the absolute y-component of S_1 defined in the wheel carrier using α_3 must coincide with the absolute y-component of S_1 defined in the handlebar using α_2;
(7) the absolute z-component of S_1 defined in the wheel carrier using α_3 must coincide with the absolute z-component of S_1 defined in the handlebar using α_2;
(8) the steer angle δ is given;
(9) the rotation of the lower rocker is given (alternatively the length of the suspension strut between T_1 and T_2 is given).

The resulting system of nine algebraic equations is now solved with a typical dataset, in order to highlight the main features of this suspension. Figure 7.42 (a) shows the duolever suspension in its nominal configuration. The (b) part of the figure shows

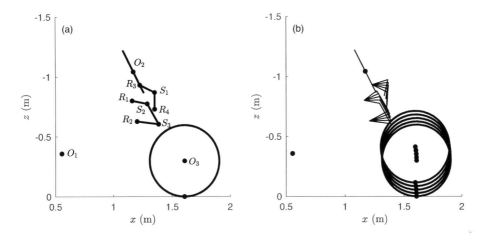

Figure 7.42: Duolever: (a) nominal configuration, (b) anti-dive effect.

the behaviour of the suspension when the steering is fixed straight ahead and the lower rocker angle is varied by 0.6 rad from the nominal configuration. The inspection of the motion of the front-tyre contact point shows an almost vertical displacement as the suspension compresses, which is related to some anti-dive behaviour—a full anti-dive configuration is achieved in the case that the direction of travel is normal to the load-transfer angle, that is, the load-transfer angle β_{lt} has the same inclination as the anti-dive line β_{sa}; Section 7.2.2. With the duolever suspension, the steering axis is determined by S_1 and S_2, and its castor angle increases when the suspension is compressed.

The bump steer behaviour of the Duolever is highlighted in Figure 7.43. While keeping the handlebar steer angle fixed at a prescribed value (10° in this example), a variation in the suspension travel (due to a bump or pothole, for example), results in a variation of the kinematic steer angle of the front wheel (herein defined as the yaw of

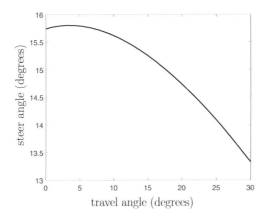

Figure 7.43: Duolever: bump steer effect at a steering angle of 10°.

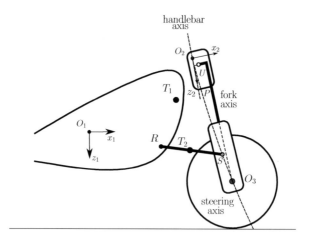

Figure 7.44: Telelever suspension scheme.

the front wheel). In a standard telescopic fork suspension, the kinematic steering angle does not vary with suspension travel when the handlebar steer angle is held fixed.

Telelever front suspension. The BMW Telelever front wheel suspension is illustrated in Figure 7.44, where R, S, P, and U^{27} denote revolute, spherical, prismatic, and universal joints respectively. A central suspension strut is linked to the trailing link at T_2 and the main chassis at T_1 [323].

As with a more conventional machine, wheel guidance is performed by the handlebars and fork legs. A trailing link attached to the main frame supports the forks and front wheel. A central strut (like a monoshock) is used for sprung-mass suspension and damping. This arrangement allows for an anti-dive characteristic using a simpler layout than the one required for the duolever. The telelever geometry can be designed so that the forks and front section respond minimally in dive under braking. The anti-dive properties of the suspension system and the vehicle's stability under braking result from an increase in the steering castor angle when the suspension compresses. With the telelever suspension the trail increases during braking, instead of decreasing as it does with traditional telescopic forks.

The Telelever is a spatial mechanism which consists of $n_B = 5$ bodies (chassis, lower fork, upper fork, handlebar, rocker between R and S), $n_R = 2$ revolute joints (one denoted by R, and one related to the handlebar steer axis between the handlebars and the chassis), $n_P = 1$ prismatic joint (between the upper and lower forks), $n_U = 1$ universal/Hooke joint, and $n_S = 1$ spherical joint. The Chebychev–Grübler–Kutzbach formula [324] shows that the system has two degrees of freedom (travel and steer):

$$n_{DOF} = 6(n_B - 1) - 5n_R - 5n_P - 4n_U - 3n_S = 2. \qquad (7.226)$$

---

[27] The universal joint between the spherical joints connecting the upper forks and the steering assembly is introduced to simulate the kinematics of the telelever using a rigid body analysis. To work properly the 'real system' relies on the structural compliance of the wishbone–fork system.

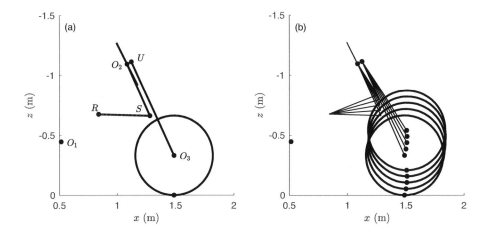

Figure 7.45: Telelever: (a) nominal configuration, (b) anti-dive effect.

A kinematics analysis can be carried out using a chassis-fixed reference frame with origin O_1, in which the x- and z-coordinates of R and T_1 are fixed, a handlebar-fixed frame with origin O_2, in which the x- and z-coordinates of U are fixed, and a lower-fork-fixed reference frame with origin O_3 (z_3 is aligned with the fork travel axis), in which the x- and z-coordinates of S and x-coordinate of U are fixed. The frame's angular positions are related using the yaw–roll–pitch (ψ–ϕ–μ) convention; (2.159).

The position and orientation of the chassis frame are given, the position and rotation of the handlebar frame are given, but the handlebar steer angle δ and the position and orientation of the wheel carrier frame are unknown. At this stage there are seven unknowns, namely δ, x_3, y_3, z_3; the coordinates of O_3, ψ_3, ϕ_3, μ_3 and as many constraints are necessary to solve the kinematic problem:

(1) the rocker length $(\boldsymbol{S} - \boldsymbol{R}) \cdot (\boldsymbol{S} - \boldsymbol{R}) = L_{RS}^2$ is constant;
(2) the rocker operates in the chassis x–z plane, that is, $(\boldsymbol{S} - \boldsymbol{R}) \cdot \boldsymbol{y}_1$ is constant;
(3) the x-component of U in the lower-fork frame coordinate system is fixed;
(4) the y-component of U in the lower-fork frame is fixed (and equal to zero);
(5) universal joint U enforces $\boldsymbol{x}_3 \cdot \boldsymbol{y}_2 = 0$;
(6) the steer angle δ is assumed given;
(7) the rotation of the rocker is given (or the length of the suspension strut between T_1 and T_2 is known).

The resulting system of seven algebraic equations is now solved for a typical dataset, in order to highlight the main features of this suspension. The nominal configuration is illustrated in Figure 7.45 (a). The (b) part of the figure shows the behaviour of the suspension when the steering is fixed straight ahead and the lower rocker angle is varied by 0.4 rad from the nominal configuration. The front-tyre contact point moves along an almost vertical path as the suspension compresses, which relates to the suspension's anti-dive properties; the same conclusion was drawn for the duolever suspension. In the case of the telelever, the steer axis passes through the spherical joint S and the

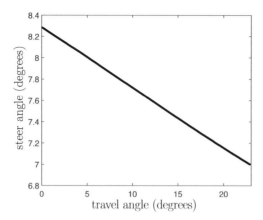

Figure 7.46: Telelever: bump steer effect at a steering angle of $10°$.

projection of the point U on the steer axis. In common with the duolever, the castor angle increases when the suspension compresses.

The bump-steer effect is illustrated in Figure 7.46. When the handlebar steer angle is fixed at a given value ($10°$ in this example), suspension travel (due to a bump or pothole) results in a variation of the kinematic steer angle of the front wheel (also defined as the yaw of the front wheel).

7.7.7   Power transmission

In motorcycles the engine power 'flows' from the engine crankshaft to a drive sprocket on a chassis-mounted gearbox, then to a driven sprocket that is fixed to the rear wheel, either through a chain or belt drive, or a drive shaft. Chain/belt drive transmissions are most common in the case of a transverse engine layout (crankshaft axis aligned with wheel spin axes), while shaft-drive transmission is most common with longitudinal engine layout (crankshaft axis in the symmetry plane of the vehicle). It was shown in Section 7.2.2 that these two different transmission layouts have some dynamic implications on the trim of the vehicle.

An analysis of the chain-drive transmission will be carried out in some detail, in order to obtain expressions for the contact angles of the upper and lower chain runs, which are required in order to apply chain forces in the correct direction in simulation models. These expressions are also necessary to compute the squat angles discussed in Section 7.2.2. A motorcycle chain/belt drive operates between the gearbox drive sprocket and a driven sprocket that is rigidly attached to the rear wheel; see Figure 7.47. When the chain is under tension, it places a secondary load on the swingarm and rear-wheel suspension thereby influencing the swingarm equilibrium angle and thus vehicle's pitch angle.

Referring to Figure 7.48, it can be seen that the chain drive functions as a tension link between the points p_6 on the gearbox sprocket and the point p_4 on the rear-wheel drive sprocket when the engine is driving the rear wheel. Under engine overrun conditions (engine braking), the tension link operates between p_7 on the gearbox sprocket

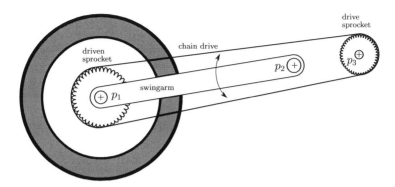

Figure 7.47: Chain drive and swingarm rear suspension. The point p_1 is the centre of the rear-wheel driven sprocket, p_2 is the swingarm pivot, and p_3 is the centre of the drive sprocket.

and p_5 on the rear-wheel sprocket. Some slack, or clearance, is allowed between driving and overrunning, and the chain is treated as elastic and damped. When building a computer model, the transmitted forces will require logic that determines whether the chain is operating under driving or overrun conditions. The points p_4 to p_7 need to be considered as moving points in the sprocket bodies, since they will rotate around their respective sprockets as the chain loading varies. When a chain section is under tension, it is perpendicular to the radial vectors defining the chain–sprocket contact points.

In the kinematic analysis it is convenient to make use of the following six vector relationships:

$$\boldsymbol{r}_{sa} = r_{sa} \begin{bmatrix} -\cos\theta \\ \sin\theta \end{bmatrix} \qquad \boldsymbol{r}_{su} = r_w \begin{bmatrix} -\sin\mu_u \\ -\cos\mu_u \end{bmatrix} \qquad \boldsymbol{r}_{du} = r_d \begin{bmatrix} -\sin\mu_u \\ -\cos\mu_u \end{bmatrix} \quad (7.227)$$

$$\boldsymbol{r}_{s0} = \begin{bmatrix} x_3 - x_2 \\ z_3 - z_2 \end{bmatrix} \qquad \boldsymbol{r}_{sl} = r_w \begin{bmatrix} \sin\mu_l \\ \cos\mu_l \end{bmatrix} \qquad \boldsymbol{r}_{dl} = r_d \begin{bmatrix} \sin\mu_l \\ \cos\mu_l \end{bmatrix}, \qquad (7.228)$$

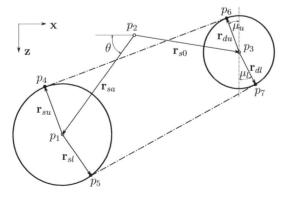

Figure 7.48: Kinematics of the chain drive and swingarm suspension. The points p_4 to p_7 represent the chain–sprocket contact points.

where r_d and r_w are the radii of the drive and driven sprockets respectively. The coordinates of point p_i are denoted (x_i, z_i); the coordinate y_i need not be considered, since this is a two-dimensional analysis. The top chain section is described by $\mathbf{r}_{s0} + \mathbf{r}_{du} - \mathbf{r}_{sa} - \mathbf{r}_{su}$, while the bottom chain section is described by $\mathbf{r}_{s0} + \mathbf{r}_{dl} - \mathbf{r}_{sa} - \mathbf{r}_{sl}$, and so

$$(\mathbf{r}_{s0} + \mathbf{r}_{du} - \mathbf{r}_{sa} - \mathbf{r}_{su}) \cdot \mathbf{r}_{du} = 0 \tag{7.229}$$

and

$$(\mathbf{r}_{s0} + \mathbf{r}_{dl} - \mathbf{r}_{sa} - \mathbf{r}_{sl}) \cdot \mathbf{r}_{dl} = 0. \tag{7.230}$$

It follows by direct calculation that

$$(x_3 - x_2 + r_{sa} \cos\theta) \sin\mu_u + (z_3 - z_2 - r_{sa}\sin\theta)\cos\mu_u - r_d + r_w = 0 \tag{7.231}$$

and

$$(x_3 - x_2 + r_{sa} \cos\theta) \sin\mu_l + (z_3 - z_2 - r_{sa}\sin\theta)\cos\mu_l + r_d - r_w = 0. \tag{7.232}$$

Equations (7.231) and (7.232) are in the form $A\sin(x) + B\cos(x) + C = 0$ and can be solved for μ_u and μ_l, in terms of the swingarm angle θ using (7.80). Once μ_u and μ_l have been determined, the points p_4, p_5, p_6, and p_7 can be computed. When the location of the chain–sprocket contact points are known, the chain/belt force direction can be fully described.

Example 7.10 Consider the chain drive arrangement given in Figure 7.48 and suppose that the origin of the analysis coordinate system is p_2. Suppose also that the drive sprocket has radius $r_d = 0.040\,\text{m}$, and is located at $x_3 = 0.113\,\text{m}$, $z_3 = 0.041\,\text{m}$. The driven sprocket has radius $r_w = 0.100\,\text{m}$, and is located by the vector $\mathbf{r}_{sa}$, which has magnitude $0.305\,\text{m}$ and inclination $\theta = 0.552\,\text{rad}$. The chain angles are determined using (7.80); (7.231) gives the angle of the upper chain run as $\mu_u = 0.155\,\text{rad}$, while (7.232) gives the angle of the lower chain run as $\mu_l = 0.463\,\text{rad}$. In addition, the co-ordinates of points p_4, p_5, p_6, p_7 are $x_4 = -0.275\,\text{m}$, $z_4 = 0.061\,\text{m}$, $x_5 = -0.215\,\text{m}$, $z_5 = 0.249\,\text{m}$, $x_6 = 0.107\,\text{m}$, $z_6 = 0.002\,\text{m}$, $x_7 = 0.131\,\text{m}$, $z_7 = 0.077\,\text{m}$.

The chain/belt force magnitude can be computed as follows. If φ_w and φ_d are respectively the wheel and gearbox sprocket angular positions, the upper chain run is in tension if $r_d\varphi_d - r_w\varphi_w > 0$. It then follows that $\delta l_u = r_d\varphi_d - r_w\varphi_w$ is the increase in separation between p_4 and p_6, and $\delta l_u K_c + \dot{\delta l}_u D_c$ determines the chain/belt force in terms of the chain/belt stiffness K_c and damping D_c. In the same way the lower chain run is in tension if $r_w\varphi_w - r_d\varphi_d > s_l > 0$, in which s_l represents the chain slack. In the overrun case the tension in the lower chain run can be determined in much the same way.

In an alternative simplified approach, when the engine torque M_e is propulsive, the upper chain run is assumed tight and the force is computed using $F_{cu} = M_d/r_d$, where $M_d = M_e\tau$ with M_d the torque at the drive shaft and τ the velocity ratio between the engine crankshaft and the drive-sprocket spin velocities. Under engine braking, the lower run is assumed tight and the related force magnitude is computed as $F_{cl} = M_d/r_d$.

7.7.8 Motorcycle multibody model

All high-fidelity motorcycle models are an evolution of the Whipple bicycle model described in Chapter 5. The new features include such things as relaxed force-generation tyres, a flexible chassis, aerodynamic forces, a rider model, front and rear suspension systems, and a chain- or shaft-drive.

A typical advanced mutibody model [290, 291] has the degree of freedom shown in Figure 7.49, with the corresponding tree structure given in Figure 7.50. The tree structure describes the way in which the various model components move relative to each other. The main frame of the motorcycle is given six degrees of freedom relative to an inertial (road-fixed) reference frame. The rotational and translation freedoms are measure with respect to the inertial n_x, n_y, n_z reference frame given in Figure 7.49. The main body has three children. The first is the swingarm that is allowed to rotate around the y-axis of a main-body-fixed reference frame at the swingarm pivot point p_6; see Figure 7.49. The rear wheel is allowed to spin around a y-axis fixed in the swingarm body at p_1. Next, the upper body of the rider is allowed to rotate around the x-axis of a main-body-fixed reference system at p_9. The third branch of the tree is a little more complicated and describes the frame flexibility, the front suspension, and the front wheel. The steered front assembly is allowed to rotate around the x'-axis of the main-body-fixed reference system x', y', z' at p_8. The z'-axis is aligned with the steer axis, with the y'-axis pointing out of the page. The front-frame compliance motion around the x' axis is usually opposed by the stiff linear rotational spring and a small parallel rotational damper. Steering occurs around the z' axis and the upper and lower parts of the front suspension can translated relative to each other along the z' axis. The front wheel rotates relative to the lower steered assembly at p_7. With this or any other model, the accuracy of the predicted behaviour will depend on effective conceptual modelling and multibody analysis, and on good parameter values [98, 200, 297, 325, 326].

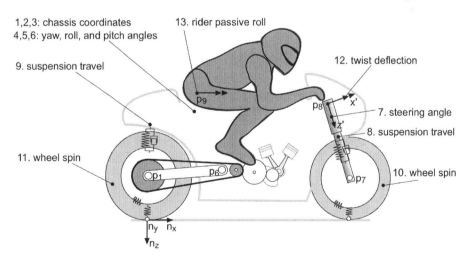

Figure 7.49: Freedoms in a typical multibody motorcycle model; the thirteen degrees of freedom are shown numbered.

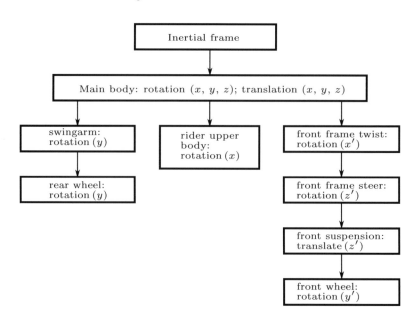

Figure 7.50: Tree structure of the multibody model depicted in Figure 7.49.

The construction of complex vehicle models is a common application of multibody codes. An example of motorcycle hands-off modes in straight running is shown in Figure 7.51 (a) for speeds from 1 to 75 m/s. The speed-dependent modes (eigenvalues) were obtained using the multibody code described in [291], and the dataset of the 1,000 cc sport motorcycle reported in [290].

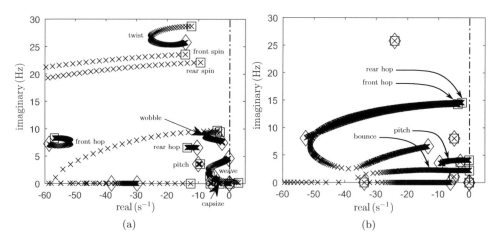

Figure 7.51: Modes of a 1,000 cc motorcycle: (a) straight running when the speed is varied from 1 (squares) and 75 m/s (diamonds); (b) straight running at 35 m/s, with the suspension damping varied from zero (squares) to the baseline value (diamonds).

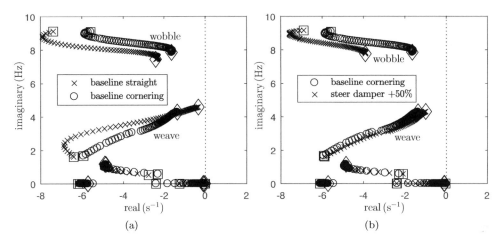

Figure 7.52: Modes of a 1,000 cc motorcycle: (a) cornering with a lateral acceleration of 0.4 g (roll angle ≈ 25°) as the speed is varied between 10 m/s (squares) and 75 m/s (diamonds); (b) cornering with a lateral acceleration of 0.4 g, with the steer damper set at baseline and 50% over baseline.

The model differential equations are a set of implicit differential equations

$$\boldsymbol{f}(\dot{\boldsymbol{x}}, \boldsymbol{x}, \boldsymbol{u}) = \boldsymbol{0}. \tag{7.233}$$

Under steady-state, these equations become

$$\boldsymbol{f}(\boldsymbol{0}, \boldsymbol{x}_0, \boldsymbol{u}_0) = \boldsymbol{0}. \tag{7.234}$$

The algebraic equations given in (7.234) can be solved to find the trim state $\boldsymbol{x}_0$ in terms of the steady-state inputs $\boldsymbol{u}_0$ as a function of the vehicle speed and lateral acceleration (under steady-state cornering). The equations in (7.233) can be linearized about the trim state to give the descriptor state-space equations

$$E\dot{\boldsymbol{x}} = A\boldsymbol{x} + B\boldsymbol{u}, \tag{7.235}$$

where

$$E = \left.\frac{\partial f}{\partial \dot{x}}\right|_{x_0,u_0} ; \qquad A = \left.\frac{\partial f}{\partial x}\right|_{x_0,u_0} ; \qquad B = \left.\frac{\partial f}{\partial u}\right|_{x_0,u_0} . \tag{7.236}$$

The speed-dependent eigenvalues of the matrix pencil $\lambda E - A$ are shown in Figure 7.51 (a).

As shown in Figure 7.51 (a), the weave mode has a natural frequency that increases with speed from 0 to 4.5 Hz; it is unstable for speeds below 7 m/s. The damping of the weave mode then increases up to 17 m/s, and then reduces as the speed increases further. It is worth noting that the weave locus appears to separate into two branches; this is caused by the rider's roll motion in the saddle [84]. The wobble mode has modal frequencies in the range 7.5–9.5 Hz, with minimum damping occurring at a speed of approximately 65 m/s. In this motorcycle the frequency reduces as the speed increases, but opposite trends are possible in other machines. Vehicles with minimum wobble-mode damping below 30 m/s are common. The capsize mode in this vehicle is always stable with the related locus on the negative x-axis. The front frame structural twist mode (the twelfth degree of freedom in Figure 7.49) has frequencies in the 25–28 Hz range. The front- and rear-tyre spin modes are related to circumferential vibrations in the tyre carcass coupled with the slip dynamics.[28] The damped natural frequency of the front-tyre spin mode decreases with speed from around 23.5 Hz (at zero speed) to approximately 22 Hz since the tyre behaves like a spring element at low speed and then like a damper when the speed increases; Section 3.8.1.

Figure 7.51 (b) focuses on the response of the four most important in-plane modes, as the front and rear suspension damping is varied from zero to the baseline value, at a speed of 35 m/s (these modes are not strongly speed dependent when the vehicle is operating in straight running). The front- and rear-wheel hop modes have low-speed frequencies of 14 and 14.5 Hz respectively, which then reduce to 8 and 6.5 Hz respectively. As the same time their damping increases as the suspension damping is increased. The bounce and pitch modes begin with frequencies of 2.2 and 4.1 Hz. The pitch frequency reduces to 3.5 Hz as the speed increases, while the bounce becomes over-damped as the damping is increased to its baseline value.

The cornering situation is considerably more complex, since in this case the in-plane and out-of-plane motions are coupled, with these interactions more prominent at large roll angles. As a consequence, several straight running modes merge together to form combined 'cornering modes' that require study. The early literature [327] discusses the existence of a modified weave mode that occurs under cornering conditions, when the suspension system plays an important role in its initiation and maintenance. To investigate the effect of suspension damping on cornering weave, [327] benchmarks several front and rear suspension dampers in laboratory experiments and riding tests. Motorcycle stability is found to be sensitive to suspension damping characteristics, and cornering weave instability is to some extent controllable through rear suspension damper settings. The influence of suspension damping on the weave mode is also studied both analytically and experimentally in [311]. A separate study [328] demonstrates, using a simple analysis, the possibility of interaction between pitch and weave modes at high speeds, where the lightly damped weave-mode natural frequency approaches that of the pitch mode. Although for straight running the coupling of in-plane and out-of-plane motions is weak, for steady-state cornering the coupling between the two modes increases with increased roll angle, indicating that the inclusion of pitch and bounce freedoms in motorcycle models is essential for handling studies involving cornering. The cornering experiments described in [311] quantify the influences of various motorcycle design parameters and operating conditions on wobble and weave. Tests with a range of motorcycles and riders were carried out for both straight running and

---

[28] When the engine inertia is included, the frequency of the rear-wheel spin mode reduces, because of the increase in the referred moment of inertia. The referred inertia changes with the engaged gear ratio, being maximum in first gear. A detailed analysis of these dynamics is reported in [304].

steady-state cornering. It is shown that reduced rear suspension damping, increased rear loading, and increased speed increase the tendency for the motorcycle to weave. As predicted by theory, the frequency of wobble varies little with speed, while that of weave increases with speed. Significant steps in the theoretical analysis of motorcycle cornering behaviour are documented in [306, 307, 329]. The model developed therein considers small perturbations about straight-running conditions, but also for the first time about steady-state cornering. The model is used to calculate the eigenvalues of the small-perturbation linearized motorcycle, where the results for straight running are consistent with the conventional wisdom. The way the weave- and wobble-mode characteristics are predicted as varying with speed is as expected, with front- and rear-suspension pitch and wheel-hop modes appearing that are almost independent of speed. Under cornering conditions, the interaction of these otherwise uncoupled modes produces more complicated modal behaviours. The cornering weave and combined wheel hop/wobble modes are illustrated, and root loci are plotted to observe the sensitivity of the results to parameter variations. Surprisingly, it is predicted that removing the suspension dampers hardly affects the stability of the cornering weave mode, contrary to the experiences of [311] and [327].

Under cornering, the wobble mode involves suspension movement and the previously speed-independent suspension-pitch and wheel-hop modes now vary markedly with speed. In general, interactions between weave and suspension modes, or wobble and suspension modes, occur when these modes are close in terms of their natural frequency. The coupling of the in-plane and out-of-plane motions also suggests the possibility of road excitation signals being transmitted into the lateral motions of the vehicle, causing steering oscillations [330].

An example of cornering modal behaviour is given in Figure 7.52 (a) for speeds between 10 m/s and 75 m/s and a lateral acceleration of 0.4 g. The equilibrium roll angle is approximately 25°, which is greater than the theoretical 22° because of the 'thick' tyres used; see Section 7.2.4. Again the dataset of [290] simulated with the multibody code [291] has been employed. Figure 7.52 (b) highlights the effect of increasing the steering damper on the cornering modes, which stabilizes the wobble whereas it destabilizes the weave mode.

In addition to the multibody code FastBike [291] used in this section, there are a number of other codes specifically designed for the analysis of motorcycle dynamics, such as BikeSim® and MotorcycleMaker®. Alternatively, the motorcycle model can be created using general-purpose multibody software such as Adams® and Virtual.Lab Motion®. While time domain simulations are possible within all of the codes mentioned above, the computation of vibration modes is sometimes laborious, since most codes cannot compute the trim configuration given the speed and acceleration as was performed in this section using [291].

7.7.9 Remarks on vibration modes

Early models such as that proposed in [2], which have no suspension and non-slipping tyres, only produce weave and capsize modes. Weave involves a fishtailing motion of the whole vehicle, with frequency typically between 2 and 5 Hz, while capsize represents a non-oscillatory rolling of the vehicle. The introduction of side-slipping tyres produces the shimmy/wobble mode, which is a steering vibration with frequency typically in the range 5–10 Hz. Modelling the relaxation behaviour of the tyres reduces significantly the stability of the wobble mode, which can now become unstable at high speeds [292]. The introduction of a twist compliance between the front frame and the chassis causes the wobble mode to become unstable at low speed (in case of high compliance) or high speed (in case of high stiffness) [84, 296]. The introduction of a frame compliance also introduces a structural mode, with frequency typically in the range 20–30 Hz. The inclusion of a passive rider roll freedom 'breaks' the weave mode locus into two branches and stabilizes the high-speed weave [84, 307, 308].

The introduction of suspension introduces two more degrees of freedom, which are associated with the bounce (primarily a vertical motion of the chassis) and pitch (mainly a pitch motion of the chassis plane) that have undamped natural frequencies in the range 2–4 Hz. Radial tyre flexibility adds two more modes to the model: the front and rear wheel-hop modes that have undamped natural frequencies in the range 12–24 Hz.

Powertrain compliance, between the engine crankshaft and the rear wheel, introduces a transmission chatter mode with a frequency of approximately 20 Hz. The transmission chatter mode, which involves an oscillation of the rear end of the vehicle, has been compared with experimental data in [301, 302]. In [331] chatter is associated with the structural mode arising from the twist compliance between the front frame and the chassis—in this case the oscillation is confined to the machine's front end. In [303] the various chatter modes are modelled and simulated in a single model.

The weave and wobble modes discussed so far are 'hands-off' modes. When modelling the effect of rider's hands on the handlebar, the wobble mode stabilizes significantly, while the weave mode destabilizes—recent supporting experimental findings are reported in [332] and [84].

In straight running, the out-of-plane modes (weave and wobble), and the in-plane modes (bounce, pitch, and hops) are decoupled. The out-of-plane freedoms include lateral translation, yaw, roll, and steer, while the in-plane freedoms comprise front- and rear-suspension travel, and front- and rear-tyre radial deflection.[29] Under cornering, the out-of-plane modes and the in-plane modes interact. In particular, the weave mode may gain a significant rear-suspension travel component, while wobble, which usually involves steer, may gain a front-suspension travel component. In addition, the in-plane modes, which are essentially independent of speed in straight running (with any variability coming from speed-dependent aerodynamic loading), become speed dependent under cornering.

We conclude with a remark relating to typical root locus patterns for a motorcycle accelerating gently from standstill. At standstill, the vehicle is unstable due to the body-capsize, and steering-capsize modes; see Section 5.4.3. Indeed, riders must put their feet on the road to prevent the vehicle from falling over. For speeds up to 6–8 m/s[30] the uncontrolled motorcycle is still unstable due to the unstable weave mode;

---

[29] An informative although dated video by Dunlop on weave and wobble vibration modes is available online at https://www.youtube.com/watch?v=z3OQTU-kE2s.
[30] This low critical weave speed range reflects typical motorcycle datasets.

in this speed range rider control is needed to stabilize the vehicle. In this low-speed range the weave frequency is typically between 0.2 and 0.4 Hz [333]. Above 6–8 m/s the vehicle enters a stable speed range that requires no active rider intervention; the machine remains stable with the rider's hands off the handlebars, and even with the rider out of the saddle! At increased speeds, the dynamic capsize mode may become slightly unstable (with hands off the handlebars), but this instability is easily stabilized by passive rider action. If the vehicle is properly designed, set up, and loaded, there should be no weave or wobble instability under normal riding conditions. At high speed inappropriate cargo distributions may promote poorly damped weave, while wobble may be triggered by wheel imbalance, rough road surfaces, or the absence of a steering damper. As a general rule, most parameters variations improving the stability of either the wobble or weave mode, while destabilizing the other. For example, a steer damper stabilizes wobble but destabilizes weave. In the same way hands-on riding introduces damping into the wobble mode but destabilizes the weave mode [332].

8

Optimal Control

8.1 Background

Optimal control problems have been studied extensively since the early part of the twentieth century. Progress in solving these problems has been driven primarily by applications in space and atmospheric flight (including: launch vehicles, interplanetary orbital transfer, and high-performance aircraft). In all of these applications the ability to solve increasingly complex optimal control problems has been made possible by advances in high-speed computing. The mathematical and computing techniques being used are now so diverse, and the range of applications of optimal control is so broad, that a comprehensive review of the subject would be nigh on impossible.

The albatross uses a variant of dynamic soaring, which is a flying strategy that can be optimized by exploiting wind gradients in the atmospheric boundary layer [334]. Dynamic soaring is exploited in unmanned aerial vehicles (UAVs) and gives them an energy-free loitering and travelling capability that prolongs their endurance [335]. Energy-optimal trajectories for UAVs equipped with solar cells are considered in [336]. These vehicles may be used to travel between given positions, within an allowed time, or for loitering in the neighbourhood of a target indefinitely. Another problem relates to the optimal level turn of a solar-powered UAV flying in the atmosphere [337]. The *Goddard problem* is a famous launch problem that was posed by Robert H. Goddard (1882–1945) in [338].

The minimal fuel thrust programming for the terminal phase of a lunar soft landing mission is studied in [339]. The optimal thrust programme consists of a period of zero thrust (free-fall) followed by full thrust until touchdown; this is a so-called *bang–bang* strategy. Controlled thrusters and moment gyroscopes are used to maintain and correct the orbit and attitude of the International Space Station. The moment gyroscopes, which are zero-propellant actuators, can be used to realign the structure. Optimal guidance trajectories can be designed using optimal control [340]. Optimal control also plays a key role in the design of low-energy, low-thrust transfers to the moon [341]. It is shown that low-energy transfers using less propellant than standard can be achieved. Optimal low-thrust transfers to geosynchronous and Molniya orbits are studied in [342]. Another iconic optimal control problem, which moved to centre stage with the introduction of the jet engine, is the *minimum-time-to-climb* problem. As with many of these problems it comes in a variety of 'flavours' [343].

Hybrid vehicles are equipped with at least two energy sources and powertrains that combine at least two modes of propulsion. While hybrid powertrains increase the complexity of the drivetrain control problem, they also increase its flexibility and in

Dynamics and Optimal Control of Road Vehicles. D. J. N. Limebeer and M. Massaro.
© D. J. N. Limebeer and M. Massaro 2018. Published in 2018 by Oxford University Press.
DOI: 10.1093/oso/9780198825715.001.0001

some sense can outperform anything achievable with a single-source drivetrain [344, 345]. Optimal control is used in [346] to optimize the powertrain energy flow and racing line of a hybrid electric vehicle (HEV) around a closed-circuit racetrack. The utility of optimal control in the solution of minimum fuel problems in HEVs is demonstrated in [347]. Fuel-optimal control strategies for HEVs are also studied in [348]. The optimal control of hybrid vehicles using direct transcription is studied in [349], where the problem is discretized and the resulting finite-dimensional optimization problem is solved using a nonlinear programming code.

The computation of numerical solutions to optimal control problems in motor racing originated in the 1950s, when simple heuristic arguments were employed to estimate lap times and optimize setup parameters [350]. The early literature related to the application of optimal control to minimum-time problems for ground vehicles includes [351], where a lane-change manoeuvre for a simple nonlinear point-mass car model is considered. Minimum-time optimization problems for a simple motorcycle model are studied for a number of curves of the Mugello circuit in [352]. Minimum-time problems for a Formula One car racing on the Barcelona and Suzuka circuits are considered in [353]. Various minimum-time and minimum-fuel problems will be discussed in detail in Chapter 9.

As with mechanics, optimal control (theory) is a mature and substantial subject that we will review only briefly, and we again hope to point the reader in the direction of the important literature. Optimal control has been covered in a wide array of excellent books including [354–360], with excellent non-specialist magazine articles given in [361, 362]. Survey articles on numerical methods for optimal control can be found in [363–365].

8.2 Fundamentals

System dynamics. An uncontrolled system may be described by an ordinary differential equation (ODE):

$$\dot{\boldsymbol{x}}(t) = \boldsymbol{f}(\boldsymbol{x}(t), t) \qquad \boldsymbol{x}(0) = \boldsymbol{x}_0 \qquad t > 0. \tag{8.1}$$

In this description we are given an initial condition $\boldsymbol{x}_0 \in \mathbb{R}^n$ and a function $\boldsymbol{f}$: $\mathbb{R}^n \times \mathbb{R} \to \mathbb{R}^n$. [1] The solution of the differential equation is a curve $\boldsymbol{x} : [0, \infty) \to \mathbb{R}^n$, which we call the 'state' of the system. The state is fully determined by $\boldsymbol{x}_0$ and $\boldsymbol{f}$.

Controlled dynamics. Suppose we include control inputs $\boldsymbol{u}(t) : [0, \infty) \to \mathbb{R}^m$, with $\boldsymbol{f} : \mathbb{R}^n \times \mathbb{R}^m \times \mathbb{R} \to \mathbb{R}^n$, then (8.1) is replaced by

$$\dot{\boldsymbol{x}}(t) = \boldsymbol{f}(\boldsymbol{x}(t), \boldsymbol{u}(t), t) \qquad \boldsymbol{x}(0) = \boldsymbol{x}_0 \qquad t > 0. \tag{8.2}$$

The state will now depend on $\boldsymbol{x}_0$, $\boldsymbol{f}$, and $\boldsymbol{u}(t)$.

---

[1] $\mathbb{R}^n$ is the space of n-dimensional real vectors.

Cost function. When the system has control inputs, the task may be to find controls that are 'best' in some sense. To do this we need to quantify 'best' and this is usually done with the help of a *performance index* (or *cost function*), which might be defined as

$$J(\boldsymbol{x}(t), \boldsymbol{u}(t), t) = m(\boldsymbol{x}(T), T) + \int_0^T l(\boldsymbol{x}(t), \boldsymbol{u}(t), t)dt, \tag{8.3}$$

which is the so-called Bolza form, where $\boldsymbol{x}(t)$ solves (8.2) for the control $\boldsymbol{u}(t)$. In (8.3), $l : \mathbb{R}^n \times \mathbb{R}^m \times \mathbb{R} \to \mathbb{R}$ and $m : \mathbb{R}^n \times \mathbb{R} \to \mathbb{R}$ are given. We call l the *running cost* (Lagrange term) and m the *terminal cost* (Mayer term). The terminal time $T > 0$ may be pre-specified, or free. In order that the problem makes sense, $m(\boldsymbol{x}(T), T)$ and $l(\boldsymbol{x}(t), \boldsymbol{u}(t), t)$ are usually chosen to be non-negative, and chosen to reflect some physical quantity whose minimization is required. The control $\boldsymbol{u}(t)$ may be required to lie in a particular set $\mathcal{U}$, such as square-integrable functions or functions of bounded magnitude.

It is noted that a problem formulated in the Bolza form (8.3) can be transformed into an equivalent problem with only a Mayer term. This is accomplished by introducing the additional variable x_0 that is described by

$$\dot{x}_0(t) = l(\boldsymbol{x}(t), \boldsymbol{u}(t), t) \tag{8.4}$$

with initial condition $x_0(0) = 0$. The performance index can now be rewritten as

$$J(\boldsymbol{x}_a(t), \boldsymbol{u}(t), t) = m(\boldsymbol{x}(T), T) + x_0(T) \tag{8.5}$$
$$= g(\boldsymbol{x}_a(T), T) \tag{8.6}$$

where $\boldsymbol{x}_a = [x_0, \boldsymbol{x}]^T$.

If the original system equations are explicitly time-varying, the state can be further augmented by the time-monitoring state

$$\dot{x}_{n+1}(t) = 1, \tag{8.7}$$

with $x_{n+1}(0) = 0$, so that neither the system of state equations nor the performance index are explicit functions of time. If the terminal time is fixed, the boundary condition

$$x_{n+1}(T) = T \tag{8.8}$$

must be imposed.

Optimal control problem. The aim of the optimal control problem is to find a control $\boldsymbol{u}^*(t)$ which minimizes the cost function. In other words, we want a $\boldsymbol{u}^*(t)$ such that

$$J(\boldsymbol{x}^*(t), \boldsymbol{u}^*(t), t) \le J(\boldsymbol{x}(t), \boldsymbol{u}(t), t) \tag{8.9}$$

for all admissible controls $\boldsymbol{u}(t)$; $\boldsymbol{u}^*(t)$ is said to be *optimal* and results in the state $\boldsymbol{x}^*(t)$. If the sign of $J(\boldsymbol{x}(t), \boldsymbol{u}(t), t)$ is reversed, the optimal control problem involves maximizing it. Important technical questions include: (i) conditions under which (8.2) has a unique solution, (ii) conditions under which optimal controls exist, and (iii) the mathematical characterization of the optimal controls, and their synthesis.

8.3 Pontryagin minimum principle (PMP)

In this section we begin by discussing first-order necessary conditions for optimality within a classical variational framework. The arguments we use parallel closely the development of the Euler–Lagrange equation (2.40) from the action integral (2.37), which we may think of as a surrogate for the performance index (8.3). In the optimal control problem we must respect the controlled system dynamics (8.2) and so we will append these dynamic 'constraints' to (8.3) as we did when dealing with holonomic constraints in (2.42). In the case that the control set is bounded, the variational paradigm runs into technical difficulty. The removal of these limitations leads to the famous *Pontryagin minimum principle*.

8.3.1 Unconstrained case

We will follow the development given in [355] and introduce the *costate* $\boldsymbol{\lambda}(t) : [0, \infty) \to \mathbb{R}^n$, which is a vector of Lagrange multipliers used to append (8.2) to (8.3) as follows:

$$J_a(\boldsymbol{x}(t), \boldsymbol{u}(t), t) = m(\boldsymbol{x}(T), T) +$$
$$\int_0^T \Big(l\left(\boldsymbol{x}(t), \boldsymbol{u}(t), t\right) + \boldsymbol{\lambda}^T \left(\boldsymbol{f}(\boldsymbol{x}(t), \boldsymbol{u}(t), t) - \dot{\boldsymbol{x}}(t)\right) \Big) \, dt. \quad (8.10)$$

For the time being we will assume that the state and controls are unconstrained.

Taking the various time dependencies as given, and integrating the last term in (8.10) by parts, gives

$$J_a(\boldsymbol{x}(t), \boldsymbol{u}(t)) = m(\boldsymbol{x}(T)) + \int_0^T \Big(l(\boldsymbol{x}, \boldsymbol{u}) + \boldsymbol{\lambda}^T \left(\boldsymbol{f}(\boldsymbol{x}, \boldsymbol{u}) - \dot{\boldsymbol{x}}\right) \Big) \, dt$$
$$= m(\boldsymbol{x}(T)) + \int_0^T \Big(l(\boldsymbol{x}, \boldsymbol{u}) + \boldsymbol{\lambda}^T \boldsymbol{f}(\boldsymbol{x}, \boldsymbol{u}) + \dot{\boldsymbol{\lambda}}^T \boldsymbol{x} \Big) \, dt - \boldsymbol{\lambda}^T \boldsymbol{x} \Big|_0^T$$
$$= m(\boldsymbol{x}(T)) - \boldsymbol{\lambda}^T \boldsymbol{x} \Big|_0^T + \int_0^T \Big(\mathcal{H} + \dot{\boldsymbol{\lambda}}^T \boldsymbol{x} \Big) \, dt, \quad (8.11)$$

where the *control Hamiltonian* is given by

$$\mathcal{H} = l(\boldsymbol{x}, \boldsymbol{u}) + \boldsymbol{\lambda}^T \boldsymbol{f}(\boldsymbol{x}, \boldsymbol{u}). \quad (8.12)$$

Assuming temporarily that T is fixed (and $\boldsymbol{x}_0$ is given), the variation in J_a corresponding to a variation $\delta \boldsymbol{u}$ in $\boldsymbol{u}$ is

$$\delta J_a = \left[\left(\frac{\partial m}{\partial \boldsymbol{x}} - \boldsymbol{\lambda} \right)^T \delta \boldsymbol{x} \right]_{t=T} + \int_0^T \left(\left(\frac{\partial \mathcal{H}}{\partial \boldsymbol{x}} \right)^T \delta \boldsymbol{x} + \left(\frac{\partial \mathcal{H}}{\partial \boldsymbol{u}} \right)^T \delta \boldsymbol{u} + \dot{\boldsymbol{\lambda}}^T \delta \boldsymbol{x} \right) dt \quad (8.13)$$

where $\delta \boldsymbol{x}$ is the variation in $\boldsymbol{x}$ in the differential equation (8.2) due to $\delta \boldsymbol{u}$.[2]

---

[2] The partial derivatives in (8.13) are vectors with the ith entry of $\frac{\partial m}{\partial \boldsymbol{x}}$ given by $\frac{\partial m}{\partial x_i}$.

At this point we can simplify (8.13) by setting

$$\dot{\boldsymbol{\lambda}} = -\frac{\partial \mathcal{H}}{\partial \boldsymbol{x}} \qquad \text{and} \qquad \frac{\partial m}{\partial \boldsymbol{x}}(T) = \boldsymbol{\lambda}(T), \tag{8.14}$$

so that

$$\delta J_a = \int_0^T \left(\left(\frac{\partial \mathcal{H}}{\partial \boldsymbol{u}}\right)^T \delta \boldsymbol{u} \right) dt. \tag{8.15}$$

As a result, the first-order necessary condition for $\boldsymbol{u}^*$ to be (locally) optimal is

$$\left.\frac{\partial \mathcal{H}}{\partial \boldsymbol{u}}\right|_{\boldsymbol{u}=\boldsymbol{u}^*} = 0 \qquad 0 \le t \le T. \tag{8.16}$$

In order to compute the optimal controls the system dynamic equations (8.2) are solved off the initial condition $\boldsymbol{x}(0) = \boldsymbol{x}_0$, while the costate equation (8.14) is solved off the given boundary condition so that

$$\dot{\boldsymbol{x}}^* = \boldsymbol{f}(\boldsymbol{x}(t), \boldsymbol{u}(t))|_{\boldsymbol{x}=\boldsymbol{x}^* \, \boldsymbol{u}=\boldsymbol{u}^*} \tag{8.17}$$

$$\dot{\boldsymbol{\lambda}}^* = -\left.\frac{\partial \mathcal{H}}{\partial \boldsymbol{x}}\right|_{\boldsymbol{x}=\boldsymbol{x}^* \, \boldsymbol{\lambda}=\boldsymbol{\lambda}^* \, \boldsymbol{u}=\boldsymbol{u}^*} \tag{8.18}$$

$$0 = \left.\frac{\partial \mathcal{H}}{\partial \boldsymbol{u}}\right|_{\boldsymbol{x}=\boldsymbol{x}^* \, \boldsymbol{\lambda}=\boldsymbol{\lambda}^* \, \boldsymbol{u}=\boldsymbol{u}^*} \tag{8.19}$$

for all $t \in [0\,T]$.[3] In order to ensure a (local) minimum, a second-order necessary condition for optimality can be derived, which is the *Legendre–Clebsch condition*

$$\left.\frac{\partial^2 \mathcal{H}}{\partial \boldsymbol{u}^2}\right|_{\boldsymbol{x}=\boldsymbol{x}^* \, \boldsymbol{\lambda}=\boldsymbol{\lambda}^* \, \boldsymbol{u}=\boldsymbol{u}^*} \ge 0 \tag{8.20}$$

for all $t \in [0\,T]$. Conditions (8.19) and (8.20) taken in combination mean that

$$\mathcal{H}(\boldsymbol{x}^*, \boldsymbol{u}, \boldsymbol{\lambda}^*, t) \ge \mathcal{H}(\boldsymbol{x}^*, \boldsymbol{u}^*, \boldsymbol{\lambda}^*, t) \tag{8.21}$$

for all $t \in [0\,T]$ and all admissible $u(t)$.

In summary, (8.17) are the state equations. Equations (8.18) are the costate equations (because they are associated with the costates $\boldsymbol{\lambda}$). The optimal control equations are given by (8.19) and (8.20), which taken together give a weak version of the PMP for unconstrained systems (8.21). Equations (8.17) and (8.18) represent a total of $2n$ ordinary differential equations with (mixed) boundary conditions $\boldsymbol{x}(0) = \boldsymbol{x}_0$ and $\boldsymbol{\lambda}(T)$ given by (8.14), with the controls derived from (8.19); these equations therefore represent a two-points-boundary-value problem (TPBVP).

---

[3] If a problem formulated in Bolza form is transformed into a problem with only a Mayer term, the additional state (8.4) attracts an additional multiplier λ_0 given by $\dot{\lambda}_0 = -\frac{\partial \mathcal{H}}{\partial x_0} = 0$. This means that λ_0 is constant and will reappear later as the *abnormal multiplier*.

8.3.2 Boundary conditions

A fixed end time is implicit in the above analysis, but this need not be the case. If the end time is *not* fixed, (8.13) must be replaced with

$$
\delta J_a = \left[\left(\frac{\partial m}{\partial x} - \lambda \right)^T \delta x + \left(\mathcal{H} + \frac{\partial m}{\partial t} \right) \delta t \right]_{t=T} +
$$

$$
\int_0^T \left(\left(\frac{\partial \mathcal{H}}{\partial x} \right)^T \delta x + \left(\frac{\partial \mathcal{H}}{\partial u} \right)^T \delta u + \dot{\lambda}^T \delta x \right) dt, \tag{8.22}
$$

which complicates slightly the boundary conditions. Comparison with (8.13) shows that the terms multiplied by δt are new. If the end time is free, we require

$$
\lambda(T) = \frac{\partial m}{\partial x}(T) \tag{8.23}
$$

which is the old condition (8.14) and the new (transversality) condition

$$
\mathcal{H}(T) = -\frac{\partial m}{\partial t}(T). \tag{8.24}
$$

In the case that $m = 0$, these conditions become

$$
\lambda(T) = 0 \tag{8.25}
$$

and

$$
\mathcal{H}(T) = 0. \tag{8.26}
$$

Other possible free and fixed end time possibilities are discussed in [355] and [360].

8.3.3 Constancy of the Hamiltonian

In Section 2.4.3 we discussed integrals of motion and showed that if the mechanics Hamiltonian (2.128) is not explicitly time dependent, $\frac{\partial \mathcal{H}}{\partial t} = 0$ and the total system energy (2.87) is conserved. In the case of the control Hamiltonian a similar property holds: if the control Hamiltonian (8.12) is not explicitly time-varying it remains constant.

Direct calculation using (8.12) gives

$$
\begin{aligned}
\frac{d\mathcal{H}}{dt} &= \left(\frac{\partial l}{\partial u} \right)^T \dot{u} + \left(\frac{\partial l}{\partial x} \right)^T \dot{x} + \lambda^T \left(\frac{\partial f}{\partial u} \dot{u} + \frac{\partial f}{\partial x} \dot{x} \right) + \dot{\lambda}^T f + \frac{\partial \mathcal{H}}{\partial t} \\
&= \left(\left(\frac{\partial l}{\partial u} \right)^T + \lambda^T \frac{\partial f}{\partial u} \right) \dot{u} + \left(\left(\frac{\partial l}{\partial x} \right)^T + \lambda^T \frac{\partial f}{\partial x} \right) \dot{x} + \dot{\lambda}^T f + \frac{\partial \mathcal{H}}{\partial t} \\
&= \left(\frac{\partial \mathcal{H}}{\partial u} \right)^T \dot{u} + \left(\frac{\partial \mathcal{H}}{\partial x} \right)^T \dot{x} + \dot{\lambda}^T f + \frac{\partial \mathcal{H}}{\partial t} \\
&= \left(\frac{\partial \mathcal{H}}{\partial u} \right)^T \dot{u} + \left(\left(\frac{\partial \mathcal{H}}{\partial x} \right)^T + \dot{\lambda}^T \right) f + \frac{\partial \mathcal{H}}{\partial t} \\
&= \frac{\partial \mathcal{H}}{\partial t};
\end{aligned} \tag{8.27}
$$

we remind the reader that partial derivatives of vector-valued quantities with respect to vectors are matrices with entry (i, j) of $\frac{\partial \boldsymbol{f}}{\partial \boldsymbol{u}}$ given by $\frac{\partial f_i}{\partial u_j}$ and so on. On an optimal trajectory (8.14) and (8.16) ensure that the control Hamiltonian satisfies $\frac{d\mathcal{H}}{dt} = 0$ when $\frac{\partial \mathcal{H}}{\partial t} = 0$; that is, $\mathcal{H}$ is constant when it is not explicitly time-varying.

8.3.4   Bounded controls

The variational approach to the establishment of necessary conditions for a control to be optimal has a number of shortcomings [360]. The first difficulty relates to the control set; the variational analysis is based on the assumption that the controls are unconstrained and of the form

$$u = u^* + \alpha \xi(t),$$

where ξ is a piecewise continuous function of time and α is a real parameter; this type of variation is not possible when the optimal control is on the boundary of the set of allowable controls. Constrained control sets are common in applications: actuators saturate, engine torques are limited, motor currents are restricted, and so on. When the maximum is achieved on a boundary point of the control set (8.19) need not be satisfied and (8.21) must be used instead. More precisely, the general conditions read [366]

$$\mathcal{H}(\boldsymbol{x}^*, \boldsymbol{u}, \boldsymbol{\lambda}^*, \lambda_0^*, t) \geq \mathcal{H}(\boldsymbol{x}^*, \boldsymbol{u}^*, \boldsymbol{\lambda}^*, \lambda_0^*, t) \qquad \text{for all } t \in [0, T] \tag{8.28}$$

for all $\boldsymbol{u}(t)$ in the admissible control set $\mathcal{U}$, or equivalently

$$\mathcal{H}(\boldsymbol{x}^*, \boldsymbol{u}^*, \boldsymbol{\lambda}^*, \lambda_0^*, t) = \min_{\boldsymbol{u}(t) \in \mathcal{U}} \mathcal{H}(\boldsymbol{x}^*, \boldsymbol{u}, \boldsymbol{\lambda}^*, \lambda_0^*, t). \tag{8.29}$$

Equations (8.28) and (8.29) differ from (8.21), because of the constant *abnormal multiplier* λ_0, which is introduced into the Hamiltonian as follows:

$$\mathcal{H} = \lambda_0 l(\boldsymbol{x}, \boldsymbol{u}) + \boldsymbol{\lambda}^T \boldsymbol{f}(\boldsymbol{x}, \boldsymbol{u}), \tag{8.30}$$

with the additional constraint that λ_0 and $\boldsymbol{\lambda}$ cannot be simultaneously zero. The $\lambda_0 = 0$ case is the *abnormal case*, because the necessary conditions for optimality are independent of the cost functional. If $\lambda_0 > 0$, we can re-normalize to get $\lambda_0 = 1$. The celebrated Pontryagin minimum principle (PMP) requires the use of the (slightly) modified Hamiltonian (8.30) together with the nontriviality condition

$$\boldsymbol{\lambda}^*(t) \text{ and } \lambda_0 \text{ are not both zero for all } t \in [0, T], \tag{8.31}$$

the Hamiltonian conditions (8.17) and (8.18), and the minimization condition (8.29). In the sequel, whenever the abnormal multiplier is not explicitly given, it is assumed that $\lambda_0 = 1$.

Example 8.1 The linear quadratic regulator (LQR) is analysed in many of the standard texts on optimal control [229, 354, 355, 367, 368]; we will show that the LQR can be derived from the PMP. In this problem we consider the dynamics

$$\dot{x}(t) = Ax(t) + Bu(t) \qquad x(0) = x_0, \tag{8.32}$$

and the performance index

$$J(x(0), u) = \frac{1}{2}x^T(T)Sx(T) + \frac{1}{2}\int_0^T \left(x^T Qx + u^T Ru\right) dt, \tag{8.33}$$

in which the weighting matrices satisfy $Q \geq 0$, $S \geq 0$, and $R > 0$. The corresponding control Hamiltonian is given by

$$\mathcal{H} = \frac{1}{2}\left(x^T Qx + u^T Ru\right) + \lambda^T\left(Ax + Bu\right). \tag{8.34}$$

The conditions for the optimal control are

$$\dot{\lambda} = -\frac{\partial \mathcal{H}}{\partial x} = -Qx - A^T\lambda \tag{8.35}$$

and

$$0 = \frac{\partial \mathcal{H}}{\partial u} = Ru + B^T\lambda \tag{8.36}$$

with the terminal boundary condition

$$\lambda(T) = Sx(T) \tag{8.37}$$

on the costate. Equations (8.32), (8.35), and (8.36) can be assembled into the following TPBVP

$$\begin{bmatrix} \dot{x} \\ \dot{\lambda} \end{bmatrix} = \begin{bmatrix} A & -BR^{-1}B^T \\ -Q & -A^T \end{bmatrix}\begin{bmatrix} x \\ \lambda \end{bmatrix} \tag{8.38}$$

with boundary conditions $x(0) = x_0$ and $\lambda(T) = Sx(T)$. The solution of (8.38) generates the open-loop optimal control

$$u^*(t) = -R^{-1}B^T\lambda^*(t). \tag{8.39}$$

The $2n \times 2n$ matrix in (8.38) is a *Hamiltonian matrix* H [55], since it satisfies $(JH)^T = JH$, where

$$J = \begin{bmatrix} 0_n & I_n \\ -I_n & 0_n \end{bmatrix}. \tag{8.40}$$

Although not particularly relevant here, the TPBVP (8.38) can be replaced by the Riccati differential equation [229]

$$\dot{P}(t) = P(t)A + A^T P(t) - P(t)BR^{-1}B^T P(t) + Q \qquad P(T) = S \tag{8.41}$$

which must be solved backwards in time from the terminal condition $P(T) = S$. The open-loop control given in (8.39) is then replaced by

$$u(t) = -R^{-1}B^T P(t)x(t), \tag{8.42}$$

which is equivalent, but in feedback form. The final cost is $J^*(x(0)) = x_0^T P(0)x_0$.

Example 8.2 The following example, which is examined in detail in [366], shows that boundary-constrained discontinuous optimal controls arise routinely. Problems of this type cannot be solved using the variational paradigm and (8.19); instead the PMP given in (8.29) must be used. This example also shows that the abnormal multiplier λ_0 may, under certain conditions, be zero.

Consider the controllable oscillatory system

$$\dot{x} = \begin{bmatrix} 0 & 1 \\ -1 & 0 \end{bmatrix} x + \begin{bmatrix} 0 \\ 1 \end{bmatrix} u \tag{8.43}$$

with $x(t_0)$ given and $x(t_f) = \mathbf{0}$ required; the end time t_f is not specified. The aim is to steer the state vector to the origin in minimum time using controls of the form $|u| \leq 1$. The control Hamiltonian is given by [4]

$$\mathcal{H} = \lambda_0 + \boldsymbol{\lambda}^T \left(\begin{bmatrix} 0 & 1 \\ -1 & 0 \end{bmatrix} x + \begin{bmatrix} 0 \\ 1 \end{bmatrix} u \right) = \lambda_0 + \lambda_1 x_2 + \lambda_2(-x_1 + u). \tag{8.44}$$

The PMP (8.29) determines that

$$u^*(t) = \begin{cases} +1 & \text{if} \quad \lambda_2(t) < 0 \\ -1 & \text{if} \quad \lambda_2(t) > 0, \end{cases} \tag{8.45}$$

with the Lagrange multipliers given by (8.18):

$$\dot{\lambda}_1 = -\frac{\partial \mathcal{H}}{\partial x_1} = \lambda_2 \tag{8.46}$$

$$\dot{\lambda}_2 = -\frac{\partial \mathcal{H}}{\partial x_2} = -\lambda_1. \tag{8.47}$$

This gives the following differential equation for the switching function:

$$\ddot{\lambda}_2 = -\lambda_2, \tag{8.48}$$

which has solution

$$\lambda_2(t) = A \sin(t - \varphi); \tag{8.49}$$

the constants A and φ are related to the initial conditions. The optimal controls are thus *bang–bang*, with fixed switching interval π. There are no boundary conditions on the multipliers, because both the initial and final states are specified [360].

When $u = -1$ the state equations (8.43) have solution

$$x_1(t) = R \cos(t - \alpha) - 1 \tag{8.50}$$

$$x_2(t) = -R \sin(t - \alpha), \tag{8.51}$$

and

$$(x_1(t) + 1)^2 + x_2^2(t) = R^2. \tag{8.52}$$

In this case the phase trajectories move on circles of radius R and centre $(-1, 0)$, with R and α determined by the initial conditions; specifically, $R = \sqrt{(x_1(t_0) + 1)^2 + x_2^2(t_0)}$ and $\alpha = \arctan\left(\frac{x_2(t_0)}{x_1(t_0)+1}\right) + t_0$.

---

[4] The time-minimization aspect of the problem is not made explicit in the treatment in [366].

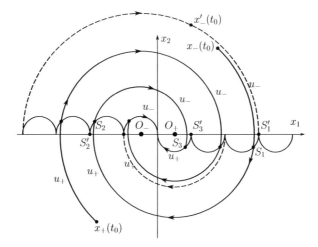

Figure 8.1: Phase trajectory for minimum-time optimal control problem.

In the case that $u = +1$ the state equations (8.43) have solution

$$x_1(t) = 1 - R\cos(t - \alpha) \tag{8.53}$$
$$x_2(t) = R\sin(t - \alpha), \tag{8.54}$$
$$\tag{8.55}$$

with

$$(x_1(t) - 1)^2 + x_2^2(t) = R^2; \tag{8.56}$$

the phase trajectories move on circles of radius R and centre $(1, 0)$. In this case $R = \sqrt{(x_1(t_0) - 1)^2 + x_2^2(t_0)}$ and $\alpha = \arctan\left(\frac{x_2(t_0)}{x_1(t_0) - 1}\right) + t_0$.

The optimal phase portraits for $u = +1$ and $u = -1$ are shown in Figure 8.1, where the phase trajectory centres O_+ and O_- are located at $(1, 0)$ and $(-1, 0)$ respectively. The optimal switching curve is made up of semicircles of unit radius, arranged along the x_1-axis in the second and fourth quadrants of the phase space (these details are covered in [366]).

An exemplar optimal phase-plane trajectory begins at the initial point $x_-(t_0)$ with the optimal control $u = -1$. The control remains negative for $t_s - 2\pi - t_0 \le \pi$ seconds; see Figure 8.2. The control then switches to $u = +1$ at S_1 in the phase plane and time $t_s - 2\pi$, and remains positive for a period of exactly π seconds. The control then

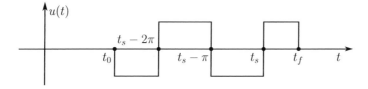

Figure 8.2: Switching control for minimum-time optimal control problem.

switches to $u = -1$ at S_2 for π seconds,[5] and then to $u = +1$ at S_3 for $t_f - t_s < \pi$ seconds. During this final interval the state approaches the origin of the state space along a circular extremal curve with radius $R = 1$; the duration of this phase is $\leq \pi$ seconds. For other initials conditions the phase-plane trajectory may begin at points such as $x_+(t_0)$ with initial optimal control $u = +1$.

Since the state vector must reach the origin of the state space at t_f, $x_1(t)$ and $x_2(t)$ must be of the form

$$x_1(t) = 1 - \cos(t - t_f) \tag{8.57}$$
$$x_2(t) = \sin(t - t_f). \tag{8.58}$$

Since S_3 is a switching point (i.e. $\lambda_2(t_s) = 0$) on the trajectory leading to the origin of the state space, the Lagrange multipliers must be of the form

$$\lambda_2(t) = A\sin(t - t_s) \tag{8.59}$$
$$\lambda_1(t) = -\dot{\lambda}_2 = -A\cos(t - t_s). \tag{8.60}$$

This means that

$$\mathcal{H} = \lambda_0 + \lambda_1 x_2 + \lambda_2(u - x_1) \tag{8.61}$$
$$= \lambda_0 - A\cos(t - t_s)\sin(t - t_f) + A\sin(t - t_s)(1 - (1 - \cos(t - t_f))) \tag{8.62}$$
$$= \lambda_0 - A\sin(t_s - t_f). \tag{8.63}$$

These arguments are essentially the same for initial conditions such as $x_+(t_0)$. Since this optimal control problem is a free end time problem, without a terminal time penalty, $\mathcal{H}(t_f) = 0$; see (8.26). The associated Hamiltonian is constant, because it is not explicitly time varying, and so $\mathcal{H}(t) = 0$. This means that $\lambda_0 = A\sin(t_s - t_f)$.

In cases of initial conditions such as $x'_-(t_0)$, which may lie anywhere on dashed semi-circular arcs with centre O_- and radius $R = 1 + 2n$, $n \geq 1$ (see the dashed semicircular arcs in Figure 8.1), the switching points will lie on the x_1-axis at points such as S'_1, S'_2, and S'_3, and in these cases $\lambda_0 = 0$. Under these circumstances the necessary conditions for optimality are independent of the running cost and the problem is deemed to be *abnormal*. In the case of initial points $x_+(t_0)$, $\lambda_0 = 0$ will occur when $x_+(t_0)$ lies below the switching curve on semicircles with centre O_+ and radius $R = 1 + 2n$, $n \geq 1$.

8.3.5   Bounded states and controls

Here we generalize the optimal control problem described in Section 8.2 to include a mixed penalty on the initial and final states, and path and boundary constraints. The

---

[5]   The semicircular subarcs such as that from S_1 to S_2 in Figure 8.1 have transit times of exactly π seconds.

objective is to determine the state $\boldsymbol{x}(t) \in \mathbb{R}^n$, the control $\boldsymbol{u}(t) \in \mathbb{R}^m$, the initial time t_0, and the terminal time $T \in \mathbb{R}$ that minimize the objective functional

$$J(\boldsymbol{x}(t), \boldsymbol{u}(t), t) = m(\boldsymbol{x}(t_0), t_0, \boldsymbol{x}(T), T) + \int_{t_0}^{T} l(\boldsymbol{x}(t), \boldsymbol{u}(t), t)dt \qquad (8.64)$$

subject to the dynamic constraint

$$\dot{\boldsymbol{x}}(t) = \boldsymbol{f}(\boldsymbol{x}(t), \boldsymbol{u}(t), t). \qquad (8.65)$$

The *path constraints* are given by

$$\boldsymbol{c}(\boldsymbol{x}(t), \boldsymbol{u}(t), t) \leq \boldsymbol{0}, \qquad (8.66)$$

while the *boundary constraints* are

$$\boldsymbol{b}(\boldsymbol{x}(t_0), t_0, \boldsymbol{x}(T), T) = \boldsymbol{0}, \qquad (8.67)$$

where

$$m : \mathbb{R}^n \times \mathbb{R} \times \mathbb{R}^n \times \mathbb{R} \longrightarrow \mathbb{R},$$
$$l : \mathbb{R}^n \times \mathbb{R}^m \times \mathbb{R} \longrightarrow \mathbb{R},$$
$$\boldsymbol{f} : \mathbb{R}^n \times \mathbb{R}^m \times \mathbb{R} \longrightarrow \mathbb{R}^n,$$
$$\boldsymbol{c} : \mathbb{R}^n \times \mathbb{R}^m \times \mathbb{R} \longrightarrow \mathbb{R}^p,$$
$$\boldsymbol{b} : \mathbb{R}^n \times \mathbb{R} \times \mathbb{R}^n \times \mathbb{R} \longrightarrow \mathbb{R}^q.$$

The associated control Hamiltonian is given by

$$\mathcal{H}(\boldsymbol{x}, \boldsymbol{\lambda}, \boldsymbol{\mu}, \boldsymbol{u}, t) = l(\boldsymbol{x}(t), \boldsymbol{u}(t), t) + \boldsymbol{\lambda}^\mathsf{T} \boldsymbol{f}(\boldsymbol{x}(t), \boldsymbol{u}(t), t)$$
$$+ \boldsymbol{\mu}^\mathsf{T} \boldsymbol{c}(\boldsymbol{x}(t), \boldsymbol{u}(t), t). \qquad (8.68)$$

The optimal control problem gives rise to the following first-order necessary conditions:

$$\dot{\boldsymbol{x}} = \frac{\partial \mathcal{H}}{\partial \boldsymbol{\lambda}}, \qquad (8.69)$$

$$\dot{\boldsymbol{\lambda}} = -\frac{\partial \mathcal{H}}{\partial \boldsymbol{x}}, \qquad (8.70)$$

$$\boldsymbol{u}^* = \arg\min_{\boldsymbol{u} \in \mathcal{U}} \mathcal{H}. \qquad (8.71)$$

The boundary conditions are given by

$$\text{either} \quad \boldsymbol{x}(t_0) \quad \text{fixed, or} \quad \boldsymbol{\lambda}(t_0) = -\frac{\partial m}{\partial \boldsymbol{x}}(t_0) + \left(\frac{\partial \boldsymbol{b}}{\partial \boldsymbol{x}}(t_0)\right)^\mathsf{T} \boldsymbol{\nu}$$

$$\text{either} \quad \boldsymbol{x}(T) \quad \text{fixed, or} \quad \boldsymbol{\lambda}(T) = +\frac{\partial m}{\partial \boldsymbol{x}}(T) - \left(\frac{\partial \boldsymbol{b}}{\partial \boldsymbol{x}}(T)\right)^\mathsf{T} \boldsymbol{\nu} \qquad (8.72)$$

$$\text{either} \quad t_0 \quad \text{fixed, or} \quad \mathcal{H}(t_0) = +\frac{\partial m}{\partial t_0} - \boldsymbol{\nu}^\mathsf{T} \frac{\partial \boldsymbol{b}}{\partial t_0}$$

$$\text{either} \quad T \quad \text{fixed, or} \quad \mathcal{H}(T) = -\frac{\partial m}{\partial T} + \boldsymbol{\nu}^\mathsf{T} \frac{\partial \boldsymbol{b}}{\partial T} \qquad (8.73)$$

$$\mu_j(t) = 0, \quad \text{when} \quad c_j(\boldsymbol{x}, \boldsymbol{u}, t) < 0, \quad j = 1, \ldots, p$$
$$\mu_j(t) < 0, \quad \text{when} \quad c_j(\boldsymbol{x}, \boldsymbol{u}, t) = 0, \quad j = 1, \ldots, p \qquad (8.74)$$
$$\mu_j(t) > 0, \quad \text{when} \quad c_j(\boldsymbol{x}, \boldsymbol{u}, t) > 0, \quad j = 1, \ldots, p$$

where $\boldsymbol{\lambda}(t) \in \mathbb{R}^n$ is the costate, $\boldsymbol{\mu}(t)$ is the path constraint multiplier, $\boldsymbol{\nu}$ is the Lagrange multiplier associated with the boundary conditions, and $\mathcal{U}$ is the admissible control set. Equations (8.69) and (8.70) are a *Hamiltonian system* derived from differentiating the (control) Hamiltonian [355, 358]. Equation (8.71) is the Pontryagin minimum principle (PMP). The conditions on the initial and final costate are given in (8.72), which need only be used when the related (initial and final) states are free.[6] The initial and final boundary conditions on the control Hamiltonian are given in (8.73), and need only be used when the initial and final times are left free. The conditions on the Lagrange multipliers of the path constraints given in (8.74) are called *complementary slackness conditions* [369–371]. The Hamiltonian system, together with its boundary conditions, transversality conditions, and complementary slackness conditions, is a *Hamiltonian boundary-value problem* (HBVP). Any solution $(\boldsymbol{x}(t), \boldsymbol{\lambda}(t), \boldsymbol{u}(t), \boldsymbol{\mu}(t), \boldsymbol{\nu})$ is an *extremal* solution and consists of the state, the costate, the controls, and any Lagrange multipliers that satisfy the boundary conditions, and any interior-point constraints on the state and costate.

8.4   Dynamic programming

An alternative approach to the solution of optimal control problems is the use of *dynamic programming*, which employs the intuitively appealing *principle of optimality* (after R. E. Bellman). In non-technical language this principle can be summarized as: 'it is better to be smart from the outset than to be stupid for a time and then become smart later'.

Following [355] we consider the optimal multi-stage decision flow graph in Figure 8.3. The first decision, made at a, gives rise to a transition to b with associated cost J_{ab}. The second decision, made at b, gives rise to a transition to c with associated cost J_{bc}. The optimal (minimum) cost is thus

$$J_{ac}^* = J_{ab} + J_{bc}, \qquad (8.75)$$

---

[6]   Suppose a system has state $\boldsymbol{x} = [y(t), z(t)]^T$, and cost function (8.64) with $m = z(T)$ and $l = 0$. The boundary conditions (8.67) are given by $\boldsymbol{b} = [y(t_0) - y_0, z(t_0) - z_0, y(T) - y_T]^T$ with y_0, z_0, y_T given, and with $z(T)$ free. The following may be computed

$$\frac{\partial m}{\partial \boldsymbol{x}}(t_0) = \begin{bmatrix} 0 \\ 0 \end{bmatrix} \quad \frac{\partial m}{\partial \boldsymbol{x}}(T) = \begin{bmatrix} 0 \\ 1 \end{bmatrix} \quad \frac{\partial \boldsymbol{b}}{\partial \boldsymbol{x}}(t_0) = \begin{bmatrix} 1 & 0 \\ 0 & 1 \\ 0 & 0 \end{bmatrix} \quad \frac{\partial \boldsymbol{b}}{\partial \boldsymbol{x}}(T) = \begin{bmatrix} 0 & 0 \\ 0 & 0 \\ 1 & 0 \end{bmatrix} \quad \boldsymbol{\mu} = \begin{bmatrix} \nu_1 \\ \nu_2 \\ \nu_3 \end{bmatrix},$$

which give the following boundary conditions on the costates (8.72)

$$\begin{bmatrix} \lambda_1(t_0) \\ \lambda_2(t_0) \end{bmatrix} = \begin{bmatrix} \nu_1 \\ \nu_2 \end{bmatrix} \quad \begin{bmatrix} \lambda_1(T) \\ \lambda_2(T) \end{bmatrix} = \begin{bmatrix} -\nu_3 \\ 1 \end{bmatrix}.$$

Only the condition $\lambda_2(T) = 1$, related to the free final state $z(t)$, is required to solve the optimal control problem as specified in (8.69), (8.70), and (8.71).

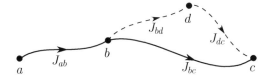

Figure 8.3: Optimal and sub-optimal paths from a to c.

because the overall path is assumed optimal. The principle of optimality asserts that if a–b–c is the optimal path from a to c, then b–c must be the optimal path from b to c. Suppose for contradiction that b–d–c is the optimal path from b to c, then

$$J_{bdc} < J_{bc}$$

and

$$J_{ab} + J_{bdc} < J_{ab} + J_{bc},$$

but this contradicts the assumed optimality of J_{ac}^*.

Bellman formulated the principle of optimality more formally as follows: 'An optimal policy has the property that whatever the initial state and initial decision are, the remaining decisions must constitute an optimal policy with respect to the state resulting from the first decision.'

8.4.1 Dynamic programming recurrence formula

Let us consider again the controllable system dynamics equation (8.2) and the performance index (8.3). The permissible controls are constrained to lie within some set $\mathcal{U}$. In order to convert this continuous-time optimal control problem into a multi-stage dynamic programming problem, we introduce a discrete approximation for the time derivative on the left-hand side of (8.2). Using the first-order explicit Euler formula one obtains

$$\frac{\boldsymbol{x}(t + \Delta t) - \boldsymbol{x}(t)}{\Delta t} \approx \boldsymbol{f}(\boldsymbol{x}(t), \boldsymbol{u}(t)) \qquad \boldsymbol{x}(0) = \boldsymbol{x}_0 \qquad t > 0. \qquad (8.76)$$

In order to simplify the notation we will no longer show time dependencies explicitly. If we suppose that $T = N\Delta t$, and introduce the shorthand $\boldsymbol{x}(k) = \boldsymbol{x}(k\Delta t)$, we can rewrite (8.76) as

$$\begin{aligned}
\boldsymbol{x}(k + 1) &= \boldsymbol{x}(k) + \Delta t \boldsymbol{f}(\boldsymbol{x}(k), \boldsymbol{u}(k)) \\
&= \boldsymbol{f}_D(\boldsymbol{x}(k), \boldsymbol{u}(k)). \qquad (8.77)
\end{aligned}$$

In a similar way we can approximate the performance index (8.3) as

$$\begin{aligned}
J(\boldsymbol{x}(t), \boldsymbol{u}(t)) &= m(\boldsymbol{x}(T)) + \int_0^{\Delta t} l(\boldsymbol{x}(t), \boldsymbol{u}(t)) dt + \int_{\Delta t}^{2\Delta t} l(\boldsymbol{x}(t), \boldsymbol{u}(t)) dt \\
&\quad + \cdots + \int_{(N-1)\Delta t}^{N\Delta t} l(\boldsymbol{x}(t), \boldsymbol{u}(t)) dt, \qquad (8.78)
\end{aligned}$$

which becomes

$$J(\boldsymbol{x}(k),\ \boldsymbol{u}(k)) \approx m(\boldsymbol{x}(N)) + \Delta t \sum_{k=0}^{N-1} l(\boldsymbol{x}(k), \boldsymbol{u}(k)) \qquad (8.79)$$

for small Δt, or more succinctly

$$J(\boldsymbol{x}(k),\ \boldsymbol{u}(k)) \approx m(\boldsymbol{x}(N)) + \sum_{k=0}^{N-1} l_D(\boldsymbol{x}(k), \boldsymbol{u}(k)). \qquad (8.80)$$

In discrete form, the optimal control problem becomes one of finding the optimal control sequence $\boldsymbol{u}^*(\boldsymbol{x}(0))$, $\boldsymbol{u}^*(\boldsymbol{x}(1)), \cdots, \boldsymbol{u}^*(\boldsymbol{x}(N-1))$.

Starting at the end of the integration period we note that

$$J_N(\boldsymbol{x}(N)) = m(\boldsymbol{x}(N)), \qquad (8.81)$$

which is the cost associated with the terminal state. If we include one earlier step, we obtain J_{N-1}, which is the cost associated with the last two steps

$$J_{N-1,N}(\boldsymbol{x}(N-1), \boldsymbol{u}(N-1)) = l_D(\boldsymbol{x}(N-1), \boldsymbol{u}(N-1)) + J_N(\boldsymbol{f}_D(\boldsymbol{x}(N-1), \boldsymbol{u}(N-1))) \qquad (8.82)$$

using (8.81). When scrutinizing (8.82), the reader will appreciate that $\boldsymbol{x}(N)$ is determined by $\boldsymbol{x}(N-1)$ and $\boldsymbol{u}(N-1)$ through the dynamics equation (8.77). In other words, the new stage cost is a function of $\boldsymbol{x}(N-1)$ and $\boldsymbol{u}(N-1)$ only. The optimal cost associated with $\boldsymbol{u}(N-1)$ is thus given by

$$J_{N-1}^*(\boldsymbol{x}(N-1)) = \min_{\boldsymbol{u}(N-1) \in \mathcal{U}} (l_D(\boldsymbol{x}(N-1), \boldsymbol{u}(N-1))$$
$$+ J_N\left(\boldsymbol{f}_D(\boldsymbol{x}(N-1), \boldsymbol{u}(N-1))\right)). \qquad (8.83)$$

The minimizing control $\boldsymbol{u}(N-1)$ depends on $\boldsymbol{x}(N-1)$, and so a feedback structure is beginning to emerge. We denote the last stage of the optimal control sequence $\boldsymbol{u}^*(\boldsymbol{x}(N-1))$.

The performance cost over the last two stages J_{N-2} can be computed as follows

$$\begin{aligned}
J_{N-2}(\boldsymbol{x}(N-2), \boldsymbol{u}(N-2), \boldsymbol{u}(N-1)) &= l_D(\boldsymbol{x}(N-2), \boldsymbol{u}(N-2)) \\
&\quad + l_D(\boldsymbol{x}(N-1), \boldsymbol{u}(N-1)) + m(\boldsymbol{x}(N)) \\
&= l_D(\boldsymbol{x}(N-2), \boldsymbol{u}(N-2)) \\
&\quad + J_{N-1}(\boldsymbol{x}(N-1), \boldsymbol{u}(N-1)). \qquad (8.84)
\end{aligned}$$

Notice that J_{N-1} is determined by $\boldsymbol{x}(N-1)$ and $\boldsymbol{u}(N-1)$, while $J_{N-2,N}$ is determined by $\boldsymbol{x}(N-2)$ and $\boldsymbol{u}(N-2)$. The optimal value of $\boldsymbol{u}(N-2)$ is determined by

$$J_{N-2}^*(\boldsymbol{x}(N-2)) = \min_{\boldsymbol{u}(N-2), \boldsymbol{u}(N-1) \in \mathcal{U}} (l_D(\boldsymbol{x}(N-2), \boldsymbol{u}(N-2))$$
$$+ J_{N-1}(\boldsymbol{x}(N-1), \boldsymbol{u}(N-1))). \qquad (8.85)$$

From the principle of optimality we deduce that irrespective of the eventual choice of $\boldsymbol{x}(N-2)$ and $\boldsymbol{u}(N-2)$, the choice of $\boldsymbol{u}(N-1)$ must be optimal in respect of the

$x(N-1)$ resulting from $x(N-2)$ and $u(N-2)$. Since we never need to renegotiate $u(N-1)$, we have that

$$J_{N-2}^*(x(N-2)) = \min_{u(N-2)\in\mathcal{U}} (l_D(x(N-2), u(N-2))$$
$$+J_{N-1}^*(f_D(x(N-2), u(N-2)))). \qquad (8.86)$$

For the three-stage process, following exactly the same reasoning, we have

$$J_{N-3}^*(x(N-3)) = \min_{u(N-3)\in\mathcal{U}} (l_D(x(N-3), u(N-3))$$
$$+J_{N-2}^*(f_D(x(N-3), u(N-3)))). \qquad (8.87)$$

Continuing backwards, repeatedly employing the principle of optimality, we arrive at the recurrence relationship

$$J_{N-k}^*(x(N-k)) = \min_{u(N-k)\in\mathcal{U}} (l_D(x(N-k), u(N-k))$$
$$+J_{N-k+1}^*(f_D(x(N-k), u(N-k)))). \qquad (8.88)$$

Summary remarks.   The recurrence relation (8.88) is a mathematical statement of the principle of optimality and forms the basis of an appealing computational procedure. As long as there is at least one feasible control sequence, this direct search procedure will find the minimum global cost, and very general systems of equations, performance criteria, and constraints can in principle be dealt with.

In the context of dynamic programming, state and control constraints make the problem easier to solve than it would otherwise be; constraints shrink the size of the search space. In the case of an unconstrained control one might have to search the space

$$-\infty < u(t) < \infty,$$

whereas in the constrained case the search interval

$$-a < u(t) < a$$

is clearly smaller.

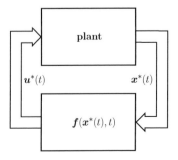

Figure 8.4: Feedback implementation of the optimal control.

The first-order necessary conditions for optimality lead to a TPBVP that generates the control as an open-loop function of time. In contrast, dynamic programming generates the control in feedback form—the control is known for every value of the state within the admissible region as shown in Figure 8.4. For the avoidance of doubt, this does not mean to imply that dynamic programming produces a functional closed-form solution such as

$$u^*(t) = f(x^*(t), t).$$

It may be possible, however, to approximate this form of solution by curve fitting.

Bellman's dynamic programming principle is one of the most powerful techniques that can be used to solve optimal control problems. This technique solves, at least in principle, a large range of important optimal control problems. However, the 'curse of dimensionality' places restriction on the utility of the methodology.

8.4.2 The Hamilton–Jacobi–Bellman equation

In our treatment of dynamic programming we approximated the original continuous-time optimal control problem with a finite-dimensional multi-stage decision process. Let us now return to (8.2) and (8.3), and consider

$$J_t(\boldsymbol{x}(t), \boldsymbol{u}(t), t) = m(\boldsymbol{x}(T)) + \int_t^T l(\boldsymbol{x}(\tau), \boldsymbol{u}(\tau), \tau) d\tau, \tag{8.89}$$

where $0 \le t < T$ and $\boldsymbol{x}_t = \boldsymbol{x}(t)$ is any admissible state vector. This performance index will depend on t, $\boldsymbol{x}_t$, and the control strategy employed on $t \le \tau \le T$. Since the cost integral has t as its lower bound, it is sometimes called the cost-to-go.

The controls that minimize (8.89) are given by

$$
\begin{aligned}
J_t^*(\boldsymbol{x}(t), t) &= \min_{\boldsymbol{u}(\tau), \, t \le \tau \le T} \left\{ m(\boldsymbol{x}(T)) + \int_t^T l(\boldsymbol{x}(\tau), \boldsymbol{u}(\tau), \tau) d\tau \right\} \\
&= \min_{\boldsymbol{u}(\tau), \, t \le \tau \le T} \left\{ m(\boldsymbol{x}(T)) + \int_t^{t+\Delta t} l d\tau + \int_{t+\Delta t}^T l d\tau \right\},
\end{aligned}
\tag{8.90}
$$

where Δt is deemed to be 'small'. By the principle of optimality

$$J_t^*(\boldsymbol{x}(t), t) = \min_{\boldsymbol{u}(\tau), \, t \le \tau \le t+\Delta t} \left\{ \int_t^{t+\Delta t} l d\tau + J_{t+\Delta t}^*(\boldsymbol{x}(t+\Delta t), t+\Delta t) \right\}, \tag{8.91}$$

where $J_{t+\Delta t}^*(\boldsymbol{x}(t+\Delta t), \boldsymbol{u}(t+\Delta t), t+\Delta t)$ is the optimal cost on $t+\Delta t \le \tau \le T$, with initial state $\boldsymbol{x}_{t+\Delta t}$.

Next, we use Taylor's theorem to expand the right-hand side of the above equation (implicit are the requisite smoothness assumptions) to obtain

$$J_t^*(\boldsymbol{x},t) = \min_{\boldsymbol{u}(\tau),\, t\leq\tau\leq t+\Delta t} \left\{ \int_t^{t+\Delta t} l\, d\tau + J_t^*(\boldsymbol{x},t) + \left[\frac{\partial J_t^*}{\partial t}(\boldsymbol{x},t)\right]\Delta t \right.$$

$$\left. + \left[\frac{\partial J_t^*}{\partial \boldsymbol{x}}(\boldsymbol{x},t)\right]^T (\boldsymbol{x}(t+\Delta t) - \boldsymbol{x}(t)) + \text{higher-order terms} \right\}.$$

$$(8.92)$$

Since Δt is small,

$$J_t^*(\boldsymbol{x},t) = \min_{\boldsymbol{u}(t)} \left\{ l\Delta t + J_t^*(\boldsymbol{x},t) + \left[\frac{\partial J_t^*}{\partial t}(\boldsymbol{x},t)\right]\Delta t \right.$$

$$\left. + \left[\frac{\partial J_t^*}{\partial \boldsymbol{x}}(\boldsymbol{x},t)\right]^T (f(\boldsymbol{x},\,\boldsymbol{u},\,t)\Delta t) \right\},$$

$$(8.93)$$

with the higher-order terms neglected. Since $J_t^*(\boldsymbol{x},t)$ is independent of $\boldsymbol{u}(t)$, it can be moved out of the minimization to obtain

$$0 = \left[\frac{\partial J_t^*}{\partial t}(\boldsymbol{x},t)\right]\Delta t + \min_{\boldsymbol{u}(t)} \left\{ l\Delta t + \left[\frac{\partial J_t^*}{\partial \boldsymbol{x}}(\boldsymbol{x},t)\right]^T (f(\boldsymbol{x},\,\boldsymbol{u},\,t)\Delta t) \right\},$$

$$= \left[\frac{\partial J_t^*}{\partial t}(\boldsymbol{x},t)\right] + \min_{\boldsymbol{u}(t)} \left\{ l + \left[\frac{\partial J_t^*}{\partial \boldsymbol{x}}(\boldsymbol{x},t)\right]^T f(\boldsymbol{x},\,\boldsymbol{u},\,t) \right\},$$

or

$$\frac{\partial J_t^*}{\partial t}(\boldsymbol{x},t) = -\min_{\boldsymbol{u}(t)} \left\{ l(\boldsymbol{x},\,\boldsymbol{u},\,t) + \left[\frac{\partial J_t^*}{\partial \boldsymbol{x}}(\boldsymbol{x},t)\right]^T f(\boldsymbol{x},\,\boldsymbol{u},\,t) \right\}, \qquad (8.94)$$

which is one form of the *Hamilton–Jacobi–Bellman* (HJB) equation. The value of $\boldsymbol{u}(t)$ that minimizes the right-hand side of (8.94) will depend on the values taken by $\boldsymbol{x}(t)$, $\frac{\partial J^*}{\partial \boldsymbol{x}}$, and t; the minimizing $\boldsymbol{u}(t)$ is an instantaneous function of the three variables $\boldsymbol{x}(t)$, $\frac{\partial J^*}{\partial \boldsymbol{x}}$, and t. Setting $t = T$ it is clear that the boundary condition on (8.94) is

$$J(\boldsymbol{x}^*(T),T) = m(\boldsymbol{x}(T),T). \qquad (8.95)$$

This is a continuous-time analogue of Bellman's recurrence relation (8.88).

Example 8.3 Page 22 [367]. Suppose we are given the system

$$\dot{x} = u$$

with performance index

$$J(x(t),u(t)) = \int_0^T (u^2 + x^2 + \frac{1}{2}x^4)dt.$$

From the HJB equation (8.94) we have

$$\frac{\partial J_t^*}{\partial t} = -\min_{u(t)} \left\{ u^2 + x^2 + \frac{1}{2}x^4 + \frac{\partial J_t^*}{\partial x}u \right\}.$$

By setting

$$\frac{\partial}{\partial u} \left\{ u^2 + x^2 + \frac{1}{2}x^4 + \frac{\partial J_t^*}{\partial x}u \right\} = 0,$$

we obtain the minimizing u given by

$$u^* = -\frac{1}{2}\frac{\partial J_t^*}{\partial x},$$

and so

$$\frac{\partial J_t^*}{\partial t} = \frac{1}{4}\left(\frac{\partial J_t^*}{\partial x}\right)^2 - x^2 - \frac{1}{2}x^4$$

is the HJB equation for this problem. Unfortunately it is rarely possible to solve the HJB equation.

8.5   Singular arcs and bang–bang control

In an important class of optimal control problems the system equations and the performance index are linear functions of the control variables. This situation may occur, for example, in minimum-time and minimum-fuel problems. When the PMP is used to determine the optimal controls for this type of problem, the solution takes the general form

$$u^*(t) = \begin{cases} a & \text{if } \Phi(t) < 0 \\ \text{unknown} & \text{if } \Phi(t) = 0 \\ b & \text{if } \Phi(t) > 0 \end{cases} \tag{8.96}$$

where a and b are, respectively, upper and lower bounds on the admissible control $u(t)$. The switching function $\Phi(t)$ is the collection of the coefficients of the control terms in the control Hamiltonian. Controls that switch between the upper and lower control limits are called *bang–bang* controls. In some problems the switching function $\Phi(t)$ is singular over a finite time interval. During this interval the Hamiltonian function is no longer an explicit function of the controls u and the PMP yields no information about the optimal control.

In order to deal theoretically with singular arcs, one requires a high-order minimum principle (HOMP), which is a generalization of the familiar PMP discussed here [372]. By using the higher derivatives of the control variations, one is able to construct new necessary conditions for optimal control problems with or without terminal constraints. It would take us too far afield to discuss these issues, but the relevant literature is available in [372] and its bibliography. The following three examples illustrate the practical application and usage of the HOMP conditions. The order of a singular arc is the lowest integer q such that the control u appears explicitly in $d^{2q}\mathcal{H}_u/dt^{2q}$ [373].

8.5.1 Example: Goddard's rocket

Optimizing the thrust programming of rockets, in order to maximize their altitude with a fixed mass of propellant, has been widely analysed for almost a hundred years. The so-called *Goddard problem* was first posed by Robert H. Goddard in his 1919 publication, 'A Method of Reaching Extreme Altitudes' [338]. In common with other classic problems such as the brachystochrone problem [361], it has become a benchmark example in optimal control that has many variants of differing levels of complexity. The Goddard problem's singular-arc structure, with the possibility of more complex sequences of singular subarcs, makes this problem particularly inviting to the theorist. As is well known, the drag plays a significant role in the switching structure of this problem. It has been shown in [374] that for the 'low-drag' case a simple bang–bang structure results. The first subarc is flown at full thrust, while the vehicle coasts on the second subarc. When atmospheric drag is introduced, the thrust programme may, under certain conditions, contain a (singular) variable-thrust subarc and a bang–singular–bang strategy results. If realist transonic drag models are used, in which there is a quick transition from a relatively low C_D (drag coefficient) in the subsonic region to a higher value of C_D, in the supersonic region, bang–singular–bang–singular–bang strategies may arise.

In order to convey some of the technical detail associated with this problem, we follow the treatment of the classical Goddard problem given in [354]. Suppose a single-stage rocket has a fixed amount of propellant, then how should the thrust be programmed, and the propellant burnt, in order to maximize its peak altitude? The equations of motion are[7]

$$\dot{v} = (T - D(v, h))/m - g \tag{8.97}$$

$$\dot{h} = v \tag{8.98}$$

$$\dot{m} = -T/c, \tag{8.99}$$

in which h is the rocket's altitude, v its velocity, m its mass, T is the (controllable) thrust, $D(v, h)$ is the drag force, c is the thrust per unit mass of fuel burnt, and g is the acceleration due to gravity. The problem is to find

$$0 \le T \le T_{\max} \tag{8.100}$$

which maximizes the final altitude following expenditure of the on-board fuel supply. The free end time performance index (in Mayer form) is

$$J = -h(t_f). \tag{8.101}$$

Since J is not explicitly time dependent, it follows from (8.24) that the control Hamiltonian satisfies $\mathcal{H}(T) = 0$, where

$$\mathcal{H} = \lambda_v \left(\frac{T - D(v, h)}{m} - g \right) + \lambda_h v - \lambda_m \frac{T}{c}. \tag{8.102}$$

---

[7] In order to avoid confusion with the control variable u, v will be used for the forward velocity in some instances.

Since $\mathcal{H}$ is not explicitly time dependent, it is constant (Section 8.3.3), and so $\dot{\mathcal{H}} \equiv 0$. In addition, $\mathcal{H}$ is linear in the controls and a bang–bang solution, possibly with singular arcs, can be anticipated.

A singular arc occurs when $\mathcal{H}_u = 0$ for a finite time interval, so that $\dot{\mathcal{H}}_u = \ddot{\mathcal{H}}_u = \cdots = 0$. In this case, realizing that $u = T$,

$$\mathcal{H}_u = \frac{\lambda_v}{m} - \frac{\lambda_m}{c} \tag{8.103}$$

and

$$\dot{\mathcal{H}}_u = \frac{\dot{\lambda}_v m - \lambda_v \dot{m}}{m^2} - \frac{\dot{\lambda}_m}{c} \tag{8.104}$$

The costate equations (8.18) are

$$\dot{\lambda}_v = \frac{\lambda_v}{m}\frac{\partial D}{\partial v} - \lambda_h \tag{8.105}$$

$$\dot{\lambda}_h = \frac{\lambda_v}{m}\frac{\partial D}{\partial h} \tag{8.106}$$

$$\dot{\lambda}_m = \frac{\lambda_v(T - D)}{m^2}. \tag{8.107}$$

Substituting (8.105) and (8.107) into (8.104) gives

$$\dot{\mathcal{H}}_u = \frac{1}{m^2}\left(\lambda_v\frac{\partial D}{\partial v} + \lambda_h\frac{D}{c} - \lambda_m m\right). \tag{8.108}$$

Along a singular arc there hold $\mathcal{H} = 0$, $\mathcal{H}_u = 0$, and $\dot{\mathcal{H}}_u = 0$. Collecting together (8.102), (8.103), and (8.108) gives

$$\begin{bmatrix} c & 0 & -m \\ \frac{\partial D}{\partial v} + \frac{D}{c} & -m & 0 \\ \frac{T-D}{m} - g & v & -\frac{T}{c} \end{bmatrix}\begin{bmatrix} \lambda_v \\ \lambda_h \\ \lambda_m \end{bmatrix} = \begin{bmatrix} 0 \\ 0 \\ 0 \end{bmatrix}. \tag{8.109}$$

The problem's *singular surface* is found by setting the determinant of (8.109) to zero. That is,

$$0 = D(v, h) + mg - \frac{v}{c}D(v, h) - v\frac{\partial D(v, h)}{\partial v}, \tag{8.110}$$

which is a function of all three state variables. In the special case [374] that

$$D(v, h) = \frac{\rho_0}{2}v^2 C_D S e^{-\beta h}, \tag{8.111}$$

in which all the new quantities are constants, (8.110) becomes

$$mg = \left(1 + \frac{v}{c}\right)D(v, h). \tag{8.112}$$

Again, the control law along the singular arc is obtained by considering $\mathcal{H}_u = 0$, $\dot{\mathcal{H}}_u = 0$, and $\ddot{\mathcal{H}}_u = 0$.[8] That is,

---

[8] The third row of (8.113) is obtained by differentiating (8.103)m^2 twice, and substituting equations (8.98)–(8.99), together with the costate equations (8.105)–(8.107).

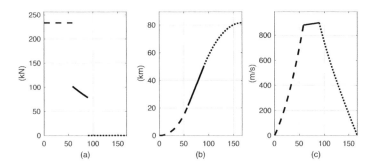

Figure 8.5: Optimized thrust profile for a V-2 ballistic missile as a function of time. Figure (a) shows the classical bang–singular–bang burn programme, Figure (b) shows the missile rising to an apogee of approximately 80 km, and Figure (c) shows the missile's speed reaching a peak during the singular subarc. The full-thrust subarc is shown dashed, the singular subarc is shown solid, and the coasting subarc is dotted. The dataset used is (in SI units): $m(0) = 12,500$, $m(t_f) = 4,539$, $g = 9.81$, $\beta = 1/90,000$, $c = 2,058$, $T_{\max} = 1.9m(0)g$, and $\rho_0 C_D S = 45,360\, g/1,828.8^2$.

$$0 = \det \left[\begin{array}{cc} c & 0 \\ \frac{\partial D}{\partial v} + \frac{D}{c} & -m \\ \left(\frac{1}{cm}\frac{\partial D}{\partial v} + \frac{D}{c^2 m}\right)(T - D) - \frac{\partial D}{\partial h} & -\frac{T}{c} \\ -m & \\ 0 & \\ \frac{T-D-mg}{c}\frac{\partial^2 D}{\partial v^2} + \frac{2T-D-mg}{c^2}\frac{\partial D}{\partial v} + \frac{TD}{c^3} + \frac{mv}{c^2}\left(c\frac{\partial D^2}{\partial v \partial h} + \frac{\partial D}{\partial h}\right) & \end{array} \right] \tag{8.113}$$

This gives the nonlinear speed-dependent feedback law in which (8.111) is employed:

$$T = D + mg + \frac{mg}{1 + 4\frac{c}{v} + 2\left(\frac{c}{v}\right)^2}\left[\frac{\beta c^2}{g}\left(1 + \frac{v}{c}\right) - 1 - 2\frac{c}{v}\right]. \tag{8.114}$$

In the case that (8.110) is never satisfied, a bang–bang thrust programme results and the missile burns all its fuel at the maximum allowed rate and then coasts to its apogee before returning to Earth. Otherwise, after the completion of the full-thrust subarc that is used to increase the speed and altitude of the missile, the singular surface characterized by (8.110) is encountered and the feedback law (8.114) is used to propel the missile along the singular subarc until its fuel is exhausted. On the third subarc the rocket coasts to its apogee prior to returning to Earth.

The performance of a supersonic ballistic missile is shown in Figure 8.5. The full-thrust phase propels the missile to an altitude of almost 20 km at a supersonic speed. A reduced-thrust singular subarc then sees the missile's altitude increased to almost 50 km. It then coasts to an apogee of over 80 km before returning to Earth.

8.5.2 Example: braking wheel

Automatic braking systems (ABS) seek to minimize the distance travelled by a braking vehicle. In this example we seek insight into the ABS design problem by analysing

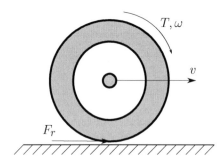

Figure 8.6: Optimal braking of a rotating wheel model.

it as an optimal control problem involving a rotating wheel of the type shown in Figure 8.6 [375].

The equations of motion involve a longitudinal force balance and a moment balance as follows

$$m\dot{v} = F_r \tag{8.115}$$

$$I\dot{\omega} = T - rF_r. \tag{8.116}$$

In these equations m is the wheel mass, v is the wheel's translational velocity, F_r is the tyre friction force, I is the wheel's spin inertia, ω is the wheel's angular velocity, r is the wheel's radius, and T is the braking torque. The tyre friction force is given by the magic formula (3.91) in which, for simplicity, the curvature factor E is set to zero. Thus

$$F_r = D\sin(C\arctan(B\kappa)) \tag{8.117}$$

in which the tyre slip is given by (3.5)

$$\kappa = \frac{\omega r - v}{v}, \tag{8.118}$$

where $v > 0$ and $v > \omega r$ (negative slip). The boundary conditions are $v(0) = v_0$, $\omega(0) = \omega_0$, $v(t_f) = \omega(t_f) = 0$, while the performance index, comprising only a Lagrange term, seeks to minimize the distance travelled while braking to a stop

$$J = \int_0^{t_f} v(\tau)d\tau; \tag{8.119}$$

the final time t_f is the time taken for the wheel to come to a standstill.

If $x_1 = v$ and $x_2 = \omega r$, (8.115) and (8.116) can be rewritten as

$$\dot{x}_1 = f(\boldsymbol{x}) \tag{8.120}$$

$$\dot{x}_2 = -\rho f(\boldsymbol{x}) + u, \tag{8.121}$$

where

$$f(\boldsymbol{x}) = \frac{F_r}{m}; \qquad \rho = \frac{mr^2}{I}; \qquad u = \frac{rT}{I}.$$

The control Hamiltonian associated with this problem is (see (8.12))

$$\mathcal{H} = x_1 + \lambda_1 f(\boldsymbol{x}) + \lambda_2 (u - \rho f(\boldsymbol{x})). \tag{8.122}$$

The costates are given by (see (8.14))

$$\dot{\lambda}_1 = -1 - (\lambda_1 - \lambda_2 \rho) \frac{\partial f}{\partial x_1} \tag{8.123}$$

$$\dot{\lambda}_2 = -(\lambda_1 - \lambda_2 \rho) \frac{\partial f}{\partial x_2}. \tag{8.124}$$

Since there is no Mayer term in the cost function, it follows from (8.24) that

$$\mathcal{H}(t_f) = 0. \tag{8.125}$$

In addition, since the Hamiltonian is not an explicit function of time (see Section 8.3.3) we conclude

$$\mathcal{H}(t) = 0; \text{ for all } t \in [0, t_f] \tag{8.126}$$

along the optimal trajectory. The PMP says

$$u^* = \underset{u}{\arg \min} \, \mathcal{H}(\boldsymbol{x}, \boldsymbol{\lambda}, u). \tag{8.127}$$

Since the applicable braking torque is bounded, the allowed control must satisfy

$$u_{\min} \leq u \leq 0. \tag{8.128}$$

It follows from (8.122) that the switching function is given by

$$\Phi = \lambda_2, \tag{8.129}$$

with the optimal control given by

$$u^* = \begin{cases} u_{\min} & \text{for } \Phi > 0 \\ 0 & \text{for } \Phi < 0 \\ u_{sing} & \text{for } \Phi \equiv 0; \end{cases} \tag{8.130}$$

the control is singular when $\Phi \equiv 0$ holds over a finite time interval. Both $\Phi = 0$ and $\dot{\Phi} = 0$ on the singular arc

$$\begin{bmatrix} 0 & 1 \\ \frac{\partial f}{\partial x_2} & \frac{\partial f}{\partial x_2} \rho \end{bmatrix} \begin{bmatrix} \lambda_1 \\ \lambda_2 \end{bmatrix} = \begin{bmatrix} 0 \\ 0 \end{bmatrix}. \tag{8.131}$$

The non-trivial solution is obtained by setting the determinant of (8.131) to zero, which gives the following condition satisfied on the singular surface

$$\frac{\partial f}{\partial x_2} = 0. \tag{8.132}$$

The slip (8.118) can be rewritten as $\kappa = x_2/x_1 - 1$ and so

$$\frac{\partial f}{\partial x_2} = \frac{\partial f}{\partial \kappa} \frac{\partial \kappa}{\partial x_2} = \frac{\partial f}{\partial \kappa} \frac{1}{x_1} = 0, \tag{8.133}$$

which means that the singular-arc condition corresponds to operating the tyre at its friction peak, where $\partial f / \partial \kappa = 0$.[9] The corresponding singular control is obtained by differentiating (8.132) with respect to time

$$\frac{d}{dt} \left(\frac{\partial f}{\partial x_2} \right) = 0. \tag{8.134}$$

By direct calculation

$$\frac{d}{dt} \left(\frac{\partial f}{\partial x_2} \right) = \frac{\partial^2 f}{\partial x_2^2} \dot{x}_2 + \frac{\partial^2 f}{\partial x_1 \partial x_2} \dot{x}_1 \tag{8.135}$$

$$= \frac{\partial^2 f}{\partial x_2^2} (u - \rho f) + \frac{\partial^2 f}{\partial x_1 \partial x_2} f \tag{8.136}$$

$$= \alpha(x) - \beta(x) u \tag{8.137}$$

where

$$\alpha(x) = \frac{\partial^2 f}{\partial x_1 \partial x_2} f - \frac{\partial^2 f}{\partial x_2^2} \rho f \tag{8.138}$$

and

$$\beta(x) = \frac{\partial^2 f}{\partial x_2^2}. \tag{8.139}$$

If $\beta(x) \neq 0$ the singular control is given by[10]

$$u_{sing} = \frac{\alpha(x)}{\beta(x)}.^{[11]} \tag{8.140}$$

It is intuitively obvious that the optimal braking strategy is to operate the tyre at the friction peak; less well known is that this strategy corresponds to the singular solution of an optimal control problem.

An example of optimal braking which follows a bang–singular–bang strategy is shown in Figure 8.7. The impulsive bang–bang torques at the beginning and end of the braking interval are used to satisfy the boundary conditions, with the remaining strategy on a singular surface that corresponds to maximal tyre friction. Observe that for the current data set $\kappa_{min} = -0.2138$, as predicted by Footnote 9.

---

[9] The peak braking force is obtained from (8.117) when $\kappa = \frac{\tan(\pi/2C)}{B}$, under the standard assumption that $C > 1$.

[10] To ensure that the extremal conditions correspond to a minimum, the so-called Legendre–Clebsch or the Kelley–Contensou conditions must hold. In the present case these conditions require $\beta(x) \leq 0$ [375].

[11] The singular control can be further simplified to $u_{sing} = f \left(\rho + \frac{x_2}{x_1} \right)$ by exploiting

$$\frac{\partial^2 f}{\partial x_1 \partial x_2} = \frac{\partial^2 f}{\partial \kappa^2} \frac{\partial \kappa}{\partial x_1} \frac{\partial \kappa}{\partial x_2}, \quad \frac{\partial^2 f}{\partial x_2^2} = \frac{\partial^2 f}{\partial \kappa^2} \left(\frac{\partial \kappa}{\partial x_2} \right)^2, \quad \frac{\partial \kappa}{\partial x_1} = -\frac{x_2}{x_1^2} \text{ and } \frac{\partial \kappa}{\partial x_2} = \frac{1}{x_1} \text{ [375].}$$

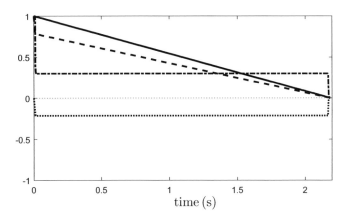

Figure 8.7: Minimum-distance braking of a wheel. The solid line represents v/v_0, the dashed line shows ω/ω_0, the dot-dash line is T/T_{min} and the dotted curve is κ. The dataset used is (in SI units): $g = 9.81$, $m = 250$, $r = 0.25$, $I = 1$, $B = 7.0$, $C = 1.6$, $D = 0.7\,mg$. The initial conditions are $v_0 = 15$, $\omega_0 = v_0/r$. The largest available braking torque is $T_{min} = -1,500$.

8.5.3 Example: Fuller's problem

The famous Fuller problem shows that even seemingly simple optimal control problems can have complex control behaviours [376], which can pose a serious challenge to numerical solution protocols. Fuller's phenomenon relates to extremals that have an infinite number of control switches in finite time (otherwise known as chattering arcs) [377].

Consider the double-integrator system with initial conditions $x_1(0)$ and $x_2(0)$

$$\dot{x}_1 = u \qquad x_1(0) = x_{10}$$
$$\dot{x}_2 = x_1 \qquad x_2(0) = x_{20}$$

with $|u(t)| \leq 1$ and performance index

$$J = \frac{1}{2} \int_0^\infty x_2^2(t)dt. \tag{8.141}$$

The aim is to control the system to the origin of the state space at minimum cost.

Using (8.12) the control Hamiltonian is given by

$$\mathcal{H}(\boldsymbol{x}, \boldsymbol{\lambda}, u) = \frac{1}{2}x_2^2 + \lambda_1 u + \lambda_2 x_1. \tag{8.142}$$

The PMP stipulates that the optimal control is given by

$$u^*(t) = \begin{cases} +1 & \text{if } \lambda_1(t) < 0 \\ -1 & \text{if } \lambda_1(t) > 0 \\ \text{unknown} & \text{if } \lambda_1(t) = 0 \end{cases} \tag{8.143}$$

with

$$\dot{\lambda}_1(t) = -\frac{\partial \mathcal{H}}{\partial x_1} = -\lambda_2(t)$$

$$\dot{\lambda}_2(t) = -\frac{\partial \mathcal{H}}{\partial x_2} = -x_2(t).$$

Since there is no Mayer term in the cost (8.141), the boundary conditions on the multipliers are $\lambda_1(t_2) = \lambda_2(t_2) = 0$; see (8.14). The relationship between the states and Lagrange multipliers is illustrated in Figure 8.8.

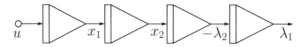

Figure 8.8: States and Lagrange multipliers for Fuller's problem.

Singular control arcs are possible on the interval $t_1 \le t \le t_2$ $(t_1 < t_2)$, if $\frac{\partial \mathcal{H}}{\partial u} = \lambda_1(t) = 0$ for a finite time. In this case it is necessary to differentiate the switching function four times before the control appears explicitly:

$$
\begin{aligned}
\tfrac{d}{dt}\tfrac{\partial \mathcal{H}}{\partial u} &= \dot{\lambda}_1(t) &= -\lambda_2(t) = 0 \\
\tfrac{d^2}{dt^2}\tfrac{\partial \mathcal{H}}{\partial u} &= -\dot{\lambda}_2(t) = x_2(t) &= 0 \\
\tfrac{d^3}{dt^3}\tfrac{\partial \mathcal{H}}{\partial u} &= \dot{x}_2(t) &= x_1(t) &= 0 \\
\tfrac{d^4}{dt^4}\tfrac{\partial \mathcal{H}}{\partial u} &= \dot{x}_1(t) &= u(t) &= 0.
\end{aligned}
\tag{8.144}
$$

It is concluded that a singular arc occurs when $x_1(t) = x_2(t) = 0$, and that the related singular control is $u = 0$. If $u \pm 1$, it follows from Figure 8.8 that

$$x_1 = \pm t + a$$

$$x_2 = \pm\frac{1}{2}t^2 + at + b$$

$$\lambda_2 = \mp\frac{1}{6}t^3 - \frac{1}{2}at^2 - bt + c$$

$$\lambda_1 = \pm\frac{1}{24}t^4 + \frac{1}{6}at^3 + \frac{1}{2}bt^2 - ct + d,$$

in which a, b, c, and d are constants of integration. These equations suggest that it is not possible to control the system to the origin $(x_1 = x_2 = \lambda_1 = \lambda_2 = 0)$ from an arbitrary point in the state space with a finite number of switches. Thus an infinite number of switches are required in finite time; switching occurs when $\lambda_1 = 0$. It can be shown [376] that the switching curve is described by [376]

$$x_2 + hx_1|x_1| = 0, \tag{8.145}$$

where

$$h = \sqrt{\frac{\sqrt{33} - 1}{24}} \approx 0.4446.$$

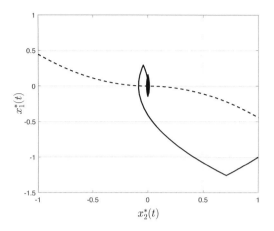

Figure 8.9: Phase plane trajectory for Fuller's problem with $x_1(0) = -1$, $x_2(0) = 1$, and final time $T = 5\,\mathrm{s}$, and the theoretical switching curve given in (8.145).

Using this strategy the origin can be reached in finite time with an infinite number of switches; the switching intervals in fact decrease in geometric progression. It can be shown [376] that the ratio between two consecutive switches t_{s1} and t_{s2} is

$$\frac{t_{s2}}{t_{s1}} = \sqrt{\frac{1 - 2h}{1 + 2h}}. \tag{8.146}$$

If one 'assumes' a control with a finite number of switches in finite time (5 s in this case), an optimization code can be used to find the optimal switching times. Figure 8.9 shows the optimal switching curve given in (8.145), together with an approximation to the optimal phase-plane trajectory in the state space. It is clear from the figure that with a pre-imposed 'finite-number-of-switches' strategy, switching does not always

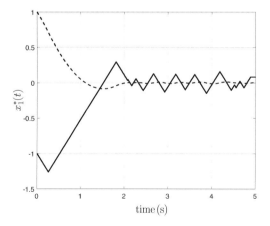

Figure 8.10: Temporal evolution of the states for the Fuller problem described in Figure 8.9.

occur on the theoretically optimal switching curve (8.145). The temporal evolution of the states is illustrated in Figure 8.10.

8.5.4 Regularization

In the special case of the linear quadratic regulator (LQR) problem, when the R matrix is singular, it is clear from (8.36) that $\frac{\partial \mathcal{H}}{\partial u} = 0$ does not specify the optimal control law and the problem is thereby singular. In the more general setting of Section 8.5, we saw that singularity can be, and often is, related to the dynamics and cost being linear in the controls. In order to 'disrupt' this singular structure, it was suggested in [378] that an index of the form

$$J(\boldsymbol{x}, \boldsymbol{u}, t) = \int_0^T \left(f_0(\boldsymbol{x}, t) + g_0(\boldsymbol{x}, t)\boldsymbol{u}(t) \right) dt \tag{8.147}$$

should be replaced by

$$J(\boldsymbol{x}, \boldsymbol{u}, t) = \int_0^T \left(f_0(\boldsymbol{x}, t) + g_0(\boldsymbol{x}, t)\boldsymbol{u}(t) + \left(\frac{\epsilon_k}{2} \boldsymbol{u}^2 \right) \right) dt, \tag{8.148}$$

where $\epsilon_k > 0$. The term $\left(\frac{\epsilon_k}{2} \boldsymbol{u}^2 \right)$ is nonlinear in the controls and is sometimes called a regularizing term, and the resulting optimal control problem is non-singular. The ϵ algorithm is:

(1) choose a starting value ϵ_1 and a nominal control $\boldsymbol{u}_1$;
(2) solve the resulting ϵ_k problem (with $k = 1$ initially); this yields a minimizing control function $\boldsymbol{u}_k(t)$;
(3) choose $\epsilon_{k+1} < \epsilon_k$, and find $\boldsymbol{u}_{k+1}(t)$, and go to step 2;
(4) terminate computation when either
  (a) ϵ_k is so small that numerical instability occurs, or
  (b) $\epsilon_k \le \sigma$ which is a small preassigned positive quantity.

This method solves a singular optimal control problem as a sequence of non-singular problems. In some cases, however, this methodology performs poorly. Alternatives that have been suggested include iterative dynamic programming [379], and the introduction of a piecewise derivative variation in the control [380]. The latter technique is motivated as a direct approach that prevents erratic oscillations that can appear in numerical solutions. It has the property that the method does not cause the approximating problems to become increasingly singular in the limit.

8.6 Numerical methods for optimal control

Most practical optimal control problems have to be solved numerically [381]. Solution can be derived (indirectly) from first-order necessary conditions such as those described in Section 8.3, or (directly) using a discrete approximation to the optimal control problem itself. One direct approach is illustrated in Section 8.4.1, where the continuous-time optimal control problem is approximated as a multi-stage decision problem. Methods that attempt to find a solution (indirectly) using first-order necessary conditions are called *indirect methods*, while methods that attempt to solve a discrete approximation to the optimal control problem itself are called *direct methods*. In an indirect approach a multiple-point boundary-value problem is solved with the boundary conditions at the endpoints and/or interior points only partially known; see for example (8.17) and (8.18), with boundary conditions $x(0) = x_0$, and (8.14). Typically there are n initial and n terminal conditions prespecified, where n is the dimension of the system state space. In a direct method an optimization problem is solved numerically by approximating the original optimal control problem with a finite-dimensional optimization problem. In this approach there is no need to provide the costate equations, although they are produced internally by most optimization routines. Further details will be provided in Section 8.6.5.

Once the solution methodology has been chosen, the problem is solved using either *explicit* or *implicit* simulation. In explicit simulation the dynamics are integrated using a time-marching technique, where the solution at a given time step is obtained from the solution of the dynamics at one or more previous time steps. This process of time marching is then repeated by making progressively better approximations to the unknown boundary conditions until the known boundary conditions are satisfied. In implicit simulation the solution of the dynamics at each time step along with the boundary conditions is obtained simultaneously without employing time marching.

8.6.1   Explicit simulation (time-marching)

In explicit simulation, typically performed by time-marching, the solution of the differential equation at a future time step is obtained using the solution at current and prior time steps. The most basic explicit simulation method is the *shooting method* [382]. In shooting an initial guess is made at the unknown boundary conditions at one end of the interval. Forward shooting refers to methods that guess the unknown initial conditions, while backward shooting refers to methods that guess the unknown final conditions.

The standard shooting method can present numerical difficulties due to instabilities in the dynamics. In order to overcome this difficulty the *multiple-shooting method* [383] was developed. In a multiple-shooting method, the time interval $[0,T]$ is divided into K subintervals $\mathcal{T}_k = [t_{k-1}, t_k]$, with $\bigcup_{k=1}^{K} \mathcal{T}_k = [0,T]$ and $t_K = T$. The shooting method is then applied over each subinterval $[t_{k-1}, t_k]$ with the initial value of $x(t_{k-1})$, $k = 2, \ldots, K$ unknown. In order to enforce continuity, the defect conditions $x(t_k^-) = x(t_k^+)$, $k = 1, \ldots, K-1$ must be recognized. These continuity conditions result in a vector-valued root-finding problem, where it is desired to satisfy the boundary conditions in combination with

$$x(t_k^-) - x(t_k^+) = 0, \quad k = 1, \ldots, K-1.$$

Despite being of higher dimension, the multiple-shooting method represents an improvement over the standard shooting method, because the shorter integration intervals reduce the sensitivity to errors of the unknown terminal conditions.

8.6.2   Implicit simulation (collocation)

In implicit simulation the boundary-value problem in (8.17) and (8.18) is again solved by dividing the solution interval $[0,T]$ into mesh intervals $\mathcal{T}_k = [t_{k-1}, t_k]$; $k = 1, \ldots, K$, where $t_K = T$ and $\bigcup_{k=1}^{K} \mathcal{T}_k = [0, T]$. In each mesh interval $\mathcal{T}_k$ the function $\boldsymbol{x}(t)$ is approximated in terms of a set of basis functions $\psi_j^{(k)}(t)$, that is,[12]

$$\boldsymbol{x}^{(k)}(t) \approx \boldsymbol{X}^{(k)}(t) = \sum_{j=1}^{J} c_j^{(k)} \psi_j^{(k)}(t). \tag{8.149}$$

This parameterization is used to integrate (8.17) and (8.18) from t_{k-1} to one or more stage times $t_{ik} \in \mathcal{T}_k$; $i = 1, \ldots, I$, where $t_{k-1} < t_{1k} < t_{2k} < \cdots < t_{(i-1),k} < t_{Ik} = t_k$ for all $i \in [1, \ldots, I]$ as follows:

$$\boldsymbol{X}(t_{ik}) = \boldsymbol{X}(t_{k-1}) + \int_{t_{k-1}}^{t_{ik}} \boldsymbol{f}(\boldsymbol{X}(\tau), \tau)d\tau, \quad (i = 1, \ldots, I). \tag{8.150}$$

The integrals given in (8.150) are then replaced by quadrature approximations of the form

$$\boldsymbol{X}_{ik} = \boldsymbol{X}_{k-1} + \sum_{j=1}^{I} A_{ij}^{(k)} \boldsymbol{f}(\boldsymbol{X}_{jk}, t_{jk}), \quad (i = 1, \ldots, I), \tag{8.151}$$

where $\boldsymbol{X}_{jk}$; $j = 1, \ldots, I$ are the values of the function approximation $\boldsymbol{X}(t)$ at the stage points t_{jk}; $j = 1, \ldots, I$ and $A_{ij}^{(k)}$; $i, j = 1, \ldots, I$ is the *integration matrix* associated with the particular integration (quadrature) rule that is used to integrate the dynamics in mesh interval $\mathcal{T}_k$. Equation (8.151) can then be rearranged in the form

$$\boldsymbol{X}_{ik} - \boldsymbol{X}_{k-1} - \sum_{j=1}^{I} A_{ij}^{(k)} \boldsymbol{f}(\boldsymbol{X}_{jk}, t_{jk}) = \boldsymbol{0}, \quad (i = 1, \ldots, I), \tag{8.152}$$

which are called *defect conditions* that must be satisfied at all stage times in each mesh interval. In addition, assuming that the solution is continuous, continuity conditions $\boldsymbol{x}(t_k^-) = \boldsymbol{x}(t_k^+)$ must be satisfied at every mesh interval interface. The goal then is to solve for the coefficients $c_j^{(k)}$ in each mesh interval. As opposed to the seemingly similar explicit simulation method and multiple shooting, in implicit simulation all of the defect equations are found *simultaneously* by solving a large-scale algebraic system of the form $\boldsymbol{f}(\boldsymbol{z}) = \boldsymbol{0}$. Implicit simulation is often referred to as *collocation*, because the defect conditions in (8.152) make the value of the function at each stage point equal to the quadrature approximation of the dynamics at those points.

8.6.3 Indirect methods

The boundary-value problem given in (8.17) and (8.18) can be solved using either explicit or implicit simulation. A typical explicit-simulation method would be *indirect shooting* or *indirect multiple-shooting*, while a typical simultaneous indirect method would be *indirect collocation*. In an indirect shooting method [382], an initial guess

---

[12] Capital letters are used to emphasize the fact that these are discretized quantities.

is made for the unknown boundary conditions at one boundary. Using this guess, together with the known initial conditions, the Hamiltonian system in (8.17) and (8.18) is integrated to the other boundary (that is, either forwards from 0 to T, or backwards from T to 0) using a time-marching method. The control at each integration step is obtained from (8.19) or (8.29). Upon reaching T, the terminal conditions obtained from the numerical integration are compared to those given in (8.14). If the integrated terminal conditions differ from the known terminal conditions by more than a specified tolerance, the unknown initial conditions are adjusted and the process is repeated until the difference between the integrated terminal conditions and the required terminal conditions is less than some specified threshold. While the simplicity of simple shooting is appealing, this technique suffers from significant numerical difficulties due to ill-conditioning. The shooting method poses particularly poor characteristics when the optimal control problem is *hypersensitive* [384] (that is, when the time interval of interest is long in comparison with the timescales of the Hamiltonian system in a neighbourhood of the optimal solution).

In an indirect collocation method, the state and costate can be parametrized using piecewise polynomials as shown in (8.149). The collocation procedure leads to a large-scale root-finding problem $\boldsymbol{F}(\boldsymbol{z}) = \boldsymbol{0}$, where the vector of unknown coefficients $\boldsymbol{z}$ consists of the coefficients of the piecewise polynomial. This system of nonlinear equations is then solved using an appropriate root-finding technique suitable for large-scale problems.

The PINS optimal control suite [385–387] is an example of a software package that employs indirect methods with state and control collocation (full collocation). The suite includes the library XOptima and the solver Mechatronix. Beginning with the system equations (8.17), the costate equations (8.18), the optimality equations (8.19), and the boundary conditions (8.14) are generated (as C++ files) by the library XOptima. A finite-difference discretization of the resulting TPBVP and its solution as a large set of algebraic equations is carried out by the solver. The constrained problem is converted into an equivalent unconstrained problem using penalty and barrier function techniques [386][13] that are applied to the state and control constraints. The control equations (8.19) are solved analytically for the controls $\boldsymbol{u}$ (in terms of the states and Lagrange multipliers) using symbolic methods, with iterative schemes employed when an analytical solution is not possible. The relationships between the controls and the states and Lagrange multipliers are used to eliminate the controls from the TPBVP. This indirect full-collocation approach has proved relatively insensitive to initial guesses on the Lagrange's multipliers. In PINS, the system equations can be written in the more general form $E\dot{\boldsymbol{x}} = \boldsymbol{f}(\boldsymbol{x}, \boldsymbol{u}, t)$, where E is a square matrix, rather than the standard $\dot{\boldsymbol{x}} = \boldsymbol{f}(\boldsymbol{x}, \boldsymbol{u}, t)$ form. This more general formulation is convenient in multibody systems when the inversion of the E matrix is problematic (e.g. it is singular, large, or has a high condition number) [352].

---

[13] With penalty functions a (possibly small) violation of the constraint is allowed, while with barrier functions there is never violation of the constraint, but a (possibly slightly) conservative solution may be obtained.

8.6.4   Direct methods

The most basic direct explicit-simulation method for solving optimal control problems is the *direct shooting method*. The control is parametrized using piecewise polynomials as in (8.149), with the parameters to be determined by optimization. The dynamics are then satisfied by integrating the differential equations using a time-marching integration method. Similarly, the cost function of (8.3) is determined using a quadrature approximation that is consistent with the numerical integrator used to solve the differential equations. The nonlinear programme (NLP) that arises from direct shooting then minimizes the cost subject to any path and interior-point constraints. Direct shooting suffers from issues similar to those associated with indirect shooting in that instabilities arise due to the long time interval over which the time-marching method is applied. Thus, in a manner similar to that for indirect methods, an alternate direct explicit-simulation method is the *direct multiple-shooting method*. Over the past two decades implicit-simulation direct-collocation methods have become the de facto standard for the numerical solution of optimal control problems.

A *direct collocation method* is an implicit simulation approach, where both the state and control are parameterized in terms of basis functions (generally piecewise polynomials) for implicit simulation. In the context of optimal control, different, fixed-order integration methods such as Runge–Kutta and Hermite–Simpson methods have been employed. In direct collocation methods the following NLP arises from the approximation of the optimal control problem:

$$\min \ F(z) \text{ subject to } \begin{cases} z_{\min} \leq z \quad \leq z_{\max} \\ \qquad g(z) \leq 0. \end{cases} \tag{8.153}$$

As a rule this NLP is large, containing tens or even hundreds of thousands of variables and constraints. The key feature of these problems, which makes them tractable, is the fact that the NLP is *sparse*. A variety of well-known NLP solvers such as SNOPT [388], IPOPT [389], and KNITRO [390] have been developed specifically for solving large sparse NLPs. In a typical fixed-order collocation method (such as a Runge–Kutta method) accuracy is improved by increasing the number of mesh intervals.

Orthogonal collocation methods.   *Pseudospectral* methods for solving optimal control problems are now widely employed. In these methods the optimization horizon is subdivided into a number of mesh intervals, with the state in each interval approximated by a linear combination of Lagrange polynomials. Collocation is then performed at Gauss-quadrature collocation points, which are the roots of orthogonal polynomials. This methodology has the advantage that Gaussian quadrature methods converge exponentially when used to approximate the integrals of smooth functions. In fixed-order algorithms such as the Runge–Kutta and Hermite–Simpson methods, accuracy is improved by adjusting the number and width of the mesh intervals. In pseudospectral methods accuracy can also be improved by adjusting the degree of the polynomial approximations. By increasing the approximating polynomial degrees in segments where the solution is smooth, it is possible to achieve convergence to the optimal solution by exploiting the convergence properties of the Gaussian quadrature. One can also add mesh boundaries at strategic locations where smoothness in the solution is lost. To see how Gaussian quadrature method works in practice, we consider the collocation methods that employ Legendre–Gauss–Radau (LGR) points, although other collocation points such as Legendre–Gauss (LG) and Legendre–Gauss–Lobatto (LGL) are possible [391].

When using pseudospectral methods the time interval $t \in [t_0, t_f]$ is transformed into the fixed interval $\tau \in [-1, +1]$ with the affine transformation $t = (t_f - t_0)\tau/2 + (t_f + t_0)/2$. The interval $[-1, +1]$ is then partitioned into a *mesh* consisting of K *mesh intervals* $\mathcal{T}_k = [T_{k-1}, T_k]$, $k = 1, \ldots, K$, where $-1 = T_0 < T_1 < \ldots < T_K = +1$. The mesh intervals have the property that $\bigcup_{k=1}^{K} \mathcal{T}_k = [-1, +1]$. Let $\boldsymbol{x}^{(k)}(\tau)$ and $\boldsymbol{u}^{(k)}(\tau)$ be the state and control in $\mathcal{T}_k$. The optimal control problem of (8.64)–(8.67) can be rewritten as follows: Minimize the cost functional

$$J = m(\boldsymbol{x}^{(1)}(-1), t_0, \boldsymbol{x}^{(K)}(+1), t_f) + \frac{t_f - t_0}{2} \sum_{k=1}^{K} \int_{T_{k-1}}^{T_k} l(\boldsymbol{x}^{(k)}(\tau), \boldsymbol{u}^{(k)}(\tau), t(\tau, t_0, t_f)) \, d\tau, \tag{8.154}$$

subject to the dynamic constraints

$$\frac{d\boldsymbol{x}^{(k)}(\tau)}{d\tau} = \frac{t_f - t_0}{2} \boldsymbol{f}(\boldsymbol{x}^{(k)}(\tau), \boldsymbol{u}^{(k)}(\tau), t(\tau, t_0, t_f)), \quad (k = 1, \ldots, K), \tag{8.155}$$

the path constraints

$$\boldsymbol{c}_{\min} \leq \boldsymbol{c}(\boldsymbol{x}^{(k)}(\tau), \boldsymbol{u}^{(k)}(\tau), t(\tau, t_0, t_f)) \leq \boldsymbol{c}_{\max}, \quad (k = 1, \ldots, K), \tag{8.156}$$

and the boundary conditions

$$\boldsymbol{b}(\boldsymbol{x}^{(1)}(-1), t_0, \boldsymbol{x}^{(K)}(+1), t_f) = \boldsymbol{0}. \tag{8.157}$$

As the state must be continuous at each interior mesh point, one requires $\boldsymbol{x}(T_k^-) = \boldsymbol{x}(T_k^+)$, $(k = 1, \ldots, K-1)$ at the interior mesh points $(T_1, \ldots, T_{K-1})$. Finally, in order to pose the collocation method as an integration scheme, (8.155) is replaced by the equivalent integral

$$\boldsymbol{x}^{(k)}(\tau) = \boldsymbol{x}^{(k)}(T_{k-1}) + \frac{t_f - t_0}{2} \int_{T_{k-1}}^{\tau} \boldsymbol{f}(\boldsymbol{x}^{(k)}(s), \boldsymbol{u}^{(k)}(s), t(s, t_0, t_f)) ds. \tag{8.158}$$

The multi-interval form of the LGR collocation method [391–400] is described as follows. First, the state of the continuous-time optimal control problem is approximated in $\mathcal{T}_k$, $k \in [1, \ldots, K]$, as

$$\boldsymbol{x}^{(k)}(\tau) \approx \boldsymbol{X}^{(k)}(\tau) = \sum_{j=1}^{N_k+1} \boldsymbol{X}_j^{(k)} \ell_j^{(k)}(\tau), \tag{8.159}$$

where $\tau \in [-1, +1]$, with

$$\ell_j^{(k)}(\tau) = \prod_{\substack{l=1 \\ l \neq j}}^{N_k+1} \frac{\tau - \tau_l^{(k)}}{\tau_j^{(k)} - \tau_l^{(k)}};$$

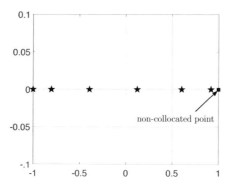

Figure 8.11: LGR collocation points with an additional non-collocated point at $+1$.

a basis of Lagrange polynomials, $[\tau_1^{(k)}, \ldots, \tau_{N_k}^{(k)}]$ are the LGR collocation points. Figure 8.11 shows the LGR points associated with a sixth-order Radau integration formula [401]. Differentiating $\boldsymbol{X}^{(k)}(\tau)$ in (8.159) with respect to τ gives

$$\frac{d\boldsymbol{X}^{(k)}(\tau)}{d\tau} = \sum_{j=1}^{N_k+1} \boldsymbol{X}_j^{(k)} \frac{d\ell_j^{(k)}(\tau)}{d\tau}. \tag{8.160}$$

The integral form of the dynamics given in (8.158), and discretizing the result with an LGR quadrature on the interval $[T_{k-1}, T_k]$, gives

$$\boldsymbol{X}_{i+1}^{(k)} = \boldsymbol{X}_1^{(k)} + \frac{t_f - t_0}{2} \sum_{j=1}^{N_k} I_{ij}^{(k)} \boldsymbol{f}(\boldsymbol{X}_j^{(k)}, \boldsymbol{U}_j^{(k)}, t(\tau_j^{(k)}, t_0, t_f)) \quad (i = 1, \ldots, N_k). \tag{8.161}$$

The following NLP then arises from the LGR discretization. Minimize the cost functional

$$J \approx m(\boldsymbol{X}_1^{(1)}, t_0, \boldsymbol{X}_{N_K+1}^{(K)}, t_f) + \sum_{k=1}^{K} \sum_{j=1}^{N_k} \frac{t_f - t_0}{2} w_j^{(k)} l(\boldsymbol{X}_j^{(k)}, \boldsymbol{U}_j^{(k)}, t(\tau_j^{(k)}, t_0, t_f)), \tag{8.162}$$

subject to defect constraints

$$\boldsymbol{X}_{i+1}^{(k)} - \boldsymbol{X}_1^{(k)} - \frac{t_f - t_0}{2} \sum_{j=1}^{N_k} I_{ij}^{(k)} \boldsymbol{f}(\boldsymbol{X}_j^{(k)}, \boldsymbol{U}_j^{(k)}, t(\tau_j^{(k)}, t_0, t_f)) = \boldsymbol{0}, \quad (i = 1, \ldots, N_k), \tag{8.163}$$

where $I_{ij}^{(k)}$, $(i = 1, \ldots, N_k, j = 1, \ldots, N_k, k = 1, \ldots, K)$ is the $N_k \times N_k$ *Legendre–Gauss–Radau integration matrix* in mesh interval $k \in [1, \ldots, K]$. The discretized path constraints

$$\boldsymbol{c}_{\min} \leq \boldsymbol{c}(\boldsymbol{X}_i^{(k)}, \boldsymbol{U}_i^{(k)}, t(\tau_i^{(k)}, t_0, t_f)) \leq \boldsymbol{c}_{\max} \quad (i = 1, \ldots, N_k) \tag{8.164}$$

and the discretized boundary conditions

$$b(\boldsymbol{X}_1^{(1)}, t_0, \boldsymbol{X}_{N_K+1}^{(K)}, t_f) = \boldsymbol{0} \tag{8.165}$$

must also be recognized. The continuity in the state at the interior mesh points $k \in [1, \dots, K-1]$ is enforced by the condition

$$\boldsymbol{X}_{N_k+1}^{(k)} = \boldsymbol{X}_1^{(k+1)}, \quad (k = 1, \dots, K-1). \tag{8.166}$$

Computationally, the constraint of (8.166) is eliminated from the problem by using the *same* variable for both $\boldsymbol{X}_{N_k+1}^{(k)}$ and $\boldsymbol{X}_1^{(k+1)}$. The NLP that arises from the LGR collocation method is then to minimize the cost function of (8.162) subject to the algebraic constraints of (8.163)–(8.165).

An example of a software package that implements LGR pseudospectral methods for the solution of nonlinear optimal control problem is GPOPS-II [402].

8.6.5   KKT vs PMP

Collocation methods have proved effective for the solution of nonlinear optimal control problems, when applied to both direct [402] and indirect [385] approaches, especially in the case of vehicle minimum-time problems [120, 125, 403–405]. When employing collocation, direct methods produce an NLP problem, which calls for an optimization solver, while indirect methods produce a large set of algebraic equations (arising from the discretization of the first-order necessary conditions), which calls for a root-finding solver. These apparently different approaches are theoretically closer than one might expect. Indeed, the Karush–Kuhn–Tucker[14] equations generated by (most) optimization routines coincide, in the limit, with the related first-order necessary conditions employed in indirect methods. In the case of direct methods, the optimization solver produces (internally) the adjoint equations, while in the case of indirect methods the adjoint equations are provided by the user.

We will now illustrate the relationship between the KKT and first-order necessary conditions in the simple case of a cost function $J = m(\boldsymbol{x}(T))$ (all cost functions can be reduced to this form; Section 8.2) subject to $\dot{\boldsymbol{x}} = \boldsymbol{f}(\boldsymbol{x}, \boldsymbol{u})$ and $\boldsymbol{x}(t_0) = \boldsymbol{x}_0$; the final time T is fixed and no inequality constraints are considered.

---

[14] The Karush–Kuhn–Tucker (KKT) conditions [406–408] generalize the Lagrange multiplier approach to the case of inequality constraints. When minimizing the cost function $F(\boldsymbol{x})$ subject to $\boldsymbol{b}(\boldsymbol{x}) = 0$ and $\boldsymbol{c}(\boldsymbol{x}) \leq 0$, the KKT conditions are

$$\frac{\partial L}{\partial \boldsymbol{x}} = \boldsymbol{F}_x + b_x^T \boldsymbol{\lambda} + \tilde{c}_x^T \boldsymbol{\mu} = 0 \tag{8.167}$$

$$\frac{\partial L}{\partial \boldsymbol{\lambda}} = \boldsymbol{b}(\boldsymbol{x}) = 0 \tag{8.168}$$

$$\frac{\partial L}{\partial \tilde{\boldsymbol{\mu}}} = \tilde{\boldsymbol{c}}(\boldsymbol{x}) = 0 \quad \tilde{\boldsymbol{\mu}} \geq 0 \tag{8.169}$$

in which $L = F + \boldsymbol{b}^T \boldsymbol{\lambda} + \tilde{\boldsymbol{c}}^T \tilde{\boldsymbol{\mu}}$ is the Lagrangian, $\boldsymbol{F}_x$ is the gradient of the cost function, and b_x and $\tilde{c}_x$ are the Jacobian matrices associated with the equality and inequality constraints respectively (the tilde indicates that only the set of active constraints needs be considered).

In the case of the indirect approach, the first variation of the augmented performance index leads to the following standard PMP set of equations:

$$\dot{\boldsymbol{\lambda}} = -\mathcal{H}_x \tag{8.170}$$

$$0 = \mathcal{H}_u \tag{8.171}$$

with boundary conditions

$$\boldsymbol{\lambda}(T) = \frac{\partial m}{\partial x}(T) \tag{8.172}$$

and control Hamiltonian

$$\mathcal{H} = \boldsymbol{\lambda}^T \boldsymbol{f}. \tag{8.173}$$

In the case of the direct approach, the discretization is carried out using (8.151). For simplicity it is assumed that $I = 1$, $A_{ij}^{(k)} = h$ (single-stage Euler method) and that time does not to appear explicitly in $\boldsymbol{f}$, thus the equations $\dot{\boldsymbol{x}} = \boldsymbol{f}(\boldsymbol{x}, \boldsymbol{u})$ are discretized as

$$\boldsymbol{X}_{k+1} = \boldsymbol{X}_k + h\boldsymbol{f}(\boldsymbol{X}_k, \boldsymbol{U}_k). \tag{8.174}$$

The resulting NLP problem has the the following set of decision variables:

$$\boldsymbol{z} = [\boldsymbol{X}_1, \boldsymbol{U}_1, \cdots, \boldsymbol{X}_K, \boldsymbol{U}_K]^T. \tag{8.175}$$

The solver generates the augmented Lagrangian

$$J_a = m(\boldsymbol{X}_K) + \sum_{k=1}^{K-1} \boldsymbol{\Lambda}_k^T \left(\boldsymbol{X}_{k+1} - \boldsymbol{X}_k - h\boldsymbol{f}(\boldsymbol{X}_k, \boldsymbol{U}_k) \right) \tag{8.176}$$

and the associated KKT conditions are

$$\frac{\partial J_a}{\partial \boldsymbol{X}_k} = -h\boldsymbol{\Lambda}_k \frac{\partial \boldsymbol{f}(\boldsymbol{X}_k, \boldsymbol{U}_k)}{\partial \boldsymbol{X}_k} - (\boldsymbol{\Lambda}_k - \boldsymbol{\Lambda}_{k-1}) = 0 \tag{8.177}$$

$$\frac{\partial J_a}{\partial \boldsymbol{U}_k} = -h\boldsymbol{\Lambda}_k \frac{\partial \boldsymbol{f}(\boldsymbol{X}_k, \boldsymbol{U}_k)}{\partial \boldsymbol{U}_k} = 0 \tag{8.178}$$

$$\frac{\partial J_a}{\partial \boldsymbol{X}_K} = \boldsymbol{\Lambda}_{K-1} + \frac{\partial m(\boldsymbol{X}_K)}{\partial \boldsymbol{X}_K} = 0. \tag{8.179}$$

Recalling that $\mathcal{H} = \boldsymbol{\lambda}^T \boldsymbol{f}$, it is evident that in the limit $h \to 0$ (8.177) converges to (8.170), in the the same way that (8.178) becomes (8.171) and (8.179) becomes (8.172). The Lagrange multipliers associated with the NLP are approximation of the costates of the indirect solution formulation. Direct approaches discretize the nonlinear optimal control problem in the beginning in order to produce an NLP, while the indirect approaches discretize the solution at the end in order to solve the TPBVP associated with the PMP equations [381].

8.6.6   Mesh refinement

The numerical solution of optimal control problems using direct transcription methods (see Section 8.6.4) involves three key steps: (a) transcribing the optimal control problem into an NLP problem using a numerical integration scheme; (b) solving the NLP; and (c) reviewing the accuracy of the solution and, if necessary, refining the mesh and re-solving the (refined) problem. The accuracy and efficiency of this process can be influenced by many things including the choice of integration scheme, the solution mesh, the mesh refinement strategy, and the problem characteristics themselves (cf. singular arcs, stiff and/or non-smooth dynamics, and/or rapidly varying path constraints).

A key aspect of efficiently generating accurate solutions to optimal control problems using collocation methods is the proper placement of the mesh segment boundaries and the collocation points within each mesh segment. The algorithmic determination of the mesh boundaries and collocation points is called *mesh refinement*. In a traditional fixed-order collocation method (trapezoid, Hermite–Simpson, or Runge–Kutta method), the degree of the polynomial in each interval is fixed and the mesh refinement determines the location of the mesh boundaries. An orthogonal collocation method has the flexibility to change the degree of the approximating polynomial (that is, the order of the method). As seen from the prior discussion of the Legendre–Gauss–Radau (LGR) method, the order of the method is changed by modifying the number of LGR points in any mesh interval. The freedom to change the degree of the approximating polynomial in a mesh interval, combined with the ability to modify the width and location of a mesh interval, are referred to, respectively, as the p and h parts of the mesh refinement.

Fixed-order (h) collocation methods have been used extensively, because they are typically implicit simulation forms of fixed-order time-marching methods, which were computationally efficient in the early days of computing. In the past decade methods that employ only p collocation (that is, a method that employs a single mesh interval and only varies the degree of the approximating polynomial) have showed promise for problems whose solutions are smooth and can be accurately approximated using relatively low-degree (that is, less than a degree 10 or 15) polynomial. The use of a pure h or a pure p method has limitations. Achieving a desired accuracy tolerance may require an extremely fine mesh (in the case of an h method), or it may require the use of an unreasonably large degree polynomial approximation (in the case of a p method).

In order to reduce significantly the size of the finite-dimensional approximation, and thus improve computational efficiency of solving the NLP arising from orthogonal collocation methods, a new class of hp orthogonal collocation mesh-refinement methods have been recently developed. As described above, in an hp mesh-refinement scheme, the number of mesh intervals, the width of each mesh interval, and the degree of the approximating polynomial within each mesh interval are allowed to vary simultaneously. The hp approach has led to several new algorithms for mesh refinement as described in [395–398]. An accurate solution to the continuous-time optimal control problem can be obtained in a much more efficient manner than that obtained using a fixed-order method, while concurrently taking advantage of the key features of a p method in segments where the solution is smooth. Various hp mesh-refinement methods have been implemented in the orthogonal collocation software package GPOPS-II [402].

8.6.7   Example: track curvature reconstruction

The previously mentioned optimal control software packages (GPOPS-II for a direct collocation method and PINS for an indirect collocation method) are now used to solve numerically the track curvature reconstruction problem [405]. The aim is to find the curvature of the closed track centre line given the (noisy) centre-line coordinates from GPS data expressed in 'rectangular' coordinates.[15]

It is convenient to model the track using a curvilinear coordinate system that follows the vehicle using the track centre-line arc length as the abscissa [255, 352]. Models of this type provide a compact way of describing the vehicle's progress and of constraining its position to remain within the track boundaries when solving optimal control problems. A curvilinear coordinate system also enables one to deal easily with a track that crosses over itself.

Figure 8.12 describes the position along the track centre line in terms of the abscissa s, which may, for example, give the distance travelled from a start/finish line. It is understood that the distance travelled is a strictly increasing function of time, and that 'time' and 'distance travelled' can be thought of as alternative independent variables. At a point A along the track, the track's radius of curvature is given by $\mathcal{R}_A$ and its curvature by $\mathcal{C}_A = 1/\mathcal{R}_A$. A vector tangent to the track at any particular point will have an orientation angle θ (see inset in Figure 8.12); θ is a function of s.

A progress increment ds along the track centre line produces increments dx and dy in the conventional rectangular inertial system:

$$dx = ds \cos \theta \tag{8.180}$$
$$dy = ds \sin \theta. \tag{8.181}$$

It follows from elementary geometry that the track curvature is given by

$$\mathcal{C} = \frac{d\theta}{ds} = \frac{d}{ds}\left(\arctan \frac{dy}{dx}\right), \tag{8.182}$$

which provides a conceptual way of finding the track curvature from GPS data expressed in rectangular coordinates.

The track centre line may be reconstructed using the centre-line curvature $\mathcal{C}$ by integrating (8.182) to find the track orientation angle θ, and then integrating equations (8.180) and (8.181) to find the track centre line in terms of the rectangular coordinates

---

[15] The latitude ϕ (rad) and longitude λ (rad) coordinates may be transformed into rectangular coordinates using the following approximate formulae:
$$x = \lambda R_0 \cos(\phi_0)$$
$$y = \phi R_0,$$
where ϕ_0 is the mean latitude of the track and $R_0 = 6,378,388$ m is the earth radius.

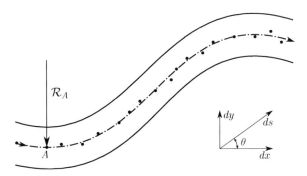

Figure 8.12: Differential geometric track model with noisy GPS centre-line coordinates.

x and y. To implement this process for closed tracks, the track curvature must satisfy the closure condition $\int_\Omega \mathcal{C} ds = 2n\pi$.[16]

A track curvature estimate can be found by solving an optimal control problem that explicitly enforces the closure condition, while solving (8.180), (8.181), and (8.182). The track centre line is described initially in terms of noisy GPS data, which will introduce errors that prevent the centre line closing on itself when simple integration is carried out.

The idea behind the track curvature optimal control problem is to minimize the reconstruction error (relative to GPS data), while enforcing a track closure constraint. Suppose that the control input is the rate of change of the centre-line curvature

$$\frac{d\mathcal{C}}{ds} = u. \tag{8.183}$$

This 'control' is then used to minimize the performance index

$$J = \int_\Omega \left((x_c - x)^2 + (y_c - y)^2 + w_u u^2\right) ds, \tag{8.184}$$

in which x_c and y_c are given GPS data for the track centre line, or estimates derived from spline interpolants. The constant w_u (in (8.184)) is used to penalize rapid changes in the track curvature, which may lead to noisy curvature estimates. In the case of a closed track, the following boundary conditions must be enforced:

$$\begin{aligned}
\mathcal{C}(s_0) &= \mathcal{C}(s_f) \\
\theta(s_0) &= \theta(s_f) \\
x(s_0) &= x(s_f) \\
y(s_0) &= y(s_f).
\end{aligned} \tag{8.185}$$

In sum, the optimal control problem is defined by the four differential equations (8.180)–(8.183) that relate x, y, θ, and $\mathcal{C}$, the cost function (8.184), and the boundary conditions (8.185).

---

[16] The left-hand side of this equation is the line integral of the track curvature around the track centre line and n is an integer (usually $-1 \le n \le 1$).

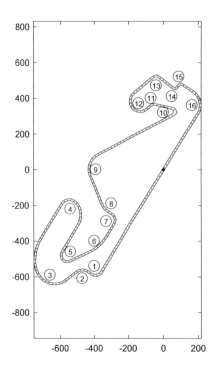
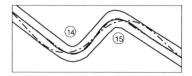

Figure 8.13: Plan view of the Circuit de Catalunya (Barcelona, Spain). The insets show magnified views of turn 10, and turns 14 and 15, and. The dash line is the centre line estimated for $w_u = 10^3$, the solid line corresponds to $w_u = 10^5$, while the dot-dashed line is for $w_u = 10^7$ respectively in (8.184). The measured centre line is shown dotted.

By way of example, we will now find an optimized curvilinear coordinate description of the centre line of the Circuit de Catalunya that is derived from GPS data. We will also study the effect of the curvature-rate weighting factor w_u. Figure 8.13 shows the measured Catalunya raceway centre line and its optimized estimate. The track curvature has been computed for three values of the curvature-rate weighting factor w_u given in (8.184). When solving the curvature optimization problem there is a trade-off between the accuracy of the centre-line estimate and the suppression of noise in the estimated track curvature. As expected, the fit between the 'true' centre line and the curvilinear approximation degrades when w_u is increased; this discrepancy is best seen in turn 10, and in turns 14 and 15, as shown on the right-hand side of Figure 8.13. Figure 8.14 illustrates this compromise, with magnified views shown for turn 10 and turns 14 and 15; the higher-accuracy centre-line fit is accompanied by increased 'activity' in the track curvature estimate.

In Section 7.3 these ideas were extended to deal with three-dimensional roads that have inclines and camber as well as corners.

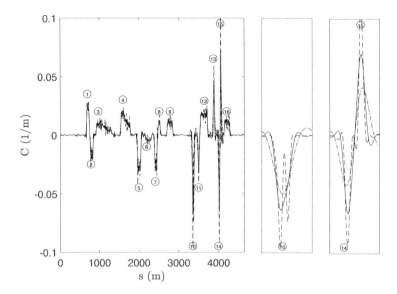

Figure 8.14: Optimized curvature estimates as a function of the elapsed distance for the centre-line descriptions given in Figure 8.13. The curve coding is consistent with Figure 8.13 for different values of w_u. The insets on the right-hand side show turn 10, and turns 14 and 15.

To conclude, we study the solution of the track curvature optimization problem for four tracks using GPOPS-II and PINS [405]. The tracks are illustrated in Figure 8.15, with the reconstruction errors given in Tables 8.1 and 8.2. The curvature weighting function used is $w_u = 10^5$. The direct collocation software GPOPS-II and the indirect collocation software PINS provide almost identical results. The numerical solution is computed with two points for each metre of track.

Table 8.1 RMS (m) of the centre line track reconstruction error of the four circuits in Figure 8.15 with a weight $w_u = 10^5$.

Circuit	PINS	GPOPS-II
Adria	1.2e-01	2.2e-01
Catalunya	1.4e-01	2.3e-01
Mugello	1.1e-01	2.0e-01
Imola	1.4e-01	2.2e-01

Table 8.2 Maximum error (m) of the centre line track reconstruction of the four circuits in Figure 8.15 with a weight $w_u = 10^5$.

Circuit	PINS	GPOPS-II
Adria	8.1e-01	5.7e-01
Catalunya	1.2e-01	1.3e-02
Mugello	4.6e-01	3.2e-01
Imola	1.1e-00	1.1e-00

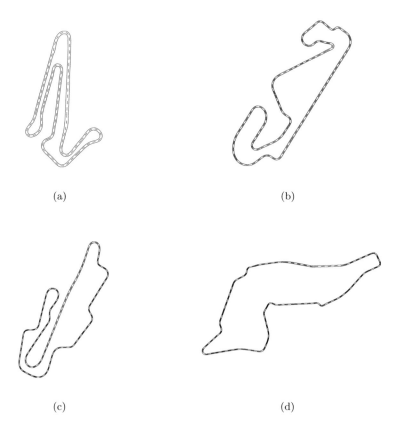

(a)   (b)

(c)   (d)

Figure 8.15: Circuits related to the errors in Tables 8.1 and 8.2: (a) Adria (Italy); (b) Catalunya (Spain); (c) Mugello (Italy); (d) Imola (Italy).

9
Vehicular Optimal Control

9.1 Background

Chapter 8 reviewed the fundamentals of solving nonlinear optimal control problems using first-order necessary conditions. Bang–bang and bang–singular–bang solution structures are common in many vehicular minimum-time and minimum-fuel problems, and are particularly common in motor racing. In order to complete a fast lap, maximum torque is used to accelerate the vehicle away from the start/finish line. Maximum braking is then applied, as late as possible, on the entry to corners, with maximum torque then reapplied, as early as possible, at the exit from corners. These bang–bang minimum-time solutions are usually accompanied by high fuel consumption, and heavy tyre and brake wear rates. Motor vehicles cannot usually be driven in this aggressive manner away from the racetrack, especially when fuel conservation, vehicle maintenance, and statutory speed limits must be considered.

This chapter focuses on some common road vehicle-related applications of optimal control. In Section 9.2 three related time- and energy-constrained vehicular optimal control problems are considered, which highlight the main features of their solution, including possible non-uniqueness and non-existence of a solution. In the interests of simplicity aerodynamic forces and power-loss effects are neglected. The first problem considers the minimum-fuel assent of a hill with constant gradient. This problem is solved analytically in order to uncover its main structural features, particularly when speed restrictions are in place. This problem is then generalized to consider a more general road profile, and is solved numerically in the case of both regenerative braking and (standard) dissipative mechanical braking.

In Section 9.3 a typical minimum-fuel driving problem with a terminal time constraint is considered for a three-dimensional road. In this case aerodynamic forces, rolling resistance, and realistic engine maps are employed together with G-G restrictions representative of typical drivers. Numerical results are given and discussed.

Minimum lap-time problems, which may be fuel constrained, are classical optimal control problems related to closed-circuit racing. There are a variety of these problems, which may have free or predetermined trajectories, and which may use dynamic or quasi-steady-state models. In the case of free-trajectory problems, the optimal trajectory is determined by the optimization solver. As a general rule, fixed-trajectory problems are faster to solve; the racing line may have been determined from race

Dynamics and Optimal Control of Road Vehicles. D. J. N. Limebeer and M. Massaro.
© D. J. N. Limebeer and M. Massaro 2018. Published in 2018 by Oxford University Press.
DOI: 10.1093/oso/9780198825715.001.0001

data, or previous free-trajectory problems. In the same way, quasi-steady models[1] are sometimes used to speed up the computation time.

One of the earliest attempts to solve minimum-lap-time problems dates back to the late 1950s [350]. In the late 1980s minimum-time simulations on a section of the Paul Ricard circuit were reported: a quasi-steady model with a pre-defined trajectory was employed and the results compared with those of three expert drivers (Andretti, Peterson, Stewart) [409]. The technique reported is based on the identification of critical points, which are followed by acceleration and preceded by braking, and was used in much subsequent research. A quasi-steady approach was also adopted in a more recent work [410], which includes a comparison with the results of a non-quasi steady model. The optimization of the roll stiffness balance is investigated in [411]. An extended quasi-steady approach is presented in [412]—the methods allow for the inclusion of some modelling features typically neglected by quasi-steady models (e.g. tyre-load dynamics, yaw dynamics, and suspension damper effects). Amongst the earliest contributions to minimum-time problems using full dynamic models are [351] and [352] for cars and motorcycles respectively. In both cases indirect optimization methods (see Section 8.6.3) are employed. Direct methods (see Section 8.6.4) were used to solve minimum-time problems for Formula One cars in [413]. More recent work relating to cars are [414], where a thermodynamic-tyre model is included, and [403] where a Formula One car with a hybrid powertrain is considered. In [415] a three-dimensional track is used, while [270] studies the interaction between the suspension system and the car's aerodynamics. Limit-handling conditions, such as 'drifting' and hand-brake manoeuvres, are studied in [120,416]. In the case of motorcycles, advanced models have been used to optimize the gearbox and investigate the effects of a counter-rotating engine [249,404]. The solution of minimum-lap-time problems with iterative optimization techniques has also been attempted.[2]

In this chapter we will consider minimum-time problems with unassigned trajectories and full dynamic models, while employing the nonlinear optimal control framework introduced in Chapter 8. In Section 9.4 we will consider a minimum-lap-time problem for a Formula One car in which a hybrid powertrain is employed. A similar minimum-lap-time problem for a racing motorcycle is considered in Section 9.5. The models employed capture the key features of four- and two-wheeled vehicle dynamics, including load transfers, suspensions, aerodynamics, slipping and relaxed tyres, and three-dimensional roads.

---

[1] In these problems the manoeuvre is divided into segments, with the vehicle operating in steady-state save for a small number of state variables such as the speed.

[2] Model predictive control is used in [417,418], for example. However this is a 'suboptimal' technique due to the 'receding-horizon' approach employed, in combination with linearized models. 'Evolutionary' approaches are reported in [419]: although a complex model can be employed, an extended computation time is required.

9.2　An illustrative example

Three problems related to the optimal control of a car that covers a prescribed distance, within a prescribed time, using minimum energy, are analysed in this section.[3]

In the first problem a car driven up an incline of constant gradient is considered; see Figure 9.1. To keep the analysis simple no braking is allowed and friction-related losses are ignored. This problem captures several of the features of vehicular minimum-

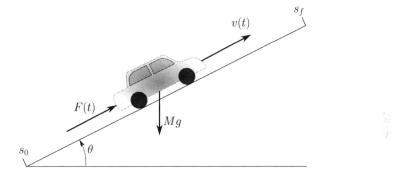

Figure 9.1: Car driving up an incline of constant gradient.

fuel optimal control problems. It is clear that fuel is required to increase the vehicle's potential energy as it ascends the incline. Also, reduced journey times require additional fuel to increase the vehicle's terminal kinetic energy. The controllably varied driving force is denoted $F(t)$, the gradient angle of the incline is θ, the mass and velocity of the car are m and $v(t)$ respectively, the acceleration due to gravity is g, and the start and end points are denoted s_0 and s_f respectively. The minimum-energy fixed-terminal-time (and fixed-distance) optimal control problem is described by

$$\min_{F(t)} E = \int_0^{t_f} F(t)v(t)\, dt \,. \tag{9.1}$$

When braking is allowed, (9.1) can again be used to measure the total energy expenditure in the case of regenerative braking (sometimes called kinetic energy recovery). When dissipative braking systems are employed, one must replace (9.1) with

$$\min_{F(t)} E = \int_0^{t_f} \max(F(t),0)v(t)\, dt, \tag{9.2}$$

since negative braking forces are associated with lost power.

In the second and third sub-problems we will consider the more general road illustrated in Figure 9.2, which is described by the cubic

---

[3]　The authors would like to recognize the contribution of Mr Gareth Pease to the development and solution of this example while he was a DPhil student at Oxford.

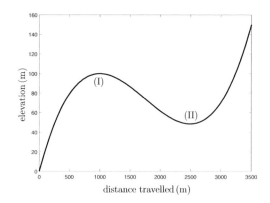

Figure 9.2: Example of an undulating road with elevation $y = 150\,\mathrm{m}$ at $x = 3{,}500\,\mathrm{m}$. Peak (I) has elevation $y = 100\,\mathrm{m}$ at $x = 1{,}000\,\mathrm{m}$, while the dip at (II) has elevation $y \approx 50\,\mathrm{m}$ at $x \approx 2{,}500\,\mathrm{m}$. The road path length is $s_f = 3{,}515\,\mathrm{m}$.

$$y(x) = \frac{27}{875{,}000{,}000}x^3 - \frac{283}{1{,}750{,}000}x^2 + \frac{202}{875}x. \tag{9.3}$$

The second problem considers the regenerative braking case with performance index (9.1), while the third problem looks at the 'lossy' braking case represented by (9.2). The first problem will be solved analytically using the PMP, while subsequent sub-problems will be solved numerically.

9.2.1   Constant-incline case

The energy-related cost function is given by (9.1), the equations of motion are given by (9.4) and (9.5), (9.6) is a velocity constraint, and the control constraint is given by (9.7).

$$\dot{v}(t) = F(t)/M - K, \tag{9.4}$$
$$\dot{s}(t) = v(t), \tag{9.5}$$
$$v(t) \le v_m(t), \tag{9.6}$$
$$0 \le F(t) \le F_m, \tag{9.7}$$

in which $K = g\sin\theta$. The problem boundary conditions are

$$s_0 = 0, \; s(t_f) = s_f, \tag{9.8}$$
$$v(0) = 0, \; v(t_f) = \text{free}. \tag{9.9}$$

We will solve this problem in its full generality, but with the velocity constraint ignored initially. The control Hamiltonian (see (8.68)) for this problem is formed by appending the dynamic state equations and the velocity inequality constraint to the cost function

$$\mathcal{H}(t) = F(t)v(t) + \lambda_v(t)(F(t)/M - K) + \lambda_s(t)v(t) + \mu(t)(v(t) - v_m(t)); \tag{9.10}$$

λ_v and λ_s are the co-states associated with the system states. The speed-related inequality constraint is appended to the Hamiltonian using the multiplier μ, which has the following properties (see (8.74)):

$$\mu(t) = 0 \quad \text{for} \quad v(t) < v_m(t), \tag{9.11}$$
$$\mu(t) < 0 \quad \text{for} \quad v(t) = v_m(t), \tag{9.12}$$
$$\mu(t) > 0 \quad \text{for} \quad v(t) > v_m(t). \tag{9.13}$$

Constraint violations increase the Hamiltonian, because in this case $\mu > 0$.

Pontryagin's minimum principle states that the optimal control (8.71) is given by

$$F^*(t) = \arg\min_{F(t)} \; \mathcal{H}(v^*, s^*, F, \lambda_v^*, \lambda_s^*, \mu^*) \tag{9.14}$$

for $t \in [0, t_f]$; the superscript $(.^*)$ denotes optimal trajectories. Since $\mathcal{H}$ is linear in the control F, bang–bang and possibly singular behaviours should be anticipated.

The first-order necessary conditions for optimality are (see (8.70))

$$\dot{\lambda}_v = -\mathcal{H}_v,$$
$$\dot{\lambda}_s = -\mathcal{H}_s. \tag{9.15}$$

The co-states are therefore given by the following ordinary differential equations

$$\dot{\lambda}_v = -(F + \lambda_s + \mu) \tag{9.16}$$
$$\dot{\lambda}_s = 0 \tag{9.17}$$

indicating that λ_s is constant. The final velocity is free, and so there is a boundary condition on $\lambda_v(t_f)$ (see (8.72)) given by

$$\lambda_v(t_f) = 0. \tag{9.18}$$

It follows from the PMP that the optimal control is given by

$$F^*(t) = \begin{cases} 0, & \text{if} \quad \Phi(t) > 0 \\ F_s(t), & \text{if} \quad \Phi(t) \equiv 0 \\ F_m, & \text{if} \quad \Phi(t) < 0 \end{cases} \tag{9.19}$$

where $\Phi = v + \lambda_v/M$ is the switching function. In the case of a singular arc, $\Phi(t) \equiv 0$ for some finite time interval, and so $\Phi(t) \equiv \dot{\Phi}(t) \equiv \ddot{\Phi}(t) \equiv \dots \equiv 0$ must also hold on this time interval. The order of the singular arc is related to the number of times Φ must be differentiated in order to have the control $u = F$ appear explicitly; see Section 8.5. In the present case $\dot{\Phi} = \dot{v} + \dot{\lambda}_v/M$, which can be rewritten using (9.4) and (9.16) as $\dot{\Phi} = -(K + (\lambda_s + \mu)/M)$. Since θ is fixed, λ_s is constant, see (9.17), and $\mu = 0$ in the case of an inactive speed constraint (as initially assumed), $\dot{\Phi}$ is constant. We therefore conclude that the control never appears following differentiation of the switching function with respect to time and thus that singular arcs are not possible.[4] The resulting control is bang–bang in character.

---

[4] If aerodynamic drag is included, singular arcs may appear. When aerodynamic drag is included, a new term of the form $-\frac{\lambda_v}{M} \frac{1}{2}\rho C_D A v^2$ appears in (9.10). The control F then appears explicitly when the switching function (which is still $\Phi = v + \lambda_v/M$) is differentiated twice; the speed is constant along the singular arc [348, 420]. Exactly the same change in solution structure appears in the Goddard rocket (see Section 8.5.1), where the inclusion of aerodynamic drag changes a bang–bang solution into a bang–singular–bang solution structure.

At the start of the optimal control interval $F^* = F_m$ (which implies $\Phi(0) < 0$), because in the case that $F^* = 0$, the car would roll down the incline according to (9.4). Since $\dot{\Phi}$ is constant, there is at most one control switch. In sum,

$$F(t) = \begin{cases} F_m, & t \in [0, \tau] \\ 0 & t \in (\tau, t_f] \end{cases} \tag{9.20}$$

with τ the thus far unknown switching time.

On the first subarc the state equations can be integrated, recognizing the boundary conditions (9.8) and (9.9), to give

$$v_1(t) = \left(\frac{F_m}{M} - K\right) t, \tag{9.21}$$

$$s_1(t) = \left(\frac{F_m}{M} - K\right) \frac{t^2}{2}. \tag{9.22}$$

On the second subarc

$$v_2(t) = K(t_f - t) + v_f \tag{9.23}$$

$$s_2(t) = -\frac{K}{2}(t - t_f)^2 + v_f(t - t_f) + s_f. \tag{9.24}$$

Equating each of the states at the switching time $t = \tau$ gives

$$v_f = \left(\frac{F_m}{M}\tau - Kt_f\right) \tag{9.25}$$

$$0 = F_m\tau^2 - 2M(Kt_f + v_f)\tau + M(Kt_f^2 + 2v_ft_f - 2s_f). \tag{9.26}$$

Eliminating v_f in (9.26) using (9.25) gives

$$0 = F_m\tau^2 - 2F_mt_f\tau + M(2s_f + Kt_f^2). \tag{9.27}$$

In the case of a switch $\tau \leq t_f$, the switching time is given by the smaller of the two roots of (9.27)

$$\tau = t_f - \sqrt{\frac{(F_m - MK)t_f^2 - 2Ms_f}{F_m}}. \tag{9.28}$$

The optimal cost is given by

$$E^* = \left(s_fK + \frac{v_f^2}{2}\right)M, \tag{9.29}$$

which is the total energy of the car as it passes the finish line.

Equation (9.28) contains useful information relating to the existence of solutions that we will now examine. Since $\tau \leq t_f$ must be real, we require $(F_m - MK)t_f^2 - 2Ms_f \geq 0$. The condition $F_m - MK > 0$ is necessary to ensure that it is possible to drive the car up the incline. The condition $(F_m - MK)t_f^2 - 2Ms_f \geq 0$ ensures that there is sufficient time for the car to cover the course when F_m is applied throughout the journey. In this limit case $(F_m/M - K)t_f^2 - 2s_f = 0$, $\tau = t_f$ and maximum drive force is required throughout the optimal control interval. The corresponding minimum final time is given by

$$t_f^{\min} = \sqrt{2s_f/(F_m/M - K)}. \tag{9.30}$$

Equations (9.21) and (9.30) give the final velocity for the minimum-time solution

$$v_f = \sqrt{2s_f(F_m/M - K)},$$

which can be substituted into (9.29) to give $E^* = F_m s_f$.

In the case that the terminal time is unconstrained, the energy (9.29) is minimized when $v_f = 0$, $E^* = MKs_f$ (the total potential energy of the car at s_f). There are many 'optimal' solutions in this case. One solutions is bang–zero, which can be derived from (9.25) and (9.26) by setting $v_f = 0$. The resulting switching time and final time are

$$\tau = \sqrt{2MKs_f/F_m(F_m/M - K)} \tag{9.31}$$

and

$$t_f^{en} = \sqrt{2F_m s_f/K(F_m - MK)} \tag{9.32}$$

respectively.

We will now consider the case when the velocity state constraint is active. In this case $v = v_m$ is constant, and $F = MK$ follows from (9.4). Referring to (9.19) we see that an active velocity constraint must occur on a singular subarc, which implies that for some finite time $\Phi(t) \equiv 0$. In the case that constraint (9.6) is active, the second derivative of the switching function is given by

$$\ddot{\Phi} = -\dot{\mu}^* = 0, \tag{9.33}$$

which implies that μ is also constant along the singular subarc; that is $\mu^* < 0$ by (9.12).

It follows, therefore, that the vehicle accelerates up to v_m on the first subarc when $\Phi < 0$, the car then holds a constant speed v_m on a second (singular) subarc when $\Phi \equiv 0$, and it eventually decelerates on a third subarc when $\Phi > 0$. The bang–singular–zero control strategy is shown in Figure 9.3, with the corresponding velocity profile shown in Figure 9.4.

We can now use purely mechanics-based arguments to find the switching times and the state trajectory. The states v_1 and s_1 on the first subarc are again described by (9.21) and (9.22). For the singular arc the velocity is constant and the driving force balances the car's weight projected along the incline. Thus

$$v_2(t) = v_m, \tag{9.34}$$

$$s_2(t) = \frac{(F_m/M - K)}{2}\tau_1^2 + v_m(t - \tau_1), \tag{9.35}$$

where τ_1 is the first switching time. For the third subarc the dynamics v_3, s_3 are again given by (9.23) and (9.24).

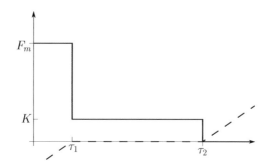

Figure 9.3: Bang–singular–zero control (solid line) and switching function Φ (dashed line).

The first switch can be calculated using (9.21) with $v_1 = v_m$ to give

$$\tau_1 = \frac{v_m}{(F_m/M - K)}. \tag{9.36}$$

Equating (9.34) and (9.35) with (9.23) and (9.24) at τ_2 gives

$$v_f = v_m - K(t_f - \tau_2) \tag{9.37}$$

$$0 = K\tau_2^2 + 2(v_m - v_f - Kt_f)\tau_2 + t_f^2 K + 2(v_f t_f - v_m \tau_1 + s_a - s_f) \tag{9.38}$$

in which

$$s_a = \frac{(F_m/M - K)}{2}\tau_1^2 \tag{9.39}$$

is the distance travelled on the first subarc.

Eliminating v_f in (9.38) using (9.37) gives the second switching time as

$$\tau_2 = t_f - \sqrt{\frac{2}{K}\left((t_f - \tau_1)v_m + s_a - s_f\right)}. \tag{9.40}$$

If this problem is to be soluble, t_f must satisfy

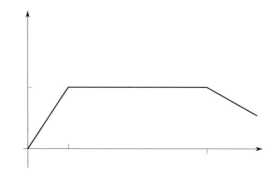

Figure 9.4: Speed profile in the speed-limited case.

$$(t_f - \tau_1)v_m - (s_f - s_a) \geq 0, \tag{9.41}$$

so that the terminal time is long enough to allow the car to cover the distance between the first switching point and the finish line at the maximum allowed velocity v_m.

It is easy to see that the energy usage is given by

$$E^* = F_m \frac{F_m/M - K}{2}\tau_1^2 + MKv_m(\tau_2 - \tau_1). \tag{9.42}$$

In the minimum-time case, that is, when $\tau_2 = t_f$, (9.40) can be solved for t_f with (9.36) and (9.39) to give

$$t_f^{\min} = \frac{v_m}{2(F_m/M - K)} + \frac{s_f}{v_m} \tag{9.43}$$

with the corresponding energy expenditure

$$E^* = M\left(v_m^2/2 + Ks_f\right). \tag{9.44}$$

Again, in the minimum-energy case with free final time, $v_f = 0$. By analogy with the case that is free of a speed constraint, there are many solutions and the minimum energy is $E^* = MKs_f$. One of these solutions is derived as follows. In the minimum-energy case t_f is free, $v_f = 0$, and (9.37) and (9.38), with (9.36) and (9.39), can be solved for t_f and τ_2 to give

$$t_f^{en} = \frac{F_m v_m}{2K(F_m - mK)} + \frac{s_f}{v_m}, \tag{9.45}$$

$$\tau_2 = t_f^{en} - v_m/K. \tag{9.46}$$

and the minimum energy is again given by

$$E^* = MKs_f. \tag{9.47}$$

Example 9.1 Consider a vehicle that has to cover 100 m on an inclined road with slope $\theta = 0.1$ rad in 10 s, starting from standstill, and with a maximum propulsive force-to-mass ratio of $\frac{F_m}{M} = 10$ with and without a speed constraint of $v_m = 50$ kph. In the case the speed constraint is neglected $t_f^{\min} = 4.71$ s (see (9.30)), $t_f^{en} = 15.05$ s (see (9.32)), and the switching time of the bang–zero solution is $\tau = 1.62$ s (see (9.28)). In the case the speed constraint is enforced, $t_f^{\min} = 7.97$ s (see (9.43)), $t_f^{en} = 15.06$ s (see (9.45)), and the switching times of the bang–singular–zero solution are $\tau_1 = 1.54$ s (see (9.36)) and $\tau_2 = 2.41$ s (see (9.40)).

9.2.2 Regenerative braking

This problem studies the use of regenerative braking (sometimes called kinetic energy recovery) on the undulating road (9.3). We make use again of performance index (9.1);

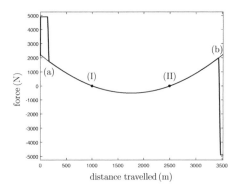

Figure 9.5: Driving and braking force with optimal regenerative braking. The dotted line represents the tangential component of the car's weight.

the problem constraints are given by (9.4)–(9.6), where the gravitational term takes the more general form[5]

$$K = g\frac{\frac{dy}{dx}}{\sqrt{1 + (\frac{dy}{dx})^2}},$$ (9.48)

and the control constraint (9.7) is replaced by

$$-F_m \leq F(t) \leq F_m.$$ (9.49)

In order to track the longitudinal position of the car, the following equation is required

$$\dot{x} = v\frac{1}{\sqrt{1 + \left(\frac{dy}{dx}\right)^2}}.$$ (9.50)

The boundary conditions are again (9.8)–(9.9). The numerical example is run with the following parameters: $M = 1,000\,\text{kg}$, $v_m = 30\,\text{m/s}$, $F_m = 500\,g$, where g is the gravitational constant, $t_f = 50\,\text{s}$ and $s_f = 3,515\,\text{m}$ which corresponds to $x_f = 3,500\,\text{m}$.

By solving numerically a discretized version of the problem, one obtains the optimal propulsive force law illustrated in Figure 9.5, which has a now familiar bang–singular–bang form. The first bang ends at (a) and is used to accelerate the vehicle as fast as possible up to the speed limit. The singular subarc from (a) to (b) holds the vehicle's speed constant, and has a magnitude that is modulated by the road gradient; it is evident that the drive force is the tangential component of the car's weight. At the road's peak and dip the propulsive force is zero; see (I) and (II). The final bang subarc is used to brake the vehicle thereby recovering the car's kinetic energy back into some form of energy store.

The car's velocity profile is given in Figure 9.6. On the first bang subarc the speed

---

[5] Recall $\sin\theta = \frac{dy}{\sqrt{dx^2+dy^2}}$ and $\cos\theta = \frac{dx}{\sqrt{dx^2+dy^2}}$.

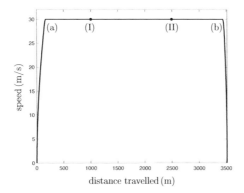

Figure 9.6: Speed characteristic with optimal regenerative braking.

accelerates up to the speed limit of $30\,\mathrm{m/s}$. The speed is then maintained at the speed limit along the singular arc between (a) and (b). Finally, kinetic energy is extracted from the car on the braking subarc in order to reduce the propulsive energy expenditure.

9.2.3 Dissipative braking

In the dissipative braking case the performance index (9.2) is employed in combination with the constraints and boundary conditions used in Section 9.2.2. The key difference in this case is that (9.2) does not provide 'credit' for regenerative braking.

The driving control strategy is more complex with dissipative braking. As is evident from Figure 9.7, the optimal force law again has a bang–singular–zero–singular–zero structure. The bang subarc, which ends at point (a), is used to accelerate the vehicle as quickly as possible up to the speed limit. The driving force on the singular subarc between (a) and (b) holds the vehicle at the speed limit by cancelling the decelerating gravitational influence on the car. At (b) the driving force is set to zero allowing the car to coast towards the brow of the hill at (I); the speed decreases on the zero subarc between (b) and (I). The car then accelerates, under the influence of gravity towards

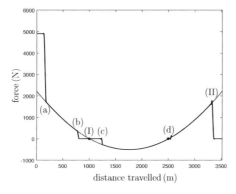

Figure 9.7: Driving and braking force with optimal dissipative braking.

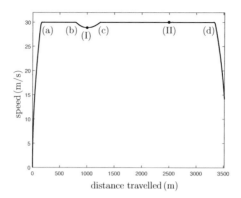

Figure 9.8: Speed characteristic with optimal dissipative braking.

the dip at (II). After the car has accelerated up to the speed limit, under the influence of gravity, a second singular subarc begins at (c) that holds the vehicle's speed on the limit, through the dip at (II), to the end of the singular subarc at (d). The car then coasts towards the finish line on the second zero subarc so as to arrive at the required $t_f = 50\,\mathrm{s}$ terminal time. The corresponding speed profile is shown in Figure 9.8.

The optimal mechanical braking power is illustrated in Figure 9.9. The area under this curve is related to the 'wasted' energy, with the equality obtained when the independent variable is time rather than the distance travelled. When comparing Figures 9.6 and 9.8, which correspond to indices (9.1) and (9.2) respectively, one observes the impact of the kinetic energy recovery incentive.

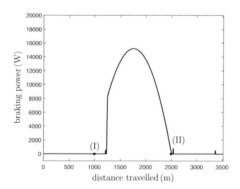

Figure 9.9: Power dissipation with a mechanical braking system.

9.3   Fuel-efficient driving

This example lends further insight into 'eco-driving' strategies that minimize fuel usage, whilst complying with a 'just-in-time' arrival time constraint [421]. In this example rolling resistance, aerodynamic drag, and a realistic powertrain are added to the ba-

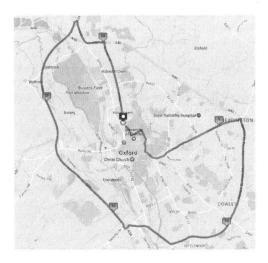

Figure 9.10: 28 km driving route around Oxford, Oxfordshire, UK.

sic vehicle model given in Section 9.2. The optimization will be performed on a real three-dimensional road in the neighbourhood of Oxford in the UK.

Track model. The 28 km route to be considered is shown in Figure 9.10, with its corresponding curvature and elevation given in Figure 9.11. In the optimal control problems considered here the route must be covered within the prescribed time of 1,800 s, whilst minimizing the fuel usage. The track modelling is carried out using the techniques described in Section 7.3—in the current application the road is modelled as a three-dimensional space curve with zero width and so road camber is not considered.

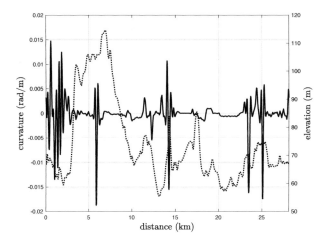

Figure 9.11: Geodesic curvature (solid) and elevation (dotted) for the route illustrated in Figure 9.10.

Vehicle model. The vehicle model used here is similar to that given in (9.4) and takes the form (see Figure 9.12)

$$\dot{v}(t) = \frac{1}{M}\left(F(t) - F_r(t)\right) \tag{9.51}$$

in which F is the tractive force, and F_r is the resistive force given by

$$F_r(t) = \frac{1}{2}\rho C_D A v(t)^2 + \mu f_N(t) - MK. \tag{9.52}$$

The first term is the aerodynamic drag, the second represents the rolling resistance (with $f_N(t) = Mg\cos\theta(t)$ normal reaction to the road surface), while the third term is the gravitational force component tangent to the road; as before, $K = g\sin\theta$.

The control input is the tractive power $p_t(t)$ available at the wheels, which is related to the tractive force by

$$F(t) = \frac{p_t(t)}{v(t)}. \tag{9.53}$$

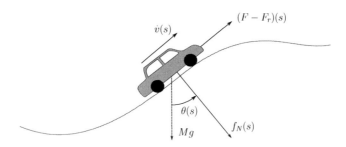

Figure 9.12: Car force system.

Powertrain model. The engine used in this study is a small 1.7 l Mercedes-Benz diesel engine with data taken from the ADVISOR software package [422]. The fuel consumption map is plotted in Figure 9.13 with a maximum torque line overlay. The fuel mass flow rate $\dot{m}_f$ is a function of the engine's angular velocity w_e and engine torque τ_e, and is approximated here by the bi-quadratic map

$$\dot{m}_f = -0.2903 + 0.0021298\, w_e + 0.00198\, \tau_e + 2.2745 \times 10^{-6}\, w_e^2$$
$$-2.2619 \times 10^{-6}\, \tau_e^2 + 4.58924 \times 10^{-5}\,\tau_e w_e.$$

The engine is linked to the wheels via a continuously variable transmission (CVT), which removes the need to constrain the engine rotational speed relative to the wheel rotational speed. The engine power is given by $p_e(t) = w_e(t)\tau_e(t)$—the tractive power p_t cannot exceed the engine power p_e available.

Optimal control problem. It is convenient to transform the problem so that the elapsed distance rather than the elapsed time is the independent variable, which makes the route information more easily utilized. This transformation is one-to-one, with both

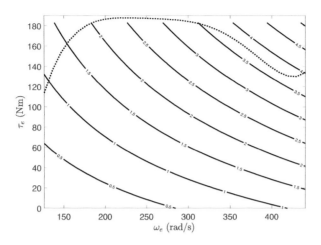

Figure 9.13: Fuel consumption map (grammes per second); the maximum torque curve is shown dotted.

quantities monotonically increasing along the test route so long as $v(t) > 0$. Changing the independent variable is achieved using

$$ds = v(s)dt. \tag{9.54}$$

The cost function is the total fuel usage along the route

$$m_f = \int_0^{s_f} \dot{m}_f \frac{ds}{v(s)}, \tag{9.55}$$

where $\dot{m}_f$ is given by (9.54). The system dynamics, given by (9.51), (9.52), and (9.53), can be written concisely as

$$\frac{dv(s)}{ds} = f(v(s), \boldsymbol{u}(s)) \tag{9.56}$$

in which the control vector[6] is

$$\boldsymbol{u}(s) = \begin{bmatrix} p_t(s) \\ \omega_e(s) \\ \tau_e(s) \end{bmatrix}. \tag{9.57}$$

There are two inequality constraints on the vehicle speed $v(s)$. The first relates to the maximum allowable lateral acceleration a_y, which is related to tyre properties

---

[6] Effectively, the only input is the tractive power p_t, but ω_e and τ_e are included in the optimal control problem in order to optimize the fuel consumption over the engine map.

(Section 7.2.3) and driving style, while the other represents the legislated speed limit $v_l(s)$

$$v(s) \leq \sqrt{\frac{a_y}{|\mathcal{C}(s)|}} \tag{9.58}$$

$$v(s) \leq v_l(s). \tag{9.59}$$

There is also a longitudinal acceleration limit

$$a_l^{min} \leq vf(v(s), \boldsymbol{u}(s)) \leq a_l^{max}, \tag{9.60}$$

which is related to the tyre and engine characteristics (Section 7.2.1). Equations (9.58) and (9.60) represents a simple rectangular G-G diagram for the vehicle. More sophisticated formulations are possible, such as an elliptical G-G characteristic.

The power balance between the powertrain and the car is modelled with the following inequality:

$$p_e(s) - p_t(s) \geq 0. \tag{9.61}$$

The tractive power is bounded by the maximum engine power p_e^{max} and the maximum braking power p_b^{max}. The engine operating point is constrained to remain on the map in Figure 9.13:

$$0 \leq \tau_e \leq \tau_e^{max}(\omega_e) \tag{9.62}$$

$$\omega_e^{idle} \leq \omega_e \leq \omega_e^{max}, \tag{9.63}$$

where $\omega_e^{idle} = 125 \, \text{rad/s}$ and $\omega_e^{max} = 450 \, \text{rad/s}$ in this example. The upper limit on the engine torque is governed by the maximum torque line—the dotted line in Figure 9.13—and is described by

$$\tau_e^{max}(\omega_e) = 1.26778 \times 10^{-12} \, \omega_e^6 - 1.12112 \times 10^{-9} \, \omega_e^5 - 1.57884 \times 10^{-8} \, \omega_e^4$$
$$+ 3.21045 \times 10^{-4} \omega_e^3 - 0.13419 \, \omega_e^2 + 22.3244 \, \omega_e - 1176.32.$$

The maximum torque line was obtained by least-squares curve-fitting over the speed range $125 \leq \omega_e \leq 450 \, \text{rad/s}$.

The boundary conditions on the state are

$$v(0) = v(s_f) = \epsilon, \tag{9.64}$$

where $\epsilon > 0$ is 'small' and ensures that the vehicle starts and ends the circuit at rest.[7] Finally, it is necessary to impose an integral constraint on the travel time

$$\int_0^{s_f} \frac{ds}{v(s)} \leq T_{max}. \tag{9.65}$$

The bound on T_{max} acts as a surrogate for the driving style; a lower arrival time constraint implies more aggressive driving. As explained in Section 9.2, the lowest value of T_{max} is related to the case when the vehicle is travelling at the maximum legislated speed, while remaining within the confines of the G-G map—requesting a smaller value for T_{max} could make the problem infeasible.

---

[7] If the relationship between distance and time is to be properly defined, $v(s) = 0$ cannot be allowed.

Table 9.1 Table of the vehicle parameters that represent a typical saloon car.

Symbol	Value
m	1,280 kg
$C_D A$	0.6 m^2
ρ	1.2 kg/m^3
a_y	0.1 g
μ	0.01
a_l^{min}	-0.1 g
a_l^{max}	+0.1 g

Solution.   The optimal control problem is solved using the data given in Table 9.1. Figure 9.14 shows the speed, speed limits, and tractive power p_t as the car drives around the given route. As one would expect, the fuel-minimum optimum strategy avoids unnecessary braking and accelerating, while operating the engine at a point

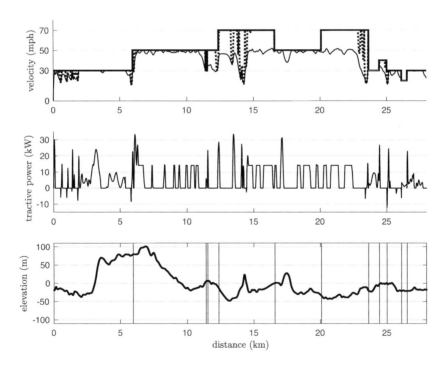

Figure 9.14: Optimal manoeuvre for a mandated arrival time of 1,800 s. The top figure shows the legislative speed limit (thick solid), road curvature speed limit (dotted), and optimal velocity (light solid). The central figure shows the tractive power. The bottom figure shows the road elevation. The vertical lines represent the different phases used for the numerical solution.

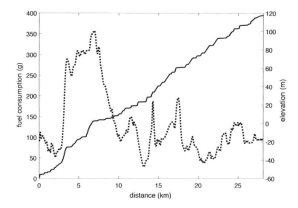

Figure 9.15: Road elevation (dotted) and accumulated fuel consumption (solid).

of high efficiency. For the most part optimal tractive power operation takes the form of bang–bang switches between zero torque and the maximum torque curve shown in Figure 9.13. The tractive power plot in Figure 9.14 shows that drive occurs primarily when the car is required to ascend hills, accelerate out of corners, or accelerate into road sections of increased mandated speed limits. Noticeable also is the 'reluctance' to make use of the 70 mph speed limit regions. The car does not accelerate to the speed limit in the first high-speed section. This is due to the large number of corners in this section, and the associated need to brake before these bends and then accelerate out of them. High-speed travel is also discouraged in the second section due to the corners at the end of it, and the fact that high-speed driving is not required for a 1,800 s transit time (a required average speed of 56 km/h).

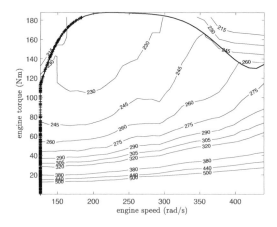

Figure 9.16: Optimal engine operating trajectory on a brake-specific fuel consumption (BSFC) map expressed in g/kWh.

Figure 9.15 shows the accumulated fuel consumption and road elevation as a function of elapsed distance. Evident are increased rates of fuel usage on inclines, and fuel saving on declines. At elapsed distances of approximately 3.15 km, 10.5 km, 13.6 km, and 17.2 km incline-rated increases in fuel consumption are evident. Conversely, very little fuel is burnt on the declines between 7 km and 10 km, 11.6 km and 13.2 km, and 17.5 km and 18.5 km.

Figure 9.16 shows the engine operating point throughout the journey on a brake-specific fuel consumption (BSFC) map for the engine employed. Contours of lower BSFC correspond to regions of fuel-efficient engine operation. The most fuel-efficient way of operating the engine is in a short-duration bang–bang fashion along the maximum torque locus. In general terms, this finding is aligned with the results in Example 9.2.

9.4 Formula One hybrid powertrain

In order to improve Formula One's 'green' credentials [423], and accelerate road-car-relevant powertrain developments, the 2014 Federation Internationale de l'Automobil (FIA) technical regulations specify a hybrid powertrain that would recover both thermal and kinetic energy [424]. The hope was that the introduction of hybrid thermal-kinetic energy recovery systems (ERS) would produce gains that would be far more significant and cost-effective than further engine and car aerodynamic developments. Indeed, contemporary lap times can be maintained using less powerful turbo-boosted engines that require two-thirds of the fuel required by predecessor cars that only used kinetic energy recovery.

In this example we will make used of a finite-width three-dimensional model of the closed-circuit 'Circuit de Spa-Francorchamps' (Spa), which will allow us to represent a number of influences known to all experienced drivers. This example is a summary of some of the research given in [255, 256, 403, 415], and is significantly more complicated than either of Examples 9.2 and 9.3.

9.4.1 Track model

The track will be modelled using the theory of *ribbons* as described in Section 7.3 and Example 7.4. The nineteen corners of the Circuit de Spa-Francorchamps are shown in Figure 9.17, while the distances to each corner from the start/finish (SF) line are given in Table 9.2. A three-dimensional representation of the track is given in Figure 7.16, which has a total elevation change of approximately 110 m.

9.4.2 Important three-dimensional influences

A vehicle's top speed will increase on downhill sections, while it will decrease on ascending inclines. It is also known from everyday experience that extended braking distances are required on descending gradients, while the opposite true on ascending ones. High-speed corners on modern roads are often banked (superelevated) in order to improve drainage and roadholding. Adverse cambering undermines roadholding and reduces the maximum achievable cornering speeds. A vehicle tends to become 'light' if it is driven fast over the brow of a hill, while it will 'squat' down on the suspension system if driven fast through a dip. This variation in apparent mass is the result of

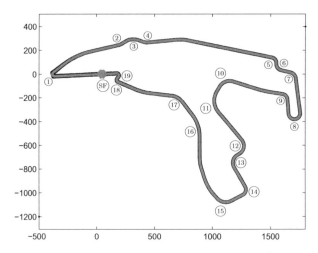

Figure 9.17: Circuit de Spa-Francorchamps showing the start/finish line and its numbered corners. The car travels around the track in a clockwise direction.

a centrifugal force, and has an effect similar to that produced by aerodynamic lift or downforce. As with all changes in downforce, this centrifugal force can impact the top speed of the vehicle. Track three-dimensionality also introduces gyroscopic moments that change the lateral and longitudinal load transfer between the tyres; see Section 7.5.2. As with all load-transfer effects, these moments may change the vehicle's top speed and handling characteristics.

We will briefly review the mechanics behind these effects. Figure 9.18 (a) shows a car travelling at speed u through a dip of radius of curvature d. It is clear that in this situation the 'effective' gravitational acceleration g_e is

$$g_e = g \cos \gamma + \frac{u^2}{d}, \qquad (9.66)$$

Table 9.2 Distances to mid corner from the start/finish line on the Circuit de Spa-Francorchamps (in metres $\pm$ 25 m).

Corner	Distance (m)	Corner	Distance (m)
1	400	11	4,075
2	1,100	12	4,525
3	1,175	13	4,650
4	1,275	14	4,925
5	2,400	15	5,200
6	2,525	16	5,875
7	2,650	17	6,200
8	3,075	18	6,725
9	3,300	19	6,800
10	3,825	SF	6,988

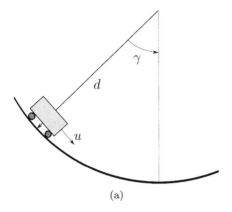 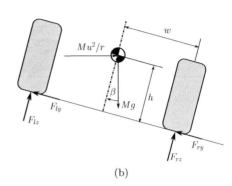

(a)   (b)

Figure 9.18: (a) Car travelling at speed u through a dip or radius of curvature d. (b) Car cornering with speed u, corner radius of curvature r, and adverse camber angle of β (the car is travelling into the page and turning to the left).

with the tangential component of gravity g_t given by

$$g_t = g \sin \gamma. \tag{9.67}$$

The speed and normal-curvature-related gravitational acceleration can have a significant influence on the normal loading of the tyres. When the car travels over the brow of a hill, the radius of curvature in (9.66) is deemed negative, and the car consequently appears 'lighter'.

Adverse cambering limits the cornering speed for the reason set out in Figure 9.18 (b) and the following single-axle analysis. Consider a car under steady-state cornering in a bend of radius r. Balancing forces normal to the road gives

$$F_{zl} + F_{zr} = M \left(g \cos \beta - \frac{u^2}{r} \sin \beta \right), \tag{9.68}$$

while balancing forces parallel to the road gives

$$F_{yl} + F_{yr} = M \left(\frac{u^2}{r} \cos \beta + g \sin \beta \right). \tag{9.69}$$

Taking moments around the right-wheel ground-contact point gives

$$F_{zl} = \frac{M}{2} \left(g \cos \beta - \frac{u^2}{r} \sin \beta \right) - \frac{Mh}{2w} \left(\frac{u^2}{r} \cos \beta + g \sin \beta \right); \tag{9.70}$$

and so

$$F_{zr} = \frac{M}{2} \left(g \cos \beta - \frac{u^2}{r} \sin \beta \right) + \frac{Mh}{2w} \left(\frac{u^2}{r} \cos \beta + g \sin \beta \right); \tag{9.71}$$

Equation (9.68) shows that adverse camber reduces the normal load on the tyres.[8] Equation (9.69) shows that adverse camber can increase the lateral forces required

---

[8] In the case of positive camber, β is deemed negative.

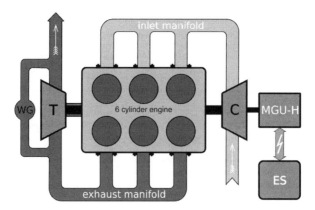

Figure 9.19: The MGU-H on the same driveshaft as the turbo-charger's turbine (T) and compressor (C). The cold inlet air is shown in blue and the hot exhaust gases are shown in red. The waste gate (WG) controls the exhaust turbine bypass. The MGU-H can be used to drive either the compressor, or charge the energy store (ES).

to hold the corner.[9] Equations (9.70) and (9.71) show that adverse cambering tends to 'exaggerate' the normal tyre load redistribution as compared with the flat road scenario, where load transfer is $\frac{Mh}{2w}\frac{u^2}{r}$; see (7.65). The negative influences of adverse cambering are thus threefold, and the top speed of the car is reduced below that achievable on a flat road. The opposite effects occur in a corner with positive camber.

9.4.3   Powertrain

The 2014 FIA technical regulations call for a 1,600 cc 4-stroke internal combustion engine (ICE) that has a fuel mass flow rate limited to 100 kg/h [424]. The regulations also specify that the ICE may be pressure charged using a single single-stage compressor linked to a single-stage exhaust turbine by a common shaft parallel to the engine crankshaft. An electric motor-generator unit (heat) (MGU-H) may be coupled directly to the same shaft. This system is illustrated in Figure 9.19. The turbine, which is driven by the exhaust gas stream, drives both the compressor and the MGU-H. The turbine is bypassed by a controllable waste gate (WG); closing the waste gate will increase the power generated by the MGU-H, but decrease the power generated by the engine. The operation of the thermal energy recovery system is summarized in Figure 9.20, which shows a plot of power from the internal combustion engine P_{IC} versus power supplied by the MGU-H; denoted P_h. In this example, when the waste gate is open the MGU-H requires 60 kW to operate the compressor and the ICE output power is boosted by 20 kW. When the waste gate is closed 40 kW is recovered from the MGU-H and the ICE output power drops to approximately 440 kW.

Operating the engine with an open waste gate is fuel inefficient; the waste gate is normally only open in race qualifying. A second motor-generator unit (kinetic) MGU-

---

[9]   For small $\beta > 0$, $\frac{u^2}{r}\cos\beta + g\sin\beta \approx \frac{u^2}{r} + g\beta > \frac{u^2}{r}$.

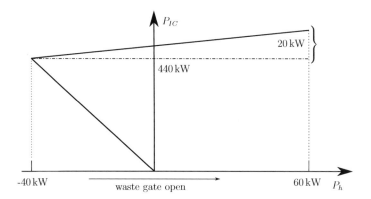

Figure 9.20: Operating regime of the 2014 engine and ERS at full power; P_h is the power absorbed by the MGU-H. Power generated by the ICE is P_{IC}. As the waste gate is opened the power generated by the ICE increases, but the electrical power used by the MGU-H also increases.

K is connected to engine crack shaft,[10] and can be used to either boost the drive to the rear wheels or recover kinetic energy into the energy store (ES) for later deployment.

There are several powertrain-related regulations that must be reflected in the minimum-time optimal control problem [424]:

(1) the car is restricted to a maximum $100\,\mathrm{kg}$ of fuel per race;
(2) the fuel mass flow rate must not exceed $0.028\,\mathrm{kg/s}$ (or $100\,\mathrm{kg/hour}$);
(3) the car's ES can store up to $4\,\mathrm{MJ}$;
(4) the ES can accept up to $2\,\mathrm{MJ}$ per lap from the MGU-K;
(5) the MGU-K can draw up to $4\,\mathrm{MJ}$ per lap from the ES;[11]
(6) power flows to and from the MGU-K is restricted to $\pm120\,\mathrm{kW}$;
(7) the power flow to and from the MGU-H is unrestricted.

We will now explain how the various constraints were set up in the optimal control calculation. The power available to the rear wheels, P_m, is constrained by the following inequality:

$$(P_{IC}^{\mathrm{max}} + P_{W_g} W_g)\dot{F} - P_{loss} + P_k - P_m \geq 0. \tag{9.72}$$

The P_{IC}^{max} term in equation (9.72) represents the power generated by the ICE under full fuelling when the waste gate is closed; the normalized fuel mass flow rate is given by $\dot{F} \in [0, 1]$. The second term in (9.72) represents the power boost resulting from opening the waste gate; $W_g \in [0, 1]$ is the waste gate control with $P_{W_g} = 20\,\mathrm{kW}$. The third term represents the engine's rotational losses and is set at a constant $40\,\mathrm{kW}$ for

---

[10] In legacy Formula One cars this motor was the central component of a kinetic energy recovery system (KERS).

[11] The energy asymmetry between rules (4) and (5) 'encourages' the use of thermal energy recovery.

illustrative purposes, P_k is the power delivered to the MGU-K (P_k is positive when the MGU-K is motoring), and P_m represents the mechanical power delivered to the rear wheels (positive values accelerate the car).

The power generated/absorbed by the MGU-H is given by

$$P_h = ((P_h^{\max} - P_h^{\min})W_g + P_h^{\min})\dot{F} \qquad (9.73)$$

and is a function of the waste gate opening and the fuel flow rate; $P_h^{\min} = -40\,\mathrm{kW}$ and $P_h^{\max} = 60\,\mathrm{kW}$.

In order to monitor the resource constraints associated with the fuel usage, and the ES and MGU-K usage, four auxiliary state variables are introduced as follows:

$$F = \int_{lap} \dot{F} ds. \qquad (9.74)$$

This state is used to monitor the fuel consumption and is constrained by $0 \le F \le 100/n_l$ kg, in which n_l is the number of laps in the race. As pointed out in rule (2) above, $\dot{F}$, when normalized, is constrained to the interval $[0, 1]$. The second auxiliary state is given by

$$\dot{E}_s = -(P_h + P_k). \qquad (9.75)$$

This state is used to monitor the stored energy and is constrained by $4\,\mathrm{MJ} \ge E_s \ge 0$ over the whole circuit. The third auxiliary state is described by

$$\dot{E}_{ES2K} = \begin{cases} P_k & P_k > 0, \ P_h > 0 \\ P_k + P_h & P_k > -P_h > 0 \\ 0 & \text{otherwise.} \end{cases} \qquad (9.76)$$

This state is used to monitor the energy supplied to the MGU-K from the ES and is constrained by $4\,MJ \ge E_{ES2K} \ge 0$. The first alternative corresponds to the case when both electrical machines are running as motors. In this case it is only the MGU-K energy usage that is 'taxed'. The second alternative corresponds to the case in which the MGU-K is motoring, but the MGU-H is generating. In this case the power generated by the MGU-H is offset against the MGU-K power requirements, with the difference monitored.

The fourth auxiliary state is used to tax energy flows from the MGU-K into the ES and is described by

$$\dot{E}_{K2ES} = \begin{cases} P_k & P_k < 0, \ P_h < 0 \\ P_k + P_h & 0 > -P_h > P_k \\ 0 & \text{otherwise;} \end{cases} \qquad (9.77)$$

this state is constrained by $0 \ge E_{ES2K} \ge -2\,MJ$. The first alternative corresponds to the case when both electrical machines are operating as generators. In this case it is only the MGU-K energy generation that is penalized. The second alternative corresponds to the case in which the MGU-K is generating, while the MGU-H is motoring. In this case the power required by the MGU-H is offset against the power generated by the MGU-K, with the difference monitored. Differentiable approximations to the

Heaviside step function are used to enforce constraints (9.76) and (9.77). Inequality constraint (9.72) in combination with the four auxiliary states (9.74) to (9.77) and their associated box constraints is used to model the power and energy constraints required by contemporary Formula One rules [424].

9.4.4   Vehicle model

The dynamic model of the car is described by the differential equations in Section 7.5. The car parameters are given in Table 9.3. An empirical tyre model of the type discussed in Chapter 3, which captures the key features of tyre behaviour, was employed in this study. The model generates forces as a function of longitudinal and lateral slips, and normal load, and includes coupling between the longitudinal and lateral forces. The relaxation properties of tyres are neglected. The specific model used was introduced in [126] and later employed in [403, 415]. The aerodynamic drag and downforce coefficients used are shown in Figure 9.21, and are used in (7.157), (7.158), and (7.159). The drag and downforce coefficients in Figure 9.21 are approximated by

$$C_D = 1.05292 - 0.0006643v - 1.018286 \times 10^{-5}v^2$$
$$C_{fL} = 1.61458 - 0.0013929v - 4.165714 \times 10^{-5}v^2$$
$$C_{rL} = 1.9354167 + 0.0083357v - 0.000124971v^2,$$

where C_D is the drag coefficient, and C_{fL} and C_{rL} are the front- and rear-axle downforce coefficients respectively.

Table 9.3 Vehicle parameters representative of a 2014 Formula One car.

Symbol	Description	Value
P_{IC}^{max}	maximum ICE power (closed waste gate)	440 kW
P_k^{max}	maximum MGU-K power	120 kW
M	vehicle mass	710 kg
I_x	moment of inertia about the x-axis	112.5 kg m^2
I_y	moment of inertia about the y-axis	450 kg m^2
I_z	moment of inertia about the z-axis	450 kg m^2
w	wheelbase	3.4 m
a	distance of the mass centre from the front axle	1.8 m
b	distance of the mass centre from the rear axle	$w - a$
h	centre of mass height	0.3 m
A	frontal area	1.5 m^2
D	roll moment distribution (fraction at the front axle)	0.5
w_f	front wheel to car centre line distance	0.73 m
w_r	rear wheel to car centre line distance	0.73 m
R	wheel radius	0.33 m
k_d	differential friction coefficient	10.5 Nm s rad^{-1}
μ_x	tyre peak longitudinal friction coefficient	1.75
μ_y	tyre peak lateral friction coefficient	1.80

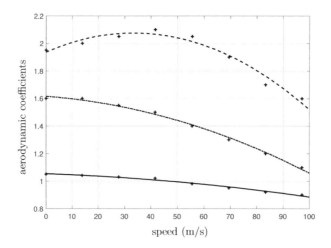

Figure 9.21: Aerodynamic coefficients. The drag coefficient C_D is the solid curve. The front-axle downforce coefficient C_{fL}, is given by the dot-dash curve. The rear-axle downforce coefficient C_{rL}, is given by the dashed curve. The $+$ symbols represent measured data points.

9.4.5   Scaling

Scaling can have a significant influence on the performance of optimization algorithms. Since convergence tolerances, and other criteria, are necessarily based on some notion of 'small' and 'large' quantities, problems with unbalanced scaling may cause difficulties. One notion of scaling is to transform the variables from their original representation, which may reflect the physical nature of the problem, to dimensionless quantities that have desirable properties in terms of the convergence of optimization algorithms. Here we take the length of the car to be the fundamental unit of length—after scaling, the car has a dimensionless length of one. In the same way, the fundamental unit of mass is the mass of the car—after scaling, the car has a dimensionless mass of one. In order to scale time, we give the car unit weight, and so a time scale of $\sqrt{g/l_0}$ is used in which l_0 is the car's length and g is the acceleration due to gravity ($\sqrt{l_0/g}$ is thus the fundamental unit of time). If the car's mass is m_0 kg, then its weight is $m_0 g$ N, which after scaling becomes unity. Once normalizing scale factors for length, mass, and time have been defined, all other scale factors can be defined in terms of these quantities. During racing the car's speed might be 90 m/s; after scaling this becomes 15.58 in dimensionless units, if $l_0 = 3.4$ m. During firm acceleration and fast cornering, the car's longitudinal and lateral tyre forces might reach 8,000 N, which becomes 1.236 in dimensionless units; the force scale factor is derived from the mass, length, and time scale factors. This scaling scheme forces the nonlinear programme's decision variables into a 'more spherical' space rather than an elongated hyperellipsoid one.

9.4.6   Non-smooth features

Nonlinear programming algorithms usually require derivative information for the functions defining the cost and constraints. For this reason non-smooth problem features have to be approximated by differentiable functions in a way that does not change significantly the problem's solution. Functions such as $\min(x, 0)$ and $\max(x, 0)$ have undefined derivatives at $x = 0$, and can be conveniently approximated using

$$\max(x, 0) \approx \frac{x + \sqrt{x^2 + \epsilon}}{2} \quad \text{and} \quad \min(x, 0) \approx -\frac{-x + \sqrt{x^2 + \epsilon}}{2}, \qquad (9.78)$$

in which ϵ is a 'small' constant. Since

$$\frac{\partial}{\partial x}\left(\frac{\pm x + \sqrt{x^2 + \epsilon}}{2}\right) = \frac{1}{2}\left(\frac{x}{\sqrt{x^2 + \epsilon}} \pm 1\right), \qquad (9.79)$$

one sees that the derivative approximations are now well defined at $x = 0$. One may also make use of $\max(a, b) = a + \max(0, b - a)$. In the case of $|x|$, one might use

$$|x| \approx \sqrt{x^2 + \epsilon};$$

in this case

$$\frac{\partial}{\partial x}\sqrt{x^2 + \epsilon} = \left(\frac{x}{\sqrt{x^2 + \epsilon}}\right),$$

with the derivative approximation at $x = 0$ again well defined. These approximations become more accurate as the value of ϵ is reduced, with values in the range $10^{-5} \leq \epsilon \leq 10^{-2}$ typical for the results given here.

9.4.7 Regularization

When the controls enter the system dynamics and performance index linearly, the possibility of bang–bang controls and/or singular arcs exists; see Section 8.5. The majority of general-purpose optimal control software codes assume a strong form of the Legendre–Clebsch necessary condition (i.e. $\nabla_{uu}\mathcal{H} > 0$, see (8.20)), and so these programs cannot be used to solve problems containing singular arcs. In order to avoid oscillatory solutions and non-convergent NLP behaviour in these problems, it is sometimes useful to introduce into the performance index small terms that are quadratic in the controls; see Section 8.5.4. These terms ensure, at least in theory, that the Pontryagin minimum principle can be used to find the optimal controls (for the perturbed problem). If these terms are 'small' relative to the elapsed time, they do not change the problem in a significant way. In our case we used an index of the form

$$J = \int_0^T (1 + \sum_{i=1}^m \epsilon_i u_i^2)dt \qquad (9.80)$$

in which the ϵ_is are small constants with the u_is the controls. This is clearly no longer a pure minimum-time problem, but is close to one if the ϵs are chosen to be sufficiently small. Instead of using J as a measure of the minimum lap time, the 'true' lap time is computed separately as an auxiliary state. For the problem analysed here, the minimum value of J, J_{opt} is of the order 110, while the computed lap time is typically $J_{opt} - 0.1\,\text{s}$; this differences is comparable with the changes obtained in J_{opt} when varying the integration mesh.

9.4.8   Computing gradients

Gradient-based methods for solving NLPs require the derivatives of the objective function and constraints with respect to the states and the controls. The most obvious way to computing these derivatives is analytically, either by hand or by using a computer algebra package such as Maple™ and Mathematica™. While the exactness of this approach is appealing, and generally results in faster rates of convergence, it is sometimes not practical to compute derivatives this way.

Another approach is to compute derivatives numerically using finite differencing. In the case of forward differencing one uses

$$\frac{df}{dx} \approx \frac{f(x + \Delta) - f(x)}{\Delta}$$

in which Δ is the step length. The difficulty with this approach is that Δ must be small enough to provide a good approximation to the derivative, but not so small that round-off errors occur when calculating the difference $f(x + \Delta) - f(x)$—this is the well-known step-size dilemma. Another approach is *complex-step differentiation*, which is both accurate and efficient [425–427]. In the case of higher-order derivatives, a more general version of the complex-step derivative approximation, called the multi-complex-step derivative approximation, can be used as a way of computing Hessians without truncation errors [428]. Another powerful approach is automatic differentiation, which computes function derivatives to the precision that one would achieve using analytic or symbolic differentiation [429]. While any of these approaches can be used, in our experience automatic and symbolic differentiation produce reliable results.

9.4.9   Problem setup

In order to solve the minimum-lap-time optimal control problem one has to assemble a performance index, which in our case is the regularized lap time (9.80), the constraints associated with the problem dynamics, the path constraints, and a constraint on the engine power.

The key kinematic equations required are: (7.101) that describes the relationship between the elapsed time and the elapsed distance, (7.95)–(7.98) that describe the evolution of the Darboux frame Euler angles, and (7.104) that describes the evolution of the car's position relative to the spine of the track. Equation (7.107) is used to describe the evolution of the vehicle's yaw angle relative to the spine.

The key dynamic equations include the longitudinal and lateral force balance equations (7.120), (7.121), and (7.117), and the yaw moment balance equation (7.122). The aerodynamic forces are given by (7.159), (7.157), and (7.158).

A number of constraint conditions have to be incorporated into the optimal control problem. A path constraint on $n(s)$ is used to ensure that the car remains on the track; the track width may be variable and thus a function of s. One also needs to include longitudinal tyre force constraints that are associated with the brakes and the differential—one possible set of constraint equations can be found in (7.137) and (7.138). The power available to the rear wheels, P_m, is constrained by inequality (9.72). The power generated/absorbed by the MGU-H is described by (9.73) and is constrained by $P_h^{\min} = -40\,\text{kW}$ and $P_h^{\max} = 60\,\text{kW}$. The fuel consumption (9.74) is constrained by

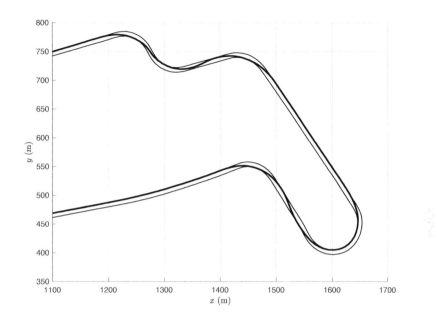

Figure 9.22: Optimized racing line for the car through turns 5 to 9.

$0 \leq F \leq 100/n_l \, kg$, in which n_l is the number of laps in the race. The stored energy (9.75) is constrained by $4 \, MJ \geq E_s \geq 0$ over each lap. The energy supplied to the MGU-K from the ES, (9.76), is constrained by $4 \, MJ \geq E_{ES2K} \geq 0$. Constraints (9.77) monitor the energy flows from the MGU-K into the ES and are constrained by $0 \geq E_{ES2K} \geq -2 \, MJ$. The tyre force constraint equations include the normal load balance condition (7.129), the roll and pitch load-transfer conditions (7.130) and (7.131), and the roll balance constraint (7.132). Path constraints are included in the problem setup to ensure that tyre normal loads don't change sign thereby producing 'non-physical' forces of attraction between the tyre and the road; see (7.136) and the surrounding discussion. The tyre slips are constrained to remain within a region of realistic tyre wear. Cyclic constraints are placed on all of the vehicle states to ensure that they do not change discontinuously across the start/finish line. In other words, while the initial and final vehicle states are free within the physical limits of the problem, they must be equal (as they cross the start/finish line).

9.4.10 Results

Figure 9.22 shows an example of the optimal racing line taken in Turns 5 to 9 at Spa. As one would expect, the optimal control algorithm seeks to maximize the radius of curvature of each corner and 'straight-line' the chicane at Turns 5 and 6. The optimal lap time on the Spa circuit is approximately 107 s with the current dataset.

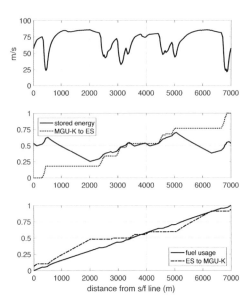

Figure 9.23: Energy usage in the 2014 ERS under racing conditions for the three-dimensional track model. The top graph shows the vehicle speed; the central graphs show the stored energy normalized by 4 MJ and the energy transfer from the MGU-K to the ES normalized by 2 MJ; the bottom graphs show the fuel usage normalized by 100/44 kg and the energy transfer from the ES to the MGU-K normalized by 4 MJ.

Figure 9.23 shows the vehicle speed, and the optimal energy usage and management, on a racing lap of Spa. It is clear from this figure that the fuel-usage constraint is active, since all the allotted fuel is consumed. The energy stored in the battery remains well within the given constraints, with the battery's state of charge the same at the start and end of the lap—this need not be the case on a qualifying lap. The energy transfer allocation from the MGU-K to the ES is fully utilized, while the energy transfer from the ES to the MGU-K remains within the allowed limit of 4 MJ.

Figure 9.24 illustrates the power flows for a racing lap of Spa. The first thing that is immediately apparent is the bang–bang nature of the engine fuelling and MGU-K utilization. Referring to the expanded view of turn 1, it is clear that on entry to the corner the MGU-K drive is cut first (at approximately 120 m), then the engine fuelling is cut and a period of regenerative braking begins (at approximately 300 m), with the brakes applied at approximately 345 m. Immediately prior to mid corner, engine fuelling is initiated, the mechanical brakes are released, and regenerative braking is terminated. On the exit of the corner the engine is fully fuelled and the MGU-K drive is initiated, both at approximately 510 m.

The next two figures illustrate the effect of a three-dimensional (3D) track as opposed to a flat (two-dimensional) one. Figure 9.25 (a) shows the influence of three-dimensionality on the racing speed. The time-difference graph shows that the car is

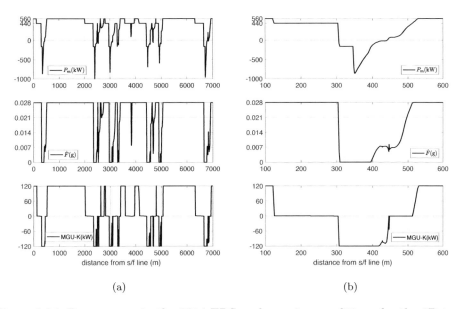

Figure 9.24: Power usage in the 2014 ERS under racing conditions for the 3D track: (a) shows the mechanical power delivered to the rear wheels (top graph), the fuel flow rate (central graph) and the MGU-K power (bottom graph); (b) is an expanded view for turn 1.

approximately two seconds faster on the 3D track. The car on the 3D track gains time in turn 1 (la Source) due to its favourable camber, which can be seen in Figure 9.25 (b).[12] One then notices that the car is quicker on the 3D track into turn 2, primarily because this is a downhill section. From the exit from turn 2 all the way to turn 5 (Kemmel Straight) the car is quicker on the 2D track, because this section is uphill. The car on the 2D track then loses time in the right–left chicane at turns 5 and 6, since this section is still uphill and the car experiences some adverse camber into turn 6. In the section between turns 10 and 12, the 3D track is noticeably faster due to the falling elevation into turn 10. The section of note is between turn 18 and the start/finish line. The car on the 3D track is faster through turn 18 due to advantageous cambering, with this speed carried through to the SF line.

We conclude this example by showing the effect of the fuel limit on the car's performance; see Figure 9.26. A full understanding of this graph requires a more detailed analysis of the car's powertrain, which is outside our consideration here. That said, it is clear that little performance is being gained by increasing the fuel quota by 10%, but there is a noticeable drop-off in the high-speed sections when the fuel quota is reduced by 10%.

---

[12] Positive camber angles represent positive (favourable) cambering on right-hand corners.

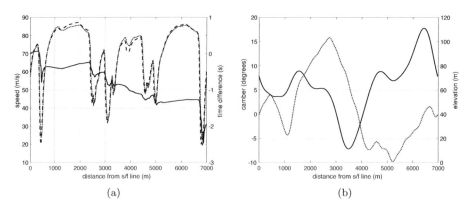

(a)   (b)

Figure 9.25: (a) Comparison between racing speed profiles for 2D (dashed) and 3D (dotted) tracks with time difference $(T_{3D} - T_{2D})$ (solid). (b) Camber (solid) and elevation (dotted) of Circuit de Spa-Francorchamps.

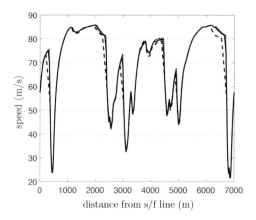

Figure 9.26: Comparison of speed profile with different fuel usage limits. The solid line shows the baseline case where $100/n_l$ kg is available per lap; n_l is the number of racing laps. The dotted line shows the case where 10% more fuel is allowed and the dashed line shows the case where 10% less fuel is allowed.

9.5 Minimum time of a racing motorcycle

In this example we consider the minimum-lap-time optimal control for a motorcycle and idealized rider. In contrast to car problems, the equations of motion of even relatively simple single-track vehicle models are too long to report explicitly. The very simple models such as Timoshenko–Young model described in Chapter 4 and the Whipple model given in Chapter 5 are too simple to be regarded as realistic. These models have various shortcomings including the absence of representative slipping force-generating tyre models, frame flexibilities, aerodynamic effects, and suspension systems. Many of these modelling issues are described in Chapter 7 and so will not be addressed again here. In the case of motorcycle system modelling it is common practice to use symbolic modelling tools such as MBSymba [430] to derive the requisite equations of motion.

Minimum-lap-time optimal control problems for motorcycles have received less attention than those for their four-wheeled counterparts, although various studies have been reported in the literature. In the late 1990s a basic model comprising eight states and three inputs was used to study optimal manoeuvring in a chicane-like track section [352]. In this case the inputs were the lateral force of the front tyre, the longitudinal force of the rear tyre, and the braking ratio. In more recent times a number of more advanced models have been used in optimal control studies [249, 385, 404]. In order to reduce solution times a number of modelling simplifications are usually necessary. These include the use of low-order polynomial representations of the suspension system (see Section 7.7.6), and simplified tyre models that capture tyre saturation and the coupling between longitudinal and lateral forces but avoid the use of full-blown magic formulae (Section 3.5).

We will consider the optimal control of a 250 cc racing motorcycle being ridden on the Aragon circuit; Figure 9.27. The vehicle parameters used are given in Table 9.4. The model has nine degrees of freedom (see Figure 7.49); the position and orientation of the chassis, the steering angle, the front- and rear-suspension travels. The vehicle model has twenty states

$$\boldsymbol{x} = [u, v, w, z, \dot{\psi}, \phi, \dot{\phi}, \mu, \dot{\mu}, \delta, \dot{\delta}, z_r, \dot{z}_r, z_f, \dot{z}_f, \hat{\alpha}_r, \hat{\alpha}_f, s, n, \xi]^T, \qquad (9.81)$$

where u, v, w are the velocities (expressed in the Darboux frame) of a fixed point in the motorcycle chassis; the elevation of the fixed point is z. The yaw rate is given by $\dot{\psi}$; ϕ and $\dot{\phi}$ are the roll angle and roll rate; μ and $\dot{\mu}$ are the pitch angle and pitch rate; δ and

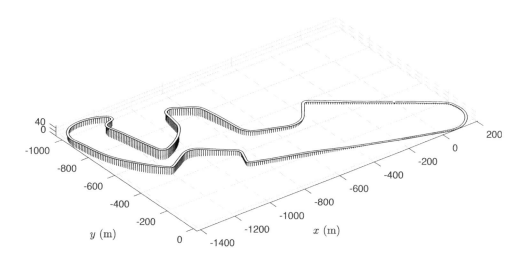

Figure 9.27: Aragon circuit.

Table 9.4 Motorcycle plus rider parameters (250 cc racing vehicle).

Symbol	Description	Value
P_{max}	engine maximum power	40 kW
M	mass	140 kg
I_x	moment of inertia about the x-axis (roll)	13 kg m^2
I_y	moment of inertia about the y-axis (pitch)	25 kg m^2
I_z	moment of inertia about the z-axis (yaw)	15 kg m^2
I_{xz}	cross moment of inertia ($I_{xz} = - \int xz\, dm$)	-5 kg m^2
w	wheelbase	1.23 m
b	distance of the mass centre from the rear axle	0.56 m
h	centre of mass height	0.63 m
k_f	front suspension stiffness	12 kN/m
k_r	rear suspension stiffness	100 kN/m
τ	velocity ratio (suspension to swingarm rotation)	0.2 m/rad
l_s	swingarm length	0.55 m
c_δ	steer damper coefficient	5 Nm s/rad
R	tyre radius	0.30 m
r	tyre crown radius	0.035 m
μ_x	tyre longitudinal friction coefficient	1.1
μ_y	tyre lateral friction coefficient	1.4
ρ	air density	1.23 kg/m^3
$C_D A$	drag coefficient in acceleration/braking	0.20/0.45

$\dot{\delta}$ are the steering angle and steering rate; z_r and $\dot{z}_r$, and z_f and $\dot{z}_f$, represent the travel of the rear and front suspensions and their travel rates; $\hat{\alpha}_r$ and $\hat{\alpha}_f$ are the rear- and front-tyre relaxed side-slip angles; and s and n describe the vehicle's road position—s describes the position along the track centre line, and n describes the lateral position of the vehicle relative to the centre line. The angle between the vehicle's central plane and track centre line is ξ; see Figure 7.20. The first fifteen states are related to the equations of motion, while the states $\hat{\alpha}_r, \hat{\alpha}_f$ are associated with the tyre relaxation (see Section 3.8)

$$\frac{\sigma_r}{V_r}\dot{\hat{\alpha}}_r + \hat{\alpha}_r = \alpha_r \tag{9.82}$$

$$\frac{\sigma_f}{V_f}\dot{\hat{\alpha}}_f + \hat{\alpha}_f = \alpha_f. \tag{9.83}$$

The front and rear relaxation lengths are given by σ_f and σ_r respectively. The front- and rear-tyre longitudinal contact velocities are given by V_f and V_r respectively. The front and rear slip angles are given by α_f and α_r, and are computed using the states $\boldsymbol{x}$ according to (3.7). The model inputs are

$$\boldsymbol{u} = [T_h, F_{xr}, F_{xf}]^T, \tag{9.84}$$

where T_h is the steering torque, and F_{xf} and F_{xr} are the front and rear longitudinal tyre forces.

In sum, the equations of motion can be written in the following descriptor system form:

$$E(\boldsymbol{x})\dot{\boldsymbol{x}} = f(\boldsymbol{x}, \boldsymbol{u}); \tag{9.85}$$

the matrix $E(\boldsymbol{x})$ may be of large dimension and/or be poorly conditioned [352].

In this example, the lateral forces are computed as follows:

$$F_{yr} = (C_\alpha^r \hat{\alpha}_r + C_\gamma^r \gamma_r) F_{zr} \tag{9.86}$$

$$F_{yf} = (C_\alpha^f \hat{\alpha}_f + C_\gamma^f \gamma_f) F_{zf}, \tag{9.87}$$

where C_α^f and C_α^r are the normalized front- and rear-tyre cornering stiffnesses, and C_γ^f and C_γ^r are the normalized front and rear camber stiffnesses. The front- and rear-tyre normal loads are given by F_{zf} and F_{zr}. Pure rolling in the longitudinal direction is assumed. The coupling between the longitudinal and lateral forces is enforced using friction ellipses:

$$\left(\frac{F_{xr}}{\mu_{xr}}\right)^2 + \left(\frac{F_{yr}}{\mu_{yr}}\right)^2 \leq F_{zr}^2 \qquad \left(\frac{F_{xf}}{\mu_{xf}}\right)^2 + \left(\frac{F_{yf}}{\mu_{yf}}\right)^2 \leq F_{zf}^2, \tag{9.88}$$

where $\mu_{x,y}$ are the friction coefficients in the longitudinal and lateral direction respectively.

A number of additional constraints must also be included. The torque at the rear wheel is limited by

$$T_r \leq \tau_p \tau_g \tau_f T_e^{\max} \eta, \tag{9.89}$$

where τ_p is the gear ratio between the engine driveshaft and the gearbox input shaft, τ_g is the gearbox ratio, τ_f is the final gear ratio between the gearbox output shaft and

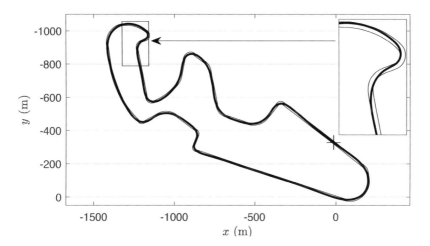

Figure 9.28: Optimal trajectory on the Aragon circuit (plan view).

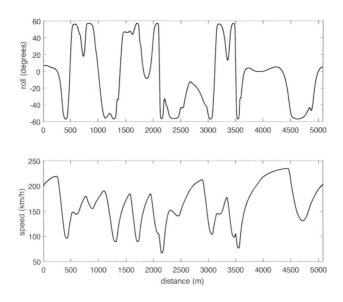

Figure 9.29: Optimal speed and roll angle for the Aragon circuit.

the rear wheel, and $T_e^{\max}$ is the maximum engine torque at the engine spin velocity ω_e (which is related to the rear wheel spin rate ω_r by $\omega_e = \tau_p \tau_g \tau_f \omega_r$). The mechanical efficiency of the powertrain is denoted η. It is also assumed that the rider engages the gear ratio that gives the maximum torque at every vehicle speed; the gear shift is assumed instantaneous. The steering angle is limited by

$$\delta \le |\delta^{\max}|. \tag{9.90}$$

The steering torque is limited, in order to recognize the rider's physical limitation

$$T_h \le |T_h^{\max}| \tag{9.91}$$

in this regard. The vehicle must be constrained to remain within the track boundaries[13]

$$n \le |n^{\max}|. \tag{9.92}$$

The front and rear tyres may be constrained to remain in contact with the road, although this is not strictly necessary (see Section 7.2):

$$F_{zr} \le 0 \qquad F_{zf} \le 0. \tag{9.93}$$

The motorcycle's optimal trajectory is given in Figure 9.28, where the optimal racing line can be observed. The corresponding optimal speed and optimal roll angle are shown in Figure 9.29; the maximum speed is 235 km/h, while the maximum roll angle is 57°. The corresponding optimal lap time is 117.9 s.

---

[13] A more precise formulation may include constraints on the lateral position of the each tyre contact point.

References

Reference entries include the page number(s) in the book where they are discussed.

[1] R. W. Hamming, *Numerical Methods for Scientists and Engineers*. New York: McGraw-Hill, 1962. See page v.

[2] F. J. W. Whipple, 'The Stability of the Motion of a Bicycle', *Quarterly Journal of Pure and Applied Mathematics*, vol. 30, pp. 312–48, 1899. See pages vi, 32, 195, 197, 212, 214, 325, and 345.

[3] A. Sharp, *Bicycles and Tricycles: An Elementary Treatise on their Design and Construction*. London, New York and Bombay: Longmans, Green, and Co., 1896, Reprinted as: *Bicycles and Tricycles: A Classic Treatise on their Design and Construction*, Dover Publications, Mineola, NY, 1977. See pages 1, 13, 15, 20, and 217.

[4] D. V. Herlihy, *Bicycle: The History*. New Haven and London: Yale University Press, 2004. See pages 1, 13, 14, 15, 17, 18, 19, 20, and 22.

[5] N. Clayton, *Early Bicycles*. Princes Risborough, Bucks, UK: Shire Publicatins, 1986. See pages 1, 17, and 19.

[6] V. Cossalter, *Motorcycle Dynamics*. Lulu, 2006. See pages 1, 177, 229, 267, and 275.

[7] T. Foale, *Motorcycle Handling and Chassis Design: The Art and Science*. Tony Foale, 2002. See pages 1 and 229.

[8] H. B. Pacejka, *Tire and Vehicle Dynamics*, 3rd edn. Oxford: Butterworth Heinemann, 2012. See pages 1, 51, 107, 123, 124, 129, 130, 133, 136, 149, 151, 152, 154, 157, 162, 163, 165, 183, and 184.

[9] G. Cocco, *Motorcycle Design and Technology: How and Why*. Vimodrone: Giorgio Nada Editore, 2005. See page 1.

[10] A. Cathcart, *The Ultimate Racers*, 1st edn. London: Gullane Children's Books, 1990. See page 1.

[11] T. P. Newcomb and R. T. Spurr, *A Technical History of the Motorcar*. Bristol and New York: A. Hilger, 1989. See pages 1, 6, 7, 9, and 11.

[12] E. Eckermann, *World History of the Automobile*. Warrendale, PA: SAE International, 2001. See pages 1 and 9.

[13] S. Parissien, *The Life of the Automobile*. London: Atlantic Books, 2013. See pages 1, 6, and 12.

[14] M. Ruiz, *The History of the Automobile*. New York: Gallery Books, 1985. See pages 1, 6, and 11.

[15] G. Genta, L. Morello, F. Cavallino, and L. Filtri, *The Motor Car: Past, Present and Future*. Dordrecht: Springer, 2014. See pages 1 and 3.

[16] D. King-Hele, 'Eramus Darwin's Improved Design for Steering Carriages—and Cars', *Notes and Records of the Royal Society of London*, vol. 56, no. 1, pp. 41–62, 2002. See pages 1 and 3.

[17] F. Freudenstein, 'Approximate Synthesis of Four-Bar Linkages', *Transaction of the ASME*, vol. 77, pp. 853–61, 1955. See page 4.

[18] G. N. Georgano, *Early and Vintage Years, 1885–1930: The Golden Era of Coach-building.* Broomall, PA: Mason Crest Publishers, 2002. See page 7.

[19] D. Clerk, *The Gas, Petrol, and Oil Engine.* London: Longmans, Green and Co., 1910. See page 7.

[20] Benz and Co., 'Fahrzeug mit Gasmotorenbetrieb', German Patent 37 435, January 29, 1886. See page 7.

[21] M. Hamer, 'Brimstone and Bicycles', *New Scientist*, pp. 48–9, 2005. See page 13.

[22] M. Strubreiter and M. Zappe, *Cyclepedia: A Tour of Classic Bicycle Designs.* London: Thames & Hudson, 2011. See page 13.

[23] P. Lallement, 'Velocipede', U.S. Patent 59 915, November 20, 1866. See pages 15 and 16.

[24] H. Kellogg, 'Bicycle', U.S. Patent 283 612, August 21, 1883. See page 17.

[25] H. J. Lawson, 'Improvements in the Construction of Bicycles and other Veloci-pedes, and in Apparatus to be Used in Connection Therewith.' Great Britain Patent 3934, September 30, 1879. See page 18.

[26] H. J. Lawson, 'Velocipede', U.S. Patent 345 851, July 20, 1886. See page 19.

[27] G. W. Pressey, 'Velocipede', U.S. Patent 233 640, October 26, 1880. See pages 18 and 19.

[28] A. Winton, 'Bicycle Frame', U.S. Patent 500,177, 1893. See page 21.

[29] M. B. Ryan, 'Folding Bicycle', U.S. Patent 518 330, April 17, 1894, first folding bicycle patent. See page 21.

[30] J. M. Starley, 'Design for a Bicycle Frame', U.S. Patent 18 820, December 18, 1888, women's bicycle frame. See page 21.

[31] E. S. Tomkins, *The History of the Pneumatic Tyre*, 1st edn. Vista, CA: Eastland Press, December 1981. See pages 26, 27, and 28.

[32] R. W. Thomson, 'Improvement in Carriage Wheels', U.S. Patent 5104, May 8, 1847. See page 26.

[33] J. B. Dunlop, 'Wheel tire for cycles', U.S. Patent 435,995, October 10, 1890. See page 26.

[34] C. K. Welch, 'Pneumatic Tire', U.S. Patent 612,981, 1898. See page 27.

[35] T. B. Jeffery, 'Wheel Tire', U.S. Patent 558,957, 1896. See pages 27 and 28.

[36] E. B. Killen, 'Improvements in or relating to Pneumatic Tyre, Rims, Wheels, and their Fitments to Road Vehicle or Chassis', British Patent 329,995, 1930. See page 29.

[37] O. J. Patin, 'Rim for Tubeless Truck Tires', U.S. Patent 2,859,792, 1958. See page 29.

[38] 'Tubeless Tires', *New York Times*, December 1954. See page 29.

[39] R. Beckadolph, 'Pneumatic Vehicle Tire', U.S. Patent 3,118,483, 1964. See page 29.

[40] J. Synge and B. Griffith, *Principles of Mechanics*, 3rd edn. New York: McGraw-Hill, 1959. See pages 30, 97, and 105.

[41] H. Goldstein, *Classical Mechanics*, 2nd edn. London: Addison-Wesley, 1980. See pages 30, 39, 44, 55, 64, 65, 66, 68, 97, and 195.

[42] H. Goldstein, C. Poole, and J. Safko, *Classical Mechanics*, 3rd edn. London: Pearson Education International, 2002. See pages 30, 67, and 299.

[43] L. D. Landau and E. M. Lifshitz, *Mechanics*, 3rd edn. Oxford: Elsevier, Butterworth-Heinemann, 1976, vol. 1. See pages 30, 43, 52, and 64.

[44] C. Lanczos, *The Variational Principles of Mechanics*. Mineola, NY: Dover, 1986. See pages 30, 33, 38, 39, 40, 42, 43, 45, and 55.

[45] V. I. Arnold, *Mathematical Methods of Classical Mechanics*, 2nd edn. New York: Springer, 1989. See pages 31, 32, and 72.

[46] E. A. Milne, *Vectorial Mechanics*. New York: Interscience, 1948. See pages 31, 36, and 90.

[47] D. D. Holm, T. Schmah, and C. Stoica, *Geometric Mechanics and Symmetry. From Finite to Infinite Dimensions*. Oxford: Oxford University Press, 2009. See page 31.

[48] J. I. Neǐmark and N. A. Fufaev, 'Dynamics of Nonholonomic Systems.' *American Math. Soc. Translations of Mathematical Monographs.*, vol. 33, pp. 330–74, 1972. See pages 31, 32, 97, 105, 106, and 196.

[49] A. A. Shabana, *Dynamics of Multibody Systems*. Cambridge: Cambridge University Press, 2013. See page 31.

[50] J. Wittenburg, *Dynamics of Multibody Systems*. Berlin: Springer, 2007. See page 31.

[51] J. G. De Jalon and E. Bayo, *Kinematic and Dynamic Simulation of Multibody Systems: the Real-time Challenge*. Berlin: Springer, 2012. See page 31.

[52] P. E. Nikravesh, *Computer-Aided Analysis of Mechanical Systems*. Englewood Cliffs, NY: Prentice-Hall, 1988. See page 31.

[53] R. M. Murray, Z. Li, and S. S. Sastry, *A Mathematical Introduction to Robotic Manipulation*. Boca Raton, FL: CRC Press, 1994. See pages 31 and 68.

[54] M. E. Carvallo, 'Théorie Du Movement Du Monocycle, part 2: Théorie De La Bicyclette.' *Journal de L'Ecole Polytechnique.*, vol. 6, pp. 1–118, 1901. See pages 32 and 195.

[55] M. Minc and H. Marcus, *A Survey of Matrix Theory and Matrix Inequalities*. Mineola, NY: Dover, 1992. See pages 47 and 355.

[56] L. Meirovitch, *Methods of Analytical Dynamics*. New York: McGraw-Hill, 1970. See page 49.

[57] J. M. Cameron and W. J. Book, 'Modeling Mechanisms with Nonholonomic Joints using the Boltzmann–Hamel Equations', *The International Journal of Robotics Research*, vol. 16, no. 1, pp. 47–59, 1997. See page 49.

[58] A. M. Bloch, J. E. Marsden, and D. V. Zenkov, 'Quasivelocities and Symmetries in Non-holonomic Systems', *Dynamical systems*, vol. 24, no. 2, pp. 187–222, 2009. See page 49.

[59] A. J. V. der Schaft and B. M. Maschke, 'On the Hamiltonian Formulation of Nonholonomic Mechanical Systems', *Reports on Mathemaical Physics*, vol. 34, no. 2, pp. 225–33, 1994. See page 62.

[60] D. M. Fradkin, 'Existence of the Dynamic Symmetries O_4 and SU_3 for All Classical Central Potential Problems', *Progress of Theoretical Physics*, vol. 37, no. 5, pp. 798–812, 1967. See page 66.

[61] P. C. W. Davies, *The Physics of Time Asymmetry*. Berkeley and Los Angeles: University of California Press, 1974. See pages 74 and 77.

[62] R. Peierls, *Surprises in Theoretical Physics*. Princeton, NJ: Princeton University Press, 1979. See page 74.

[63] R. Feynman, *The Character of Physical Law*. Cambridge, MA: MIT, 1967. See page 74.

[64] R. C. Tolman, *The Principles of Statistical Mechanics*. Oxford: Oxford University Press, 1938. See page 74.

[65] K. Hutchison, 'Is Classical Mechanics Really Time-Reversible and Deterministic?' *The British Journal for the Philosophy of Science*, vol. 44, no. 2, pp. 307–23, June 1993. See page 75.

[66] S. H. Strogatz, *Nonlinear Dynamics and Chaos: With Applications to Physics, Biology, Chemistry, and Engineering*. Boston, MA: Addison-Wesley, 1994. See page 75.

[67] R. Kubo, 'The Fluctuation-Dissipation Theorem', *Reports on Progress in Physics*, vol. 29, no. 1, pp. 255–84, 1966. See pages 77 and 78.

[68] A. O. Caldeira and A. J. Leggett, 'Path Integral Approach to Brownian Motion', *Physica A*, vol. 121, pp. 587–616, 1983. See page 78.

[69] R. Zwanzig, 'Nonlinear Generalized Langevin Equations', *Journal of Statistical Physics*, vol. 9, no. 3, pp. 215–20, 1973. See page 78.

[70] G. W. Ford and M. Kac, 'On the Quantum Langevin Equation', *Journal of Statistical Physics*, vol. 46, no. 5-6, pp. 803–10, 1987. See page 78.

[71] M. C. Smith, 'Synthesis of Mechanical Networks: The Inerter', *IEEE Trans. Automatic Control*, vol. 47, no. 10, pp. 1648–62, 2002. See pages 80, 237, 263, and 264.

[72] A. Ruina, 'Nonholonomic Stability Aspects of Piecewise Holonomic Systems', *Reports on Mathematical Physics*, vol. 42, pp. 91–100, 1998. See page 83.

[73] C. Carathéodory, 'Der Schlitten', *Zeitschrift für Angewandte Mathematik und Mechanik*, vol. 13, no. 2, pp. 71–6, 1933. See page 87.

[74] O. M. O'Reilly, 'The Dynamics of Rolling Disks and Sliding Disks', *Nonlinear Dynamics*, vol. 10, pp. 287–305, 1996. See page 97.

[75] L. A. Pars, *A Treatise on Analytical Mechanics*. London: Heinemann, 1965. See page 97.

[76] E. J. Routh, *Advanced Part of a Treatise on the Dynamics of a System of Rigid Bodies*, 6th edn. London: McMillan Co., 1905. See pages 97, 105, and 106.

[77] A. M. Bloch, J. E. Marsden, and D. V. Zenkov, 'Nonholonomic Dynamics', *Notices of the AMS*, vol. 52, no. 3, pp. 324–33, 2005. See page 97.

[78] A. M. Bloch, P. S. Krishnaprasad, J. E. Marsden, and R. M. Murray, 'Nonholonomic Mechanical Systems with Symmetry', *Arch. Rational Mech. Anal.*, vol. 136, pp. 21–99, 1996. See page 97.

[79] H. K. Moffatt, 'Euler's Disk and its Finite-Time Singularities', *Nature*, vol. 404, pp. 833–34, 2000. See page 103.

[80] O. Reynolds, 'On Rolling-Friction', *Philos. Transact. of the Royal Soc. of London*, vol. 166, pp. 155–74, 1876. See page 107.

[81] G. Becker, H. Fromm, and H. Maruhn, *Schwingungen in Automobillenkungen*. Berlin: Verlag M. Krayn, 1931, p.43. See page 107.

[82] D. Sensaud de Lavaud, 'Dandinement, Shimmy et Pseudo-Shimmy d'une Voiture Automobile', Communications à l'Académie des sciences, Tech. Rep., 1927. See page 107.

[83] NACA, 'Papers on Shimmy and Rolling Behaviour of Landing Gears', NACA, Tech. Rep. TM 1365, 1954, english translation of the papers presented at the Stuttgart conference, 16–17 October 1941. See pages 107 and 158.

[84] V. Cossalter, R. Lot, and M. Massaro, 'The Influence of Frame Compliance and Rider Mobility on the Scooter Stability', *Vehicle System Dynamics*, vol. 45, no. 4, pp. 313–26, 2007. See pages 107, 183, 325, 343, and 346.

[85] M. Plöchl, J. Edelmann, B. Angrosch, and C. Ott, 'On the Wobble Mode of a Bicycle', *Vehicle System Dynamics*, vol. 50, no. 3, pp. 415–29, 2012. See pages 107, 183, and 222.

[86] G. Magnani, J. Papadopoulos, and N. Ceriani, 'On-Road Measurements of High Speed Bicycle Shimmy, and Comparison to Structural Resonance', in *Proceeding of the IEEE International Conference on Mechatronics*, Vicenza, 2013. See pages 107, 183, and 222.

[87] I. J. M. Besselink, 'Shimmy of Aircraft Main Landing Gears', PhD dissertation, TU Delft, 2000. See pages 107, 142, and 183.

[88] L. Martin, V. Jacques, and Y. Martin-Siegfried, 'Study of Shimmy in Industrial Context', in *Proc. of the International Forum on Aeroelasticity and Structural Dynamics (IFASD)*, Saint Petersburg, 2015. See page 107.

[89] H. Fromm, 'Seitenschlupf und Fuhrungswert des Rollenden Rades', Bericht der Lilienthalgesellschaft fur Luftfahrtforschung, Tech. Rep. NACA TM 1365 (english translation), 1941. See page 115.

[90] E. Fiala, 'Seitenkra fte am Rollenden Luftreifen', *VDI Zeitschrift*, vol. 96, 1954. See pages 115 and 136.

[91] G. Freudenstein, 'Luftreifen bei Schrg-und Kurvenlauf', *Deutsche Kraftfahrzeug Forschung und Str. Verk. techn*, vol. 152, 1961. See pages 115 and 126.

[92] H. B. Pacejka, 'The Wheel Shimmy Phenomenon', PhD dissertation, TU Delft, 1966. See pages 115, 142, and 224.

[93] F. Frank, 'Theorie des Reifenschraglaufs', PhD dissertation, Braunschweig, 1965. See page 115.

[94] K. Guo, 'The Effect of Longitudinal Force and Vertical Load Distribution on Tire Slip Properties', *SAE*, no. 945087, 1994. See page 115.

[95] Y. Zhang and Y. Jingang, 'Static Tire/Road Stickslip Interactions: Analysis and Experiments', *IEEE/ASME Transactions on Mechatronics*, vol. 19, no. 6, pp. 1940–50, 2013. See page 115.

[96] H. B. Pacejka, 'Spin: Camber Turning', *Vehicle System Dynamics*, vol. 43, no. S1, pp. 3–17, 2005. See pages 122 and 123.

[97] A. Higuchi, 'Transient Response of Tyres at Large Wheel Slip and Camber', PhD dissertation, TU Delft, 1997. See pages 129 and 154.

[98] V. Cossalter, A. Doria, R. Lot, N. Ruffo, and M. Salvador, 'Dynamic Properties of Motorcycle and Scooter Tires: Measurement and Comparison', *Vehicle System Dynamics*, vol. 39, no. 5, pp. 329–52, 2003. See pages 129, 157, and 341.

[99] P. Zegelaar, 'The Dynamic Response of Tyres to Brake Torque Variations and Road Unevennesses', PhD dissertation, TU Delft, 1998. See page 129.

[100] E. Bakker, L. Nyborg, and H. B. Pacejka, 'Tire Modeling for Use in Vehicle Dynamics Studies', *SAE*, no. 870421, 1987. See page 130.

[101] E. Bakker, H. B. Pacejka, and L. Lidner, 'A New Tire Model with an Application in Vehicle Dynamics Studies', *SAE*, no. 890087, 1989. See page 130.

[102] H. B. Pacejka and E. Bakker, 'The Magic Formula Tyre Model', *Vehicle System Dynamics*, vol. 21, no. S1, pp. 1–18, 1993. See page 130.

[103] P. Bayle, J. F. Forissier, and S. Lafon, 'A New Tyre Model for Vehicle Dynamics Simulation', *Automotive Technology International*, 1993. See page 130.

[104] I. J. M. Besselink, A. J. C. Schmeitz, and H. B. Pacejka, 'An Improved Magic Formula/Swift Tyre Model that can Handle Inflation Pressure Changes', *Vehicle System Dynamics*, vol. 48, no. S1, pp. 337–52, 2010. See pages 130, 137, and 156.

[105] H. B. Pacejka and I. J. M. Besselink, 'Magic Formula Tyre Model with Transient Properties', *Vehicle System Dynamics*, vol. 27, no. S1, pp. 243–49, 1997. See pages 130 and 152.

[106] H. B. Pacejka, *Tyre and Vehicle Dynamics*, 1st edn. Oxford: Butterworth Heinemann, 2002. See page 130.

[107] H. B. Pacejka, *Tyre and Vehicle Dynamics*, 2nd edn. Oxford: Butterworth Heinemann, 2006. See page 130.

[108] H. B. Pacejka, 'Study of the Lateral Behaviour of an Automobile Moving upon a Flat Level Road', Cornell Aeronautical Laboratory, Tech. Rep. YC-857-F-23, 1958. See page 136.

[109] H. B. Pacejka, *Mechanics of Pneumatic Tires*, Washington D.C., 1981, ch. Analysis of Tire Properties. See page 136.

[110] D. Arosio, F. Braghin, F. Cheli, and E. Sabbioni, 'Identification of Pacejka's Scaling Factors from Full-Scale Experimental Tests', *Vehicle System Dynamics*, vol. 43, no. S1, pp. 457–74, 2005. See page 136.

[111] M. Massaro, V. Cossalter, and G. Cusimano, 'The Effect of the Inflation Pressure on the Tyre Properties and the Motorcycle Stability', *Proceedings of the Institution of Mechanical Engineers, Part D: Journal of automobile engineering*, vol. 227, no. 10, pp. 1480–88, 2013. See pages 137 and 157.

[112] V. Cossalter, A. Doria, E. Giolo, L. Taraborrelli, and M. Massaro, 'Identification of the Characteristics of Motorcycle and Scooter Tyres in the Presence of Large Variations in Inflation Pressure', *Vehicle System Dynamics*, vol. 52, no. 10, pp. 1333–54, 2014. See pages 137, 152, and 157.

[113] T. D. Gillespie, *Fundamentals of Vehicle Dynamics*. Warrendale, PA: SAE, 1992. See pages 137, 277, and 287.

[114] J. Y. Wong, *Theory of Ground Vehicles*. Hoboken, NJ: John Wiley & Sons, 2001. See page 137.

[115] W. R. Allen, T. J. Rosenthal, and J. P. Chrstos, 'Vehicle Dynamics Tire Model for both Pavement and Off-Road Conditions', *SAE Technical Paper*, no. 970559, pp. 27–38, 1997. See page 137.

[116] C. Harnisch, B. Lach, R. Jakobs, M. Troulis, and O. Nehls, 'A New Tyre-Soil Interaction Model for Vehicle Simulation on Deformable Ground', *Vehicle System Dynamics*, vol. 43, no. S1, pp. 384–94, 2005. See page 137.

[117] C. Senatore and C. Sandu, 'Off-Road Tire Modeling and the Multi-Pass Effect for Vehicle Dynamics Simulation', *Journal of Terramechanics*, vol. 48, no. 4, pp. 265–76, 2011. See page 137.

[118] P. Coyne, *Roadcraft: The Police Driver's Handbook.* London: Stationery Office, 2007. See page 138.

[119] P. Mares, *Motorcycle Roadcraft: The Police Rider's Handbook.* London: Stationery Office, 2013. See page 138.

[120] D. Tavernini, M. Massaro, E. Velenis, D. Katzourakis, and R. Lot, 'Minimum Time Cornering: The Effect of Road Surface and Car Transmission Layout', *Vehicle System Dynamics*, vol. 51, no. 10, pp. 1533–47, 2013. See pages 138, 290, 383, and 392.

[121] S. de Technologie Michelin, Ed., *The Tyre Grip*, Clermont-Ferrand, 2001. See page 138.

[122] W. Horne and R. Dreher, 'Phenomena of Pneumatic Hydroplaning', NASA, Tech. Rep. D-2056, 1963. See page 138.

[123] NCHRP, 'Guide for Pavement Friction', National Cooperative Highway Research Program, NCHRP Project 01-43 108, 2009. See page 138.

[124] H. Grogger and M. Weiss, 'Calculation of the Hydroplaning of a Deformable Smooth-Shaped and Longitudinally-Grooved Tire', *Tire Science and Technology*, vol. 25, no. 4, pp. 265–87, 1997. See page 139.

[125] A. J. Tremlett and D. J. N. Limebeer, 'Optimal Tyre Usage on a Formula One Car', *Vehicle System Dynamics*, vol. 54, no. 10, pp. 1448–73, 2016. See pages 139, 141, 142, and 383.

[126] D. P. Kelly, *Lap Time Simulation with Transient Vehicle and Tyre Dynamics.* Cranfield University School of Engineering, 2008, PhD Thesis. See pages 139, 141, and 415.

[127] P. Fevrier and G. Fandard, 'A New Thermal and Mechanical Tyre Model for Handling Simulation', *11th International VDI congress, Tyres-Chassis-Road, Hanover*, 2007. See page 139.

[128] B. Lorenz, B. Persson, G. Fortunato, M. Giustiniano, and F. Baldoni, 'Rubber Friction for Tire Tread Compound on Road Surfaces', *Journal of Physics Condensed Matter*, vol. 25, no. 9, 2013. See page 139.

[129] K. Grosch, 'The Relation Between the Friction and Visco-Elastic Properties of Rubber', *Proceedings of the Royal Society of London. Series A, Mathematical and Physical Sciences.*, vol. 274, no. 1356, pp. 21–39, 1963. See page 139.

[130] M. Williams, R. Landel, and J. Ferry, 'The Temperature Dependence of Relaxation Mechanisms in Amorphous Polymers and Other Glass-Forming Liquids', *Journal of the Americal Chemical Society*, vol. 77, no. 14, pp. 3701–7, 1955. See page 139.

[131] D. Moore, 'Friction and Wear in Rubbers and Tyres', *Wear*, vol. 61, no. 2, pp. 273–82, 1980. See page 139.

[132] P. Haney, *The Racing & High-Performance Tire*. Warrendale, PA: SAE International, 2003. See page 139.

[133] B. N. J. Persson and E. Tosatti, 'Qualitative Theory of Rubber Friction and Wear', *The Journal of Chemical Physics*, vol. 112, no. 4, pp. 2021–29, 2000. See page 139.

[134] S. Clark, *The Pneumatic Tire*. Washington D.C.: U.S. Department of Transportation, National Highway Traffic Safety Administration, 2006. See page 139.

[135] A. Schallamach, 'Recent Advances in Knowledge of Rubber Friction and Tire Wear', *Rubber Chemistry and Technology*, vol. 41, no. 1, pp. 209–44, 1968. See page 139.

[136] D. F. Moore, *The Friction of Pneumatic Tyres*. Amsterdam: Elsevier Scientific, 1975. See page 139.

[137] A. L. Browne and L. E. Wickliffe, 'Parametric Study of Convective Heat Transfer Coefficients at the Tire Surface', *Tire Science and Technology*, vol. 8, no. 3, pp. 37–67, 1980. See page 141.

[138] B. v. Schlippe and R. Dietrich, 'Das Flattern eines Bepneuten Rades. Bericht 140 der Lilienthal Gesellschaft', NACA, Tech. Rep. TM 1365, 1941. See pages 142 and 149.

[139] L. Segel, 'Force and Moment Response of Pneumatic Tires to Lateral Motion Inputs', *Transactions ASME, Journal of Engineering for Industry*, vol. 88B, 1966. See page 142.

[140] R. S. Sharp and C. J. Jones, 'A Comparison of Tyre Representations in a Simple Wheel Shimmy Problem', *Vehicle System Dynamics*, vol. 9, no. 1, pp. 45–57, 1980. See page 142.

[141] J. P. Maurice and H. B. Pacejka, 'Relaxation Length Behaviour of Tyres', *Vehicle System Dynamics*, vol. 27, no. S1, pp. 339–42, 1997. See page 152.

[142] A. Higuchi and H. B. Pacejka, 'The Relaxation Length Concept at Large Wheel Slip and Camber', *Vehicle System Dynamics*, vol. 27, no. S1, pp. 50–64, 1997. See pages 152 and 154.

[143] E. J. H. de Vries and H. B. Pacejka, 'Motorcycle Tyre Measurememts and Models', *Vehicle System Dynamics*, vol. 29, no. S1, pp. 280–98, 1998. See page 152.

[144] J. P. Maurice, M. Berzeri, and H. B. Pacejka, 'Pragmatic Tyre Model for Short Wavelength Side Slip Variations', *Vehicle System Dynamics*, vol. 31, no. 2, pp. 65–94, 1999. See page 154.

[145] A. Schmeitz, I. Besselink, and S. Jansen, 'TNO MF-SWIFT', *Vehicle System Dynamics*, vol. 45, no. S1, pp. 121–37, 2007. See pages 154 and 156.

[146] P. v. d. Jagt, 'The Road to Virtual Vehicle Prototyping', PhD dissertation, TU Eindhoven, 2000. See page 155.

[147] G. Rill, 'First Order Tire Dynamics', in *III European Conference on Computational Mechanics, Solids and Coupled Problems in Engineering*, 2006. See page 155.

[148] W. Hirschberg and H. Rill, G. Weinfurter, 'Tire Model TMeasy', *Vehicle System Dynamics*, vol. 45, no. S1, pp. 101–19, 2007. See page 155.

[149] P. Fvrier and G. Fandard, 'A New Thermal and Mechanical Tire Model for Handling Simulation', *VDI Berichte*, pp. 261–275, 2007. See page 156.

[150] M. Grob, O. Blanco-Hague, and F. Spetler, 'TaMeTirE's Testing Procedure Outside Michelin', in *4th International Tyre Colloquium*, 2015. See page 156.

[151] N. Trevorrow and R. Gearing, 'MuRiTyre – Real Time Simulation of Distributed Contact Patch Forces, Moments and Temperatures', in *4th International Tyre Colloquium*, 2015. See page 156.

[152] F. Farroni, D. Giordano, M. Russo, and F. Timpone, 'TRT: Thermo Racing Tyre a Physical Model to Predict the Tyre Temperature Distribution', *Meccanica*, vol. 49, no. 3, pp. 707–23, 2014. See page 156.

[153] F. Farroni, M. Russo, R. Russo, and F. Timpone, 'A Physical-Analytical Model for a Real-Time Local Grip Estimation of Tyre Rubber in Sliding Contact with Road Asperities', *Proceedings of the Institution of Mechanical Engineers, Part D: Journal of Automobile Engineering*, vol. 228, no. 8, pp. 955–69, 2014. See page 156.

[154] M. Gipser, 'FTire – The Tire Simulation Model for All Applications Related to Vehicle Dynamics', *Vehicle System Dynamics*, vol. 45, no. S1, pp. 139–51, 2007. See page 157.

[155] M. Gipser, R. Hofer, and P. Lugner, 'Dynamical Tyre Forces Response to Road Unevennesses', *Vehicle System Dynamics*, vol. 27, no. S1, pp. 94–108, 1997. See page 157.

[156] C. Oertel, 'On Modeling Contact and Friction Calculation of Tyre Response on Uneven Roads', *Vehicle System Dynamics*, vol. 27, no. S1, pp. 289–302, 1997. See page 157.

[157] C. Oertel and A. Fandre, 'Tire Model RMOD-K 7 and Misuse Load Cases', *SAE*, no. 2009-04-20, 2009. See page 157.

[158] J. A. Tanner, R. H. Daugherty, and H. C. Smith, 'Mechanical Properties of Radial-Ply Aircraft Tires', *SAE*, no. 2005-01-3438, 2005. See page 157.

[159] A. E. Dressel, 'Measuring and Modeling the Mechanical Properties of Bicycle Tires', PhD dissertation, University of Wisconsin-Milwaukee, 2013. See page 157.

[160] I. G. Salisbury, D. J. N. Limebeer, A. Tremlett, and M. Massaro, 'The Unification of Acceleration Envelope and Driveability Concepts', in *24th International Symposium on Dynamics of Vehicles on Road and Tracks (IAVSD)*, Graz, Austria, 2015. See pages 168 and 222.

[161] W. T. Koiter and H. B. Pacejka, 'Skidding of Vehicle due to Locked Wheels', *Proceedings of the Institute of Mechanical Engineers*, vol. 183(8), pp. 3–18, 1968. See page 168.

[162] W. F. Milliken and D. L. Milliken, *Race Car Vehicle Dynamics*. Warrendale, PA: SAE, 2005. See page 172.

[163] H. Radt and D. V. Dis, 'Vehicle Handling Responses using Stability Derivatives', *SAE Technical Paper*, no. 960483, 1996. See page 172.

[164] J. Yi, J. Li, J. Lu, and Z. Liu, 'On the Stability and Agility of Aggressive Vehicle Maneuvers: A Pendulum-Turn Maneuver Example', *IEEE Transactions on Control Systems Technology*, vol. 20(3), pp. 663–676, 2012. See page 172.

[165] J. Slotine and W. Li, *Applied Nonlinear Control.* Englewood Cliffs, NJ: Prentice Hall, 1991. See page 173.

[166] A. Tremlett, F. Assadian, D. Purdy, N. Vaughan, A. Moore, and M. Halley, 'Quasi-Steady-State Linearisation of the Racing Vehicle Acceleration Envelope: a Limited Slip Differential Example', *Vehicle System Dynamics*, vol. 52, no. 11, pp. 1416–42, 2014. See pages 173 and 304.

[167] S. Timoshenko and D. Young, *Advanced Dynamics.* New York: McGraw-Hill, 1948. See page 173.

[168] K. J. Åström, 'Limitations on Control System Performance', *European Journal of Control*, vol. 6, no. 1, pp. 2–20, 1980. See pages 178 and 216.

[169] J. D. G. Kooijman, J. P. Meijaard, J. M. Papadopoulos, A. Ruina, and A. L. Schwab, 'A Bicycle can be Self-Stable without Gyroscopic or Caster Effects', *Science*, vol. 332, no. 6027, pp. 339–42, 2011. See pages 179 and 197.

[170] R. Van Der Valk and H. B. Pacejka, 'An Analysis of a Civil Aircraft Main Gear Shimmy Failure', *Vehicle System Dynamics*, vol. 22, no. 2, pp. 97–121, 1993. See page 183.

[171] J. P. Den Hartog, *Mechanical Vibrations.* Mineola, NY: Dover, 1985. See page 183.

[172] M. Olley, 'Road Manners of the Modern Car', *Proceedings of the Institution of Automobile Engineers*, vol. 41, pp. 147–82, 1946. See pages 184 and 228.

[173] P. Bandel and C. D. Bernardo, 'A Test for Measuring Transient Characteristics of Tires', *Tire Sci. Technol.*, vol. 17, no. 2, pp. 126–137, 1989. See page 188.

[174] Z. Sheng and K. Yamafuji, 'Postural Stability of a Human Riding a Unicycle and Its Emulation by a Robot', *IEEE Trans. Robotics and Automation*, vol. 13, no. 5, pp. 709–720, 1997. See pages 193 and 194.

[175] Y. Naveh, P. Z. Bar-Yoseph, and Y. Halevi, 'Nonlinear Modeling and Control of a Unicycle', *Dynamics and Control*, vol. 9, pp. 279–96, 1999. See page 194.

[176] A. Schoonwinkel, 'Design and Test of a Computer Stabilized Unicycle', PhD dissertation, Stanford University, Stanford, USA, 1987. See page 194.

[177] D. W. Vos and A. H. von Flotow, 'Dynamics and Nonlinear Adaptive Control of an Autonomous Unicycle (Theory and Experiment)', in *29th IEEE Conference on Decision and Control*, 1990. See page 194.

[178] R. Corona, A. P. Aguiar, and J. J. Gaspar, 'Control of Unicycle Type Robots Tracking, Path Following and Point Stabilization', in *In Proc. of IV Jornadas de Engenharia Electrónica e Telecomunicações e de Computadores*, 2008. See page 194.

[179] D. V. Zenkov, A. M. Bloch, and J. E. Marsden, 'Stabilization of the Unicycle with Rider', in *38th IEEE Conference on Decision and Control*, 1999. See page 194.

[180] Y. Naveh, P. Z. Bar-Yoseph, and Y. Halevi, 'Nonlinear Dynamics of Unicycles in Leader Follower Formation', *Commun Nonlinear Sci Numer Simulat*, vol. 14, pp. 4204–19, 2009. See page 194.

[181] R. S. Sharp, 'On the Stability and Control of Unicycles', *Proc. R. Soc. A*, vol. 466, pp. 1849–69, 2010. See page 194.

[182] F. Klein and A. Sommerfeld, *Über die Theorie des Kreisels*. Teubner, Leipzig, 1910, See particularly: Chapter IX, Section 8, 'Stabilität des Fahrrads' pp. 863-84. See pages 195, 197, and 213.

[183] J. P. Meijaard, J. M. Papadopoulos, A. Ruina, and A. L. Schwab, 'Historical Review of Thoughts on Bicycle Self-Stability', Cornell University, Tech. Rep., 2011. See pages 196 and 197.

[184] D. E. H. Jones, 'The Stability of the Bicycle', *Physics Today*, vol. 23, no. 4, pp. 34–40, 1970. See pages 196, 197, and 212.

[185] R. D. Roland, 'Computer Simulation of Bicycle Dynamics', *Proc. ASME Symposium Mechanics and Sport*, pp. 35–83, 1973. See page 196.

[186] E. Döhring, 'Stability of Single-Track Vehicles', *Institut für Fahrzeugtechnik, Technische Hochschule Braunschweig, Forschung Ing.-Wes.*, vol. 21, no. 2, pp. 50–62, 1955. See pages 196 and 324.

[187] J. P. Meijaard, J. M. Papadopoulos, A. Ruina, and A. L. Schwab, 'Linearized Dynamics Equations for the Balance and Steer of a Bicycle: a Benchmark and Review', *Proc. R. Soc. A*, vol. 463, pp. 1955–82, 2007. See pages 196, 198, 203, 204, and 212.

[188] K. J. Åström, R. E. Klein, and A. Lennartsson, 'Bicycle Dynamics and Control.' *IEEE Control Systems Magazine*, vol. 25, no. 4, pp. 26–47, 2005. See pages 196, 212, and 218.

[189] R. E. Klein, 'Using Bicycles to Teach System Dynamics.' *IEEE Control Systems Magazine*, vol. 6, no. 4, pp. 4–9, 1989. See page 197.

[190] J. D. G. Kooijman, A. L. Schwab, and J. P. Meijaard, 'Experimental Validation of a Model of an Uncontrolled Bicycle', *Multibody System Dynamics*, vol. 19, no. 1-2, pp. 115–32, 2008. See page 212.

[191] A. L. Schwab, J. P. Meijaard, and J. M. Papadopoulos, 'A Multibody Dynamics Benchmark on the Equations of Motion of an Uncontrolled Bicycle.' in *Proceedings of the ENOC-2005*, 2005. See page 213.

[192] R. A. Wilson-Jones, 'Steering and Stability of Single-Track Vehicles', *Proc. Auto. Div. Institution of Mechanical Engineers*, pp. 191–9, 1951. See pages 214 and 223.

[193] J. B. Hoagg and D. S. Bernstein, 'Nonminimum-Phase Zeros—Much to do About Nothing—Classical Control—Revisited Part II.' *IEEE control systems*, vol. 27, no. 3, pp. 45–57, 2007. See page 217.

[194] V. Cossalter, M. Lot, M. Massaro, and M. Peretto, 'Motorcycle Steering Torque Decomposition', in *World Congress on Engineering*, London, 2010. See page 217.

[195] J. Fajans, 'Steering in Bicycles and Motorcycles', *American Journal of Physics*, vol. 68, no. 7, pp. 654–9, 2000. See pages 218 and 219.

[196] D. J. N. Limebeer, R. S. Sharp, and S. Evangelou, 'The Stability of Motorcycles under Acceleration and Braking', *Journal of Mechanical Engineeering Science*, vol. 215, no. 9, pp. 1095–109, 2001. See page 222.

[197] D. G. Wilson, *Bicycling Science*. Cambridge, MA: The MIT Press, 2004. See page 222.

[198] C. Juden, 'Shimmy', *Cycletouring*, vol. June/July, pp. 208–9, 1988. See pages 222 and 223.

[199] M. Massaro and D. J. Cole, 'Neuromuscular-Steering Dynamics: Motorcycle Riders vs. Car Drivers.' in *ASME 2012 5th Annual Dynamic Systems and Control Conference*, October 2012, pp. 217–24. See pages 223, 312, and 314.

[200] V. Cossalter, A. Doria, R. Lot, and M. Massaro, 'The Effect of Riders Passive Steering Impedance on Motorcycle Stability: Identification and Analysis', *Meccanica*, vol. 46, pp. 279–92, 2011. See pages 223, 313, 314, and 341.

[201] T. Wakabayashi and K. Sakai, 'Development of Electronically Controlled Hydraulic Rotary Steering Damper for Motorcycles', in *International Motorcycle Safety Conference, IFZ*, Munich, 2004, pp. 1–22. See page 223.

[202] R. Lot and M. Massaro, 'The Kick-Back of Motorcycles: Experimental and Numerical Analysis', in *Proceedings of the ECCOMAS Thematic Conference, Multibody Dynamics*, 2007. See page 223.

[203] A. G. Thompson, 'Optimum Damping in a Randomly Excited Non-Linear Suspension', *Proceeding of the Institution of Mechanical Engineers*, vol. 184.2A, no. 8, pp. 169–78, 1969-70. See page 229.

[204] J. C. Dixon, *Tires, Suspension, and Handling*. Warrendale, PA: SAE, 1996. See page 229.

[205] R. S. Sharp and D. A. Crolla, 'Road Vehicle Suspension System Design - A Review', *Vehicle System Dynamics*, vol. 16, no. 3, pp. 167–92, 1987. See page 229.

[206] L. Segel, 'An Overview of Developments in Road Vehicle Dynamics: Past, Present and Future', in *IMechE Conference Transaction; Vehicle Ride and Handling*, vol. 9, Birmingham, 1993, pp. 1–12. See pages 229 and 298.

[207] D. J. Cole, 'Fundamental Issues in Suspension Design for Heavy Road Vehicles', *Vehicle System Dynamics*, vol. 35, no. 4-5, pp. 319–60, 2001. See page 229.

[208] D. Cao, X. Song, and M. Ahmadian, 'Editors Perspectives: Road Vehicle Suspension Design, Dynamics, and Control', *Vehicle System Dynamics*, vol. 49, no. 1-2, pp. 3–28, 2011. See page 229.

[209] M. J. Griffin, 'Discomfort from Feeling Vehicle Vibration', *Vehicle System Dynamics*, vol. 45, no. 7-8, pp. 679–98, 2007. See page 229.

[210] D. Hrovat, 'Survey of Advanced Suspension Developments and related Optimal Control Applications', *Automatica*, vol. 33, no. 10, pp. 1781–817, 1997. See page 229.

[211] H. E. Tseng and D. Hrovat, 'State of the Art Survey: Active and Semi-Active Suspension Control', *Vehicle System Dynamics*, vol. 53, no. 7, pp. 1034–62, 2015. See page 229.

[212] B. Mashadi and D. A. Crolla, 'Influence of Ride Motions on the Handling Behaviour of a Passenger Vehicle', *Proceeding of the Institution of Mechanical Engineers Part D: Journal of Automobile Engineering*, vol. 219, no. 9, pp. 1047–58, 2005. See page 229.

[213] M. Mitschke, 'Influence of Road and Vehicle Dimensions on the Amplitude of Body Motions and Dynamic Wheel Loads (Theoretical and Experimental Vibration Investigations)', *SAE Transactions*, vol. 70, 1962. See pages 229 and 248.

[214] D. Robson and C. Dodds, 'Stochastic Road Inputs and Vehicle Response', *Vehicle System Dynamics*, vol. 5, no. 1-2, pp. 1–13, 1976. See page 229.

[215] D. Cebon and D. Newland, 'Artificial Generation of Road Surface Topography by the Inverse F.F.T. Method', *Vehicle System Dynamics*, vol. 12, no. 1-3, pp. 160–5, 1983. See page 230.

[216] M. Shinozuka and G. Deodatis, 'Simulation of Stochastic Processes by Spectral Representation', *Applied Mechanics Reviews*, vol. 44, no. 4, pp. 191–204, 1991. See page 230.

[217] D. E. Newland, *An Introduction to Random Vibrations and Spectral Analysis*. London and New York: Longman, 1984. See pages 230 and 250.

[218] A. Hać, 'Suspension Optimization of a 2-DOF Vehicle Model using a Stochastic Optimal Control Technique', *Journal of sound and vibration*, vol. 100, no. 3, pp. 343–57, 1985. See page 230.

[219] F. Tyan, Y. F. Hong, S. H. Tu, and W. S. Jeng, 'Generation of Random Road Profiles', *Journal of Advanced Engineering*, vol. 4, no. 2, pp. 1373–8, 2009. See page 230.

[220] C. W. Campbell, 'Monte Carlo Turbulence Simulation using Rational Approximations to von Kármán Spectra', *AIAA journal*, vol. 24, no. 1, pp. 62–6, 1986. See page 230.

[221] K. M. A. Kamash and J. D. Robson, 'The Application of Isotropy in Road Surface Modelling', *Journal of Sound and Vibration*, vol. 57, no. 1, pp. 89–100, 1978. See page 231.

[222] K. Bogsjö, 'Coherence of Road Roughness in Left and Right Wheel-Path', *Vehicle System Dynamics*, vol. 46, no. S1, pp. 599–609, 2008. See page 231.

[223] T. Takahashi and H. B. Pacejka, 'Cornering on Uneven Roads', *Vehicle System Dynamics*, vol. 17, no. S1, pp. 469–80, 1988. See page 232.

[224] J. K. Hedrick and T. Butsuen, 'Invariant Properties of Automotive Suspensions', *Proc Instn Mech Engrs*, vol. 204, pp. 21–7, 1990. See page 235.

[225] D. C. Youla and M. Saito, 'Interpolation with Positive-Real Functions', *Journal of The Franklin Institute*, vol. 284, no. 2, pp. 77–108, 1967. See page 235.

[226] G. Zames, 'Feedback and Optimal Sensitivity: Model Reference Transformations, Multiplicative Seminorms, and Approximate Inverses', *IEEE Transactions on Automatic Control*, vol. 26, no. 2, pp. 301–20, 1981. See page 235.

[227] M. C. Smith, 'Achievable Dynamic Response for Automotive Active Suspensions', *Vehicle System Dynamics*, vol. 24, no. 1, pp. 1–33, 1995. See pages 235 and 253.

[228] C. Yue, T. Butsuen, and K. Hedrick, 'Alternative Control Laws for Automotive Active Suspensions', in *Proceedings of the IEEE American Control Conference*, Atlanta, GA., 1988. See page 236.

[229] M. Green and D. J. N. Limebeer, *Linear Robust Control*. Englewood Cliffs, NJ: Prentice Hall, 1995. See pages 237, 238, 239, 252, 253, and 355.

[230] R. W. Newcomb, *Linear Multiport Synthesis*. New York: McGraw-Hill, 1966. See page 237.

[231] B. D. O. Anderson and S. Vongpanitlerd, *Network Analysis and Synthesis*. Englewood Cliffs, NJ: Prentice-Hall, 1973. See page 237.

[232] S. Skogestad and I. Postlethwaite, *Multivariable Feedback Control: Analysis and Design*. New York: John Wiley & Sons, 2005. See page 239.

[233] F. Scheibe and M. C. Smith, 'Analytical Solutions for Optimal Ride Comfort and Tyre Grip for Passive Vehicle Suspensions', *Vehicle System Dynamics*, vol. 47, no. 10, pp. 1229–52, 2009. See page 243.

[234] D. J. N. Limebeer and B. D. O. Anderson, 'An Interpolation Theory Approach to H^∞ Controller Degree Bounds', *Linear Algebra and Its Applications*, vol. 98, pp. 347–86, 1988. See page 253.

[235] A. H. Sayed, T. Kailath, H. Lev-Ari, and T. Constantinescu, 'Recursive Solutions to Rational Interpolation Problems via Fast Matrix Factorization', *Integral Equations and Operator Theory*, vol. 20, pp. 84–118, 1994. See page 253.

[236] O. Brune, 'Synthesis of a Finite Two-Terminal Network whose Driving-Point Impedance is a Prescribed Function of Frequency', *J. Math. Phys.*, vol. 10, pp. 191–236, 1931. See page 264.

[237] R. Bott and R. J. Duffin, 'Impedance Synthesis Without Use of Transformers', *J. Appl. Phys.*, vol. 20, p. 816, 1949. See page 264.

[238] M. Reichert, 'Die kanonisch und bertragerfrei realizierbaren Zweipolfunktionen zweiten Grades (Transformerless and Canonic Realization of Biquadratic Immittance Functions)', *Arch. Elek. bertragung*, vol. 23, pp. 201–8, 1969. See page 264.

[239] J. Z. Jiang and M. C. Smith, 'Regular Positive Function and Five Element Network Sytnthesis for Electrical and Mechanical Networks', *IEEE Trans. Automatic Control*, vol. 56, no. 6, pp. 1275–90, 2011. See page 264.

[240] J. Z. Jiang and M. C. Smith, 'Series-Parallel Six-Element Synthesis of Biquadratic Impedances', *IEEE Trans on Circuits and Systems*, vol. 59, no. 11, pp. 2543–54, 2012. See page 264.

[241] T. H. Hughes and M. C. Smith, 'On the Minimality and Uniqueness of the Bott–Duffin Realization Procedure', *IEEE Trans on automatic control*, vol. 59, no. 7, pp. 1858–73, 2014. See page 264.

[242] T. H. Hughes, 'Why RLC Realizations of Certain Impedances Need Many More Energy Storage Elements than Expected', *IEEE Transactions on Automatic Control*, vol. 62, no. 9, pp. 4333–46, 2017. See page 264.

[243] C. Papageorgiou and M. C. Smith, 'Positive Real Synthesis using Matrix Inequalities for Mechanical Networks: Application to Vehicle Suspension', *IEEE Transactions on Control technology*, vol. 14, no. 3, pp. 423–35, 2006. See page 265.

[244] A. Sharma and D. J. N. Limebeer, 'Motorcycle Suspension Design using Matrix Inequalities with Passivity Constraints', *Vehicle System Dynamics*, vol. 50, no. 3, pp. 377–93, 2012. See page 265.

[245] R. J. Guyan, 'Reduction of Stiffness and Mass Matrices', *AIAA Journal*, vol. 3, no. 2, p. 380, 1965. See page 266.

[246] R. R. Craig and M. C. C. Bampton, 'Coupling of Substructures for Dynamic Analyses', *AIAA Journal*, vol. 6, no. 7, pp. 1313–9, 1968. See page 266.

[247] R. R. Craig, 'Coupling of Substructures for Dynamic Analysis: An Overview', in *41st Structures, Structural Dynamics, and Materials Conference and Exhibit*, 2000. See page 266.

[248] J. Katz, *Race Car Aerodynamics: Designing for Speed.* Cambridge, MA: Bentley, 2006. See page 267.

[249] M. Massaro and E. Marconi, 'The Effect of Engine Spin Direction on the Dynamics of Powered Two Wheelers', *Vehicle System Dynamics*, vol. 56, no. 4, pp. 604–20, 2018. See pages 268, 392, and 423.

[250] G. Mastinu and M. Plöchl, Eds., *Road and Off-Road Vehicle System Dynamics Handbook.* Boca Raton, FL: CRC Press, 2014. See pages 269 and 273.

[251] G. Genta, *Motor Vehicle Dynamics: Modeling and Simulation.* Singapore: World Scientific, 1997. See page 277.

[252] A. B. W. Kennedy, *The Mechanics of Machinery.* New York: Macmillan, 1893. See page 283.

[253] R. Goldman, M. El-Gindy, and B. Kulakowski, 'Rollover Dynamics of Road Vehicles: Literature Survey', *Heavy Vehicle Systems*, vol. 8, no. 2, pp. 103–41, 2001. See page 287.

[254] C. C. Chou, R. W. McCoy, and J. Le, 'A Literature Review of Rollover Test Methodologies', *International journal of vehicle safety*, vol. 1, no. 1-3, pp. 200–37, 2005. See page 287.

[255] G. Perantoni and D. J. N. Limebeer, 'Optimal Control of a Formula One Car with Variable Parameters', *Vehicle Systems Dynamics*, vol. 52, no. 5, pp. 653–78, 2014. See pages 290, 303, 386, and 409.

[256] G. Perantoni and D. J. N. Limebeer, 'Optimal Control of a Formula One Car on a Three-Dimensional Track. Part 1: Track Modelling and Identification', *ASME Journal of Dynamical Systems, Measurement, and Control*, vol. 137, no. 5, 2015. See pages 290, 294, and 409.

[257] E. Kreyszig, *Differential Geometry.* Mineola, NY: Dover, 1991. See page 290.

[258] T. J. Willmore, *An Introduction to Differential Geometry.* Mineola, NY: Dover, 2013. See page 290.

[259] D. J. Struik, *Lectures on Classical Differential Geometry*, 2nd edn. Mineola, NY: Dover, 2003. See page 290.

[260] J. J. Koenderink, *Solid Shape.* Cambridge, MA: MIT Press, 1990. See page 293.

[261] L. Segel, 'Theoretical Prediction and Experimental Substantiation of the Response of the Automobile to Steering Control', *Proceedings of the Institution of Mechanical Engineers: Automobile Division*, 1956. See page 298.

[262] L. Segel, 'On the Lateral Stability and Control of the Automobile as Influence by the Dynamics of the Dynamics of the Steering System', *Journal of Engineering for Industry*, 1966. See page 298.

[263] H. Dugoff, P. Fancher, and L. Segel, 'An Analysis of Tire Traction Properties and their Influence on Vehicle Dynamic Performance', *SAE Technical Paper*, no. 700377, 1970. See page 298.

[264] S. Yang, Y. Lu, and S. Li, 'An Overview on Vehicle Dynamics', *International Journal of Dynamics and Control*, vol. 1, no. 4, pp. 385–95, 2013. See page 298.

[265] A. Tremlett, M. Massaro, D. Purdy, E. Velenis, F. Assadian, A. Moore, and M. Halley, 'Optimal Control of Motorsport Differentials', *Vehicle System Dynamics*, vol. 53, no. 12, pp. 1772–94, 2015. See page 304.

[266] J. Reimpell, H. Stoll, and J. W. Betzler, *The Automotive Chassis: Engineering Principles*, 2nd edn. Warrendale, PA: SAE, 2001. See page 304.

[267] M. Raghavan, 'Number and Dimensional Synthesis of Independent Suspension Mechanisms', *Mechanism and Machine Theory*, vol. 31, no. 8, pp. 1141–53, 1996. See page 304.

[268] D. L. Cronin, 'MacPherson Strut Kinematics', *Mechanism and Machine Theory*, vol. 16, no. 6, pp. 631–44, 1981. See page 304.

[269] K. Chen and D. G. Beale, 'Base Dynamic Parameter Estimation of a MacPherson Suspension Mechanism', *Vehicle System Dynamics*, vol. 39, no. 3, pp. 227–44, 2003. See page 304.

[270] M. I. Masouleh and D. J. N. Limebeer, 'Optimizing the Aero–Suspension Interactions in a Formula One Car', *IEEE Transactions on Control Systems Technology*, vol. 24, no. 3, pp. 912–27, 2016. See pages 309, 310, and 392.

[271] Fédération Internationale de l'Automobile (FIA), 'Formula One Technical Regulations', 2017. See page 312.

[272] D. J. Cole, A. J. Pick, and A. M. C. Odhams, 'Predictive and Linear Quadratic Methods for Potential Application to Modelling Driver Steering Control', *Vehicle System Dynamics*, vol. 44, no. 3, pp. 259–84, 2006. See page 312.

[273] R. S. Sharp, 'Optimal Preview Speed-Tracking Control for Motorcycles', *Multibody System Dynamics*, vol. 18, no. 3, pp. 397–411, 2007. See page 312.

[274] R. S. Sharp, 'Motorcycle Steering Control by Road Preview', *Journal of Dynamic Systems, Measurement and Control, Transactions of the ASME*, vol. 129, no. 4, pp. 373–81, 2007. See page 312.

[275] S. D. Keen and D. J. Cole, 'Application of Time-Variant Predictive Control to Modelling Driver Steering Skill', *Vehicle System Dynamics*, vol. 49, no. 4, pp. 527–59, 2011. See page 313.

[276] S. Rowell, A. Popov, and J. Meijaard, 'Application of Predictive Control Strategies to the Motorcycle Riding Task', *Vehicle System Dynamics*, vol. 46, no. S1, pp. 805–14, 2008. See page 313.

[277] M. Massaro, 'A Nonlinear Virtual Rider for Motorcycles', *Vehicle System Dynamics*, vol. 49, no. 9, pp. 1477–96, 2011. See page 313.

[278] W. Hoult, 'A Neuromuscular Model for Simulating Driver Steering Torque', PhD dissertation, University of Cambridge, 2008. See pages 313 and 314.

[279] N. Kim and D. J. Cole, 'A Model of Driver Steering Control Incorporating the Driver's Sensing of Steering Torque', *Vehicle System Dynamics*, vol. 49, no. 10, pp. 1575–96, 2011. See page 313.

[280] M. Massaro and R. Lot, 'Application of Laplace Transform Techniques to Non-Linear Control Optimization', in *Proceedings of the ECCOMAS Thematic Conference, Multibody Dynamics*, 2007. See page 313.

[281] A. J. Pick and D. J. Cole, 'Dynamic Properties of a Driver's Arms Holding a Steering Wheel', *Proceedings of the Institution of Mechanical Engineers, Part D: Journal of Automobile Engineering*, vol. 221, no. 12, pp. 1475–1486, 2007. See pages 313 and 314.

[282] M. Plöchl and J. Edelmann, 'Driver Models in Automobile Dynamics Application', *Vehicle System Dynamics*, vol. 45, no. 7-8, pp. 699–741, 2007. See page 315.

[283] J. Kooijman and A. L. Schwab, 'A Review on Bicycle and Motorcycle Rider Control with a Perspective on Handling Qualities', *Vehicle System Dynamics*, vol. 51, no. 11, pp. 1722–64, 2013. See pages 315 and 326.

[284] C. J. Nash, D. J. Cole, and R. S. Bigler, 'A Review of Human Sensory Dynamics for Application to Models of Driver Steering and Speed Control', *Biological Cybernetics*, vol. 110, no. 2-3, pp. 91–116, 2016. See page 315.

[285] J. P. Meijaard, 'The Loop Closure Equation for the Pitch Angle in Bicycle Kinematics', in *Bicycle and Motorcycle Dynamics 2013 Symposium on the Dynamics and Control of Single Track Vehicles*, Narashino, Japan, November, 11-13 2013. See page 318.

[286] M. L. Psiaki, 'Bicycle Stability: A Mathematical and Numerical Analysis', Master's thesis, Physics Department, Princeton University, 1979. See page 318.

[287] T. R. Kane, 'Fundamental Kinematical Relationships for Single-Track Vehicles', *International Journal of Mechanical Sciences*, vol. 17, no. 8, pp. 499–504, 1975. See page 318.

[288] V. Cossalter, A. Doria, and R. Lot, 'Steady Turning of Two-Wheeled Vehicles', *Vehicle System Dynamics*, vol. 31, no. 3, pp. 157–81, 1999. See page 318.

[289] G. Frosali and F. Ricci, 'Kinematics of a Bicycle with Toroidal Wheels', *Communications in Applied and Industrial Mathematics*, 2012. See page 318.

[290] R. S. Sharp, S. Evangelou, and D. J. N. Limebeer, 'Advances in the Modelling of Motorcycle Dynamics', *Multibody System Dynamics*, vol. 12, no. 3, pp. 251–283, 2004. See pages 321, 325, 341, 342, and 345.

[291] V. Cossalter, R. Lot, and M. Massaro, 'An Advanced Multibody Code for Handling and Stability Analysis of Motorcycles', *Meccanica*, vol. 46, pp. 943–58, 2011. See pages 321, 325, 341, 342, and 345.

[292] R. S. Sharp, 'The Stability and Control of Motorcycle', *Journal of Mechanical Engineering Science*, vol. 13, no. 5, pp. 316–29, 1971. See pages 324, 325, and 346.

[293] R. S. Sharp, 'The Influence of Frame Flexibility on the Lateral Stability of Motorcycles', *J. Mech. Eng. Sci.*, vol. 16, no. 2, pp. 117–20, 1974. See page 324.

[294] G. Roe and T. Thorpe, 'A Solution of the Low-Speed Wheel Flutter Instability in Motorcycles', *J. Mech. Eng. Sci.*, vol. 18, no. 2, pp. 57–65, 1976. See pages 324 and 325.

[295] R. S. Sharp and C. J. Alstead, 'The Influence of Structural Flexibilities on the Straight Running Stability of Motorcycles', *Vehicle System Dynamics*, vol. 9, no. 6, pp. 327–57, 1980. See page 325.

[296] P. T. J. Spierings, 'The Effects of Lateral Front Fork Flexibility on the Vibrational Modes of Straight-Running Single-Track Vehicles', *Vehicle System Dynamics*, vol. 10, no. 1, pp. 21–35, 1981. See pages 325 and 346.

[297] V. Cossalter, A. Doria, M. Massaro, and L. Taraborrelli, 'Experimental and Numerical Investigation on the Motorcycle Front Frame Flexibility and its Effect on Stability', *Mechanical Systems and Signal Processing*, vol. 60-61, pp. 452–71, 2015. See pages 325 and 341.

[298] R. S. Sharp, 'Vibrational Modes of Motorcycles and their Design Parameter Sensitivities', in *Proc. Int Conf. Vehicle NVH Refinement*, Birmingham, 1994, pp. 107–21. See page 325.

[299] K. Lake, R. Thomas, and O. Williams, 'The Influence of Compliant Chassis Components on Motorcycle Dynamics: An Historical Overview and the Potential Future Impact of Carbon Fibre', *Vehicle System Dynamics*, vol. 50, no. 7, pp. 1043–52, 2012. See page 325.

[300] R. S. Sharp and D. J. N. Limebeer, 'A Motorcycle Model for Stability and Control Analysis', *Multibody System Dynamics*, vol. 6, no. 2, pp. 123–42, 2001. See page 325.

[301] V. Cossalter, R. Lot, and M. Massaro, 'The Chatter of Racing Motorcycles', *Vehicle System Dynamics*, vol. 46, no. 4, pp. 339–53, 2008. See pages 325 and 346.

[302] R. Lot, V. Cossalter, and M. Massaro, 'The Significance of Powertrain Characteristics on the Chatter of Racing Motorcycles', in *Proc. of ASME 11th Biennal Conference on Engineering Systems Design and Analysis*, 2012, pp. 607–13. See pages 325 and 346.

[303] M. Massaro, V. Cossalter, R. Lot, and A. Croce, 'The Motorcycle Chatter', in *Proceedings, Bicycle and Motorcycle Dynamics 2013 Symposium on the Dynamics and Control of Single Track Vehicles, 11–13 November 2013, Narashino, Japan*, 2013. See pages 325 and 346.

[304] M. Massaro, R. Sartori, and R. Lot, 'Numerical Investigation of Engine-to-Slip Dynamics for Motorcycle Traction Control Applications', *Vehicle System Dynamics*, vol. 49, no. 3, pp. 419–32, 2011. See pages 325 and 344.

[305] M. Massaro, R. Lot, and V. Cossalter, 'On Engine-to-Slip Modelling for Motorcycle Traction Control Design', *Proceedings of the Institution of Mechanical Engineers, Part D: Journal of Automobile Engineering*, vol. 225, no. 1, pp. 15–27, 2011. See page 325.

[306] C. Koenen and H. B. Pacejka, 'Vibrational Modes of Motorcycles in Curves', in *Proc. Int. Motorcycle Safety Conf.*, vol. II. Washington: Motorcycle Safety Foundation, 1980, pp. 501–43. See pages 325 and 345.

[307] C. Koenen and H. Pacejka, 'The Influence of Frame Elasticity, Simple Rider Body Dynamics, and Tyre Moments on Free Vibrations of Motorcycles in Curves', in *Proc. 7th IAVSD Symp. Dynamics Vehicles Roads Railway Tracks*, Cambridge, 1981, pp. 53–65. See pages 325, 345, and 346.

[308] T. Nishimi, A. Aoki, and T. Katayama, 'Analysis of Straight Running Stability of Motorcycles', in *Proc. 10th Int. Technical Conf. Experimental Safety Vehicles*, Oxford, 1985, pp. 1080–94. See pages 325, 326, and 346.

[309] T. Katayama, A. Aoki, T. Nishimi, and T. Okayama, 'Measurements of Structural Properties of Riders', *SAE Technical Paper*, no. 871229, 1987. See page 326.

[310] H. Imaizumi, T. Fujioka, and M. Omae, 'Rider Model by Use of Multibody Dynamics Analysis', *SAE Japan*, vol. 17, no. 1, pp. 75–7, 1996. See page 326.

[311] D. Weir and J. Zellner, 'Lateral-Directional Motorcycle Dynamics and Rider Control', in *SAE*, no. 780304. Warrendale, PA: SAE, 1978, pp. 7–31. See pages 326, 344, and 345.

[312] T. Katayama, A. Aoki, and T. Nishimi, 'Control Behaviour of Motorcycle Riders', *Vehicle System Dynamics*, vol. 17, no. 4, pp. 211–29, 1988. See page 326.

[313] J. K. Moore, J. D. G. Kooijman, A. L. Schwab, and M. Hubbard, 'Rider Motion Identification During Normal Bicycling by means of Principal Component Analysis', *Multibody System Dynamics*, vol. 25, no. 2, pp. 225–44, 2011. See page 326.

[314] K. R. Cooper, 'The Effects of Aerodynamics on the Performance and Stability of High Speed Motorcycles', in *Proc. 2nd AIAA Symp. Aerodynamics Sport Competition Automobiles*, Los Angeles, 1974. See page 326.

[315] K. R. Cooper, 'The Effect of Handlebar Fairings on Motorcycle Aerodynamics', *SAE Technical Paper*, no. 830156, 1983. See page 327.

[316] P. Bridges and J. Russell, 'The Effect of Topboxes on Motorcycle Stability', *Vehicle System Dynamics*, vol. 16, no. 5-6, pp. 345–54, 1987. See page 327.

[317] Y. Araki and K. Gotou, 'Development of Aerodynamic Characteristics for Motorcycles Using Scale Model Wind Tunnel', *SAE Technical Papers*, no. 2001-01-1851, 2001. See page 327.

[318] Y. Takahashi, Y. Kurakawa, H. Sugita, T. Ishima, and T. Obokata, 'CFD Analysis of Airflow Around the Rider of a Motorcycle for Rider Comfort Improvement', *SAE Technical Papers*, no. 2009-01-1155, 2009. See page 327.

[319] M. Angeletti, L. Sclafani, G. Bella, and S. Ubertini, 'The Role of CFD on the Aerodynamic Investigation of Motorcycles', *SAE Technical Paper*, no. 2003-01-0997, 2003. See page 327.

[320] L. Scappaticci, G. Risitano, M. Battistoni, and C. Grimaldi, 'Drag Optimization of a Sport Motorbike', *SAE Technical Papers*, no. 2012-01-1171, 2012. See page 327.

[321] T. Van Dijck, 'Computational Evaluation of Aerodynamic Forces on a Racing Motorcycle during High Speed Cornering', *SAE Technical Papers*, no. 2015-01-0097, 2015. See page 327.

[322] D. Fintelman, H. Hemida, M. Sterling, and F.-X. Li, 'A Numerical Investigation of the Flow Around a Motorbike when Subjected to Crosswinds', *Engineering Applications of Computational Fluid Mechanics*, vol. 9, no. 1, pp. 528–42, 2015. See page 327.

[323] G. Heyl and J. Seidl, 'Motorcycle Having a Front Wheel Suspension', US Patent 7,784,809, 2010. See pages 333 and 336.

[324] G. Gogu, 'Mobility of Mechanisms: A Critical Review', *Mechanism and Machine Theory*, vol. 40, no. 9, pp. 1068–97, 2005. See pages 334 and 336.

[325] M. Da Lio, A. Doria, and R. Lot, 'A Spatial Mechanism for the Measurement of the Inertia Tensor: Theory and Experimental Results', *ASME Journal of dynamic systems, measurement, and control*, vol. 121, no. 1, pp. 111–16, 1999. See page 341.

[326] A. Doria, M. Formentini, and M. Tognazzo, 'Experimental and Numerical Analysis of Rider Motion in Weave Conditions', *Vehicle System Dynamics*, vol. 50, no. 8, pp. 1247–60, 2012. See page 341.

[327] G. Jennings, 'A Study of Motorcycle Suspension Damping Characteristics', in *SAE*, no. 740628, 1974. See pages 344 and 345.

[328] R. S. Sharp, 'The Influence of the Suspension System on Motorcycle Weave-Mode Oscillations', *Vehicle System Dynamics*, vol. 5, no. 3, pp. 147–54, 1976. See page 344.

[329] C. Koenen, 'The Dynamic Behaviour of Motorcycles when Running Straight Ahead and when Cornering', PhD dissertation, Delft University of Technology, 1983. See page 345.

[330] D. J. N. Limebeer, R. S. Sharp, and S. Evangelou, 'Motorcycle Steering Oscillations Due to Road Profiling', *ASME J. Applied Mechanics*, vol. 69, no. 6, pp. 724–39, 2002. See page 345.

[331] R. S. Sharp and Y. Watanabe, 'Chatter Vibrations of High-Performance Motorcycles', *Vehicle System Dynamics*, vol. 51, no. 3, pp. 393–404, 2013. See page 346.

[332] M. Massaro, R. Lot, V. Cossalter, J. Brendelson, and J. Sadauckas, 'Numerical and Experimental Investigation of Passive Rider Effects on Motorcycle Weave', *Vehicle System Dynamics*, vol. 50, no. S1, pp. 215–27, 2012. See pages 346 and 347.

[333] S. Fujii, T. Kishi, and T. Sano, 'Low Speed Weave Mode Measurement of Motorcycles in Slalom Running', in *ASME 2012 5th Annual Dynamic Systems and Control Conference joint with the JSME 2012 11th Motion and Vibration Conference*, 2012, pp. 613–17. See page 347.

[334] J. A. Wilson, 'Sweeping Flight and Soaring Albatrosses', *Nature*, vol. 257, pp. 307–8, 1975. See page 348.

[335] M. Deittert, A. Richards, C. A. Toomer, and A. Pipe, 'Engineless Unmanned Aerial Vehicle Propulsion by Dynamic Soaring', *Journal of Guidance, Control and Dynamics*, vol. 32, no. 5, pp. 1446–57, 2009. See page 348.

[336] A. T. Klesh and P. T. Kabamba, 'Solar-Powered Aircraft: Energy-Optimal Path Planning and Perpetual Endurance', *Journal of Guidance, Control and Dynamics*, vol. 32, no. 4, pp. 1320–8, 2009. See page 348.

[337] D.-M. Ma, J.-K. Shiau, Y.-J. Su, and Y.-H. Chen, 'Optimal Level Turn of Solar-Powered Unmanned Aerial Vehicle Flying in Atmosphere', *Journal of Guidance, Control and Dynamics*, vol. 33, no. 5, pp. 1347–56, 2010. See page 348.

[338] R. H. Goddard, 'A Method of Reaching Extreme Altitudes'. Washington: Smithsonian Miscellaneous Collection, 1919, vol. 71, no. 2. See pages 348 and 367.

[339] J. S. Meditch, 'On the Problem of Optimal Thrust Programming for a Lunar Soft Landing', *IEEE Transactions on Automatic Control*, vol. 9, no. 4, pp. 477–84, 1964. See page 348.

[340] N. S. Bedrossian, S. Bhatt, W. Kang, and I. M. Ross, 'Zero-Propellant Maneuver Guidance', *IEEE Control Systems Magazine*, vol. 29, no. 5, pp. 53–73, 2009. See page 348.

[341] G. Mingotti, F. Topputo, and F. Bernelli-Zazzera, 'Low-Energy, Low-Thrust Transfers to the Moon', *Celestial Mechanics and Dynamical Astronomy*, vol. 105, no. 1-3, pp. 61–74, 2009. See page 348.

[342] J. T. Betts, 'Optimal Low-Thrust Orbit Transfers with Eclipsing', *Optimal Control Applications and Methods*, vol. 36, pp. 218–40, 2014. See page 348.

[343] A. E. Bryson, N. Desai, and W. C. Hoffman, 'Energy-State Approximation in Performance Optimization of Supersonic Aircraft', *Journal of Aircraft*, vol. 6, no. 6, pp. 481–8, 1969. See page 348.

[344] M. Ehsani, Y. Gao, and J. M. Miller, 'Hybrid Electric Vehicles: Architecture and Motor Drives', *Proceedings of the IEEE*, vol. 95, no. 4, pp. 719–28, 2007. See page 349.

[345] S. G. Wirasingha and A. Emadi, 'Classification and Review of Control Strategies for Plug-In Hybrid Electric Vehicles', *IEEE Transactions on Vehicular Technology*, vol. 60, no. 1, pp. 111–22, 2011. See page 349.

[346] R. Lot and S. A. Evangelou, 'Lap Time Optimization of a Sports Series Hybrid Electric Vehicle', in *Proceedings of the World Congress on Engineering*, vol. III, London, 2013. See page 349.

[347] N. Kim, S. Cha, and H. Peng, 'Optimal Control of Hybrid Electric Vehicles Based on Pontryagin's Minimum Principle', *IEEE Transactions on Control Systems Technology*, vol. 19, no. 5, pp. 1279–87, 2011. See page 349.

[348] A. Sciarretta, M. Back, and L. Guzzella, 'Optimal Control of Parallel Hybrid Electric Vehicles', *IEEE Transactions on Control Systems Technology*, vol. 12, no. 3, pp. 352–63, 2004. See pages 349 and 395.

[349] L. V. Perez and E. A. Pilotta, 'Optimal Power Split in a Hybrid Electric Vehicle using Direct Transcription of an Optimal Control Problem', *Mathematics and Computers in Simulation*, vol. 79, pp. 1959–70, 2009. See page 349.

[350] H. Scherenberg, 'Mercedes-Benz Racing Design and Cars Experience', *SAE Transactions*, no. 580042, pp. 414–20, 1958. See pages 349 and 392.

[351] J. P. M. Hendrikx, T. J. J. Meijlink, and R. F. C. Kriens, 'Application of Optimal Control Theory to Inverse Simulation of Car Handling', *Vehicle System Dynamics*, vol. 26, no. 6, pp. 449–61, 1996. See pages 349 and 392.

[352] V. Cossalter, M. D. Lio, R. Lot, and L. Fabbri, 'A General Method for the Evaluation of Vehicle Manoeuvrability with Special Emphasis on Motorcycles', *Vehicle System Dynamics*, vol. 31, no. 2, pp. 113–35, 1999. See pages 349, 379, 386, 392, 423, and 425.

[353] D. Casanova, *On Minimum Time Vehicle Manoeuvring: The Theoretical Optimal Lap*. Cranfield University School of Engineering, 2000. See page 349.

[354] A. E. J. Bryson and Y.-C. Ho, *Applied Optimal Control: Optimization, Estimation, and Control*. New York: Hemisphere Publishing, 1975. See pages 349, 355, and 367.

[355] D. E. Kirk, *Optimal Control Theory: An Introduction*. Englewood Cliffs, NJ: Prentice-Hall, 1970. See pages 349, 351, 353, 355, and 360.

[356] D. G. Hull, *Optimal Control Theory for Applications*. New York: Springer-Verlag, 2003. See page 349.

[357] F. L. Lewis and V. L. Syrmos, *Optimal Control*, 2nd edn. New York: John Wiley & Sons, 1995. See page 349.

[358] M. A. Athans and P. L. Falb, *Optimal Control: An Introduction to the Theory and Its Applications*. Mineola, NY: Dover, 2006. See pages 349 and 360.

[359] D. P. Bertsekas, *Dynamic Programming and Optimal Control (Vol. II)*. Belmont, MA: Athena Scientific, 2007. See page 349.

[360] D. Liberzon, *Calculus of Variations and Optimal Control Theory: A Concise Introduction.* Princeton, NJ: Princeton University Press, 2012. See pages 349, 353, 354, and 356.

[361] H. J. Sussmann and J. C. Willems, '300 Years of Optimal Control: From the Brachystochrone to the Maximum Principle', *IEEE Control Systems Magazine*, vol. 17, no. 3, pp. 32–44, 1997. See pages 349 and 367.

[362] A. E. Bryson, 'Optimal control—1950 to 1985.' *IEEE Control Systems Magazine*, vol. 13, no. 3, pp. 26–33, 1996. See page 349.

[363] J. T. Betts, 'Survey of Numerical Methods for Trajectory Optimization', *Journal of guidance, control, and dynamics*, vol. 21, no. 2, pp. 193–207, 1998. See page 349.

[364] A. V. Rao, 'A Survey of Numerical Methods for Optimal Control', *Advances in the Astronautical Sciences*, vol. 135, no. 1, pp. 497–528, 2009. See page 349.

[365] B. A. Conway, 'A Survey of Methods Available for the Numerical Optimization of Continuous Dynamic Systems', *Journal of Optimization Theory and Applications*, vol. 152, no. 2, pp. 271–306, 2012. See page 349.

[366] L. S. Pontryagin, V. G. Boltyanskii, R. V. Gamkrelidze, and E. F. Mishchenko, *The Mathematical Theory of Optimal Processes.* New York, London: John Wiley & Sons, 1962. See pages 354, 356, and 357.

[367] B. D. O. Anderson and J. B. Moore, *Linear Optimal Control.* Englewood Cliffs, NJ: Prentice-Hall, 1971. See pages 355 and 365.

[368] A. P. Sage and C. C. White, *Optimum Control Systems*, 2nd edn. Englewood Cliffs, NJ: Prentice Hall, 1977. See page 355.

[369] M. S. Bazaraa, H. D. Sherali, and C. M. Shetty, *Nonlinear Programming: Theory and Algorithms*, 3rd edn. Hoboken, NJ: Wiley-Interscience, 2006. See page 360.

[370] D. Bertsekas, *Nonlinear Programming.* Belmont, MA: Athena Scientific Publishers, 2004. See page 360.

[371] S. Boyd and L. Vandenberghe, *Convex Optimization.* Cambridge: Cambridge University Press, 2004. See page 360.

[372] A. J. Krenner, 'The High Order Maximal Principle and Its Application to Singular Extremals', *SIAM J. Control and Optimization*, vol. 15, no. 2, pp. 256–93, 1977. See page 366.

[373] H. Kelley, R. Kopp, and H. Moyer, *Singular Extremals, Topics in Optimization.* New York: Academic Press, 1967, pp. 63–103. See page 366.

[374] H. S. Tsien and R. C. Evans, 'Optimal Thrust Programming for a Sounding Rocket', *Journal of the American Rocket Society*, vol. 21, no. 3, pp. 99–107, 1951. See pages 367 and 368.

[375] P. Tsiotras and C. Canudas de Wit, 'On the Optimal Braking of Wheeled Vehicles', in *Proceedings of the American Control Conference*, Chicago, Illinois, 2000, pp. 569–73. See pages 370 and 372.

[376] A. T. Fuller, 'Study of an Optimum Non-Linear Control System', *Journal of Electronics and Control*, vol. 15, no. 1, pp. 63–71, 1963. See pages 373, 374, and 375.

[377] M. I. Zelikin and V. F. Borisov, *Theory of Chattering Control: with Applications to Astronautics, Robotics, Economics, and Engineering.* Boston: Springer Science & Business Media, 2012. See page 373.

[378] D. H. Jacobson, S. B. Gershwin, and M. M. Lele, 'Computation of Optimal Singular Controls', *IEEE Trans. Automatic Control*, vol. 15, no. 1, pp. 67–73, 1970. See page 376.

[379] S. A. Dadebo and K. B. McAuley, 'On the Computation of Optimal Singular Controls', in *Proceedings of the 4th IEEE Conference of Control Applications*, 1995, pp. 150–5. See page 376.

[380] A. L. Schwartz, 'Theory and Implementation of Numerical Methods Based on Runge–Kutta Integration for Solving Optimal Control Problems', PhD dissertation, Electrical Engineering and Computer Sciences, University of California at Berkley, 1996. See page 376.

[381] J. T. Betts, *Practical Methods for Optimal Control and Estimation using Nonlinear Programming*, 2nd edn. Philadelphia, PA: SIAM, 2001. See pages 376 and 384.

[382] H. B. Keller, *Numerical Solution of Two Point Boundary Value Problems.* Philadelphia, PA: SIAM, 1976. See pages 377 and 378.

[383] J. Stoer and R. Bulirsch, *Introduction to Numerical Analysis.* New York: Springer-Verlag, 2002. See page 377.

[384] A. V. Rao and K. D. Mease, 'Dichotomic Basis Approach to Solving Hyper-Sensitive Optimal Control Problems', *Automatica*, vol. 35, no. 4, pp. 633–42, 1999. See page 379.

[385] E. Bertolazzi, F. B. Biral, and M. Da Lio, 'Symbolic-Numeric Efficient Solution of Optimal Control Problems for Multibody Systems', *Journal of computational and applied mathematics*, vol. 185, no. 2, pp. 404–21, 2006. See pages 379, 383, and 423.

[386] E. Bertolazzi, F. Biral, and M. Da Lio, 'Real-Time Motion Planning for Multibody Systems', *Multibody System Dynamics*, vol. 17, pp. 119–39, 2007. See page 379.

[387] F. Biral, E. Bertolazzi, and P. Bosetti, 'Notes on Numerical Methods for Solving Optimal Control Problems', *IEEJ Journal of Industry Applications*, vol. 5, no. 2, pp. 154–66, 2016. See page 379.

[388] P. E. Gill, W. W. Murray, and M. A. Saunders, 'SNOPT: An SQP Algorith for Large-Scale Constrained Optimization', *SIAM Review*, vol. 47, no. 1, pp. 99–131, 2005. See page 380.

[389] A. Wächter and L. T. Biegler, 'On the Implementation of an Interior-Point Filter Line-Search Algorithm for Large-Scale Nonlinear Programming', *Mathematical Programming*, vol. 106, pp. 25–57, 2006. See page 380.

[390] R. H. Byrd, J. Nocedal, and R. A. Waltz, *Large-Scale Nonlinear Optimization.* Boston, MA: Springer, 2006, ch. Knitro: An Integrated Package for Nonlinear Optimization, pp. 35–59. See page 380.

[391] D. Garg, M. A. Patterson, W. W. Hager, A. V. Rao, D. A. Benson, and G. T. Huntington, 'A Unified Framework for the Numerical Solution of Optimal Control Problems Using Pseudospectral Methods', *Automatica*, vol. 46, no. 11, pp. 1843–51, 2010. See page 381.

[392] D. Garg, M. A. Patterson, C. L. Darby, C. Francolin, G. T. Huntington, W. W. Hager, and A. V. Rao, 'Direct Trajectory Optimization and Costate Estimation of Finite-Horizon and Infinite-Horizon Optimal Control Problems via a Radau Pseudospectral Method', *Computational Optimization and Applications*, vol. 49, no. 2, pp. 335–58, 2011. See page 381.

[393] D. Garg, W. W. Hager, and A. V. Rao, 'Pseudospectral Methods for Solving Infinite-Horizon Optimal Control Problems', *Automatica*, vol. 47, no. 4, pp. 829–37, 2011. See page 381.

[394] S. Kameswaran and L. T. Biegler, 'Convergence Rates for Direct Transcription of Optimal Control Problems Using Collocation at Radau Points', *Computational Optimization and Applications*, vol. 41, no. 1, pp. 81–126, 2008. See page 381.

[395] C. L. Darby, W. W. Hager, and A. V. Rao, 'An *hp*–Adaptive Pseudospectral Method for Solving Optimal Control Problems', *Optimal Control Applications and Methods*, vol. 32, pp. 476–502, 2011. See pages 381 and 385.

[396] C. L. Darby, W. W. Hager, and A. V. Rao, 'Direct Trajectory Optimization Using a Variable Low-Order Adaptive Pseudospectral Method', *Journal of Spacecraft and Rockets*, vol. 48, no. 3, pp. 433–45, 2011. See pages 381 and 385.

[397] M. A. Patterson, W. W. Hager, and A. V. Rao, 'A *ph* Mesh Refinement Method for Optimal Control', *Optimal Control Applications and Methods*, vol. 36, pp. 398–421, 2014. See pages 381 and 385.

[398] F. Liu, W. W. Hager, and A. V. Rao, 'Adaptive Mesh Refinement for Optimal Control Using Discontinuity Detection and Mesh Size Reduction', in *IEEE Conference on Decision and Control*, Los Angeles, California, 2014. See pages 381 and 385.

[399] M. A. Patterson and A. V. Rao, 'Exploiting Sparsity in Direct Collocation Pseudospectral Methods for Solving Continuous-Time Optimal Control Problems', *Journal of Spacecraft and Rockets,*, vol. 49, no. 2, pp. 364–77, 2012. See page 381.

[400] C. C. Francolin, W. W. Hager, and A. V. Rao, 'Costate Approximation in Optimal Control Using Integral Gaussian Quadrature Collocation Methods', *Optimal Control Applications and Methods*, vol. 36, p. 38197, 2014. See page 381.

[401] M. Abramowitz and I. Stegun, *Handbook of Mathematical Functions with Formulas, Graphs, and Mathematical Tables*. Mineola, NY: Dover, 1965. See page 382.

[402] M. A. Patterson and A. V. Rao, 'GPOPS - II: A Matlab Software for Solving Multiple-Phase Optimal Control Problems Using *hp*–Adaptive Gaussian Quadrature Collocation Methods and Sparse Nonlinear Programming', *ACM Transactions on Mathematical Software*, vol. 41, no. 1, 2014. See pages 383 and 386.

[403] D. J. N. Limebeer, G. Perantoni, and A. V. Rao, 'Optimal Control of Formula One Car Energy Recovery Systems', *International Journal of Control*, vol. 87, no. 10, pp. 2065–80, 2014. See pages 383, 392, 409, and 415.

[404] S. Bobbo, V. Cossalter, M. Massaro, and M. Peretto, 'Application of the Optimal Maneuver Method for Enhancing Racing Motorcycle Performance.' *SAE International Journal of Passenger Cars—Mechanical Systems*, vol. 1, no. 1, pp. 1311–18, 2009. See pages 383, 392, and 423.

[405] N. Dal Bianco, F. Biral, E. Bertolazzi, and M. Massaro, 'Comparison of Direct and Indirect Methods for Minimum Lap Time Optimal Control Problems', *Vehicle System Dynamics*, 2018. See pages 383, 386, and 389.

[406] W. Karush, 'Minima of Functions of Several Variables with Inequalities as Side Constraints', Master's thesis, Dept. of Mathematics, Univ. of Chicago, 1939. See page 383.

[407] H. W. Kuhn and A. W. Tucker, 'Nonlinear Programming', in *Proceedings of 2nd Berkeley Symposium*. Berkeley: University of California Press, 1951. See page 383.

[408] T. H. Kjeldsen, 'A Contextualized Historical Analysis of the Kuhn–Tucker Theorem in Nonlinear Programming: The Impact of World War II', *Historia Mathematica*, vol. 27, pp. 331–61, 2000. See page 383.

[409] D. Metz and D. Williams, 'Near Time-Optimal Control of Racing Vehicles', *Automatica*, vol. 25, no. 6, pp. 841–57, 1989. See page 392.

[410] D. Brayshaw and M. Harrison, 'A Quasi Steady State Approach to Race Car Lap Simulation in order to Understand the Effects of Racing Line and Centre of Gravity Location', *Proceedings of the Institution of Mechanical Engineers, Part D: Journal of Automobile Engineering*, vol. 219, no. 6, pp. 725–39, 2005. See page 392.

[411] D. Brayshaw and M. Harrison, 'Use of Numerical Optimization to Determine the Effect of the Roll Stiffness Distribution on Race Car Performance', *Proceedings of the Institution of Mechanical Engineers, Part D: Journal of Automobile Engineering*, vol. 219, no. 10, p. 114151, 2005. See page 392.

[412] T. Völkl, M. Muehlmeier, and H. Winner, 'Extended Steady State Lap Time Simulation for Analyzing Transient Vehicle Behavior', *SAE International Journal of Passenger Cars—Mechanical Systems*, vol. 6, no. 1, 2013. See page 392.

[413] D. Casanova, R. S. Sharp, and P. Symonds, 'Minimum Time Manoeuvring: The Significance of Yaw Inertia', *Vehicle System Dynamics*, vol. 34, no. 2, pp. 77–115, 2000. See page 392.

[414] D. P. Kelly and R. S. Sharp, 'Time-Optimal Control of the Race Car: Influence of a Thermodynamic Tyre Model', *Vehicle System Dynamics*, vol. 50, no. 4, pp. 641–62, 2012. See page 392.

[415] D. J. N. Limebeer and G. Perantoni, 'Optimal Control of a Formula One Car on a Three-Dimensional Track Part 2: Optimal Control', *ASME Journal of Dynamical Systems, Measurement, and Control*, vol. 137, no. 5, 2015. See pages 392, 409, and 415.

[416] D. Tavernini, E. Velenis, R. Lot, and M. Massaro, 'The Optimality of the Hand-brake Cornering Technique', *Journal of Dynamic Systems, Measurement, and Control*, vol. 136, no. 4, 2014. See page 392.

[417] J. P. Timings and D. J. Cole, 'Minimum Maneuver Time Calculation Using Convex Optimization', *Journal of Dynamic Systems, Measurement, and Control*, vol. 135, no. 3, 2013. See page 392.

[418] J. Timings and D. J. Cole, 'Robust Lap-Time Simulation', *Proceedings of the Institution of Mechanical Engineers, Part D: Journal of Automobile Engineering*, vol. 228, no. 10, pp. 1200–16, 2014. See page 392.

[419] M. Maniowski, 'Optimisation of Driver Actions in RWD Race Car Including Tyre Thermodynamics', *Vehicle System Dynamics*, vol. 54, no. 4, p. 52644, 2016. See page 392.

[420] A. Sciarretta and L. Guzzella, 'Fuel-Optimal Control of Rendezvous Maneuvers for Passenger Cars', *Automatisierungstechnik*, vol. 53, no. 6, pp. 244–50, 2005. See page 395.

[421] G. Pease, D. J. N. Limebeer, and P. Fussey, 'Diesel Engine Optimisation, with Emissions Constraints, on a Prescribed Driving Route', in *IEEE Africon*, 2017. See page 402.

[422] T. Markel, A. Brooke, T. Hendricks, V. Johnson, K. Kelly, B. Kramer, M. O. Keefe, S. Sprik, and K. Wipke, 'ADVISOR : A Systems Analysis Tool for Advanced Vehicle Modeling', *Journal of Power Sources*, vol. 110, pp. 255–66, 2002. See page 404.

[423] A. Trabesinger, 'Power Games', *Nature*, vol. 447, June 2007. See page 409.

[424] Fédération Internationale de l'Automobile (FIA), '2014 Formula One Technical Regulations', Tech. Rep., 2014. See pages 409, 412, 413, and 415.

[425] W. Squire and G. Trapp, 'Using Complex Variables to Estimate Derivatives of Real Functions', *SIAM Review*, vol. 40, no. 1, pp. 110–2, 1998. See page 418.

[426] J. R. R. A. Martins, P. Sturdza, and J. J. Alonso, 'The Connection Between the Complex-Step Derivative Approximation and Algorithmic Differentiation', *AIAA paper*, no. 0921, 2001. See page 418.

[427] J. R. R. A. Martins, P. Sturdza, and J. J. Alonso, 'The Complex-Step Derivative Approximation', *ACM Transactions on Mathematical Software*, vol. 29, no. 3, p. 24562, 2003. See page 418.

[428] G. Lantoine, R. P. Russell, and T. Dargent, 'Using Multicomplex Variables for Automatic Computation of High-Order Derivatives', *ACM Transactions on Mathematical Software*, vol. 38, no. 3, 2012. See page 418.

[429] A. Griewank and A. Walther, *Evaluating Derivatives: Principles and Techniques of Algorithmic Differentiation*. Philadelphia, PA: SIAM, 2008. See page 418.

[430] R. Lot and M. Massaro, 'A Symbolic Approach to the Multibody Modeling of Road Vehicles', *International Journal of Applied Mechanics*, vol. 9, no. 5, 2017. See page 423.

Main Index

Freudenstein's equation, 4
front-wheel drive, 161
frozen-time eigenvalues, 168, 170, 173,
   222
Fuller's problem, 373, 375
   chattering arcs, 373
   phenomenon, 373

Gauss quadrature methods, 380
generalized coordinates
   see coordinates, 37
generalized force, 42
generalized momenta, 61
generalized plant, 237, 252
generalized velocities, 37
geosynchronous orbit, 348
Goddard problem, 348, 367
Goddard rocket, 367, 395
GPOPS-II, 383, 386, 389
gyroscopic moment, 107, 152, 179, 184,
   190, 195, 196, 288, 301

Hamilton's canonical equations, 59, 62
   see also equations of motion, 62
Hamilton's principle, 43
Hamilton–Jacobi–Bellman equation,
   364, 365
Hamiltonian boundary-value problem,
   360
Hamiltonian matrix, 47, 355
Hamiltonian vector, 65
heave mode, 234
Heaviside step function, 415
Hessian matrix, 418
higher-order interpolation conditions,
   253
homogeneity, 52, 54
hybrid powertrain, 348
hydrogen atom, 64
hyper-sensitive, 379

implicit simulation, 377
inclined turntable, 90
independent vs. live, 276
inerter, 243, 263
inertial frame, 33

inscribed angle theorem, 164
integrals of the motion, 52
integration matrix, 378
IPOPT, 380
irreversibility paradox, 77
ISO 8608:2016, 229
isotropy, 52
   spatial, 57

Jacobi identity, 63
Jacobian matrix, 383
just-in-time constraint, 402

Karush–Kuhn–Tucker conditions, 383
Kelley-Contensou condition, 372
Kepler problem, 64
kinematic steering angle, 176
kinetic energy, 41
kinetic energy recovery, 399, 402
kingpin compliance, 183
KNITRO, 380

Lagrange equations, 43
   see also equations of motion, 43
Lagrange multipliers, 38, 43, 44, 178,
   383
Lagrange polynomials, 380
Langevin equation, 78, 79
Laplace–Runge–Lenz vector, 65
Legendre transformation, 59
Legendre–Clebsch condition, 372, 417
Legendre–Gauss points, 381
Legendre–Gauss–Lobatto points, 381
Legendre–Gauss–Radau points, 381
Legendre–Gauss–Radau integration matrix,
   382
Levi–Civita symbol, 66
linear fraction transformation, 238
linear momentum
   conservation, 54
   definition, 33
linear quadratic (LQ) preview control,
   312
linear quadratic regulator (LQR), 355
Loschmidt's objection, 77
Lyapunov equation, 240

Person Index

Printed and bound by CPI Group (UK) Ltd, Croydon, CR0 4YY